Hydraulics for Pipeliners

VOLUME 1: FUNDAMENTALS

SECOND EDITION

Hydraulics for Pipeliners

VOLUME 1: FUNDAMENTALS

SECOND EDITION

C. B. Lester

Gulf Publishing Company
Houston, London, Paris, Zurich, Tokyo

Hydraulics for Pipeliners
Volume 1: Fundamentals, Second Edition

Library of Congress Cataloging-in-Publication Data
Lester, C. B.
Hydraulics for pipeliners / C. B. Lester. — 2nd ed.
p. cm.
Includes bibliographical references and index.
Contents: v. 1. Fundamentals
ISBN 0-88415-400-9 (alk. paper)
1. Pipelines—Fluid dynamics. I. Title.
TJ935.L44 1994
621.8′672—dc20 94-3403
CIP

Printed on Acid-Free Paper (∞)

10 9 8 7 6 5 4 3 2 1

Gulf Publishing Company
Book Division
P.O. Box 2608 □ Houston, Texas 77252-2608

Contents

Introduction: Pipeline Hydraulics in a Computer Environment

Introducing Computer Computation Templates

This introduction is an example of an author's preface that got out of hand; it is a little large to be set in small type ahead of the table of contents.

Also, author's prefaces usually aren't read. This introduction is an important part of this book, and it should be read, primarily because it introduces Computer Computation Templates—**CCT**s—which are the major improvement in this edition of *Hydraulics for Pipeliners* over the original edition.

Background

I was just ten years in the business, chief engineer of a single–project big–inch U.S. pipeline company, when I sat down and wrote:

> "For some time it has been apparent that there is a need in the pipeline industry for a textbook covering fluid mechanics as applied by working pipeliners. An engineer coming into the pipeline business has no ready reference he may study and become technically well grounded in pipeline hydraulics within a reasonable time."

I was right, at the time; and I set out to produce such a book. But as the book was written, over a period of four years, I learned as well as explained; and I wrote of many things required of working pipeliners that had very little or nothing to do with hydraulics. But, the name was already there. I am stuck with it. This is *Hydraulics for Pipeliners: Second Version*.

Hydraulics for Pipeliners has had an interesting, even colorful, life. It was first published in English language as a series in a U.S. trade magazine, and then in book form in one edition of two printings; but I have copies of (unauthorized) translations in French and German; I was offered, but didn't take, (I regretfully look back at that refusal) a Japanese copy; and, in Yakutsk, in the outback of eastern Siberia, I was asked to autograph a copy in Russian. I have heard that it was translated into Spanish and Italian, but I do not know this.

It has been photocopied innumerable times, because when it went out of print in the original English edition there was no other way to obtain a copy. The library of a friendly competitor once had five photocopies bound in hard cover. When I lost my last copy for about five years, a friend lent me one of his two photocopies, which I used as a reference until my own copy showed up in a storage box where it had no business being.

I think it was a pretty good book for it's time; it made me a lot of friends; and I'm proud of it. For what I was, for what knowledge and experience I had at the time, it was the best I could do.

Now, much later in life, with a wider if not greater knowledge and a much wider experience, it's time to remake it.

The Computing Environment of Hydraulics for Pipeliners: Original Edition

The original *Hydraulics for Pipeliners* reflected the computational environment of most engineering fields in the post–WWII era in the United States—mechanical calculators, slide rules, graph paper, curve templates, and colored pencils.

The feeling at the time was: *Any engineering problem can be reduced to a straight line on the proper kind of graph paper if the proper variables are selected.* This was very nearly true.

Also: *The second place can be trusted but the third place is a guess.* This, also, was very nearly true.

The kind of environment shapes the kind of work. The study of transient fluid flow—especially transient flow in pipeline networks—was typical. The equations describing such kind of flows could be written, but they couldn't be solved. There were no classical solutions to these kinds of equations and manual calculation of numerical solutions was out of the question. Professor Bergeron filled an entire book[1] with graphical constructions of solutions to water–hammer problems, but these were so complicated that no one without the exceptional patience and drafting ability of Professor Bergeron could have used them.

Computers have changed the routine, as well as the exceptional, ways of the world, including the ways of the world of pipeline hydraulics. *Hydraulics for Pipeliners* could have been rewritten almost any time in the last 20 years—the HP–35 came on the market in 1972—to take advantage of the improved tools available for doing mathematical computations. Why it wasn't, and why it is being rewritten at the present time, is the subject of the following.

The Computing Environment of Hydraulics for Pipeliners: Second Edition

The Age of Numerical Solutions

The fundamental equations of the flow of viscous fluids —the Navier–Stokes equations mentioned in Chapter 10—usually yield sets of non–linear partial differential equations that have no closed solution and must be either solved numerically or approximated by some kind of empirical method.

Applied fluid mechanics (hydraulics) is a science of approximations. But, in these days of powerful desk–top computers, and of hand–held calculators almost equally powerful if not so capable of handling large quantities of data, being a science of approximations is not too bad. And, consider the following quotation from Fazarinc's[2] extraordinary paper given before the mathematics panel of the American Association for the Advancement of Science in 1986:

> "When the (infinitesimal) calculus finally gets taught, its limitations are rarely discussed, and when (they are) encountered, are often dismissed as special cases. The myth of (the) calculus' power propagates into our research and industrial organizations, and (it) is responsible for considerable loss of time in fruitless attempts to apply it to real–life scientific or engineering problems....
>
> "The fact is that *we cannot solve algebraic equations of order higher than four.*
>
> "Consequently *we cannot solve differential equations of order higher than four, either.*
>
> "Furthermore, *we can solve only a handful of special non–linear differential applications.*
>
> "It is therefore fair to say that *calculus is only applicable to linear problems of order four or less.* This leaves out the vast majority of interesting problems....
>
> "To put it bluntly, let us not teach our youngsters how to derive a differential equation that we cannot solve anyway and the computer does not understand. Teach them instead how to express the problem in the finite difference form. This makes the problem ready for direct computer solution, in most cases with immediate graphic feedback.
>
> "At the same time, this approach circumvents the pitfalls of complex derivations and removes the abstraction of infinitesimal calculus from the problem."

Dr. Fazarinc's discourse points the way to the future of fluid mechanics, and, of course, of pipeline hydraulics. Substitute the word *pipeliners* for the word *youngsters* in the quotation from Dr. Fazarinc and you will have a strong recommendation for future work. Locate a copy of this reference and study it well.

Computing With Handheld Calculators

The Scientific Calculators

The early hand calculators appeared in the late 1960's, and they were a wonder: they would do chain calculations *including additions and subtractions*. For the slide rule world, this was a miracle.

Very shortly thereafter, in the early 1970's, came the first scientific calculators, typified by the **RPN** (**R**everse **P**olish **N**otation) HP–35, which not only performed chain calculations but had all of the transcendental functions of a good log–log slide rule under a button. Real chain calculations were now possible: $\sin^2 x + \cos^2 x$ really did equal 1! These machines made life easier but no new problems were solved.

The Programmable Calculators

Then came the HP–65, quickly followed by the HP–67 and HP–97, and, ultimately the HP–41 which, with all its power and add–ons, became a cottage industry by itself. These machines had two things going for them: (1) they could store programs for reuse, and (2) they could run iterative programs which offered numerical solution of non–linear equations.

The original machines had very small memories and so could only run very small programs. The later ones, with add–ons, could have several KB of RAM, in addition to several KB of calculator ROM already in the machine, and could run substantial programs.

These machines did create a minor storm in the routine work of pipeline hydraulics: they could use iterative techniques to solve the Colebrook–White function. This one program, in one step, freed pipeliners from having to carry around a paper copy of the Moody diagram, and pipeliners loved it.

These machines had a major drawback, however; they used a key–per–function version of an assembly language that was not easy to use and, for the machines without a printer, listing a program was impossible and it was easy to get lost between the lines.

And, what is really worse, once a program had been written it was almost impossible to read it because there is no way to *comment* it. There were forms to use in hand–coding, of course, but very few of us used them because it was easier, and quicker, to code directly to the machine.

As a result, the hand–held calculator simplified life but it didn't advance the science very much.

Computing with Desktop Computers

Procedural Languages

There were workstations available well before the IBM PC became available—the HP–9830 and HP–9845 were good examples—and the powerful HP Technical BASIC language in ROM used by these machines facilitated writing, debugging, and running some rather lengthy programs. A program to plot the profile and allowable–pressure lines of a pipeline route is a typical example of a program that was not difficult technically but was complicated, made sense of a large amount of input data, and took a lot of the detail work off a pipeline designer.

Also, there were many programs to solve transient flow problems written and so numerical solution of water–hammer in long pipelines became common.

The early ROM BASIC and BASICA of the IBM PC machines and the GWBASIC supplied with PC clones, and the similar BASIC interpreters for the Apple designs, did not provide the easy development environment the workstations offered; and even when FORTRAN, PASCAL, C and C++, and other compilers became available, the desktops did not become workstations. *True*BASIC, by Kemeny and Kurtz, who implemented the first BASIC as a timeshare interpreter on the Dartmouth computer network in 1965, improved on the friendly BASIC but remained a compiled language. Microsoft's *Quick*Basic version 4.5, which offered a compiler that, during the developmental phase of a problem, acted like an interpreter, offered a development environment very similar to that of HP technical basic, except the MAT (matrix) functions were missing, and rapidly became a kind of standard language for the individual engineer who needed to write a single program quickly to do one specific thing. *Q*Basic, a subset of *Quick*Basic, is packaged with most PC type machines.

But what is important for all kinds of computers using procedural languages is that the equations must be written in the form of a string of ASCII characters that can be entered from the keyboard of a typewriter or of a telex machine—which was the base for the 7–bit ASCII code—*and they are unreadable*.

As a result, a reasonably complex computer program, if not highly commented by the author, including not only line–by–line comments but comments on what data the program expects as inputs, what it will do with these data, and what values it will place in its output, *is not understandable*.

As a consequence, until recently computers have done very little to advance the teaching, or learning, of science except as they afford the user the opportunity to code his own solutions to a problem or to use the solution coded by another placing his trust in the source code hoping it is truly applicable to his problem. Either way is not very conducive to understanding the problem: The first may teach the user a lot about computer programming, and the second may offer a quick answer, but neither advances understanding any more than learning by rote.

Books written on technical subjects that include coded solutions to specific problems usually use FORTRAN because FORTRAN is the technical language of choice of academia—the eighth edition of Streeter and Wylie's *Fluid Mechanics*,[3] coded in BASICA, is a prominent exception—and the programs in these books are neither readable nor understandable without a step-by-step examination of each line of code, which, of course, no one will do unless he has an immediate need for that particular piece of code.

Program Recipe Books

Happily, there are better ways available, most of them along the lines of Professor Fazarinc's philosophy that numerical solutions to problems are readily available and, what is of primary importance, in the real world numerical answers are what is required.

Among the first *computer cookbooks* were Volumes I and II of *Basic Scientific Subroutines*.[4,5] These books offered solutions to common scientific problems in BASIC programs, each provided with a good set of descriptions of the input variables, the output values, and comments on what the program was intended to do.

These were followed by the now-famous *Numerical Recipe* series, the first written being in FORTRAN-77[6] with a subset of Pascal in an appendix, followed by a specialized edition in Pascal,[7] a specialized edition in C[8] and, the most recent, a text written in QuickBasic 4.5.[9] Diskettes with programs, demonstration drivers, and examples of use for each language became available over time.

This class of book offered, first, a statement of the problem in the terms of mathematical formulas rendered in the usual symbolic form, followed by the commented computer code in the selected language which, with associated example books and diskettes, allowed the user to run problems he might otherwise not try. While many of the programs were of no interest to a pipeliner, some of them were, and they could be used as written or with minor changes. The Levenberg-Marquardt solution to the non-linear least squares problem developed by the Argonne National Laboratory published in both book and computer-readable tape form in MINPACK-1,[10] while not exactly a recipe book, provided an instructive lesson on numerical solutions to (sometimes) very difficult problems. MINPACK-1 is in the public domain and both the book and tape can be obtained from the Argonne National Laboratory.

Still, the troublesome requirement remained: To be sure you understood a program you had to dissect it by hand, line-by-line and—in effect—prepare your own set of comments that were meaningful to you in your field of work.

The recipe books were a great advance, and there are now many of them. I still did not want to write a college text kind of reference book about pipeline hydraulics, however, and I did not rewrite *Hydraulics for Pipeliners*.

Symbolic Calculating Engines

The really useful computer aides, however, had to await the arrival of the canned *solvers*. These programs allowed the user to enter an equation in an accepted procedural language, give values to the inputs, and obtain a numerical answer in the output.

Solvers appeared in some handheld calculators in the mid-1980's, but except for the one-time solution of non-linear equations in a single variable, these did not do much to advance the state of the art.

The most important additions to the computer tool kit, however, were *Macsyma*,[11] *Maple*,[12] *MathCad*,[13] and *Mathematica*[14]—there are probably others—which not only have the ability to evaluate all kinds of linear and non-linear equations but had the added, most important, ability to solve large systems of simultaneous equations that are non-linear in one or more equations in one or more variables. These software suites offered the user the ability to code a problem using the ordinary symbols of mathematics, give numeric values to the known variables, and solve for the unknowns. *All without writing a single line of procedural code.*

They also offered symbolic solution (as differs from numeric solution) of some problems (*MathCad* has a sub-set of *Maple*'s symbolic processor), so that if a problem *did have* a closed solution it could be found and used instead of an iterative solution to a specific numeric problem.

These symbolic processors will also simplify, evaluate, manipulate, integrate, and differentiate expressions symbolically, manipulate matrices, etc.

This also meant that a book, such as this one, could be written that would offer *understandable* patterns for computer algorithms to solve specific problems, and this is the philosophy for, finally, rewriting *Hydraulics for Pipeliners*:

Present the best available data and methods, and integrate these into understandable, workable, solutions to the problems of working pipeliners.

Computer Computation Templates

A Computer Computation Template, or **CCT**, as developed in the following, is *a readable pattern for a computer algorithm* for solving a problem.

CCT No. 0-0 is an annotated sample. The major sections of a **CCT** are:

1. A description of the application
2. References
3. Input data, subdivided into
   - factors and constants of proportionality
   - names, values, and dimensions of variables
   - estimated values of unknowns (seeds)
4. The procedure...equation(s) to be solved... written in the usual mathematical symbols.
5. Results

A **CCT** is, then, a pattern describing a solution to a problem. A procedure can be written in any procedural language to implement the pattern: first write a calling sequence to call the procedure and pass the inputs to it, write the procedure, and write an output module to receive the answer. If one of the software suites with a programming language of its own is used (*Mathematica, Macsyma,* and *Maple* have their own languages), an implementation can be written that is more readable and, therefore, more understandable, than code written in one of the general procedural languages.

Or, a **CCT** can be used for what it is—a file developed in *MathCad* to solve a problem using *MathCad's* engine: change an input item, punch the *calculate* button, and a new answer appears in the results.

All **CCT**s are written in *MathCad* terminology:

:= definition (assignment) operator: assigns a value to a variable; the value may be a number or a combination of numbers and *previously defined* variables

= bold face equal: establishes equality of expressions but does *not* initiate valuation

= equal: initiates evaluation of an expression

Three examples are included herein. **CCT** No. 0-1 is a very simple pattern written to solve the API 2509C shrinkage equation. The Δ°API function G is set in the form of a *MathCad* range variable, taking on values from 0 °API to 100 °API in steps of 10 °API. The values of G are written in the first output table, the values of S_G are written in the second table, and the results are plotted on the accompanying graph. New values for S_G, and a new curve, can be generated essentially instantly by changing the value of C in the input section.

CCT No. 0-2 is a more complicated pattern showing an iterative solution with the *MathCad root* function. This **CCT** solves the Colebrook–White function for a range of 10 values of Q expressed in mbd. The output tables contain Q_i, R_i, and f_i, respectively, and the plot shows the function as calculated. As in all **CCT**s, varying any of the input values quickly produces a new output.

CCT No. 0-3 is a very complex pattern which solves a problem set up by a five–leg, four–node network with one source and two sinks with only three head values and the flows out of the source and into the sinks known. The problem: to find the five unknown flowrates.

CCT No. 0-3 uses the powerful *MathCad solve block* procedure—based on the Levenberg–Marquardt solution to non–linear least squares problems presented in the previous reference[10]—to solve a system of five simultaneous equations of which three are simple and linear and two are complex and highly non–linear. Since this kind of problem can be quite ill conditioned, the *given–minerr*() procedure is used. Solutions obtained with the *given–minerr*() procedure are not precise; if the output of **CCT** No. 0-3 is put back in the procedure rearranged to calculate the input variables, the results will not be exact but *they will be much more accurate than the input data available for a real life problem.*

For well behaved functions the *given–find MathCad* procedure, which only gives up searching after the error vector is smaller than some established tolerance, is used.

Conclusion

There are many **CCT**s included in this book, and others will be made available separately. Used either as a template or as a file called by the *MathCad* engine, however, the intent is the same—describe the inputs, the equations, and the expected outputs of a problem in pipeline hydraulics.

I sincerely hope this kind of presentation will improve the awareness, understanding, and use of mathematical aides in pipeline hydraulics. *The size of some projects is so great that the use of anything but the best procedures acting on the best information available is technical malpractice.*

References

[1] Bergeron, L., Du Coup de Bélier en Hydraulic Au Coup de Foudre en Électricité. Paris: Dunod, 1950.

[2] Fazarinc, Zvonko. "A Viewpoint of Calculus." *Hewlett–Packard Journal*, pp 38-40, March 1987.

[3] Streeter, V. L., and Wylie, B. E., *Fluid Mechanics*, 8th. ed. New York: McGraw–Hill, Inc., 1986.

[4] Ruckdeschel, F. R., *BASIC Scientific Subroutines, Vol. I.* Peterborough, NH: BYTE/McGraw–Hill, 1981.

[5] Ruckdeschel, F. R., *BASIC Scientific Subroutines, Vol. II.* Peterborough, NH: BYTE/McGraw–Hill, 1981.

[6] Press, W. H., Flannery, B. P., Teukolsky, S. A., and Vetterling, W. T., *Numerical Recipes: The Art of Scientific Computing*, 1st. ed. New York: Cambridge University Press, 1986.

[7] Press, W. H., Flannery, B. P., Teukolsky, S. A., and Vetterling, W. T., *Numerical Recipes in Pascal: The Art of Scientific Computing*, rev. ed. New York: Cambridge University Press, 1989.

[8] Press, W. H., Flannery, B. P., Teukolsky, S. A., and Vetterling, W. T., *Numerical Recipes in C: The Art of Scientific Computing*, 2nd. ed. New York: Cambridge University Press, 1992.

[9] Sprott, J. C., *Numerical Recipes: Routines and Examples in BASIC*, 1st. ed. New York: Cambridge University Press, 1991.

[10] MINPACK–1, *A Package of FORTRAN Subprograms for the Numerical Solution of Systems of Nonlinear Equations and Nonlinear Least Squares Problems*, ANL-80-74. Argonne, Illinois: Argonne National Laboratory, 1980

[11] *Macsyma*, a suite of software published by Macsyma, Inc., Arlington, MA.

[12] *Maple*, a suite of software published by Waterloo Maple Software, Waterloo, Canada.

[13] *Mathcad*, a suite of software published by MathSoft, Inc., Cambridge, MA.

[14] *Mathematica*, a suite of software published by Wolfram Research, Champaign, IL.

Sample

Computer Computation Template

CCT No. 0-0: Short description of the application.

References: Includes chapter number, and equation, table, and/or figure numbers, as apply.

Input

Factors — Factors are usually constants of proportionality, but may be application–specific.

Input Data — Input data, not including *seed variables* (see below).

Variables — Variable definitions include units of measurement.

Equations

Seed Variables — *Seed Variables* are the initial *guesses* required for iterative solutions.

Given — *Given* indicates the system of equations following is part of a MathCad *solve block*. Not required for simple numeric evaluation or for *root* function solutions.

[Equations written in standard mathematical notation].

Results———————————

Find() — The MathCad *find*() function is used to obtain a solution to a system of well behaved equations.

Minerr() — The MathCad m*inerr*() function is used to obtain a solution to an ill conditioned system of equations.

Otherwise — Result of evaluation of a function, obtained either directly or iteratively by the MathCad *root* function.

Computer Computation Template

CCT No. 0-1: Evaluate API 2509 C formula for shrinkage of a light hydrocarbon when mixed with a large quantity of a heavy hydrocarbon.

References: Chapter 2. Equation 2-21. Figure 2-10.

Input

Variables

$C := 5$ Volume percent concentration of lighter component

$G := 0, 10..100$ Range of difference in API degrees between lighter and heavier component frp, 0 to 100 in steps of 10.

Equations

$S_G := 0.00214 \cdot C^{-0.0704} \cdot G^{1.75}$

Results

G	S_G
0	0
10	0.11
20	0.36
30	0.73
40	1.22
50	1.80
60	2.47
70	3.24
80	4.09
90	5.02
100	6.04

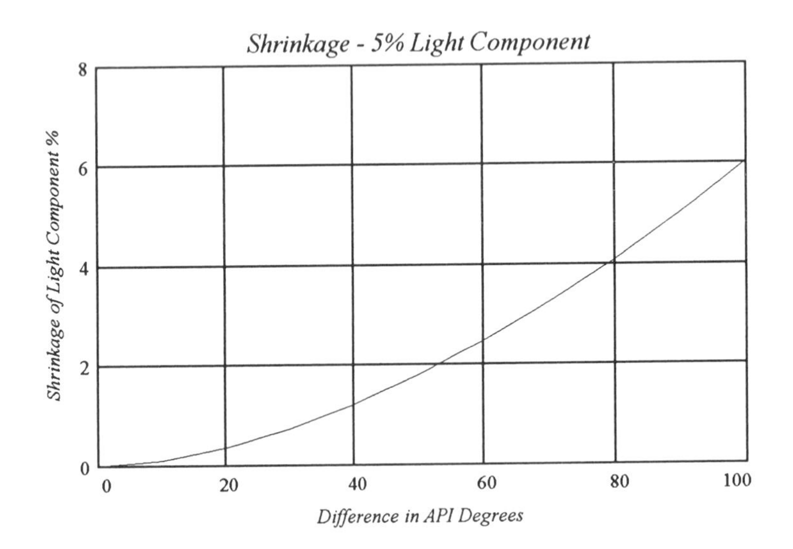

Computer Computation Template

CCT No. 0-2: Solves Colebrook–White Function iteratively using MathCad *root* function.

References: Chapter 13. Table 13-4.

Input

Factors $C_1 := 92.241232$ $\quad i := 1, 2..10$

Input Data $d := 35.25$ in $\quad \nu := 7.0$ cSt $\quad e := 0.0007$ in

Variables $Q_i := 50000 + 25000 \cdot i$ bpd $\quad R_i := C_1 \cdot \frac{Q_i}{d \cdot \nu}$

Seed $f := 0.01$

Equations

$$f_i := root\left[\frac{1}{\sqrt{f}} + 2 \cdot log\left[\left(\frac{\frac{10 \cdot e}{d}}{37}\right) + \frac{2.51}{R_i \cdot \sqrt{f}}\right], f\right]$$

Results

Q_i	R_i	f_i
75000	28037	0.0239
100000	37382	0.0223
125000	46728	0.0213
150000	56074	0.0204
175000	65419	0.0198
200000	74765	0.0192
225000	84111	0.0187
250000	93456	0.0183
275000	102802	0.0180
300000	112147	0.0177

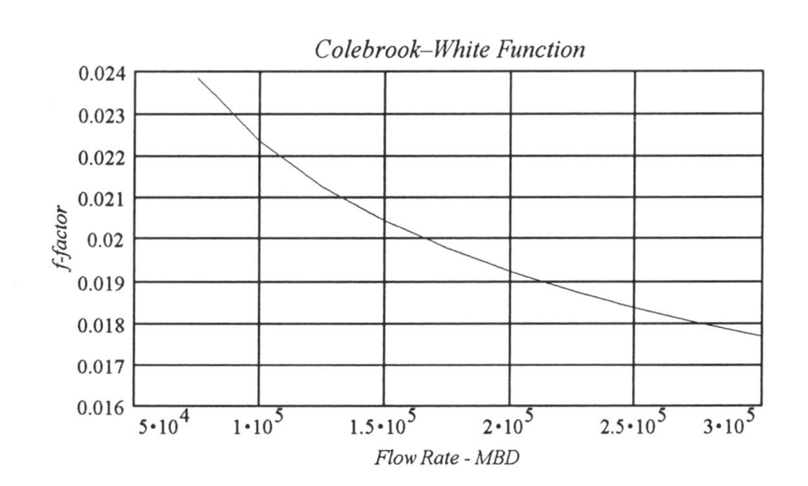

Computer Computation Template

CCT No. 0-3: Calculate flow rates for five leg, four node network with one source and two sinks. Use solve-block *given–minerr*() procedure to solve five simultaneous equations of which two are highly non–linear.

References: Chapter 16. Figure 16-1 (general case). Figure 16-11 (case specific).

Input

Factors $C_1 := 92.241232$ $C_2 := 0.02647271$ $C_3 := 0.01147629$ $K := \frac{3.6}{2.302585}$ $k := \frac{2.0}{2.302585}$

Variables $L_1 := 3.0$ kft $L_2 := 3.5$ kft $L_3 := 1.0$ kft $L_4 := 1.0$ kft $L_5 := 1.5$ kft

$d_1 := 12.0$ in. $d_2 := 10.0$ in. $d_3 := 6.0$ in. $d_4 := 8.0$ in $d_5 := 12.0$ in

$Qa := 50000$ bpd $Qb := 15000$ bpd $Qc := 35000$ bpd $\nu := 7.0$ cSt

Seed Variables $Q_1 := 40000$ bpd $Q_2 := 10000$ bpd $Q_3 := 20000$ bpd $Q_4 := 20000$ bpd $Q_5 := 15000$ bpd

Equations

Ha, Hb and Hc known, find: Q_1, Q_2, Q_3, Q_4, Q_5.

given

$$Q_1 + Q_3 = Qa$$
$$Q_1 - Q_2 - Q_4 = Qb$$
$$Q_4 + Q_5 = Qc$$

$$\left[C_2 \cdot L_1 \cdot \left[\frac{4}{\left(K \cdot \ln\left(\frac{C_1 \cdot Q_1}{8 \cdot d_1 \cdot \nu}\right)\right)^2}\right] \cdot \frac{Q_1^2}{d_1^5} + C_2 \cdot L_2 \cdot \left[\frac{4}{\left(K \cdot \ln\left(\frac{C_1 \cdot Q_2}{8 \cdot d_2 \cdot \nu}\right)\right)^2}\right] \cdot \frac{Q_2^2}{d_2^5}\right] - C_2 \cdot L_3 \cdot \left[\frac{4}{\left(K \cdot \ln\left(\frac{C_1 \cdot Q_3}{8 \cdot d_3 \cdot \nu}\right)\right)^2}\right] \cdot \frac{Q_3^2}{d_3^5} = 0$$

$$\left[C_2 \cdot L_4 \cdot \left[\frac{4}{\left(K \cdot \ln\left(\frac{C_1 \cdot Q_4}{8 \cdot d_4 \cdot \nu}\right)\right)^2}\right] \cdot \frac{Q_4^2}{d_4^5} - C_2 \cdot L_5 \cdot \left[\frac{4}{\left(K \cdot \ln\left(\frac{C_1 \cdot Q_5}{8 \cdot d_5 \cdot \nu}\right)\right)^2}\right] \cdot \frac{Q_5^2}{d_5^5}\right] - C_2 \cdot L_2 \cdot \left[\frac{4}{\left(K \cdot \ln\left(\frac{C_1 \cdot Q_2}{8 \cdot d_2 \cdot \nu}\right)\right)^2}\right] \cdot \frac{Q_2^2}{d_2^5} = 0$$

Results

$$minerr(Q_1, Q_2, Q_3, Q_4, Q_5) = \begin{bmatrix} 37860 \\ 8884 \\ 12140 \\ 13976 \\ 21024 \end{bmatrix} \begin{matrix} \text{bpd} \\ \text{bpd} \\ \text{bpd} \\ \text{bpd} \\ \text{bpd} \end{matrix}$$

Part I
Characteristics of Pipeline Fluids

Part I, consisting of Chapters 1–9, is dedicated to, first, a review of units of measurement and of the precision and accuracy that can be expected in measurements and calculations; and, second, to the characteristics of fluids, usually from the standpoint of those characteristics that affect fluid flow although, for completeness, certain other characteristics important when the fluid is being stored or becomes involved in custody transfer are included.

Considerable emphasis is placed on the development of methods for calculating, or estimating, the value of fluid characteristics and on ways of estimating, or predicting, the effect of changes in pressure or temperature.

Tabulations of important formulas and equations, and, where appropriate, Computer Computation Templates, are provided.

1

Units, Conversion Factors, Accuracy, and Precision

Introduction

The first edition of *Hydraulics for Pipeliners* was based almost entirely on the *gravitational*, or *engineer's*, foot–pound–second (FPS) system of units. There were, of course, some *exceptions*: the usual unit of kinematic viscosity was the *Saybolt Universal Second,* reported in seconds, which was converted to *centistokes,* which have the dimensions of mm^2/s; the usual unit of capacity was the *oil barrel*, which is 42 U.S. gallons of 231 in^3; density was reported in terms of *API gravity*, which is a nonlinear function of relative density 60/60°F, etc. Not a consistent approach, but not unusual for the U.S. oil industry at that time. And it worked.

But it was tedious to convert these kinds of formulas to accept other input or output units with a high degree of numeric consistency (note the word is *consistency*, and not *accuracy*) because the tool the engineering trade had for multiplying, dividing, and taking roots and powers, in those days was the ten–inch slide rule. Slide rules did not offer a realistic probability of repeating a calculation to more than three significant figures; and if the calculation required forming more than one or two intermediate results, the validity of the third figure in the answer was seriously in question. The ten–inch rule was sometimes called a *one percent tool* because reliable results could be expected to only two significant figures.

The result was that each pipeliner had his own set of formulas for the units he preferred. These formulas had carefully calculated constants, usually determined by computations made with mechanical ten–bank calculators and six–place log tables, designed to yield output in terms of units he understood from input in units he preferred.

As an example, my own set of formulas for pipeline friction loss was designed to produce answers in terms of *psi/kft*. This isn't now, and wasn't then, a very useful design parameter; but I didn't realize that at the time. I understood it, I designed and operated pipelines with it, and I was happy with it. I really didn't need anything better.

Today we all need better things. When the first edition was being written most of the world's existing pipelines had been constructed in the U.S., but already the Canadians were beginning to build pipelines of some stature, starting with the Trans–Mountain system; U.S. and European oil companies were building significant pipelines in Venezuela; and TAPLine and the original IPC pipelines had been built in the Middle East. The first product pipeline in Europe had just been built from Le Havre to Paris, and NATO and the U.S. JCA were beginning to construct their systems of pipelines for the distribution of military fuels in Western Europe.

It is noteworthy that almost without exception the designers of these pipelines still worked primarily with FPS units. With miscellaneous special units, such as those previously described, added in at their pleasure, they satisfied themselves—they didn't have to satisfy anyone else.

There often was no consistency between units of the same company. For example, two U.S. companies used the Darcy–Weisbach f–factor kind of flow equation for crude lines but regressed to the Hazen–Williams C–factor waterline kind of calculation for products lines.

These arrangements worked if you were comfortable with their idiosyncrasies and aware of their limitations, but it was nearly impossible to explain them to anyone else, much less justify why they were chosen.

Now pipelines are built in every corner of the world, and the FPS system of units, despite all the adjustments made to it by the pipeline industry, is no longer the only system available.

Today we must have a consistent set of rules and formulations for solving pipeline problems, together with conversion factors to allow inputting the data in hand, in whatever consistent or inconsistent units, while yielding answers in the desired terms.

This is possible today because we have standards and conversion factors accepted worldwide, and because we have handheld calculators and desktop computers that have computing capacities orders of magnitude greater than we had in the past.

But while the problems of using miscellaneous, rather than a consistent set of, units have mostly gone away with the availability of these new computing tools, another problem—the *appearance of great accuracy* presented by 10- or even 15-digit output of simple problems—has appeared and it threatens to distort the judgment of pipeliners everywhere. *Consistency* has been mistakenly understood as *accuracy*.

Systems of Units

In *gravitational* systems of units, as compared to *absolute* systems, the factor included as a multiplier or divisor in the unit definitions is the acceleration due to gravity, usually denoted as g.

Gravitational systems tend to be more intuitive than absolute systems. The units in the gravitational FPS system are the *foot*, the *pound–force*, and the *second*. Gravitational FPS systems are really satisfactory for engineers educated in the FPS system. They have grown up with pounds and feet and seconds and they have a certain instinct about these units that serves them well in their profession.

These FPS units, however, are not instinctively understood by those educated under one (there *are* several) of the metric systems. The preference for gravitational systems of units, as compared to absolute (or other) systems, does persist in the metric world, however; and until ISO decided that gravitational systems were out, and absolute systems were in, those working in the metric system had no trouble with systems based on the *meter*, the *kilogram–force*, and the *second*.

Before proceeding we must understand the difference between *gravitational* and *absolute* systems of units; why gravitational systems have long been used by engineers; and why the *United States Customary System (USCS)* (now sometimes called the *inch–pound–second* system), based on the *yard* and *pound* as defined by the *U.S. Bureau of Standards*, is a gravitational system.

Three basic dimensions are necessary to make up a complete dimensional system applicable to mechanics and, thereby, hydraulics problems.

Two of these dimensions are *always* taken as *length* and *time*. Length is taken as the inch or foot, or as the centimeter or meter. Time is (almost) universally taken as the *second*.

The third dimension may be taken as any number of things. In the astrophysical system of units it is taken as the acceleration due to gravity. Commonly, however, the third dimension is taken as either *mass* or *force*, and these two properties are then related by Newton's equivalency $F = Ma$. If *weight* is used instead of mass the equation becomes $Fg = Wa$. In both equations a is *acceleration* in terms of length/time². In the second equation g is the special *acceleration due to gravity*.

Engineering work (before ISO) was almost universally carried out in what is called the *gravitational* system, in which the third dimension is taken as *force*. A gravitational FPS system, for instance, has as its third unit the *pound–force*, and so it is necessary to invent and name a new unit of *mass* which will be accelerated 1 ft/s² by the uniform application of 1 pound–force. In the U.S. this new unit is called the *slug*, and is numerically equal to *g pounds–force*. The internationally accepted value for the acceleration due to gravity is 9.806 650 E+00 m/s², *exactly*, and the conversion from feet to meters is 3.048 000 E–01, *exactly*, and the value of 1 slug is 3.217 405 E+01 pounds to seven significant figures.

Scientific work, as distinguished from engineering work, is commonly carried out in what is called an *absolute* system of units in which the third dimension is taken as *mass* instead of *force*. An absolute FPS system has as its third unit the *pound–mass*, and it is necessary to invent and name a new unit of force which will accelerate a 1 pound–mass by 1 ft/s² when applied uniformly. In the English language world this unit is called the *poundal*, and is equal to 1 *pound–mass/g* which equals 3.108 095 E–02 pounds to seven significant figures.

The advantage of gravitational systems is that they are instinctively understood from common measurements. We all know what a pound is because we can *weigh* it: a pound–mass weighs a pound under conditions of free–fall gravity and, for all practical purposes, it weighs a pound anywhere near the surface of the earth. A pound–mass exerts a pound–force when put in the weighing basket of a scale, and a pound–force will accelerate a slug 1 ft/s^2. Pressures reported in pounds–force/in^2 are intuitively understood because we can visualize a one pound weight sitting on top of a 1 in^2 piston that is pressurizing a closed vessel. Pressure given in pounds/in^2 is instinctively understood: poundals/in^2 is not.

On the other hand, the advantage of absolute systems is found in theoretical work. Equations written in an absolute system seem to be cleaner, and are probably more intuitively understood, than equations written in a gravitational system. The factor g, acceleration due to gravity, doesn't seem to get in the way of more important factors so often.

This kind of factor–of–proportionality problem is not peculiar to mechanical systems of units. In the metric MKSA system, in which the fourth unit is the *ampere*, the unit of electrical current, one must choose between a *rationalized* and a *non–rationalized* kind of system, the difference between these systems being where a factor 4π is inserted in the table of units.

The problem of g in mechanical problems arises from the Newtonian equivalency. In electromagnetism the 4π problem arises from the constants of proportionality in Ampere's and Coulomb's equations. Coulomb's equation includes the factor $4/\pi$ to take care of the divergence of electrostatic lines of force, whereas Ampere's equation deals with electromagnetic lines of force, which are closed and by definition do not diverge, and the factor is not needed.

Rationalization can be carried to almost ridiculous extremes, as in the Gaussian system of CGS units. This system defines special electrostatic and electromagnetic units which make both theoretical manipulations and computations much simpler than with any other system of units. It is excellent for teaching—and for learning—but it is difficult to handle in practical problems. One's instincts fail when certain units are defined as being multiplied, or divided, by c, the speed of light, or the *square* of this value, as well as 4π.

The SI System of Units

The SI system of units has had much written about it. SI is the internationally accepted abbreviation for the *Système International d'Unités* (International System of Units) promulgated by the *Conférence Générale des Poids et Mesures* (General Conference of Weights and Measures). The CGPM, which is supported by many nations, is responsible for the resolution of all problems pertaining to the metric system, and is the parent organization of the *Bureau International des Poids et Mesures* (International Bureau of Weights and Measures) which is charged with maintaining the fundamental standards of measurement, for making comparisons with the standards of others, and for performing research toward the development of new standards.

The SI system grew out of a *rationalized absolute MKS* system of units, first proposed by Giorgi in 1935, which would add one of the fundamental electromagnetic units (ampere, coulomb, ohm, volt) for a fourth unit to arrive at a system of units which would be suitable for both mechanical and electromagnetic work.

In 1950, the *ampere*, the unit of electric current flow, was accepted by the CGPM as the fourth dimension and the MKSA system came into being. In 1954, two more base units were added, the *kelvin*, as the base unit for temperature, and the *candela*, as the base unit for luminous intensity. This six unit system was officially designated as the SI system by the CGPM in 1960. A seventh unit, the *mole*, the base unit of substance, was added to SI in 1971.

The SI system is divided into three classes of units called *base* units, *supplementary* units, and *derived* units.

The *base* units are those shown in Table 1-1. The use of capital and lower case letters is in accordance with SI procedures. The first letter of *symbols* derived from a person's name is capitalized; any other letters in symbols are in lower case. The *names* of units, as contrasted to *symbols* of units, however, are never capitalized; *viz.* 1 pascal = 1 Pa.

Table 1-1
SI Base Units

Quantity	Unit	Symbol
length	meter	m
mass	kilogram	k
time	second	s
electric current	ampere	A
thermodynamic temperature	kelvin	K
amount of substance	mole	mol
luminous intensity	candela	cd

The work of defining the seven base units is almost done. As this is being written only one of them, the kilogram, remains defined in terms of an artifact (the mass of the international kilogram held by the BIPM in Sèvres, France). The other six units have been related to natural phenomena which are considered unvarying and can be measured very accurately.

Table 1-2 lists the *supplementary* units. There are only two of these. The *radian*, symbol *rad*, is the unit of plane angle, and the *steradian*, symbol *sr*, is the unit of solid angle. These are purely geometrical units, defined exactly in their definition.

All other SI units are called *derived* units. At the time this was written there were four *named* derived units applicable to hydraulics. These are shown in Table 1-3. Some common derived units, not yet given names, are listed in Table 1-4.

All of the *base*, *supplementary* and *derived* units are *coherent* which, in SI terminology, is understood as that property of a system of units in which there is a one–to–one correspondence among base units and derived units. One newton of force will accelerate one kilogram of mass at a rate of one meter per second squared; one pascal of pressure is one newton per square meter; etc. Prefixes of 100, 10, 1/10 and 1/100 (hecto, deca, deci, and centi) are frowned on; the mantissa of the number quantifying a value should be chosen between 0.1 and 1000 so that prefixes representing powers evenly divided by three (k, kilo, thousand and M, mega, million are examples) are applicable.

It is coherence, as well as the fact that SI is an absolute rather than a gravitational system, which acts to inhibit the application of SI to most engineering, and pipeline, problems.

Table 1-2

SI Supplementary Units

Quantity	Unit	Symbol
plane angle	radian	rad
solid angle	steradian	sr

Table 1-3

SI Named DerivedUnits

Quantity	Unit	Symbol	Formula
force	newton	N	kg·m/s²
pressure, stress	pascal	Pa	N/m²
energy, work	joule	J	N·m
power	watt	W	J/s

Table 1-4

Other Derived Units

Quantity	Units	Formula
acceleration	meters per second squared	m/s²
angular acceleration	radians per second squared	rad/s²
angular velocity	radians per second	rad/s
area	square meter	m²
density, mass	kilogram per meter cubed	kg/m³
moment of force	newton meter	N·m
specific heat capacity	joules per kilogram kelvin	J/(kg·K)
specific volume	cubic meter per kilogram	m³/kg
thermal conductivity	watts per meter kelvin	W/(m·K)
velocity	meters per second	m/s
viscosity, dynamic	pascal second	Pa/s
viscosity, kinematic	square meter per second	m²/s
volume	cubic meter	m³

The inherent inflexibility of SI makes for difficulties everywhere. The derived unit for *velocity*, for instance, is the meter/second, and this is not a reasonable unit for calibration of the speedometer of an automobile. The unit of *density* is kilogram/cubic meter, making the density of water 1000 instead of 1.000 which is instinctively a more proper reference value. And the unit of pressure is the *pascal*, 1 newton/square meter, which is such a tiny unit that either the *megapascal* (1 MPa = 10^6 pascals), a large unit, or the *kilopascal* (1 kPa = 10^3 pascals), a small unit, is recommended for common use.

However, the *bar*, a unit of pressure which has been in use for more than 100 years, equal to 10^5 pascals, is only accepted for limited use, such as reporting the value of physical measurements. The fact that it is intuitively a reasonable size of unit for pressure in the absolute MKS system, being only a little larger than the gravitational (engineer's) 1 kg/cm² unit, and only a little less than 1 standard atmosphere (1 *atm* = 760 mm Hg), is given no credence: there is no *coherent* prefix in SI for 10^5 so the bar is not acceptable in the usual case.

SI is a moving target. I followed it with interest for years, and finally gave up when it appeared that changes were being made seemingly for change's sake.

The original edition of ASTM E380-68[1] reporting on SI said "The preferred metric unit of pressure and stress is the newton per square meter. However, in some countries using cgs and MKS units, measurements are given in kilograms–force per square centimeter or kilograms–force per square millimeter. Where the inclusion of such units is desirable in the interest of international cooperation, the kilogram–force per square centimeter (or per square millimeter) may be used providing the metric unit (newtons per square meter) is also given in parentheses..." Taken from ASTM E380-72[2] (the same publication, four years later): "The SI unit of pressure and stress is the pascal (newton per square meter). The use of *metric engineering units* (italics are mine) (the kilogram–force per square centimeter, kilogram–force per square millimeter, bar, torr, etc.) is discouraged and should be discontinued." And, four *more* years later, ASTM E380-76[3] was saying: "The SI unit of pressure is the pascal (newton per square meter) and with proper prefixes is applicable to all such measurement. Old metric gravitational units for pressure and stress such as kilogram–force per square centimeter (kg_f/cm^2) shall not be used. Widespread use has been made of other non–SI units such as the bar and torr for pressure, but this is strongly discouraged. The *millibar* is widely used in meteorology; *this usage will continue for the present in order to permit meteorologists to communicate easily within their profession* (again the italics are mine). The kilopascal should be used in presenting data to the public."

At that point I lost interest, and I'm not sure what has happened to the bar in recent years. However, I do wish pipeliners had as much clout in the scientific community as the meteorologists. Maybe we could have had one or two SI units of our own, too.

There may come a time when engineers are educated in the SI system of units and will intuitively understand it. They will instinctively convert 1 meter H_2O of head to 9806 pascals, and will have stored in their memory the standard conditions for measuring crude oil and refined products, these being 288.15 K, 101.325 kPa.

However, this book is being prepared in a less perfect time and the described ideal conditions do not exist. Therefore, with all due respect to the CGPM and what they are trying to do, the metric system used herein will be the mixed–up gravitational metric system engineers have used for the past century or so, brought up to date with the latest CGPM figures as to fundamental units, conversion between systems of units, and the like. Where I refer to standard documents, or to the published work of others, if pressures are reported in kPa, or stresses in MPa, I will use the pascal as a unit of pressure or stress.

Selected Metric System

The metric system selected for use in this book is a gravitational MKS system using as fundamental (i.e., base) units the meter, the kilogram–force, and the second.

It will be understood that the *kg* used to measure pressure and stress is the *kilogram–force*, and that pressures reported in *kg/cm²* or *kg/mm²* are really in *kilogram–force/cm²* and *kilogram–force/mm²*. And I will use these kinds of compound units as advantageous: the kg/cm² is best for measuring pipeline pressures, but the kg/mm² is better for reporting pipe yield strengths, for instance.

Densities will be related to water at maximum density as unity, *i.e.*, 1 dm³ of H_2O at 4°C will be assumed to weigh 1.0000 kg. Except when working with commercial measurements of quantities of crude oils and products in which case it will be assumed to be 999.012 kg/m³ at 60°F.

A liter will be assumed to be 1 dm^3; the fact that between 1901 and 1964 1 liter was 1.000 028 dm^3 is of no interest to a practicing pipeliner.

Where original references are in bars or torrs, these will be used in the text. The *atm* (the normal atmosphere is recognized world–wide as 760 *torr*, with torr being defined as 1 mm Hg at °C), equal to 1.013 25 bars, a common unit of measurement in some European countries, will *not* be used even if found in an original reference because it can be confused with the *atü* (technical atmosphere) which, in some countries, is used to mean 1 kg_f/cm^2.Some countries codified both the *atm* and the *atü*, and this really was confusing.

And I will use the °C whether the C stands for *centigrade* or *Celsius*, instead of *kelvins* except when looking at a situation in which the use of absolute temperatures is mandatory.

Selected FPS System

The FPS system I will use is, of course, the gravitational FPS system.

Pressures and stresses will be in *pounds–force/in²*, which are usually abbreviated as *psi*. If very high pressures are involved, or when discussing *stress*, I will use the *ksi*, which is the abbreviation for *thousands of pounds–force/in²*.

[In MKS there is *ksc* for *kilograms–force/cm²*. This isn't an accepted abbreviation but, since SI doesn't like the unit, anyway, they shouldn't be too upset about it. And if I make reference to kPa × 100, I will use the *bar*.]

Linear dimensions will be in decimal fractions or multiples of the *foot*.

Weights will be in *pounds* unless *tons* are indicated, and I will use the 2000–pound *short ton,* ST, the (approximately) 2204–pound *metric ton*, MT, and the 2400–pound *long ton,* LT, as best applies, but we will always identify them; we will not use the confusing *tonne*.

Temperatures will be in *°F* for *Fahrenheit* unless absolute temperatures are involved in which case we will use *°R* for *Rankin*.

Common Values

All the usual pipeline terms will be maintained as they are in use today.

Product volumes will be in *cubic meters* or *oil barrels*. When discussing commercial measurement of quantities I will also have reference to the *metric ton* and *long ton* because use of these units is legally mandated in certain jurisdictions.

Flow rates will generally be in *cubic meters per hour* (m^3/h) or in *oil barrels per day* (bpd) except when dealing with transient flow when rates will be measured in *cubic meters per second* (cms or m^3/s) or *cubic feet per second* (cfs or ft^3/s). I will not touch the (almost) antiquated MTA (*metric tons per annum*) or LTA (*long tons per annum*). These are journalistic, or legal, not pipeline, terms.

Viscosities are reported in many ways, and we will provide conversions for most of them in the proper place. Flow calculations end up using kinetic viscosities in CGS or FPS units, as applies, but these are almost always converted from viscometer readings in some other kind of unit.

The same comment applies to calorific values; without regard to how they are measured or reported they almost always end up being used as some multiple of *kcal/kg* or *Btu/lb*.

And, of course, the world oil industry's favorite anachronism, the *°API*. The °API never enters into a calculation without being converted into another kind of number, but as a *scale* for measuring *relative densities* (or *specific gravities*, if you will) it is probably written down in more places in any given day than any other number of its kind. Since there will be some emphasis on chemical (as distinguished from refined hydrocarbon product) pipelines, reference will also be made to *°Baumé*, similar to °API, which is another anachronism used widely in the world chemical industry.

SI really doesn't like *horsepower*, but pipeliners do, and so I will have reference to the *hp* used in most English–language countries, defined as 550 ft–lb/s, and to both the French *cheval–vapeur*, *cv*, and the German *pferdestärke, ps* which are equally 75 m–kg/s. Since SI wants to talk about power in terms of *watts*, I will agree with SI that 1 hp = 746 watts, and that 1 cv or ps = 735.5 watts. If a manufacturer wants to deliver me a 2000–kW motor I will know it is really a 2681–hp machine.

And, finally, there are the special parameters—usually a characteristic of a process or a substance—which are mostly of oil industry origin and use. Among these are *cloud point*, *pour point*, *flash point*, *BS&W*, *Reid vapor pressure*, and the like. Since there are generally no alternatives to these I will use them as I find them.

Conversion Factors

Much of the work of the CGPM and BIPM over the years developing definitions and conversion factors has been outstanding diplomatically as well as technically.

Example: For years the British thought the inch was 25.399 978 mm long while the Americans thought it was 25.400 050 8 mm in length. In 1969, however, the parties got together under the auspices of the CGPM and the yard was defined as 9.144 E–01 m, *exactly*; and this makes the foot 3.048 E–01 m, *exactly*, and thus the inch is 2.54 E–02 m, *exactly*. And the world now agrees, as engineers always did, that one inch is 25.4 mm, and while this may not be a great technical achievement it is a diplomatic achievement of high rank and does give official recognition to something which, for practical purposes, had long been an accepted fact.

For those who do not understand the difficult diplomacy of such an agreement we have included Table 1-5. Taken from Orlando,[4] this table shows some of the foot measures used in different countries in relation to the SI defined foot of 304.8 mm.

Table 1-5
Approximate Equivalents of the Foot

Country	Name	mm
Mexico	Pie	279
Netherlands	Voet	283
Poland	Stopa	288
Sweden	Fot	296
Switzerland	Fuss	305
Russia	Øyt	304
Great Britian	Foot	305
Germany	Rheinisher Fuss	313
Denmark	Fod	314
Austria	Weiner Fuss	316
Norway	Fod	314
France	Pied	325
Italy	Piede Librando	513
SI (exact)	Foot	304.8

We pointed out earlier that a liter is now 1 dm^3. CGPM really doesn't like the liter, but it accepts it for liquid measurement. It absolutely doesn't accept multiples or sub–multiples, however; thus a kiloliter is 1 m^3, and a milliliter is 1 cm^3, but SI doesn't like that, either, because the divider of 100 required is not *coherent*.

There is an exact equivalency between the pound and the kilogram, insofar as the U.S. Congress is concerned, because in 1895 it established that the pound was 453.592 427 7 grams, exactly, and the U.S. Bureau of Standards uses this definition. SI relates 1 pound–force to 1 newton by the factor 4.448 221 615 260 5, *exactly*; and dividing this factor by the SI standard gravity of 9.806 650 m/s^2, *exactly*, and multiplying by 1000 to convert to grams, we arrive at 453.592 4 grams which is the same as the U.S. Congress' value to the seventh significant figure. The reciprocal relation gives 1 kg = 2.204 623 lb.

Essentially all of the conversion factors used herein will be derived and used in the form X.YYYY E±ZZ, which is scientific notation with one number and four decimal places in the mantissa. To derive these factors, identities accurate to at least six, and usually seven, significant figures have been used. Such kind of precision is usually not justified (commercial measurement of product volume is an important exception), but is used herein strictly to allow comparison between the results of calculations to assess the *consistency* thereof.

Exact Conversions

Many of the conversions are *exact*, or are defined by the CGPM, the U.S. Bureau of Standards, or some other body of similar standing, as *being* exact. The CGPM has decided that 9.806 650 m/s^2 is the acceleration due to the *standard* gravity g and that is all there is to it. The value of g at your location is doubtless a little different.

Other units, some of which result from multiplying or dividing two exact factors, may not be exact within the five figures used. For instance, the legal definition of the pound required 14 significant figures; and while this may make sense legally it is nonsense from an engineering standpoint, being tantamount to being able to specify the weight of the cargo of a 250 000–dwt tanker to a precision of considerably less than one pound.

Most of the necessary conversion factors can be derived from the equivalencies in Table 1-6 which lists a few of the important relationships between MKS, FPS and hybrid units. To convert centistokes, which have the dimensions $(cm^2/s)\cdot(1/100)$ to FPS units, multiply 1/100 by $1/(30.48)^2)$ to yield the factor 1.0764 E–05. To convert oil barrels to cubic meters we form the product $42\cdot(231/12^3)\cdot(0.30.48^3)$ which yields 1.5899 E–01; or, the inverse, 1 m^3 = 6.2898 barrels.

Table 1-6
Fundamental Equivalents

1 inch	=	2.5400 E-02	*	meters
1 foot	=	1.2000 E+01	*	inches
1 mile	=	5.2800 E+03	*	feet
1 liter	=	1.0000 E-03	*	cubic meters
1 U. S. gallon	=	2.3100 E+02	*	cubic inches
1 U.K. gallon	=	2.7742 E+02		cubic inches
1 oil barrel	=	4.2000 E+01	*	U. S. gallons
1 pound	=	4.5359 E-01		kilograms
1 kilogram	=	2.2046 E+00		pounds
1 gravity	=	9.8067 E+00		meters/second^2
	=	3.2174 E+01		feet/second^2
1 year	=	3.6500 E+02	*	days
	=	8.7600 E+03	*	hours

Note: The asterisk (*) indicates the equivalency is exact.

Conversions Accepted as Being Exact

Conversion factors described in the above *are* exact; they are the product or quotient of other values *known* or *specified* to be exact.

There are other conversion factors that are *defined* as exact. A good example here is the conversion of °C to °F, or the conversion between °API and *relative density 60/60°F*. The conversion is the definition, itself, and is *accepted* as being exact.

Quasi–Exact Conversions

There are other conversions which are taken from *quasi–exact* definitions established by authoritative sources. An example here is the ASTM procedure for converting flow times through Saybolt viscometers to kinematic viscosity in centistokes. These are not exact conversions—and the ASTM makes no claim that they are—but they are the best available and are usually considered as *quasi–exact*.

Precise Conversions of Imprecise Values

Lastly there are what appear to be very precise conversions of imprecise values.

One of these is the *bulk modulus* of liquids, which is defined as

$$K = -(1/V)(dp/dV) \quad (1\text{-}1)$$

The absolute value of bulk modulus can be measured with an accuracy no better than five percent under the best kind of conditions, and even though there is an easy conversion from ksc to psi accurate to five significant figures it is not technically correct to state that a bulk modulus measured as being 190 000 psi is the same as 13 359 ksc.

However, in the world of calculators and computers it is often proper to make such a conversion if the converted number is to be used in arriving at an intermediate result in a calculation because to do otherwise may destroy the numerical integrity of the number and so the *consistency* of the calculation. We will see in Chapter 2 the use of 16– and 17–digit intermediate results to assure six places in the result.

Oil Field Conventions

There are also oil field conventions, in addition to all of the kinds of measurements and conversions described previously, where the appearance of exactness is a commercial necessity.

Thus, a crew takes a good quality steel tape, measures a large oil tank with it, and calculates the volumes to be contained in it for each one–eighth inch of level to the nearest one–hundredth barrel of capacity for each such level; a pair of gagers using an ordinary commercial steel tape and plumb bob agree on the opening and closing liquid levels to the nearest one–eighth inch, opening and closing temperatures to the nearest one–half degree Fahrenheit, opening and closing densities to within one–tenth degree API, and the content of basic sediment and water to the nearest one–tenth of one percent; run tickets are computed using five– or six–place correction tables and two numbers, representing the computed opening and closing volumes in the tank, are thus determined; the amount of product delivered to or from the tank is calculated by subtracting one from the other; and this number, reported to the nearest one–hundredth barrel, is reported as the volume of product bought or sold, *exactly*. Note that the calibration, measurement, correction and computation procedures do not aim at exactness, and no one believes they are exact, but the results are *presumed* to be exact. And to do otherwise would be to invite commercial chaos, because no one *ever* buys, or sells, *approximately* any amount of a fungible product if said product has a *precise unit price*.

Errors and Mistakes

Hand–held calculators, and computers of any kind, can give one a sense of accuracy which is not justified because these devices normally compute in terms of 10 or 15 digits and, if not restrained, will produce output values of such precision. Some of the better workstations use a 64–bit internal representation of floating–point numbers and, while it is good to have such accuracy available to minimize the tendency of intermediate results to introduce propagation errors into the final answer, output displayed or printed to the decimal precision represented by a 64–bit binary number is overkill of the worst kind.

For engineering purposes it is not logical to design a device to a precision higher than that to which it can be built. If a parameter can be measured to an accuracy of one digit in the third place, computations to a precision of one in the fourth place are justified, and even one in the fifth or sixth place if intermediate results are involved in the calculation; but using more figures is a delusion and can yield ludicrous answers. I have seen reports in which pipeline flows were calculated, and reported, to six significant figures *just because that was the way the computer output the numbers.*

But computers have no judgment, and sometimes they are just *wrong*: many software languages will give $\sqrt{4}$ as 1.9999999999, and while this is certainly close enough, it is not correct: $\sqrt{4} = 2$.

What calculators and computers can give us, which we never had before, is *consistency*. A calculation can be made over and over again, by different people at different times in different places, and, if the same input data is used in each case, each will get the same result. And this is, if nothing else, intellectually comforting, because if a calculation can be repeated, exactly, under different circumstances, the calculation *procedure* is probably correct. Whether the procedure yields the *correct answer* to the problem is another matter.

Consistency is a matter of handling *errors* in a uniform manner so as to obtain a uniform output; there is no way to handle a *mistake*.

Like the computer that couldn't calculate the square root of four, a mistake is *wrong*. Mistakes result from improperly applied, or incorrect, formulas or procedures (if we were writing on computers we would have said *algorithms*).

The GIGO of computerese, which means *garbage in–garbage out*, is in common use. It describes the results of inputting wrong data to the computer.

There ought to be an acronym GIJO, meaning *good input–junk output*, to describe the case where the data *input is correct* but the *output is not*.

And it is in this context we make the distinction between *errors* and *mistakes*.

Errors result from mistakes (or errors, if you will) in the *data*; *mistakes* result from errors (or mistakes, if it makes more sense to you) in the *procedure*.

We should use the computing capacity of calculators and computers to check the numeric consistency of our calculations, but not make the mistake of believing all those numbers. The validity of most of the computations made with calculators or computers today is not sensibly better than those we made years ago with the ten–inch slide rule*: there are, however, more problems we can solve.*

There are, of course, exceptions; but the general rule is that the *data*, by which is meant the variables we input, and the *procedures*, by which is meant the formulas and algorithms we plug into our calculators and computers, don't justify the consistency, which is a kind of numerical precision, routinely available today.

Some of the time it is just the input data which doesn't justify such digital overkill and some of the time it is only the procedure which is fundamentally incorrect.

We should take a hard look at the confusing world of accuracy, precision, repeatability, reproducibility, and consistency; understand them for what they are so we can recognize them when we see them; and adjust our pipeline world accordingly.

Accuracy, Precision, Repeatability, and Reproducibility

It is important to understand the difference between *accuracy* and *precision*. The definitions of ASTM E177,[5] are very good:

> *Precision* of a measurement process refers to the degree of mutual agreement between the individual measurements of the process.
>
> *Accuracy* refers to the degree of agreement of such measurements with an accepted reference level of the property of the material measured.

Accuracy applies to conversion factors, for instance; and the fact that conversion factors with only five significant figures have been selected for use herein implies that there is a certain inaccuracy in them unless, as in some case, exact conversion is possible with five or fewer significant figures.

Precision can be considered as describing the results of measurements of the properties of a material and, by extension, of calculations using such results. ASTM E177 says "Precision may be stated in the term of an *index of precision* in the form of $\pm a$, where a is some positive number...In general, the larger the index, the less precise the process...Thus, these indexes are really direct measures of imprecision."

Certain standards differentiate between *repeatability* and *reproducibility* as different kinds of *precision*.

ASTM D1298[6] says:

> 9.1 The following criteria should be used for judging the acceptability of results:
>
> 9.1.1 *Repeatability*—Duplicate results by the same operator should be considered suspect if the results differ by more than the following amounts:
>
> (lists of acceptable values)
>
> 9.1.2 *Reproducibility*—The results submitted by each of two laboratories should not be considered suspect unless the results differ by more than the following amounts:
>
> (lists of acceptable values)

There are many variables, most of them derived from measured physical quantities, that enter into pipelining calculations. Each kind of variable and, sometimes, each variable, has special characteristics which make a varying precision of measurement intrinsic to the variable.

Length, for instance, can be measured to a very great precision, although on pipelines fourth–order surveying accuracy is usually accurate enough and a precision no better than about 1:2000 or about 0.05% can be expected. *And this is the most precise parameter apt to enter into a pipeline calculation.*

The most *imprecise* parameter ordinarily encountered is the previously mentioned bulk modulus of liquids. Not only is the measurement difficult to make, but it is both temperature and pressure sensitive, which makes the selection of the value to use in calculations more an act of judgment than of technical fact.

Vapor pressures are also notable in that the measurement of vapor pressure is a tedious, difficult, procedure that can be easily performed improperly without the operator knowing it. The result is that the vapor pressure determined is well removed from the actual vapor pressure. This kind of error in a parameter is more in the nature of a *mistake*, however, than an *error*, even though the imprecision applies to an input variable.

Other parameters, especially measurements of the characteristics of fluids such as density, viscosity, and pour point, can technically be quite precise but generally aren't. It is very seldom that an operator will take all of the painstaking little steps set out in the procedure for the determination of pour point, for instance. This is another case where the spirit might be willing, but the flesh could care less. We should always beware of values derived from measurements of the characteristics of fluids, unless we have great faith in the source thereof.

It is often necessary to estimate the effect of an error in one or more variables on the integrity of the answer. The problem where only two variables are to be combined to obtain a result is relatively simple. In cases where three or more variables are involved we must either go into probability theory and statistical methods or, what is simpler and often just as effective, assume all the errors are concentrated in only two of the variables, which means, of course, that we assume the values of all other variables are precisely correct. Davis[7] quotes a rule of thumb attributed to a nineteenth century physical chemist named Ostwald which says "A variable error may be neglected if it is less than one–tenth of a large error, and often indeed if it is less than one–fifth." By use of this rule—which has a very sound mathematical basis—it is usually possible to pick out of any group of variables the variable, or pair of variables, which may tend to error so much more rapidly than other variables that it is not necessary to consider possible or probable errors in the other variables at all.

Marshall[8] set down a simple set of rules that are also mathematically well founded. The rules assume two variables that have true values of X and Y. The errors in the measured values of X and Y are n and m. The measured value of X has the higher percent of error.

Addition

Fraction of error in the result = $n+Y(m-n)/(X+Y)$

Rule: When two or more numbers are added, the sum cannot be in error by a larger percent than the percent of error in the most erroneous number.

Subtraction

Fraction of error in the result = (nX–mY)/(X+Y)

Rule: If one inaccurate number is subtracted from a nearly equal number, the fraction of error in the resulting difference may be very great. If one of the numbers is small compared to the other, the fraction of error in the difference approaches the fraction of error in the larger number.

Multiplication

Fraction of error in the result = n+m

Rule: When two inaccurate numbers are multiplied, the error in the product is equal to the algebraic sum of the individual errors.

Division

Fraction of error in the result = (n–m)/(1+m)

Rule: When one inaccurate number is divided by another, the maximum possible fraction of error in the quotient is equal to the sum of the fractions of error in the divisor and dividend.

Exponentiation (Error confined to the number)

Fraction of error in the result = nP (P is the exponent)

Rule: When an inaccurate number is raised to an accurate power, fractional or otherwise, the result will have a percent of error equal to the percent of error in the number multiplied by the power.

Exponentiation (Error confined to the exponent.)

Fraction of error in the result = $X^{Pn}-1$

Rule: When a correct number is raised to a power which is in error, the fraction of error in the result may be very high or it may be very low. Each situation should be analyzed separately.

These are relatively simple rules and formulas based on a single determination of the variables and a single calculation involving only two variables. When more than one—perhaps several dozen—determinations of the value of the variables are available, it is necessary to go into statistical techniques to arrive at an appropriate answer. A good set of instructions for using statistical methods on hydraulic measurements can be found in API 2534.[9] While this standard is specifically applicable only to measurement of hydrocarbons by turbine meters, the statistical methods so well described therein are applicable to any homogeneous, statistically valid set of measurements.

At which point it might be well to introduce an observation attributed to Dr. William E. Deming having to do with statistics generally:

> "Without a homogeneous population, a statistician's calculations by themselves are an illusion, if not a delusion."

Which is an example of garbage in–garbage out *without* a computer.

And a paraphrase of another quote from ANSI Z1.2[10] is in order:

> "In the choice of action limits for indicating when to look for assignable causes of variation, an attempt is made to strike an economic balance with respect to the net consequences of two kinds of *errors* that may occur in practice; namely, *looking* for errors that *do not* exist, and *not looking* for errors that *do* exist."

While it is important to develop an instinct that tells us that the second kind of error may be in play and should be pursued, it is just as important to develop a feel for the kinds of errors that, if they do exist, will be so minor as to be inconsequential in the final result.

As a final thought, here is a paraphrase of a warning by Littlewood and Strigini[11] on the validity of complex computations of any kind:

> "Despite rigorous and systematic testing, most large programs contain some residual bugs when delivered. The reason for this is the complexity of the source code....
>
> "A program can yield the wrong answer because the particular inputs that triggered the problem had not been used during the test phase when defects could be corrected....
>
> "The situation responsible for such inputs may even have been *misunderstood* or *unanticipated.*

> "The designer either *correctly* programmed the *wrong* reaction, or just *completely failed to take the situation into account.*
>
> "This kind of bug is the most difficult to eradicate."

The inputs which lead to the kind of wrong answer that Littlewood and Strigini describe may be perfectly good, valid, input items, but the program had never seen that kind of combination of inputs, or *combination of values* of inputs, and so generated a wrong answer.

All of which leads to two observations on inputs:

- It is sometimes more difficult to establish the proper conditions for making the measurement of an input parameter than to make the measurement itself. Measuring discharge pressure of an out of balance centrifugal pump is an example of a nearly impossibly difficult measurement. Measuring viscosity at flowing temperature is another.
- It is also philosophically difficult to take the correct decisions when selecting inputs to use, given a wide range of possible inputs for several variables.

For instance, friction loss calculations—the seminal calculation of all pipelining—are usually taken as isothermal in nature. Only when the characteristics of the flowing fluid vary widely with temperature, or when the temperature varies widely along the length of the pipeline, is temperature taken into account.

Yet, calculations based on a *fixed* average temperature, and an *unchanging* design fluid, a procedure followed too often, will *by definition* be in error when the temperature rises above or drops below the average temperature because the characteristics of the design fluid will vary as the temperature varies.

Given these realities, it can be understood how a pipeline design based on a design fluid taken as the most difficult (highest density, highest viscosity) product the pipeline is ever apt to encounter, and a flowing temperature based on the lowest ground temperature ever recorded in the region, is *over designed* as to pipe tonnage and *under designed* as to pumping horsepower with the result that a terribly uneconomic pipeline is built. Pipelines conceived and built against such assumptions cost too much and operate inefficiently.

The procedure described is common, but it is incorrect. It is the result of a major failure in the mental process of establishing the inputs for the very large number of decisions which must be taken to realize a pipeline.

Throughout this book there will be reminders of the practicality of making certain calculations, as well as consideration of the precision to which they can be made, considering the way the variables involved are obtained.

And, in all cases, we will assume it is understood that pipelining is a *craft*, and that it is not, and never was, a *science*.

Standards

Throughout this book are quotations from standards promulgated by authorities such as the American Society for Testing Materials (ASTM), the American Petroleum Institute (API), and the American National Standards Institute (ANSI).

These references to standards are for two completely different reasons.

The most obvious is that a standard is a *standard*, whatever one may think of it. Sometimes it must be used as a matter of legal or commercial acceptance.

The other is that the *thinking* behind a standard, which quite often is summarized and included in the standard as complementary information, is thinking of the very best kind. If you can satisfy a committee of your peers that some fact or method belongs in a standard, you are probably right and the fact or method can be trusted.

I have not tried to keep references to standards up to date as of the latest publication date of this book. Laying aside the fact that to do so would be costly, the facts are that standards change slowly, for the most part; and a quotation from or citation of a standard published in (say) 1977 is not apt to be measurably different from a similar extract from a (say) 1987 edition.

A major exception to this thesis—that most standards change slowly, if at all—is *Petroleum Measurement Tables, Volume Correction Factors* (Volumes I through XI), API Standard 2540 (also IP 200, ASTM D1250, ISO R91, and ANSI/ASTM D1250), which went in a set of books and came out a set of computer programs in Fortran IV. This situation is discussed in detail in Chapter 2.

Also, the API *Petroleum Measurement Standards*, now a set of books filled with three–hole punched pamphlets and booklets, is in flux at this writing and has been so for many years, and I may quote from an API standard with an incorrect identification number because standards once moved into the *Petroleum Measurement Standards* lose their numerical identity. They usually maintain their titles, approximately if not exactly.

References

1. ASTM E380-68. *Metric Practice Guide*. Philadelphia: American Society for Testing Materials.
2. ASTM E380-72. *Metric Practice Guide*. Philadelphia: American Society for Testing Materials.
3. ASTM E380-76. *Metric Practice Guide*. Philadelphia: American Society for Testing Materials.
4. Orlando, R. "Building Standards Development in Sweden and the Metric Building World." NBS Publication No 504, *Metric Dimensional Coordination–The Issues and Precedent*. Washington: National Bureau of Standards, 1978.
5. ASTM E177. *Use of the Terms Precision and Accuracy as Applied to Measurement of a Property of a Material.* Philadelphia: American Society for Testing Materials.
6. ASTM D1298. *Density, Relative Density (Specific Gravity), or API Gravity of Crude Petroleum and Petroleum Products by the Hydrometer Method.* Philadelphia: American Society for Testing Materials.
7. Davis, L. M. "Error Analysis for Assessing Accuracy of Liquid Measurement." *Pipeline Engineer*, Mar. 1968.
8. Marshall, R. B. *Measurements in Electrical Engineering*, 2nd ed. Cincinnati: John S. Swift & Co., 1948.
9. API 2534. *Measurement of Liquid Hydrocarbons by Turbine Meter Systems*. Washington: American Petroleum Institute.
10. ANSI Z1.2. *Control Chart Method of Analyzing Data.* New York: American National Standards Institute.
11. Littlewood, Bev. and Strigini, Lorenzo. "The Risks of Software." *Scientific American*, Vol. 267, No. 5., Nov. 92, page 62.

2

Density, Relative Density, Specific Gravity, and API Gravity

What Is Density?

Albert Batik, writing editorially in the July 1977 issue of *ASTM Standardization News*, discussed the difficulty of agreeing on a standard definition of the most simple things and concepts. He gives as an example the word *lot*, which 15 ASTM committees have given 25 definitions; and he mentions that there are 17 definitions of *viscosity* written by 13 technical committees.

He should have looked further. *Glossary of ASTM Definitions, 1973*, lists 25 definitions for *density*, and if to these are added definitions of density that are modified by an adjective or an adjectival phrase, the total number of definitions of density rises to 51. All of which only serves to point out the difficulty of naming fundamental properties that are not described by a single accepted definition of some recognized authority, which is itself accepted as the final arbiter and issuer of definitions.

In the absence of such arbiter the following definition is offered:

> Density, relative density, specific gravity and API gravity are in the nature of measures of the quantity of a substance expressed in terms of its mass per unit volume.

This definition, perhaps in a few more carefully chosen words, will suffice for most ordinary work. However, as soon as any serious *commercial* work, or any kind of *precise* work, for that matter, is contemplated, we must look more thoroughly into these parameters and try to really understand what they mean in the context in which they are used.

Commercial Density

Density is certainly the most important commercial characteristic of crude oils and refined products that can be determined without performing some kind of chemical analysis.

It is almost universal practice to buy, sell, or transfer custody of petroleum products by volume measure, i.e., (gallons, barrels, cubic meters), but since the volume of these liquids varies considerably with temperature, it is necessary to stipulate in the documents of trade just how the volume amount of the transaction is to be measured. Almost without exception the volume is presumed to be measured at a *standard temperature*; and if it is *not in fact* measured at the standard temperature, the volume at the temperature of measurement is corrected to yield the volume the liquid *would have had if it had been measured at the standard temperature*.

There is an important difference between commercial and technical density buried in that last sentence:

- Commercial density is *always* specified as at a *stated temperature*, and this is almost always a *standard temperature*.
- Technical density, which is the density of pipeline hydraulics and related fields, is the density at whatever temperature it happens to be measured.

The factor used to correct the measured volume to the standard volume is called the *volume correction factor*, usually abbreviated *VCF*, which is the ratio of the density of the liquid at the temperature of measurement to the density of the substance at the standard temperature. The method used to calculate the density of the substance at the standard temperature, having in hand its density at the measured temperature, involves the *coefficient of thermal expansion*, usually abbreviated as the Greek α.

Density as a commercial characteristic of a petroleum product is usually determined and used to four or five decimal places.

Technical Density

Density is the most important *technical* characteristic of a liquid flowing in a pipeline. Pressure, pressure loss, horsepower, and quantity pumped, all vary directly with density. Some other operating measures vary with density in a secondary way.

When density is used as a technical characteristic of a crude or refined product, four decimal places are always sufficient. Quite often three places are satisfactory; and, sometimes, a density accurate to two places is perfectly acceptable.

Measurement of Density

Commercial measurements of petroleum must have the characteristics of commercial exactness, i.e., though a measurement is known to only a certain precision, it is presumed to be exact, and the world's oil industry has agreed long ago how density, relative density, and API gravity are to be measured, the instruments to be used to measure them, the method of reporting them, and the precision with which we can expect the measurements to be made by persons skilled in the trade making such measurements on a routine, day–to–day basis.

The standard in its present form is *API Manual of Petroleum Measurements,* Chapter 9.1,[1]which also carries the ANSI/ASTM D1298 standard number. This standard is usually called by its ASTM designation.

There are routinely available methods for determining density to six or more decimal places, but these methods have no place in ordinary pipelining and such precision is seldom required even for commercial purposes.

Density may be expressed as *absolute density*, by which is meant the mass of a unit volume of a substance, or *relative density*, which is understood to be the ratio of the mass of a substance to the mass of an equal volume of a reference material.

Relative density is often called *specific gravity*, though the term is being replaced with *relative density* in much of the modern literature and in newly issued standards. In the oil business relative density is often expressed as *API gravity*, reported in *degrees* and defined as

$$Sp\ Gr = \frac{141.5}{131.5 + API^{\circ}} \tag{2-1}$$

Density of non–petroleum liquids are sometimes reported in *degrees Baumé* defined, for liquids lighter than water, as

$$Sp\ Gr = \frac{140}{130 + B\acute{e}^{\circ}} \tag{2-2}$$

For liquids heavier than water another *Baumé* scale, with 145 as the modulus in both the numerator and the denominator of the equation, is used.

For liquids, the reference material is almost always water; for gases it is almost always air.

It is necessary to specify the conditions of temperature and, for gases, pressure at which the referenced mass applies. It is also necessary to specify whether the mass is measured as if buoyed by the air—under some certain conditions of air pressure and temperature, of course—or if it is measured as if in a vacuum (*in vacuo*).

Values of the density of water at several temperatures used in world standards are shown in Table 2-1. Note the values of absolute density are given to six decimal places, and those of relative density only to five places, though each selection is from the *Smithsonian Tables* which are the standard for such things.

ASTM D1298 describes the use of a glass hydrometer, properly calibrated, to measure density, relative density, or API gravity of a petroleum liquid having a RVP of 26 psia or less.

Table 2-1
Density of Water

Temperature °C	Absolute Density[1] kg/cm³	Relative Density[2] kg/cm³
0	999.842	999.87
4	999.973	1000.00
15	999.099	999.13
15.56[3]	999.012	999.04
20	998.203	998.23

[1]Density of water at 3.78°C=999.87
[2]Density of water at 4.00°C=1000.00
[3]15.56°C=60°F

It is interesting to note the definitions in the standard as to the value of the output of the measurements:

6. Definitions

6.1 *density*—for the purpose of this method, the mass (weight in vacuo) of liquid per unit volume at 15°C.

6.2 *relative density (specific gravity)*—for the purpose of this method, the ratio of the mass of a given volume of liquid at 15°C (60°F) to the mass of a given volume of water at the same temperature. When reporting results, explicitly state the standard reference temperature, for example, relative density (specific gravity) 60/60°F.

6.3 *API gravity*—a special function of relative density (specific gravity) 60/60°F, represented by

API gravity, deg = (141.5/sp gr 60/60°F)–131.5

No reference to temperature is required, since 60°F is included in the definition.

6.4 *observed values*—values observed at temperatures other than the specified reference temperature. *These values are only hydrometer readings, not density, relative density (specific gravity), or API gravity at that other temperature.*

The phrase "The sample has an API gravity at 37°F at 45.6 degrees" is not meaningful: by definition there is no "API at 37°F." D1298 reinforces this statement later on in the text where it says

10. Calculations and Report

Note 9—Hydrometer scale readings at temperatures other than the calibration temperatures (15°C or 60°F) should not be considered as more than scale readings since the hydrometer bulb changes with temperature.

There are ways to correct for the expansion of glass hydrometers to a certain extent. These are discussed under *Petroleum Measurement Tables* herein.

The *repeatability* of measurements on opaque liquids (the most usual case) is 0.0006 density units, 0.0006 relative density (specific gravity) units, or 0.2 degrees API. *Reproducibility* is 0.0015 density units, 0.0015 relative density (specific gravity) units, or 0.5 degrees API.

There are other methods available. ASTM D1217, for instance, describes use of the Bingham–type pycnometer. This is a laboratory method yielding densities and relative densities to a *repeatability* of 0.00002, a *reproducibility* of 0.00003, and an *accuracy* (compared to theoretical values) of 0.00003. These are considerably better than the hydrometer method of D1298 but the method is a research–type test and is not practical for routine commercial—or technical—determinations.

Effect of Temperature and Pressure on Density

Both temperature and pressure affect the density of a liquid.

The effect of *temperature* is a first–order effect and is important commercially. It is also important technically: density is one of the three factors entering into the one–dimensional continuity equation $\rho_1 v_1 A_1 = \rho_2 v_2 A_2$, and if $\rho_1 \neq \rho_2$, and $A_1 = A_2$ or $A_1 / A_2 = Constant$, the *volume* passing point 1 is not equal to the *volume* passing point 2 even though (in non–transient conditions) the *mass* passing the two points is equal. Precise flow calculations, especially on very long pipelines on which input and output flowing temperatures may differ greatly, must consider this effect.

The effect of *pressure* is a second order effect, and is commercially important only when a product volume measured under one pressure must be converted to an equivalent volume at another pressure. This is important when metering light petroleum fractions, but usually does not enter into custody transfer operations involving crude oils. The variation of density with pressure is important in the generation of pressure surges and the like during transient flow conditions, and calculations made toward understanding these phenomenon make use of the concept of density as a function of pressure.

Density as a Function of Temperature

The definition of the *coefficient of thermal expansion* of a liquid is

$$\alpha = \left(\frac{1}{V}\right)\left(\frac{dV}{dt}\right) \quad (2\text{-}3)$$

where α = coefficient of thermal expansion
V = volume at any temperature
t = temperature

In this equation the units of the two volume variables cancel out so that the dimensions of α are $1/t$, i.e., the coefficient of thermal expansion of a certain oil may be given as 0.000500/°F or, preferably, 500×10^{-6}/°F.

Determination of commercially acceptable coefficients of thermal expansion for crudes and products has been a long process. The *Petroleum Measurement Tables* are the result.

Petroleum Measurement Tables–Background

The *American Petroleum Institute* (Washington) and the *Institute of Petroleum* (London) historically have been the sponsors of the *Petroleum Measurement Tables*. These are the commercial guides for buying and selling crude oils and refined products around the world, and provide uniform, accepted ways, of converting densities and volumes measured at one temperature to densities and volumes at another (usually a standard) temperature.

They also contain, or contain references to, some of the best information available on the thermal behavior of crude oils and refined products, and so are important technically as well.

Petroleum Measurement Tables–1953 Edition

The first edition of the *Petroleum Measurement Tables* was derived from Circular 410, *National Petroleum Oil Standards*, published by the U.S. National Bureau of Standards in 1936, and the *Tables for the Measurement of Oil*, issued by the British Institute of Petroleum in 1945. This first edition, issued in 1953, went through more than a dozen printings before being retired in 1980.

The *Tables* were originally issued in three versions:

- The *American* version, based on U.S. gallons, oil barrels, short tons, degrees Fahrenheit, degrees API and specific gravity 60/60°F ;
- The *British* (sometimes called the *Imperial*) version, based on imperial gallons, oil barrels, long tons, degrees Fahrenheit, degrees API and specific gravity 60/60°F; and
- The *Metric* version, based on liters, cubic meters, metric tons, degrees Celsius, and density 15°C.

All three were founded on measurement of *weights in air*, with corrections for the buoyancy of air based on a standard atmosphere with a density of 0.00122 g/cm³, and for brass weights having a density of 8.1 g/cm³, both at 0°C.

For the convenience of those operating in countries where measurements must be based on *weights in vacuo* the *Tables* provided conversion factors for densities 15°C in the range of densities from 0.5000 to 1.1000. For these conversions, densities are presumed to be given to four decimal places and conversion factors are given to five places.

The *Petroleum Measurement Tables* (hereinafter abbreviated *Tables*) were a monumental piece of work. Containing 58 tables in the three editions, they were essentially nothing but hundreds of pages of tabulations of factors of one kind or another, each factor the result of an individual calculation. The tables themselves ranged in size from a single page to monumental Table 53 which ran to some 250 pages. Over 250 000 IBM punch cards, weighing over two tons, were required for calculating and proofing the *Tables*.

Petroleum Measurement Tables–1980 Edition

The *Tables* were unchanged, except for minor additions and some errors found and corrected, until 1980 when a completely new edition was issued.

There were several reasons for the new edition, but primarily they were developed because of work done by Downer and Inkley[2] which indicated that the expansion coefficients used to develop the old *Tables* in the range of densities appropriate to some crude oils were incorrect for several crudes of then–current economic importance. Their investigations, involving 15 crudes, 13 heavy fuel oils, and 8 other refined products, found that almost without exception crudes expanded approximately eight percent more than predicted by the old *Tables*. The coefficients for products were found to be more accurate. Downer and Inkley's data are summarized in Table 2-2.

Table 2-3 summarizes the coefficients of expansion for the eight *Groups* used in calculating the old *Petroleum*

Table 2-2
Mean Expansion Coefficients
Crude Oils, Fuel Oils, and Other Products

Material	Sp Gr 60/60°F	Mean Expansion Coefficient
CRUDE OILS		60–100°F
Abu Dhabi Marine (Zakum)	0.8256	0.000527
Abu Dhabi Land (Murban)	0.8285	0.000521
Barrow Island (Australia)	0.8399	0.000499
Barrow Island (Australia) Repeat	0.8400	0.000502
Libyan Sarir (Tobruk Export)	0.8430*	0.000495+
Libyan Sarir (Tobruk Export) Repeat	0.8431*	0.000498+
Iraq Mediterraneann Grade A (Kirkuk)	0.8451	0.000490
Iranian Light Export Blend	0.8568	0.000486
Kuwait Export Blend	0.8709	0.000462
Nigerian Forcados Blend	0.8705	0.000478
Iranian Heavy Export Blend	0.8732	0.000476
Tia Juana Medium	0.8908	0.000448
Alaskan North Slope Prudhoe Bay	0.8914	0.000454
Alaskan North Slope Prudhoe Bay Repeat	0.8914	0.000454
Nigerian Bonny Medium Blend	0.8988	0.000453
FUEL OILS		140–170°F
Commercial Blend A	0.9108*	0.000411
Abu Dhabi Land (Murban)	0.9217*	0.000411
Libyan Sarir (Tobruk Export)	0.9250*	0.000409
Commercial Blend B	0.9346*	0.000404
Barrow Island (Australia)	0.9347	0.000405
Barrow Island (Australia) Repeat	0.9347	0.000405
Abu Dhabi Marine (Umm Shaif)	0.9362*	0.000402
Iranian Light Export Blend	0.9473	0.000389
Iranian Heavy Export Blend	0.9558*	0.000387
Iraq Mediterranean Grade A (Kirkuk)	0.9567*	0.000384
Iraq Mediterranean Grade A (Kirkuk) Repeat	0.9567	0.000384
Tia Juana Medium	0.9645	0.000379
Kuwait Export Blend	0.9651*	0.000383
OTHER PRODUCTS		60–100°F
Light Car, Cracked Gasoline	0.6811	0.000707
Desulphurized Straight Run Gasoline	0.7346	0.000662
Kerosine A	0.7900	0.000515
Kerosine B	0.7912	0.000538
Diesel Oil A	0.8340	0.000477
Diesel Oil B	0.8480	0.000464
Diesel Oil C	0.8527	0.000470
Lubricating Oil (SAE 10)	0.8754	0.000419

* Determined by linear expansion from higher temperature.
+ Mean expansion coefficient 100-140°F

Table 2-3

Reduction of Volume to 60°F Against Specific Gravity 60/60°F From Petroleum Measurement Tables (1953)

Group No.	Range of Group Specific Gravity 60/60°F	Coefficient of Expansion per °F at 60°F	Corresponding Specific Gravity 60/60°F
7	0.6112 to 0.6275	0.00090	0.6193
6	0.6276 to 0.6417	0.00085	0.6360
5	0.6418 to 0.6722	0.00080	0.6506
4	0.6723 to 0.7238	0.00070	0.6953
3*	0.7239 to 0.7753	0.00060	0.7468
2	0.7754 to 0.8498	0.00050	0.8063
1	0.8499 to 0.9659	0.00040	0.9218
0	0.9660 to 1.0760	0.00035	1.0291

* All blends of gasoline and benzene are considered to fall in Group 3; when the presence of benzene is uncertain, the oil shall be classified in Group 3 if the specific gravity is more than 0.7750 and the 50% distillation recovery point is less than 293°F.

Measurement Tables. Note there are distinct differences between the Table 2-2 and Table 2-3 values for certain *Groups*.

In Table 2-3 coefficients of expansion are presumed to run from 0.00035/°F to 0.00090/°F at 60°F. The step between Group 0 and Group 1, Group 5 and Group 6, and between Group 6 and Group 7, is 0.00005; the step between all other groups is 0.00010. Products are all deemed to belong to Group 3.

These factors were commercially accepted as *exact* though, of course, they were not; and since there were eight different coefficients of expansion—with a discrete change in the value of the coefficient between each of them—a change in specific gravity of 0.0001 in moving from one group to another could result in a change in the coefficient of expansion of 0.00010. The *Tables* were commercially practical; in light of the data of Downer and Inkley they were no longer technically defensible.

The new *Petroleum Measurement Tables* were a joint effort of the *API Committee on Petroleum Measurement* (COPM), the *Joint API/ASTM Committee on Static Petroleum Measurement* (COSM), and the *Physical Properties Working Group* (PPWG) of the COSM. They put together a database of 2278 measurements on 349 samples involving some 600 measurements on 124 crude samples, and the remainder involved a dozen or more miscellaneous refined products, preponderantly gasolines, jet fuels and medium oils. There were also some 107 measurements on 17 samples of lubricating oils, and 77 measurements on miscellaneous products. There were other samples, and other measurements were available, but the database retained was judged sufficient to provide the bases for dividing the data into five distinct populations for further study—one crude group and four refined products groups—and further development was based on these five statistical populations.

The new edition of the *Tables* was first announced to the oil industry in general by Hankinson, Segers, Buck and Gielzecki in late 1979[3] and issued by the API as Standard 2540 in August of 1980. It also was adopted as IP 200, ASTM D1250, ANSI/ASTM D1250, and ISO R91.

In the U.S. the complete set of the *Tables* was issued as *Manual of Petroleum Measurement Standards*, Chapter 11.1, Volumes I–XIV, as part of the API Measurement Coordination series.

The 1980 edition was reaffirmed in 1987. The only major challenge to the *Tables*, a dispute about the selection of the standard measurement temperature for hot climates, was settled by West, Hankinson, and Downer[4] in 1989.

While all the printed tables, or equivalents thereof, of the old *Tables* are contained in the 14 volumes of Chapter 11.1, the single most important volume is Volume X,

Background, Development and Computer Documentation, which describes what is contained in the other 13 volumes and how it got there.

Volume X is extraordinary, as standards go, in that it is, first, a well written, concise history of how the new *Tables* were developed and, second, nearly 400 pages of computer code and associated documentation.

And the standard is the computer code, not the printed tables derived from the code. This is thought to be the first time computer code, rather than the output of the code, has been given the status of a world–wide standard.

For each of the tables (there are 15 of them) there is a section containing

- A statement of the reason for the table,
- A statement as to whether it applies to *generalized crude oils*, *generalized products*, or to *individual and special applications* (for liquids for which thermal expansion coefficients are known *a priori*), and
- Subroutines in Fortran IV to generate the table.

The subroutines, and their associated documentation, are such that anyone reasonably skilled in programming who has access to a computer capable of handling 32-bit integer arithmetic can generate code which, in turn, will generate measurement tables that will *exactly* match other tables generated by the same subroutines anywhere in the world.

This kind of repeatability is assured not only by the use of integer arithmetic, which most all machines will handle identically up to the limit of their word size, but also by the use of division and multiplication algorithms capable of working to 16- or 17-digit precision to assure a minimal introduction of computational errors into the intermediate results.

Probably equally as important is that the truncating and rounding decisions are in the program logic and not in the output stream; in other words, the subroutines will not print out a full string of 17 decimal numbers just because the computer holds them internally.

But the best of Volume X is *History and Development of Petroleum Measurement Tables (1980)*, pp. 73–85. For these pages alone every pipeliner should have a copy of Volume X in his library.

Here we find the equations, and references to the sets of data, used to derive the computer programs and, thus, the *Tables*, and thereby a set of identities and functions for α and VCF to satisfy at one time the requirements of both commercial and technical pipelining.

Once again the best thinking, and the best data, in the industry has ended up in a *Standard*.

Commercial Calculations

We start with the previously given definition for the coefficient of thermal expansion

$$\alpha = \left(\frac{1}{V}\right)\left(\frac{dV}{dt}\right) \tag{2-4}$$

where α = coefficient of thermal expansion
V = volume at any temperature
t = temperature

The form of equation for α in terms of easily measured parameters is

$$\alpha = \alpha_T + \beta\,\Delta t \tag{2-5}$$

where α_T = α at the base temperature T
β = a function of α independent of temperature

From Equations 2-4 and 2-5

$$\left(\frac{1}{V}\right)\left(\frac{dV}{dt}\right) = \alpha_T + \beta\Delta t \tag{2-6}$$

where $\Delta t = t - T$

We can rearrange and integrate between T and t to give

$$\log_e\left(\frac{V}{V_T}\right) = \alpha_T\,\Delta t + \left(\frac{\beta}{2}\right)\Delta t^2 \tag{2-7}$$

A study of the selected database by the PPWG showed that

$$\beta = k\,\alpha_T^{\,2} \tag{2-8}$$

where k = a temperature independent constant

The developers decided on the basis of studies of data in hand that the best all around value for k was 1.6. Substituting the computed value for β in Equation 2-7 we arrive at the following equation for *VCF* which is valid for any liquid of known α:

$$VCF = \left(\frac{V_T}{V}\right) = EXP\left[-\alpha_T \Delta t\left(1 + 0.8\,\alpha_T \Delta t\right)\right] \tag{2-9}$$

where t = any temperature
T = base temperature

It was determined that the coefficient of expansion at the base temperature is related to the density at the base temperature by

$$\alpha_T = \left(\frac{K_0 + K_1 \rho_T}{\rho_T^{\,2}}\right) \tag{2-10}$$

Table 2-4 summarizes the minimum and maximum values of ρ, the values of K_0 and K_1, and the values of α_T, for the five groups making up the population of the measurements studied.

We can rearrange these data in the form of five straight lines on an X–axis of $= (1/\rho_T{}^2) \times 10^6 [\mathrm{kg/cm^2}]$ against a Y–axis of $\alpha_T \times 10^6 (\mathrm{F^\circ})^{-1}$. In other words, plots of the reciprocal of density against the coefficient of thermal expansion. Figure 2-1 is in this form.

The data for crude oils was extrapolated to cover the range from 0 °API to 100 °API, thus a range of specific gravities of 0.6112 to 1.0760. The α_T and K_0 values for the crude oil group were used to produce a straight line relationship between α_T and $1/\rho^2$ over this range and the *generalized crude oils* group was defined. This is shown graphically in Figure 2-2.

The data from the fuel oils, jet fuels and kerosines, and gasolines groups were combined as shown in Figure 2-3 to yield the *generalized products* group. The fuel oils data was used for the range 0–37 °API, the jet fuels and kerosines data from 37 to 50 °API, and the gasolines data finished out the group from 50 to 85 °API.

The discontinuity at 50 °API is handled by making a straight–line interpolation between the value of α_T at 48 °API (0.7885 sp gr) on the jet fuels and kerosines curve and 52 °API (0.7710 sp gr) on the gasolines curve in the general form of $\alpha_{60} = A + B/\rho_{60}^2$ using $A = -0.00186840$ and $B = 1489.0670$ as the two constants.

Once again, no one really believes that α_T in the range 48>°API>52 lies on this line; but if both parties to a

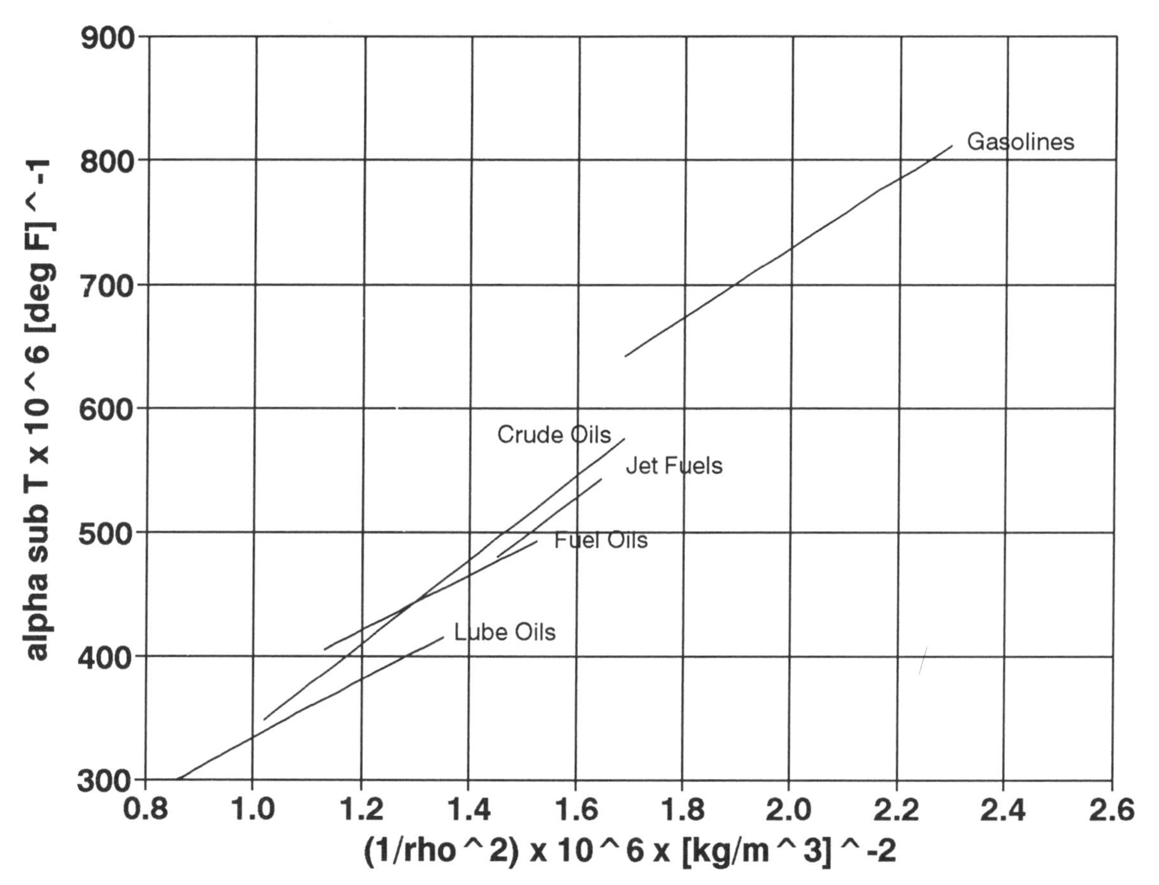

Figure 2-1. Coefficients of expansion for five statistically homogeneous groups of petroleum hydrocarbons.

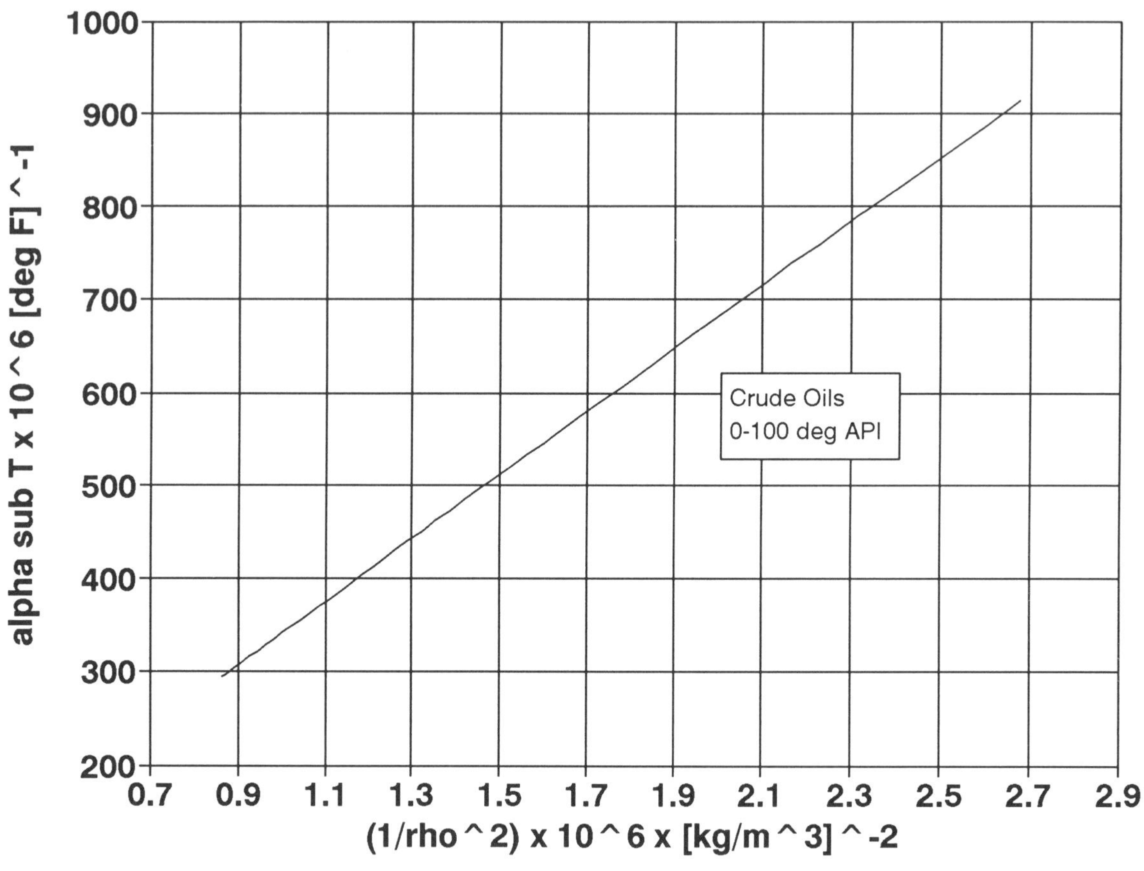

Figure 2-2. Illustrating base for generalized crude oils tables.

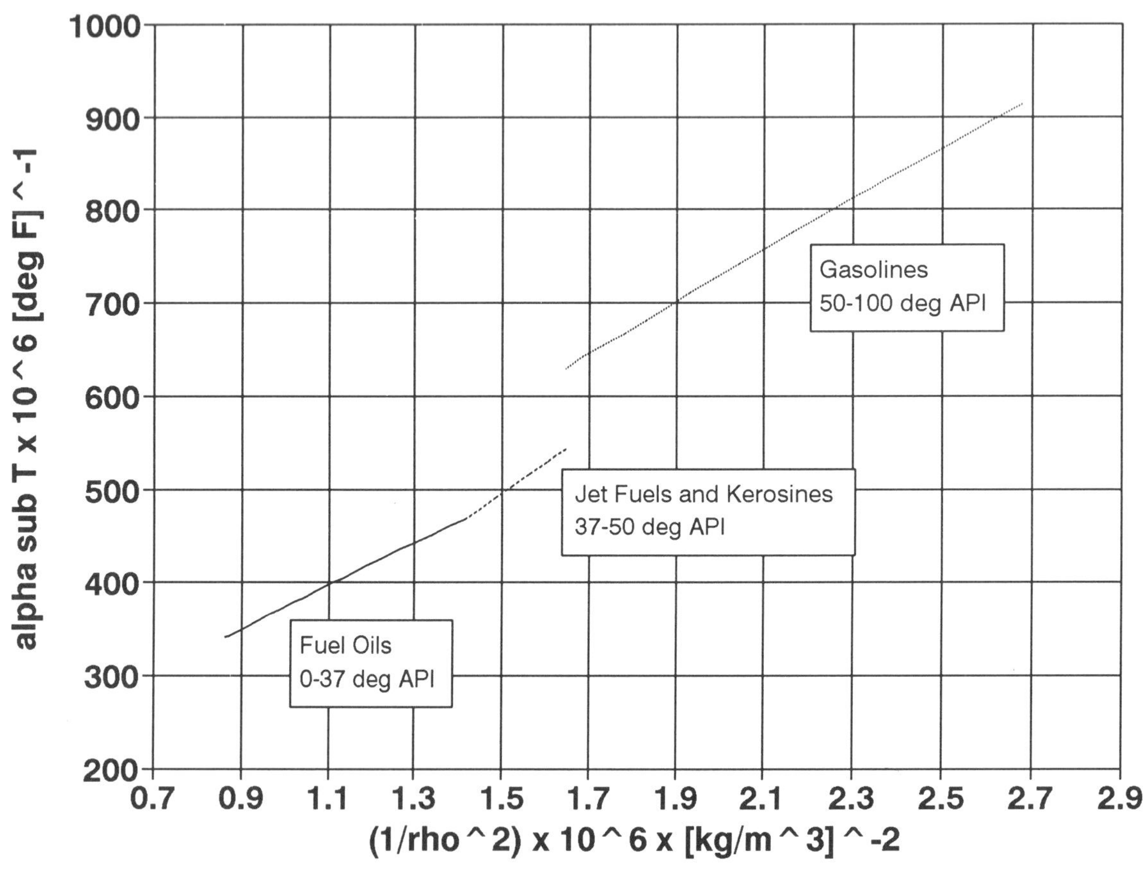

Figure 2-3. Illustrating bases for generalized product tables.

transaction agree that it does, then *commercially it does*. *Technically,* it does *not*.

These data are all incorporated in the Fortran IV subroutines of Chapter X. These calculations are precise:

- The density of water at 60°F is taken as 999.012 kg/m², and the 141.5 factor of the formula for °API (see later herein) is taken as (141360.1980/1000).
- If the specific gravity (relative density) is measured with a glass hydrometer a correction factor is applied to provide for the effect of the expansion of the glass with temperature. The correction factor for temperatures given in Fahrenheit is

$$HYC = 1.0 - 0.00001278(t - 60) - 0.0000000062(t - 60)^2 \quad (2\text{-}11)$$

For temperatures given in Celsius the correction is

$$HYC = 1.0 - 0.000023(t - 15) - 0.00000002(t - 15)^2 \quad (2\text{-}12)$$

These are taken from the *Report on the Development, Construction, Calculation, and Preparation of the ASTM–IP Petroleum Measurement Tables* (1960).

- Contrary to the values in the 1953 *Tables*, all measurements in the 1980 *Tables* are presumed to be in vacuo.

In addition, the term *specific gravity* was discontinued and replaced with *relative density*. Specific gravity, by definition in the 1953 *Tables*, was presumed measured against weights in air; relative density, by definition in the 1980 *Tables*, is presumed measured against weights in vacuo.

This distinction is sometimes important in commercial measurements, especially in countries where weights are presumed to be measured in vacuo. It doesn't mean much technically.

Technical Calculations

We can commence our development of tools for use in working with *technical density*, as differs from the *commercial density* considered in some detail in the preceding three sections, by agreeing on what technical density is *not*. Technical density is *not* necessarily a highly accurate measurement of the quantity of mass of a substance, but it is a very important characteristic which varies with the *kind* of substance, and with the effects of temperature and pressure *on* the substance, and has a great deal to do with how pipelines work.

The considerations of commercial density were given over to the *very* accurate, *repeatable*, determination of ρ, the density of a substance at some temperature; of ρ_T, the density of the substance at some standard temperature *T;* and of α_T, the coefficient of expansion of the substance at *T;* and the volume correction factor, *VCF*.

For technical purposes we need to be able to compute these same values in a *reasonably* accurate, *consistent* way.

Note the use of the word *reasonably* as a replacement for *very*, and *consistent* for *repeatable*.

We can start with the statement from ASTM D1298 cited previously to the effect that the *observed* density, relative density, or API gravity, is only an *observation*; it is *not* density, relative density, or API gravity. And, for commercial purposes this, of course, is true; the standard says so. But for most technical purposes the accuracy of observations of density, relative density or API gravity of a substance is good enough; and if we want to design a new pipeline, or predict the performance of an existing pipeline, the accuracy of the observed values, corrected to the operating temperature of the pipeline, *whether this is the stated standard temperature for the test or not*, is sufficiently accurate.

The plot of sp gr 60/60°F–vs–API gravity in Figure 2-4 shows the relationship of the two values, and for many uses a similar plot, to a larger and more

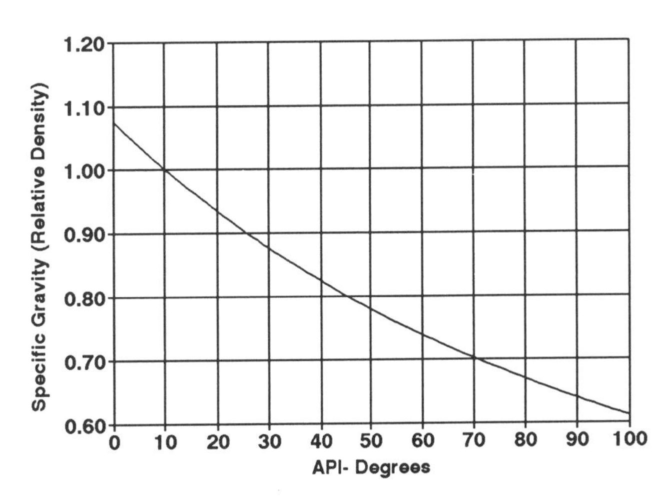

Figure 2-4. Spgr 60/60F -vs- API.

accurately divided set of axes, is sufficient for determining the flowing density of an oil from an observation of API gravity.

Simplifications apply to the other factors we can measure or compute. Equation 2-10 is an example:

$$\alpha_T = \left(\frac{K_0 + K_1 \rho_T}{\rho_T^{\ 2}} \right) \tag{2-10}$$

This can be rewritten in the form

$$\alpha_T = \frac{K_0}{\rho_T^2} + \frac{K_1}{\rho_T} \tag{2-13}$$

From Table 2-4 we know that for crude oils K_0 equals 341.0957 or, for our purposes here, 341, and K_1 equals zero, so that the equation for the coefficient of expansion of crude to an accuracy sufficient for all technical use is

$$\alpha = \frac{341}{\rho_{60}^2} \tag{2-14}$$

This equation is plotted in Figure 2-5. Equations in the form of Equation 2-13 and or 2-14 can be written for the four classes of refined products as well. For measurements using °C and kg/m³, K_0 = 613.9723 so a

Table 2-4

Tabulation of Product Classifications, Classification Ranges, and Thermal Expansion Coefficients

Product Classification	Product Classification Density Ranges kg/m³		Parameters for Formula for Coefficient of Thermal Expansion		Coefficient of Thermal Expansion at 60°F $\alpha_T = \frac{K_o + K_1 \rho_T}{\rho_T^{\ 2}}$
	Minimum	Maximum	K_0	K_1	
Crude Oils	770		341.0957	0	575.30 E-06
		990	341.0957	0	348.02 E-06
Gasolines	657		192.4571	0.2438	816.95 E-06
		770	192.4571	0.2438	641.23 E-06
Jet Fuels	785		330.3010	0	536.01 E-06
		825	330.3010	0	485.29 E-06
Fuel Oils	812		103.8720	0.2701	490.17 E-06
		1075	103.8720	0.2701	341.14 E-06
Lube Oils	861		144.0427	0.1896	414.51 E-06
		940	144.0427	0.1896	364.72 E-06

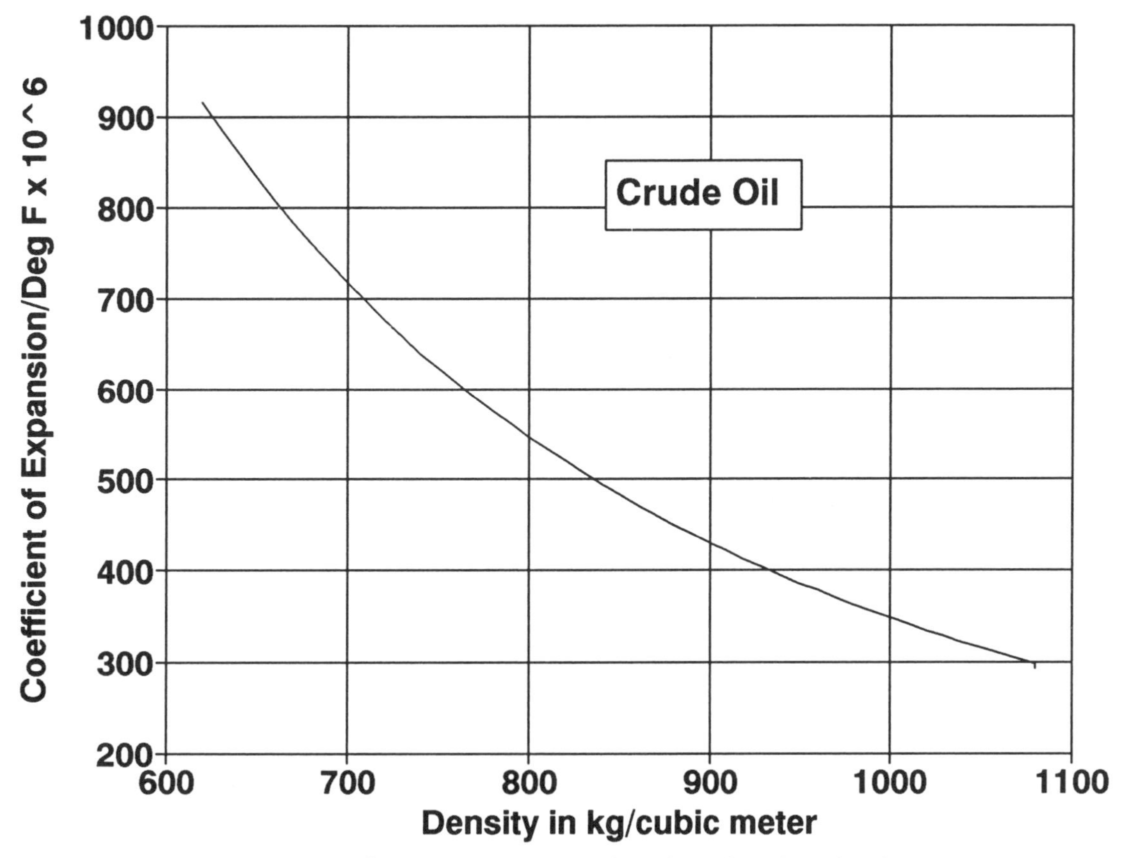

Figure 2-5. Coefficients of expansion plotted against density.

numerator of 614 will be sufficiently accurate for m–k–s calculations.

The volume correction factor defined in Equation 2-9 is

$$VCF = \left(\frac{V_T}{V}\right) = EXP\left[-\alpha_T \Delta t\left(1 + 0.8\,\alpha_T \Delta t\right)\right] \quad (2\text{-}9)$$

This can also be written as

$$VCF = \frac{\rho}{\rho_T} \quad (2\text{-}15)$$

Either formulation will produce a usable value for *VCF*.

We will sometimes be interested in the relationship between the *molecular weight* of a crude or of a product and its density. There is no rigorous direct correlation, of course, because the molecular weight of a compound substance is a well defined function of molecular weights of the individual fractions making up the substance. Without a knowledge of what the fractions are, and the percentages in which they appear, any direct calculation of molecular weight is impossible. E. R. Robinson,[5] however, working with various inputs to develop a set of formulas and associated graphs for use in determining the density of *spiked crudes*,[6] developed two relationships in terms of density that allow at least *estimating* the molecular weight of a crude or product. The relations are:

$$\begin{aligned}
&\text{For } 500 \le D \le 750\\
&\quad MW = -886.37 + 5.1671D - \left(9.79 \times 10^{-3}\right)D^2 + \left(6.3477 \times 10^{-6}\right)D^3\\
&\text{For } 750 \le D \le 1050\\
&\quad MW = -1524.4 + 4.4958D - \left(4.25 \times 10^{-3}\right)D^2 + \left(1.6667 \times 10^{-6}\right)D^3
\end{aligned} \quad (2\text{-}16)$$

Figure 2-6 is a plot of molecular weight against density at 15°C from these two relations. The curves in the vicinity of D = 750 kg/m³ have been smoothed.

We will, later on, be deeply interested in the effect of temperature on water. This is a quite complex relation

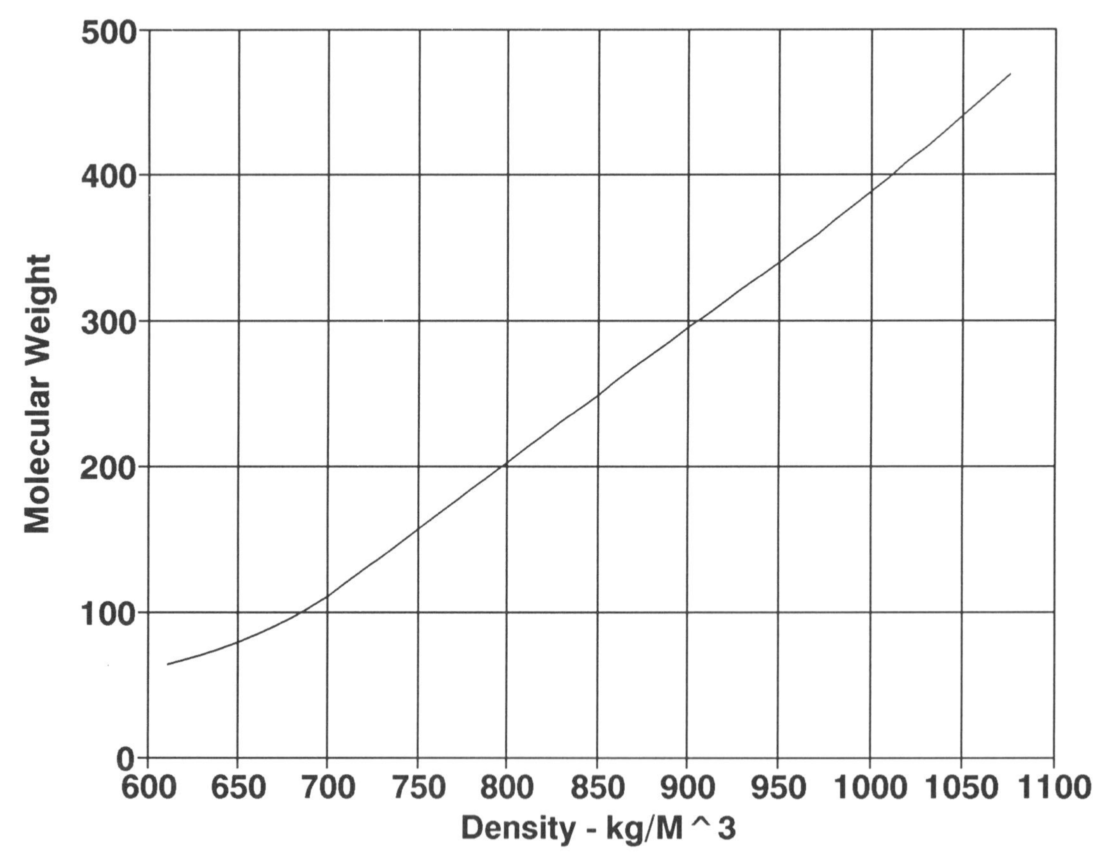

Figure 2-6. Relationship between molecular weight and density.

because the thermal coefficient of expansion varies considerably with temperature. But hydrostatic testing of oil pipelines is usually done with high pressure water; and more and more the high pressure water lines of the world are looking like big-inch oil pipelines; and it is necessary that we have good data on water in our tool kit.

Gray[7] includes an equation based on the Smithsonian Tables that gives the coefficient of expansion of water at atmospheric pressure as a function of temperature in °C.

$$\alpha \times 10^6 = -64.268 + 17.0105t - 0.20369t^2 + 0.0016048t^3 \quad (2\text{-}17)$$

This is much handier than a set of tables since it programs easily into a calculator or computer and is then instantly available. Figure 2-7 is a plot of this equation. Note the coefficient of expansion passes through zero at the point of maximum density at 4°C.

Density as a Function of Pressure

Where we consider the effect of pressure on density and, thereby, volume, we will use the *compressibility* of the product, defined as

$$\kappa = -\left(\frac{1}{V}\right)\left(\frac{dV}{dp}\right) \quad (2\text{-}18)$$

where κ = compressibility
V = volume at any pressure
p = pressure

The units of dV/V cancel, so compressibility is always reported in terms of reciprocal pressure or stress. Thus the compressibility of water at ordinary temperatures and pressures is about 0.0000032 psi or 3.2×10^{-6} psi. The *bulk modulus* is the inverse of compressibility and is reported as a pressure or stress, i.e., the bulk modulus of water is about 310,000 psi.

Commercial Calculations

We are commercially interested in the effect of pressure on density any time pressure affects the accuracy of the volume number written on the sales or custody transfer ticket.

Crudes are for the greatest part traded at atmospheric pressures. This was the case with products, as well, until

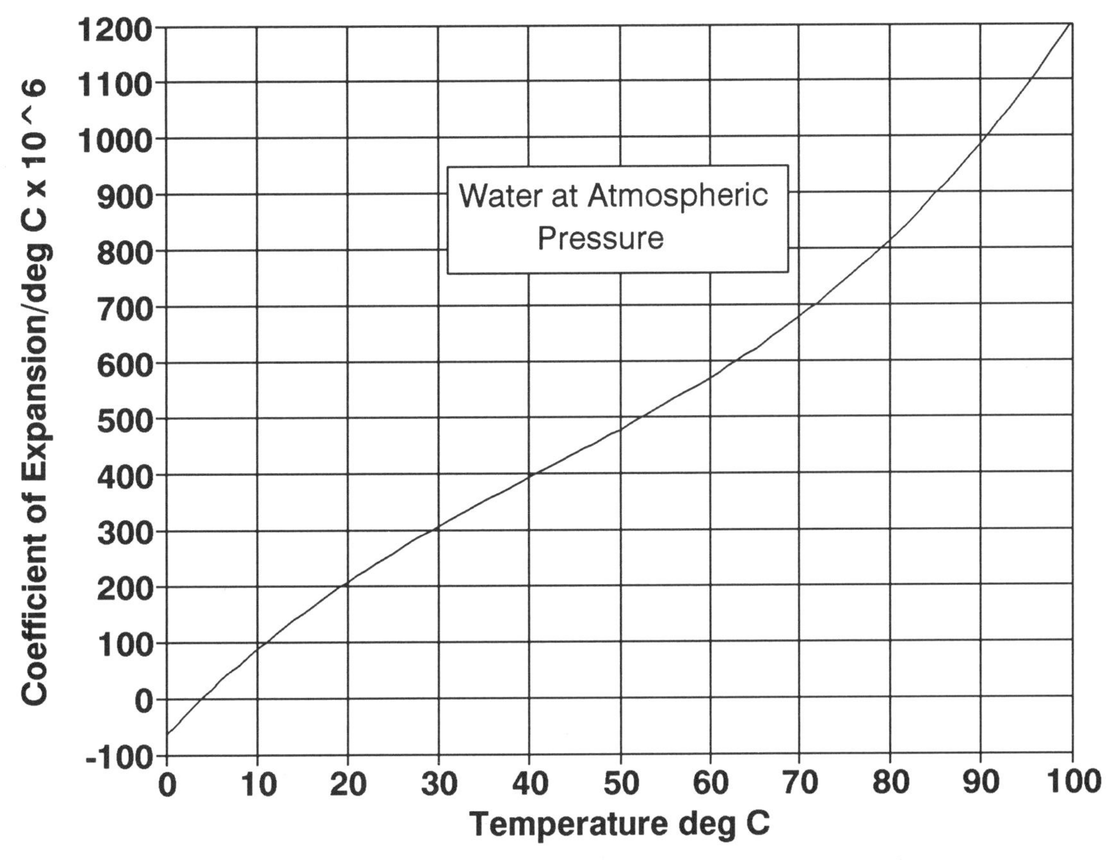

Figure 2-7. Coefficients of expansion plotted against temperature for water.

it became technically possible to ship LPG and other light petroleum fractions by pipeline, which brought up the *desirability*, if not the *necessity*, of maintaining sufficient pressure on the liquid so that it would not flash into a multi–phase mixture of liquid, vapor, and gas and in doing so make accurate volume measurements impossible.

The problem was that the light ends of the petroleum barrel were quite compressible, and metering the volume of light products at pipeline pressures produced troublesome errors.

Considerable work was done on this problem, and the first issue of API 1101[8] contained the famous Table II, *Compressibility Factors per Pound per Square Inch,* compiled by M. L. Barrett, Jr. These data were based on work by Jacobson, Ambrosius, Dashiell and Crawford,[9] a summary of which appeared in a set of curves in API 1101.

The present–day equivalents of the compressibility data in API 1101 are in the API *Manual of Petroleum Measurement Standards*, Chapter 11.2.1, 11.2.1M, 11.2.2, and 11.2.2M. The two Chapters 11.2.1 cover the range 0–90° API and, in the M edition, 638–1074 kg/m³; the two Chapters 11.2.2 cover the lighter hydrocarbons in the range 0.350–0.637 relative density 60/60°F and, in the M edition, 350–637 kg/m³.

Both sets of data are available from API in the form of 9–track, 1600 bpi tapes for inclusion as look–up tables in computer routines working up commercial data.

Technical Calculations

Table II provided a method of correcting for the compressibility of the product when measured at high pressures, but the data were still in tabular form; to use it you had to have a copy of the table and look up the data you needed in the table. This, at a time when computers were beginning to be used fairly widely in the pipeline industry, was a worrisome problem.

Techo[10] undertook to reduce Table 6 of the *Petroleum Measurement Tables*, 1953, (Reduction of Volume to 60°F Against API Gravity at 60°F) and the above mentioned Table II from API 1101, 1960, to equation form. In the case of Table II Techo produced two equations, both equally valid over the range of densities from 21° to 80° API at temperatures from 20°F to 120°F. I personally prefer Techo's Equation 6, which is

$$\beta = 28.328 + (1.07946 \times 10^{-3})t^2 + [(1.02286 \times 10^{-2}) + (5.1659 \times 10^{-7})t^2] \times (°API)^2 \quad (2\text{-}19)$$

Figure 2-8 is a plot of Dr. Techo's Equation 6. Note that it only covers the range 20°>API>80° whereas Table II covers the entire range of 0°>API>100°. Techo states the deviation of Equation 6 over the narrower range is always less than three in the third place; and he gives an equation that is cubic in the squares of the temperature and °API which has a maximum error of 1.5 in the last place.

Robinson[5] drew on data from IP Table 54 to develop the following equation for compressibility in *bars* as a function of density in kg/m³:

$$C = EXP\left(\begin{array}{l} 1.38315 + 0.00343804 \times t \\ \quad -3.02909 \times \ln\left(\frac{\rho_{15}}{1000}\right) \\ \quad -0.0161654 \times t \times \ln\left(\frac{\rho_{15}}{1000}\right) \end{array}\right) \quad (2\text{-}20)$$

where C = compressibility ($\times\ 10^{-5}$ bars)
ρ_{15} = density at 15°C
t = temperature in °C

Figure 2-9 is a plot of Robinson's equation. This covers the range from 600 to 1000 kg/m³, which is approximately the same range as 0 to 100 °API. The range of accuracy is from 0 to 35 bar.

Either the Techo or Robinson equation is satisfactory for technical work; the accuracy available is always better than the accuracy required.

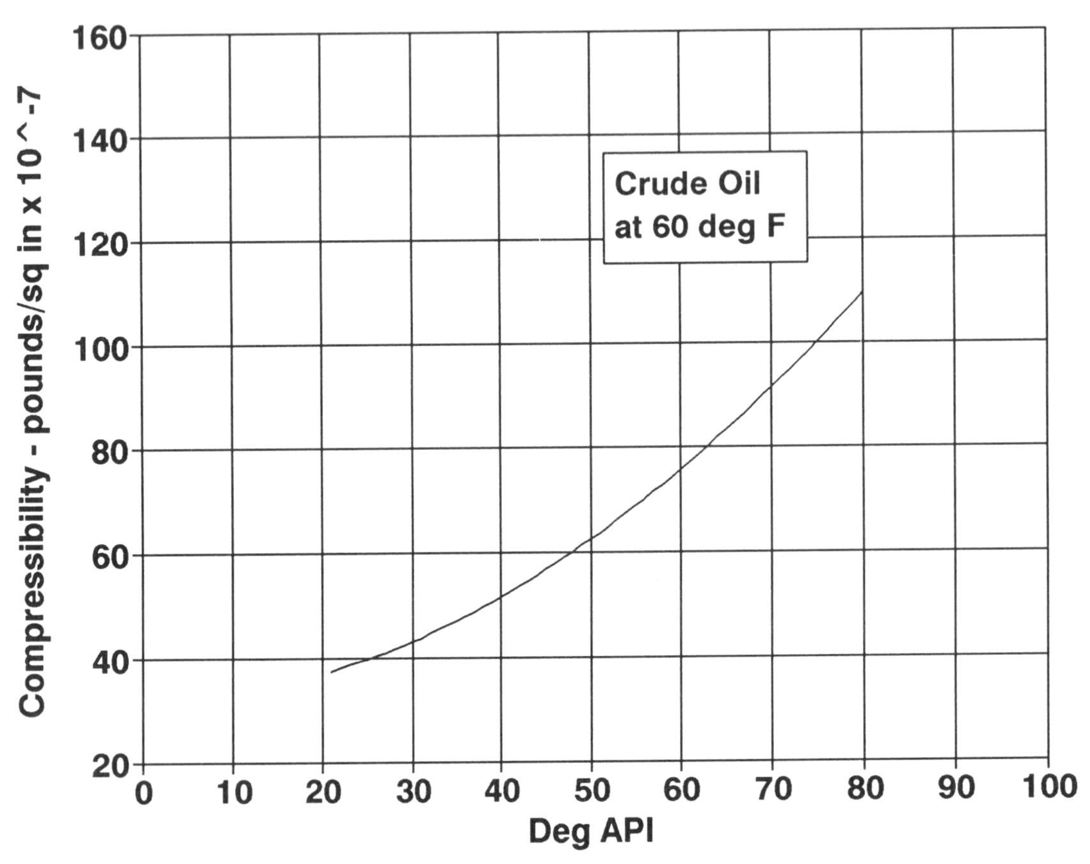

Figure 2-8. Compressibility plotted against relative density in deg API (Techo Equation 6).

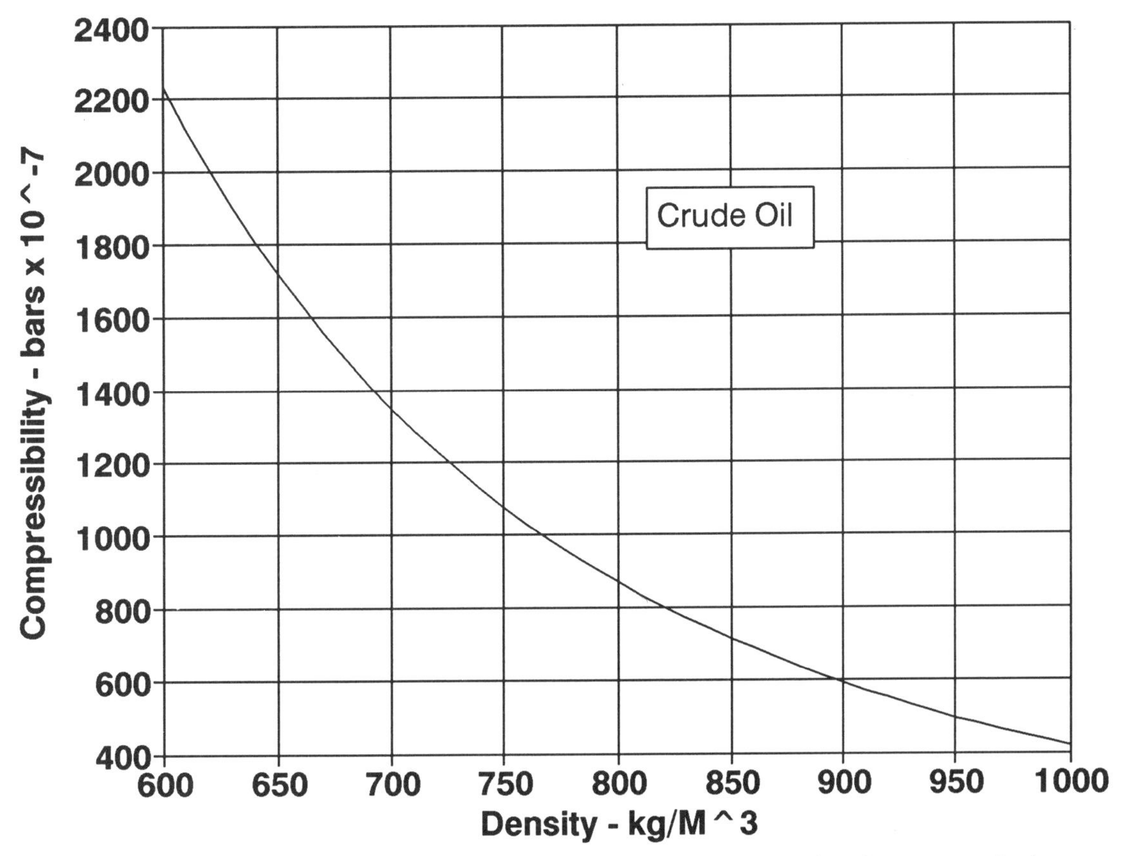

Figure 2-9. Compressibility plotted against density in kg/M ^ 3 (Robinson equation).

The compressibility of water is relatively constant with reasonable changes in pressure but does vary somewhat with temperature. It is difficult to find acceptable values for the compressibility of water with relation to either pressure or temperature, and the bases of Table 2-5 were taken from Streeter[11] who, in turn, compiled the table from *ASCE Manual of Engineering Practice*, No. 25, Hydraulic Models, 1942, long out of print. The original table was in terms of bulk modulus; the values of compressibility were calculated for Table 2-5.

Shrinkage

A *perfect* solution can be defined roughly as one in which X volume parts of liquid A mixed with Y volume parts of liquid B will produce (X+Y) volumes parts of a solution of A and B.

Any kind of solution made up of X weight parts of solution A and Y weight parts of solution B will produce (X+Y) weight parts of a solution of A and B and its density will be AX + BY.

A mixture of a liquid light hydrocarbon and a heavier crude oil is *a kind of solution*, but it is not a *perfect solution*: a solution made up of 1 barrel of propane and 100 barrels of 30 °API crude oil will not produce 101 barrels of propane–crude mixture. No pounds, kilograms, or whatever kind of weight measure will be lost; but some part of the barrel of propane will, for practical purposes, have disappeared.

When accurate volumetric metering of high–pressure pipelines became available, it immediately proved what had been suspected for years: a small volume of a light product mixed with a large volume of a heavy crude, mixes with the crude in such way that it is impossible to volumetrically balance a pipeline; i.e., $\rho_1 v_1 A_1 = \rho_2 v_2 A_2$ but $\rho_1 \neq \rho_2$ and therefore $A_1 v_1 = Q_1 \neq A_2 v_2 = Q_2$.

There was a movement, quite active for a while, to measure such movements by weight, because 1 pound of a light product + 100 pounds of crude *does* equal 101 pounds of mixture: $\rho_1 v_1 A_1$ *does equal* $\rho_2 v_2 A_2$. This movement foundered, in the end, though in recent years a form of *weight* measure–called *mass* measurement–of

Table 2-5

Density, Compressibility and Bulk Modulus Of Water

Temperature °F	Specific Weight γ lb/ft³	Density ρ slugs/ft³	Bulk Modulus K psi	Compressibility β psi x 10^{-6}
32	62.42	1.940	293000	3.41
40	62.43	1.940	294000	3.40
50	62.41	1.940	305000	3.28
60	62.37	1.938	311000	3.22
70	62.30	1.936	320000	3.13
80	62.22	1.934	322000	3.11
90	62.11	1.931	323000	3.10
100	62.00	1.927	327000	3.06
110	61.86	1.923	331000	3.02
120	61.71	1.918	333000	3.00
130	61.55	1.913	334000	3.00

flowing quantities, as distinguished from total quantity delivered over a period of time, has come into popularity in some fields. Mass measurement has become especially popular for metering ethylene, which reacts in peculiar ways in a pipeline: it can enter the line as a liquid; flow for a while as a multi–phase mixture of gas, vapor, and liquid; and exit the pipeline as any one of the three phases, including the liquid phase in which it first entered the line. Systems that use this kind of measurement electrically integrate the varying output of the mass flowmeter to determine total quantity flowed over time.

Nevertheless, with the advent of crude cocktails, a mixture of a crude and one or more light hydrocarbons, it became important to discover a way of handling mixtures of light hydrocarbons with crudes: (1) to measure both components by volumetric techniques at the points of receipt, (2) measure the output by volumetric techniques at the delivery point(s), and (3) compute the amounts received and delivered in such way that neither of the shippers who delivered product into the system, nor the purchaser(s) of the output, would gain or lose because of measurement inconsistencies.

After considerable study of a sizable database the industry came up with the procedures described in API Bulletin 2509C[12] to allow the computation, with an acceptable accuracy, of the reduction in volume of a mixture of a light and a heavy hydrocarbon in terms of the inputted volume of the light fraction. An empirical formula for this *shrinkage* given in the Bulletin is:

$$S = 0.0000214C^{-0.0704}G^{1.75} \tag{2-21}$$

where S = shrinkage factor, as decimal fraction of the lighter component volume
C = concentration, in liquid volume percent, of lighter component in mixture
G = gravity difference, °API

The data used to develop this formula were such that the formula is most applicable at or near 60°F and 100 psig.

Figure 2-10 is a reproduction of a plot included in API 2509C generated using Equation 2-21.

API 2509C is not the last word on this problem, though it is, at this writing in 1994, the only *official* word.

But Robinson, previously cited, used the Costwald equation of state with input from API and IP standards to look further into the effect of temperature and pressure on shrinkage. He points out API 2509C is only applicable near standard conditions (15°C and 1.013 bar), and no methods are available to predict the densities of mixtures (thus the shrinkage) at higher pressures and temperatures found in most pipelines. While API *indicates* shrinkage increases with temperature and decreases with pressure,

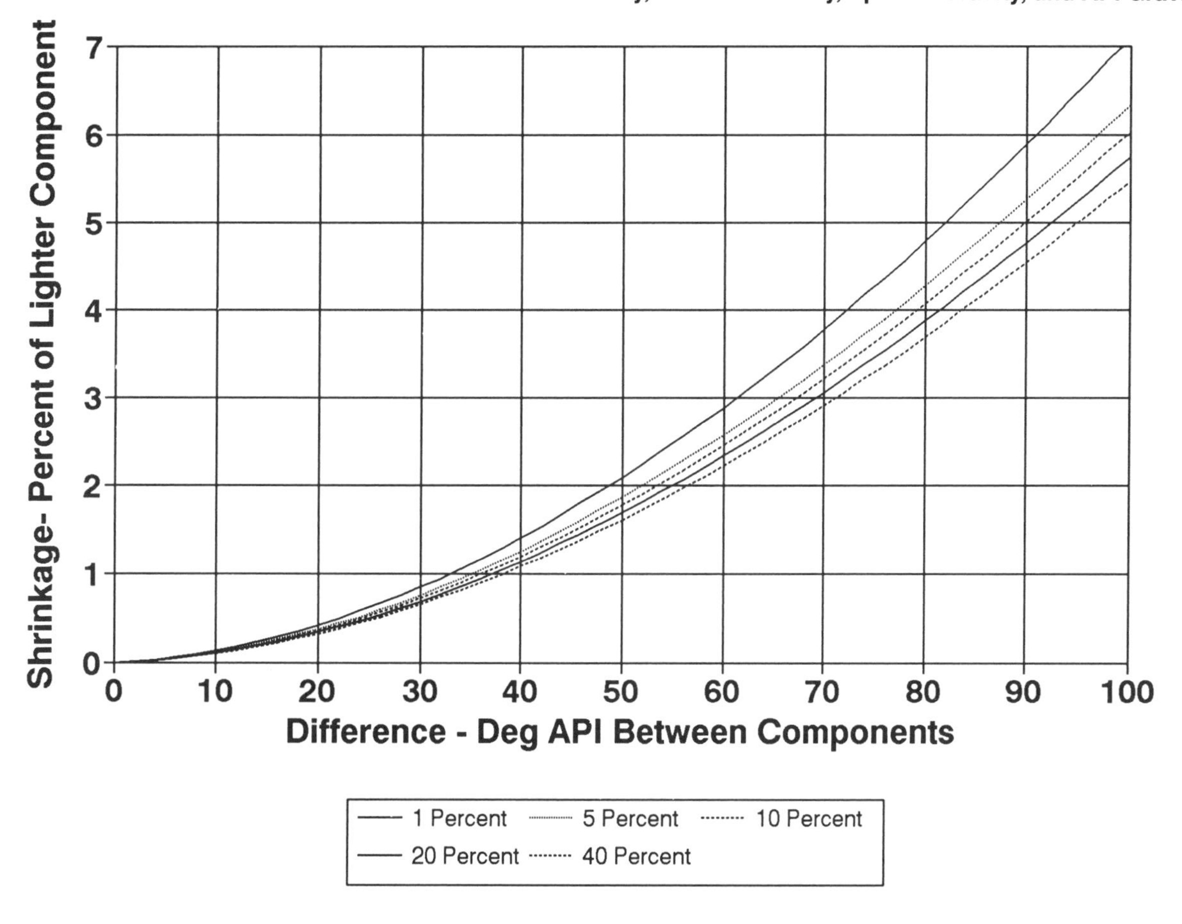

Figure 2-10. Shrinkage as a function of percentage admixture and difference in deg API.

Robinson's work *proves* it. His Figures 6 and 7, neither reproduced here, show that a 0.25 mole fraction propane component at 0 bar would shrink 7% at 0°C, but nearly 13% at 50°C; and that the 7% shrinkage at 0 bar would drop to approximately 6% at 40 bar. It is evident the effect of varying temperature is much greater than that of varying pressure under pipeline conditions.

In 1991 Ashcroft, et al.,[13] noting that there had been no published data on measured, as differed from calculated, shrinkage for over 30 years, reported on work done on another set of samples covering seven crudes and eight *spikes* (light fractions). They made their measurements at atmospheric pressure and at 15°C and 25°C. Their oscillating tube densitometer was capable of producing an accuracy of $\pm 5 \times 10^{-6}$ g/cm³ and of detecting changes in the order of $\pm 5 \times 10^{-7}$ g/cm³. Miscellaneous matter in the crudes reduced overall accuracy to the order of $\pm 1 \times 10^{-4}$ g/cm³, but this allowed determination of shrinkage factors to within ±0.1.

The equation Ashcroft et al. developed is

$$S = 2.940 \times 10^{-6} (100 - C)^{0.892} G^{2.34} \tag{2-22}$$

where S = shrinkage, vol% of spike
C = concentration of spike, vol%
G = gravity difference °API

The equation applies within the temperature range 15°C and 25°C at a pressure of 0 psig.

This equation is of the same form as Equation 2-21 but understates shrinkage for lesser differences between the densities of the components and overstates it for larger differences. Figure 2-11 illustrates this difference for a 5% spike.

Ashcroft et al. reported on work done by Childress and Grove[14] to estimate the effect of pressure on shrinkage, and developed a formula to estimate its effect in the range 0 to 100 psig.

$$S_p = S_{100}(1 + 0.000210(100 - P)) \tag{2-23}$$

where S_{100} = shrinkage at 100 psig, API equation
S_p = shrinkage at P psig, corrected

The correction is probably well within the range of accuracy of the original equation, but it does indicate the way pressure effects shrinkage.

In the end, shrinkage is something which, if it is very important, should be measured, not computed, and the procedure of API 2509C is specific in all the necessary technical details. It is just a little difficult to assemble the equipment.

Working Equations

Calculators

The main working equations for density and specific gravity are summarized on Table 2-6. All of these are amenable to solution by calculators.

Computers

There are seven Computer Computation Templates provided in this chapter. **CCT** No. 2-1 includes both the °API and °Bé –vs– specific gravity equations. **CCT** Nos. 2-2 and 2-3 cover thermal expansion of hydrocarbons, including both the coefficient of thermal expansion α and the volume correction factor *VCF*. **CCT** Nos. 2-4 through 2-7 provide for estimating molecular weight of hydrocarbons, computing the thermal coefficient of expansion of water, the isothermal compressibility of hydrocarbons using the Techo method, and shrinkage of light hydrocarbons in *spikes* using the API 2509C method.

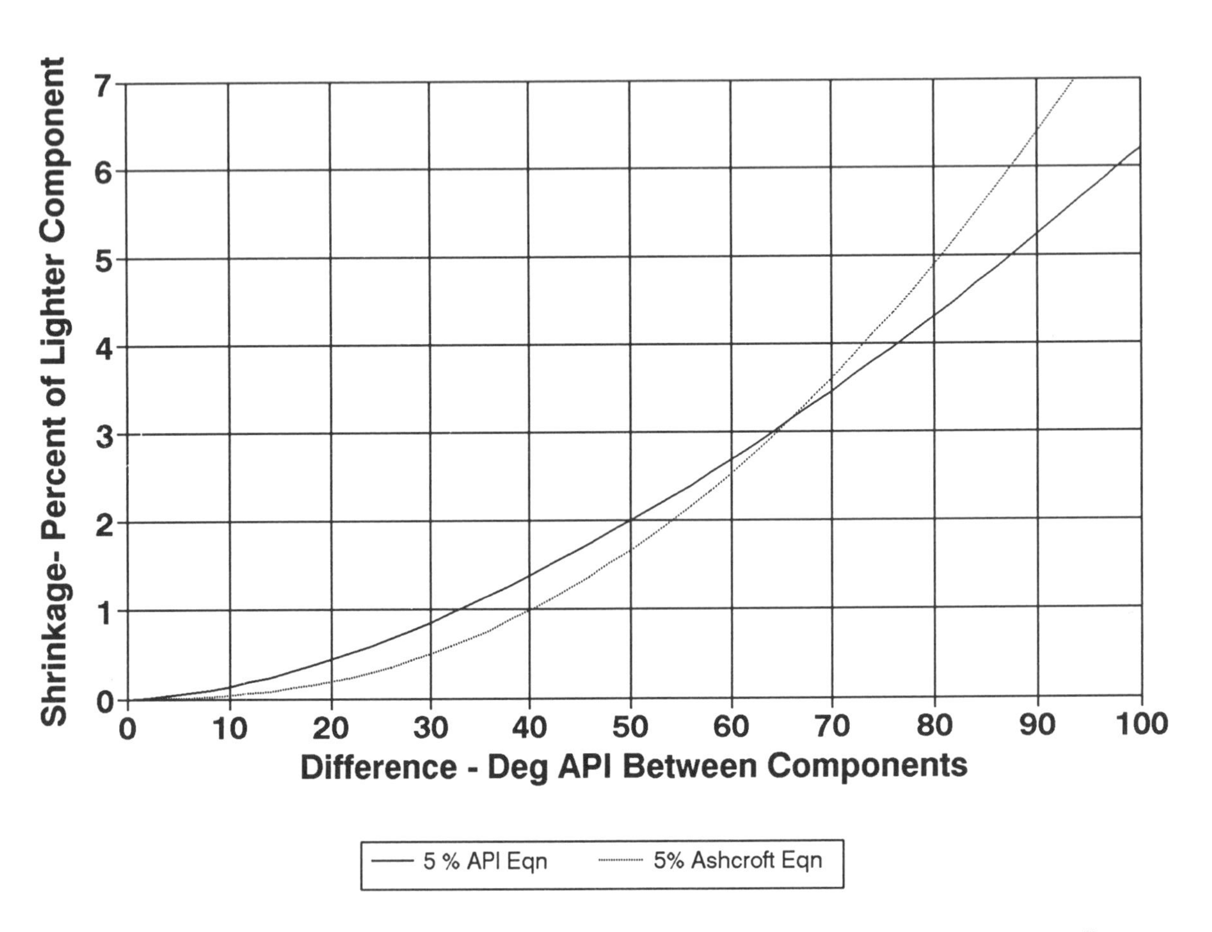

Figure 2-11. Comparison of API 2509C and Ashcroft shrinkage correlations for a 5% spike.

Table 2-6
Reference and Working Equations for Density, Relative Density Specific–, API°–, and Baumé° Gravity

°API–vs–Specific Gravity

$$Sp\ Gr = \frac{141.5}{131.5 + API^{\circ}}$$

$$API^{\circ} = \frac{141.5}{SpGr} - 131.5$$

°Baumé–vs–Specific Gravity

$$Sp\ Gr = \frac{140}{130 + B\acute{e}^{\circ}}$$

$$Be'^{\circ} = \frac{140}{SpGr} - 130$$

Coefficient of Thermal Expansion of Oils

Definition $\alpha = \left(\frac{1}{V}\right)\left(\frac{dV}{dt}\right)$

Working Definition $\alpha = \alpha_T + \beta\,\Delta t$

API/ASTM Definition $\alpha_T = \left(\frac{K_0 + K_1\rho_T}{\rho_T^{\,2}}\right)$

$$\alpha_T = \frac{K_0}{\rho_T^2} + \frac{K_1}{\rho_T}$$

API/ASTM Equation for Crude Oil $\alpha = \frac{341}{\rho_{60}^2}$

Volume Correction Factor–Definitions

Definition $VCF = \frac{\rho}{\rho_T}$

API/ASTM Definition =

$$VCF = \left(\frac{V_T}{V}\right) = EXP\left[-\alpha_T\,\Delta t(1 + 0.8\,\alpha_T\,\Delta t)\right]$$

Coefficient of Thermal Expansion of Water

Gray, from Smithsonian Tables for Water

$$\alpha \times 10^6 = -64.268 + 17.0105t - 0.20369t^2 + 0.0016048t^3$$

Compressibility of Hydrocarbons

Techo's Equation for Hydrocarbon Liquids

$$\beta = 28.328 + (1.07946 \times 10^{-3})t^2 + \left[(1.02286 \times 10^{-2}) + (5.1659 \times 10^{-7})t^2\right] \times (^{\circ}API)^2$$

Shrinkage of Light Hydrocarbons Mixed With Heavier Hydrocarbons

API 2509C Method $S = 0.0000214C^{-0.0704}G^{1.75}$

Ashcroft Method $S = 2.940 \times 10^{-6}(100 - C)^{0.892}G^{2.34}$

Molecular Weight of Hydrocarbons–Robinson Method

For $500 \le D \le 750$

$$MW = -886.37 + 5.1671D - (9.79 \times 10^{-3})D^2 + (6.3477 \times 10^{-6})D^3$$

For $750 \le D \le 1050$

$$MW = -1524.4 + 4.4958D - (4.25 \times 10^{-3})D^2 + (1.6667 \times 10^{-6})D^3$$

References

1 API Manual of Petroleum Measurement Standards, Chapter 9.1 (previously API Std 2547). Hydrometer Test Method for Density, Relative Density (Specific Gravity) or API Gravity of Crude Petroleum and Liquid Petroleum Products by Hydrometer Method. Washington: American Petroleum Institute, 1987. (Also ANSI/ASTM D1298).

2 Downer, L. and Inkley, F. A., "Thermal Expansion Coefficients Need Update to Combat Errors in Crude Oil Measuring," *Oil and Gas Journal*, June, 1972.

3 Hankinson, R. W., Segers, R. G., Buck, T. K., and Gielsecki, F. P. "Revision of Petroleum Measurement Tables Adopted." *Oil and Gas Journal*, Dec. 24, 1979, pp. 64–70.

4 West, K. J., Hankinson, R. W., and Downer, L. "Petroleum Measurement Tables of 1980 Reaffirmed." *Oil and Gas Journal*, Aug. 7, 1989, pp. 63–67.

5 Robinson, E. R. "Calculate Density of Spiked Crudes." *Hydrocarbon Processing*, May, 1983.

6 Author's Note: A spiked crude is a crude that is mixed with (usually) a small fraction of its volume of a much lighter product (LPG, condensate, etc.) to either improve the value of the crude or to facilitate the sale and shipment of the light product.

7 Gray, J. C. "How Temperature Affects Pipeline Hydrostatic Testing." *Pipeline and Gas Journal*, Dec., 1976.

8 API Standard 1101. *Measurement of Petroleum Liquid Hydrocarbons by Positive Displacement Meter*. First Edition, August, 1960. Washington: American Petroleum Institute.

9 Jacobson, W. E., Ambrosius, E. E., Dashiell, J. W., and Crawford, C. L. "Compressibility of Liquid Hydrocarbons." Proc. API 25 [IV] 39–41 (1945).

10 Techo, Dr. Robert. "Volumetric Corrections for Pipe, Products Due to Pressure, Temperature." *Pipeline and Gas Journal*. July, 1975.

11 Streeter, Victor L. *Fluid Mechanics*, 3rd ed. New York: McGraw–Hill Book Company, 1962, page 533.

12 API Bulletin 2509C, *Volumetric Shrinkage Resulting From Blending Volatile Hydrocarbons With Crude Oils*, American Petroleum Institute, Washington, D. C., 1967 (R 1987).

13 Ashcroft, S. J., Booker, D. R., and Turner, J. C. R., "Find Shrinkage of Spiked Crudes." *Hydrocarbon Processing*, October, 1991, pp. 109–111.

14 Childress, H. M., and Grove, M. B., *The Petroleum Engineer*, pp. D35–47, December 1955.

Computer Computation Template

CCT No. 2-1: API gravity and Baume gravity -vs- specific gravity. Finds and graphs five points above and below specified value of API or Baume gravity.

References: Chapter 2. Equations 2-1 and 2-2.

Input

Factors $i := 90, 92 .. 110$

Input Data $A := 40$ °API $B := 50$ °Be

Variables $A_i := \frac{i}{100} \cdot A$ $B_i := \frac{i}{100} \cdot B$

Equations

$$SpGrAPI_i := \frac{141.5}{A_i + 131.5} \qquad SpGrBe_i := \frac{140.0}{B_i + 130.0}$$

Results

A_i	$SpGrAPI_i$
36.0	0.845
36.8	0.841
37.6	0.837
38.4	0.833
39.2	0.829
40.0	0.825
40.8	0.821
41.6	0.817
42.4	0.814
43.2	0.810
44.0	0.806

API Degrees -vs- SpGr 60/60 deg F

SpGr 60/60 deg F

0.86 0.854 0.848 0.842 0.836 0.83 0.824 0.818 0.812 0.806 0.8

36 36.8 37.6 38.4 39.2 40 40.8 41.6 42.4 43.2 44

API Degrees

B_i	$SpGrBe_i$
45.0	0.800
46.0	0.795
47.0	0.791
48.0	0.787
49.0	0.782
50.0	0.778
51.0	0.773
52.0	0.769
53.0	0.765
54.0	0.761
55.0	0.757

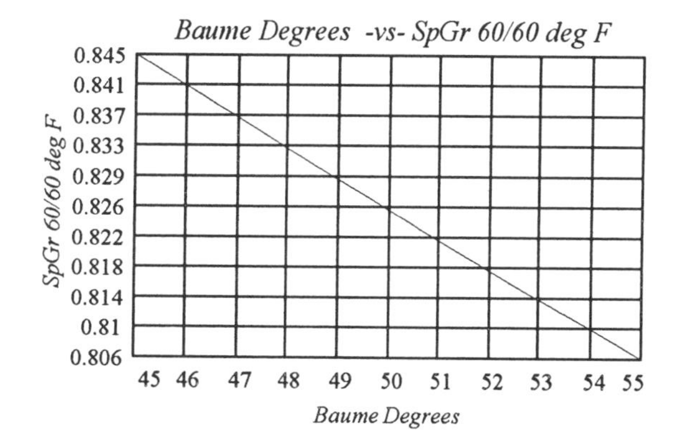

Computer Computation Template

CCT No. 2-2: α_T—Coefficient of thermal expansion for hydrocarbons. Method based on API/ASTM Petroleum Measurement Tables. Finds and graphs five points above and below specified value of density. Example is for crude oil. For other petroleum hydrocarbons select values for K_o and K_1 from Table 2-4.

References: Chapter 2. Equation 2-10. Table 2-4.

Input

Factors $K_0 := 341.0957$ $H_2O=1.000$ $K_1 := 0$ $i := 90, 92 .. 110$

Input Data $\rho_T := 0.800$

Variables $\rho_{T_i} := \frac{i \cdot \rho_T}{100}$

Equations

$$\alpha_{T_i} := \left[\frac{K_0}{\left(\rho_{T_i}\right)^2} + \frac{K_1}{\rho_{T_i}} \right]$$

Note: α_T is in terms of 1/1 000 000; i.e., α_T = 603.974 = 0.000603974

Results

ρ_{T_i}	α_{T_i}
0.720	657.978
0.736	629.681
0.752	603.171
0.768	578.301
0.784	554.938
0.800	532.962
0.816	512.266
0.832	492.753
0.848	474.334
0.864	456.929
0.880	440.464

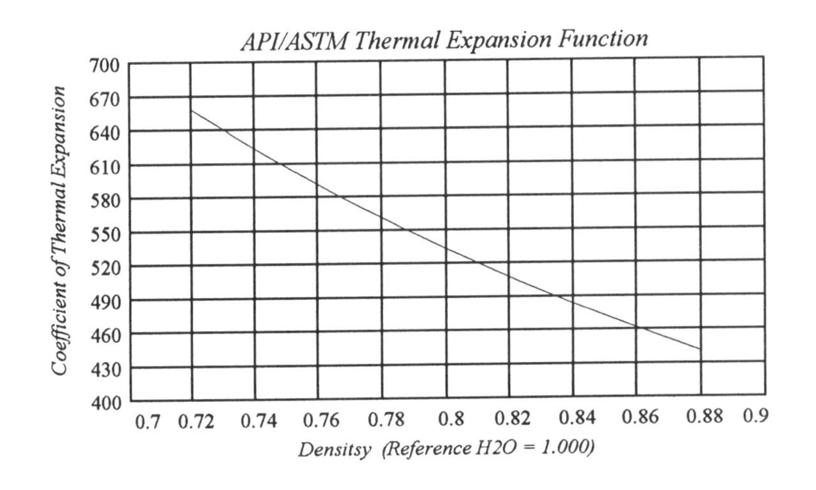

Computer Computation Template

CCT No. 2-3: VCF—volume correction factor for thermal expansion of hydrocarbons. Method used is based on the API/ASTM Petroleum Measurement Tables. Values of α_T, the coefficient of thermal expansion at base temperature *T*, are generated from the equation given in CCT 2-02. Note Δt is positive, thus the VCF obtained applies to temperatures above 60 F, the base temperature; for temperatures below 60 F, use a negative Δt. Example is for crude oil. For other petroleum hydrocarbons select values for K_0 and K_1 from Table 2-4.

References: Chapter 2. Equations 2-9 and 2-10. Table 2-4.

Input

Factors $K_0 := 341.0957$ $K_1 := 0$ $i := 0, 1 .. 10$ $H_2O = 1.000$

Input Data $\rho_T := 0.835$

Variables $\Delta t := 10$ deg F $\Delta t_i := i \cdot \Delta t$ $\alpha_T := \left[\frac{K_0}{(\rho_T)^2} + \frac{K_1}{\rho_T}\right]$

Note: α_T is in terms of 1/1 000 000; i.e., 603.974=0.000603974

Equations

$$VCF_i := \exp\left[-\frac{\alpha_T}{1000000} \cdot \Delta t_i \cdot \left(1 + 0.8 \cdot \frac{\alpha_T}{1000000} \cdot \Delta t_i\right)\right]$$

Results

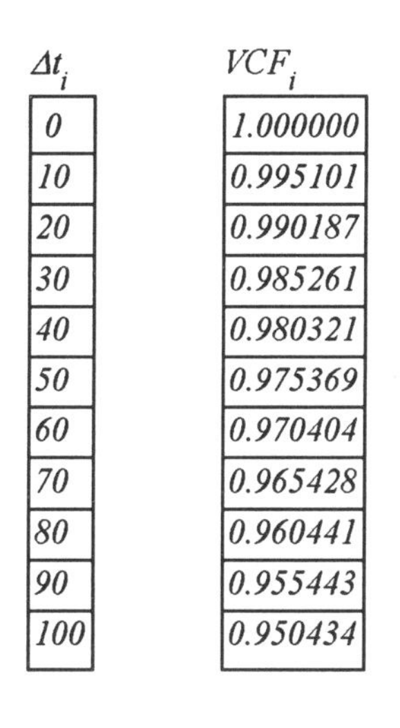

Δt_i	VCF_i
0	1.000000
10	0.995101
20	0.990187
30	0.985261
40	0.980321
50	0.975369
60	0.970404
70	0.965428
80	0.960441
90	0.955443
100	0.950434

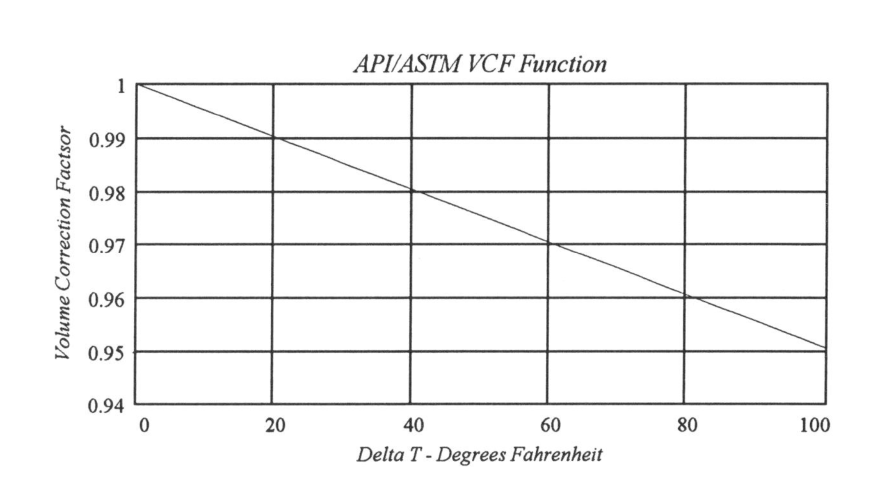

Computer Computation Template

CCT No. 2-4: MW—molecular weight of hydrocarbons using Robinson's method. The equation used is valid for D–the density referred to H_2O as 1000–between 750 and 1050, which will accommodate most crude oils. For lighter hydrocarbons use the first of equations (2-17), set D to 500, and adjust i and d.

References: Chapter 2. Equation 2-16. Figure 2-6.

Input

Factors $i := 30, 31 .. 42$

Input Data $D := 750$

Variables $d := \frac{D}{30}$ $\qquad d_i := i \cdot d$

Equations

$$MW_i := -1524.4 + 4.4958 \cdot d_i - 4.25 \cdot 10^{-3} \cdot (d_i)^2 + 1.6667 \cdot 10^{-6} \cdot (d_i)^3$$

Results

d_i	MW_i
750	160
775	183
800	206
825	228
850	250
875	272
900	294
925	317
950	340
975	364
1000	388
1025	413
1050	440

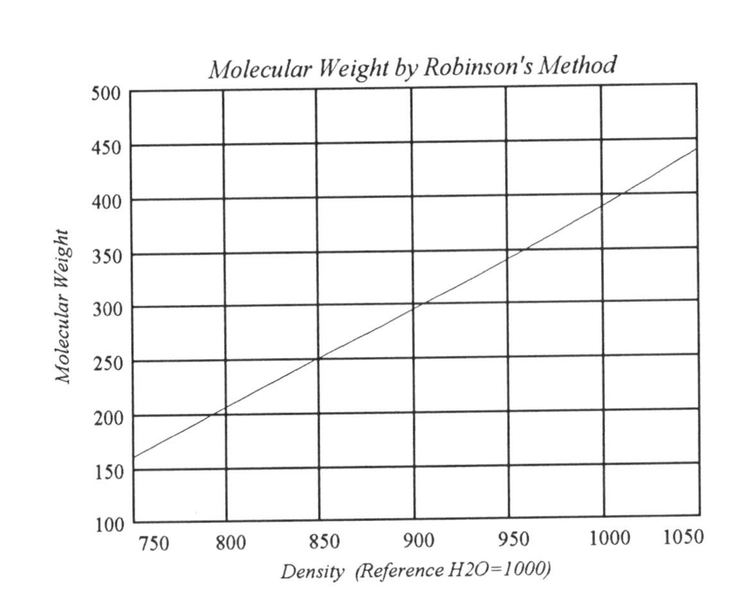

Computer Computation Template

CCT No. 2-5: α—Coefficient of thermal expansion of water. Based on Gray's analysis of the Smithsonian Tables data. Temperatures are in degrees Celsius.

References: Chapter 2. Equation 2-17. Figure 2-7.

Input

Factors $i := 0, 5 .. 100$

Variables $t := 1$ $\quad t_i := i \cdot t$

Equations

$$\alpha_i := -64.268 + 17.010 \cdot t_i - 0.20369 \cdot \left(t_i\right)^2 + 0.0016048 \cdot \left(t_i\right)^3$$

Results

t_i	α_i
0	-64
5	16
10	87
15	150
20	207
25	259
30	306
35	350
40	393
45	435
50	478
55	522
60	570
65	622
70	679
75	743
80	815
85	895
90	987
95	1089
100	1205

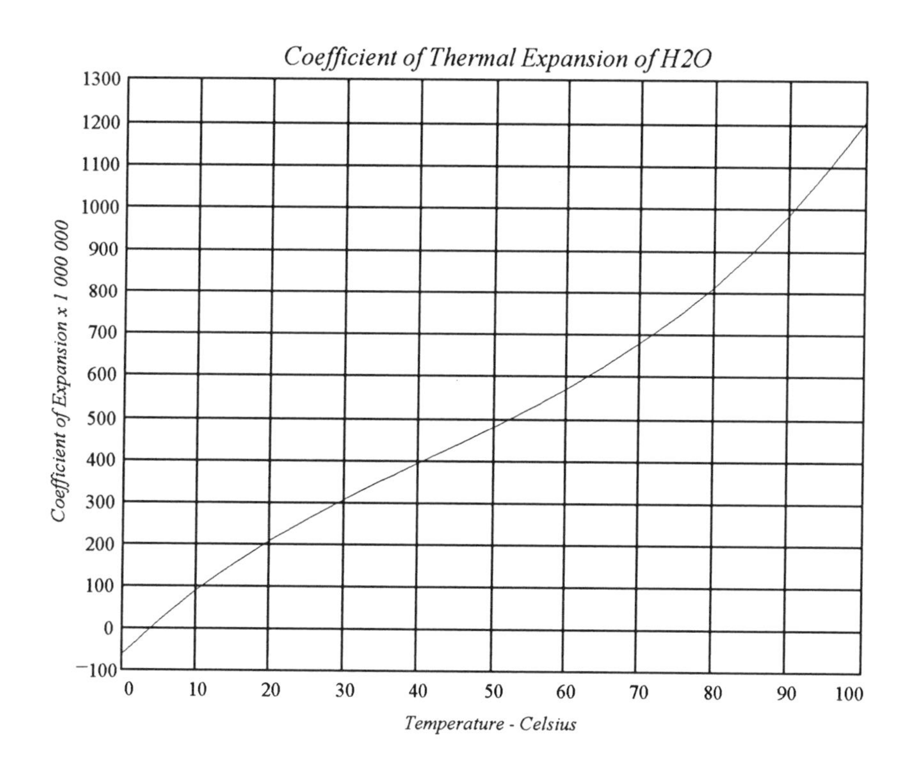

Computer Computation Template

CCT No. 2-6: β—Compressibility of hydrocarbons using Techo's method.

References: Chapter 2. Equation 2-19. Figure 2-8.

Input

Factors $i := 20, 25 .. 80$

Input Data $t := 60$ deg F Note: Change t to obtain β at temperatures other than 60 °F.

Variables $API_i := 1 \cdot i$

Equations

$$\beta_i := 28.328 + \left(1.07956 \cdot 10^{-3}\right) \cdot t^2 + \left[\left(1.02286 \cdot 10^{-2}\right) + \left(5.1659 \cdot 10^{-7}\right) \cdot t^2\right] \cdot \left(API_i\right)^2$$

Results

API_i	β_i
20	37.0
25	39.8
30	43.1
35	47.0
40	51.6
45	56.7
50	62.4
55	68.8
60	75.7
65	83.3
70	91.4
75	100.2
80	109.6

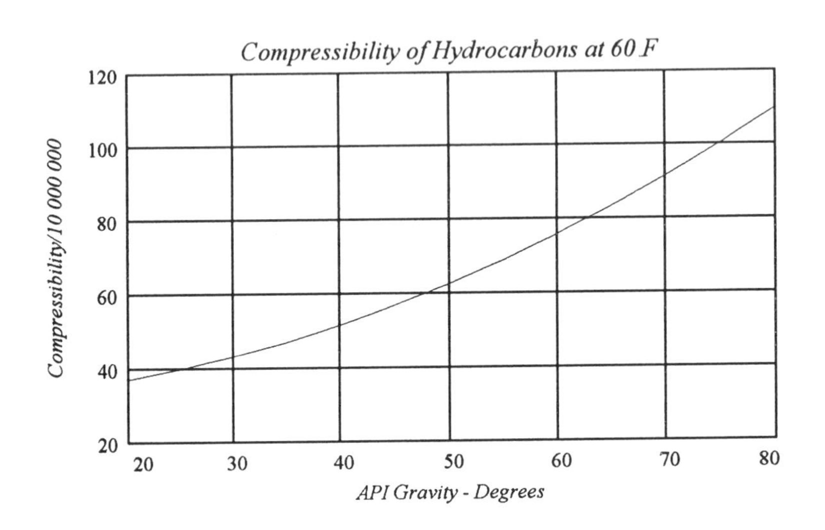

Computer Computation Template

CCT No. 2-7: API 2509C Formula for Shrinkage of Light Hydrocarbon Mixed With a Large Amount of a Heavy Hydrocarbon.

References: Chapter 2. Equation 2-21. Figure 2-10.

Input

Input Data $C := 5$ volume percent concentration of lighter component

Variables $G := 0, 10 .. 100$ range of difference in API degrees between lighter and heavier component

Equations

$$S_G := 0.00214 \cdot C^{-0.0704} \cdot G^{1.75}$$

Results

G	S_G
0	0
10	0.11
20	0.36
30	0.73
40	1.22
50	1.80
60	2.47
70	3.24
80	4.09
90	5.02
100	6.04

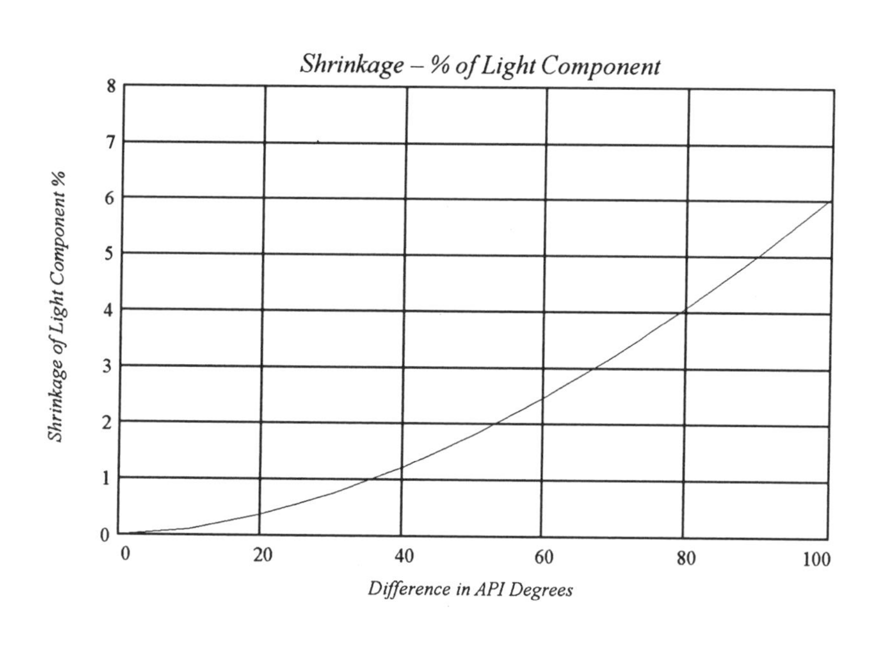

3

Newtonian Viscosity

What Is Viscosity?

One of the best descriptions of viscosity is given in a rather old (1958) British edition of the Encyclopedia Britannica in the first paragraph of the item on viscosity.

> When a fluid, which may be a liquid or a gas, is set in motion by any system of forces and then left to itself, it comes to rest by virtue of an internal friction which tends to resist the sliding of one part of the fluid over another. This internal friction is known as *viscosity*. If a state of steady motion is maintained by any system of forces the viscosity of the fluid tends to oppose the motion and leads to the dissipation of energy as heat, just as does the friction between solid bodies.

This is a concise, non–mathematical definition, and provides a better insight as to what viscosity really *is* than any number of precise formulations.

Unfortunately, hydraulics is a science—for the most part—and the definitions of parameters which enter into hydraulics problems should be mathematically precise. Therefore the following description of viscosity is a more appropriate starting point for this chapter.

> Viscosity is a measure of the shearing stress necessary to induce a unit velocity shear gradient in a substance.
>
> A fluid is said to exhibit *Newtonian* behavior when the *rate of shear* is proportional to the *shear stress*, i.e., the *coefficient of viscosity is a constant*. It is so called because Newton, in considering the imaginary fluid that, in the Cartesian world system, would have filled all of space, assumed that it would have this property.
>
> A fluid is said to exhibit *non–Newtonian* behavior when an increase or decrease in the rate of shear is *not* accompanied by a proportional increase or decrease in the shear stress, i.e., *the coefficient of viscosity varies with the rate of shear*.
>
> Viscosity is measured by either (1) measuring the shear force required to produce a given shear gradient, or (2) measuring the time required for a given volume of liquid to flow through a capillary or restriction. Tests of type (1) yield the *absolute viscosity*; type (2) tests produce the *kinematic viscosity*.
>
> Within ordinary limits viscosity is an *intensive* property of a substance, which means a large sample of a Newtonian fluid will have the same viscosity as a small sample. This is not necessarily true for non–Newtonian fluids.

Absolute Viscosity

Absolute viscosity—sometimes called the *dynamic* viscosity—can be visualized by the two–sliding–plates analogy shown in Figure 3-1.

Consider that there are two plates, plane and parallel and separated by a distance x, each of area A which is so great that the boundary effects of the edges of the plates are of secondary importance. The upper plate is free to

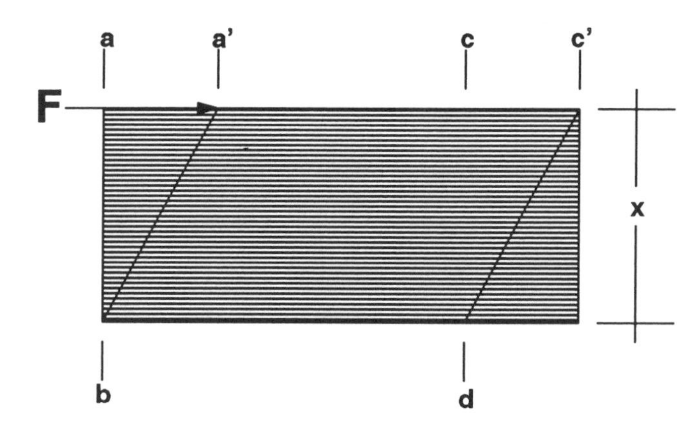

Figure 3-1. Parallel plate analogy.

move horizontally, but is restrained from moving in the vertical direction and this restraint is assumed to be without horizontal friction. There is a liquid separating the two plates which adheres to both plates. This fluid is completely homogeneous and Newtonian in behavior.

Then assume that a force F is applied to the upper plate, and that after the plate has reached a uniform velocity it advances a distance $a–a' = c–c'$ as time flows from $t = 0$ to $t = T$. The velocity of the upper plate is therefore $(a–a')/T = (c–c')/T$.

The infinitely thin layer of fluid in contact with the upper plate attains a constant velocity which is the same as that of the upper plate; the thin layer of fluid in contact with the lower plate does not move because the plate itself does not move. The fluid, being homogeneous, takes a uniform deformation over the time interval T of $(a–a')/x = (c–c')/x$, and the rate of deformation is $(a–a')/Tx = (c–c')/Tx$.

Since $(a–a')/T$ is the velocity v of the upper plate produced by the force F, the unit rate of deformation is v/x, and it can be shown that under Newtonian conditions $F \propto A(v/x)$ or, dividing by A, $F/A \propto v/x$.

The term F/A has the dimensions of a stress, called the *shearing stress*, and is usually symbolized as τ. Since *proportionality* has been established, *equality* may be established by adding a *constant of proportionality*. Thus

$$\tau = \frac{F}{A} = \frac{\mu(a-a')}{Tx} = \mu\frac{dv}{dx} \tag{3-1}$$

This equation is called Newton's *equation of viscosity*. Sometimes it is called the *rheological equation*, or the *constitutive equation*. It is usually given in transposed form, viz.:

$$\mu = \left(\frac{\tau}{\frac{dv}{dx}}\right) = \left(\frac{\tau}{\lambda}\right) \tag{3-2}$$

In this form it defines μ in terms of τ, the *shear stress*, and dv/dx or γ, which is called the *shear rate*. The coefficient μ is called the *coefficient of viscosity*, *absolute viscosity*, *dynamic viscosity* or, sometimes, simply *viscosity*.

The dimensions of absolute viscosity in absolute systems of measurements are $ML^{-1}T^{-1}$; in gravitational systems the dimensions are FTL^{-2}.

The most commonly used unit of measurement is the *poise*, named after Poiseuille, who first developed the rational equation for streamline flow of viscous liquids. The poise may be expressed as dyne·sec/cm² or as g/cm·sec. The poise is a rather large unit, and the *centipoise*, abbreviated cP and equal to 0.01 poise, is usually used in measurements and calculations.

The SI unit of absolute viscosity is the pascal·second (Pa·s) which, since a pascal is a newton/m², is equal to N·s/m². To compute the SI unit multiply poises by 0.1 or centipoises by 0.001.

English units are little used. The most common, in the gravitational system, is (lb·s)/ft². To obtain this unit multiply cP by 2.0087 E–05.

It is interesting to note that pure water at 20.2°C = 68.4°F has a viscosity of 1.00 cP, which means that the viscosity of any fluid measured in cP is the viscosity of the fluid relative to water at 68.4°F.

Air at 18°C = 64.4°F has a viscosity of 0.0183 cP = 183 μP.

Kinematic Viscosity

In situations where accelerating or decelerating motion is involved the determining parameter is not the absolute viscosity but the absolute viscosity divided by density. This parameter is called the *kinematic viscosity*.

The significance of this second viscosity parameter can be quickly seen if it is understood that stirred air, which has an *absolute viscosity* of some 0.000183 poise, comes to rest much more quickly that stirred water, which has an *absolute viscosity* of about 0.01056 poise, nearly six times "more viscous" than the air. However, air has a *kinematic viscosity* of some 0.1630 stoke, whereas the *kinematic viscosity* of water is only about 0.0101 stoke, a ratio of 16:1, and so when compared on the basis of *kinematic viscosity* air is the "more viscous" of the two substances. The ratios of inertial forces to viscous friction forces are markedly different for the two cases, and stirred air settles down much more quickly than stirred water. And if the concept of water having a lower viscosity than air seems to be an out–of–the–ordinary concept, consider that mercury has a kinematic viscosity of 0.00115 stoke, about a tenth that of water and less than a hundredth that of dry air.

The usual representation for the equivalence between absolute and kinematic viscosity is as follows:

$$\nu = \frac{\mu}{\rho} \tag{3-3}$$

where ν = kinematic viscosity
μ = absolute viscosity
ρ = density

The dimensions of kinematic viscosity in absolute systems of measurements are L^2T^{-1}, and the dimensions are the same in gravitational systems because neither force nor mass enters into the dimensions.

The most commonly used unit of measurement is the *stoke*, named after Sir George Gabriel Stokes, a British scientist and mathematician, who was a pioneer in the study of the theory of viscous flow. The stoke has dimensions cm^2/s, and conversion into fps units is obtained by dividing by $30.48^2 = 9.2903$ E+02 to obtain ft^2/s, or by $2.54^2 = 6.4516$ to obtain in^2/s.

The usual unit of measurement is the *centistoke*, abbreviated *cSt*, which is 0.01 stoke.

Note that the centistoke, which uses the *centi* prefix, is not SI. However, a centistoke is 1 mm^2/s, which *is* SI; and it is now common to refer to kinematic viscosities in millimeters squared per second rather than centistokes.

Viscosity–Temperature Relations

The viscosity of liquids *decrease* with increasing temperature; the viscosity of gases *increase*. The general relations are

$$\text{liquids} \quad \log_e \mu = A + BT \tag{3-4}$$

$$\text{gases} \quad \mu = CT^{1/2} \tag{3-5}$$

Equation 3-4 is often seen in a form called Andrade's equation, after Andrade, which is

$$\mu = A \exp B / T \tag{3-6}$$

Andrade's equation is the basis for most calculations having to do with viscosity–temperature relations of liquids.

The procedures in ASTM D 341, *Standard Viscosity–Temperature Charts for Liquid Petroleum Products*, provide the most commonly used way to investigate the viscosity–temperature relationships of liquid hydro-carbons. These so–called *Vis–Temp Charts*, one of which is reproduced in Figure 3-2, are so constructed that for temperatures in the range between the cloud point at low temperatures, and the initial boiling point at high temperatures, two viscosity–temperature points are sufficient to establish a straight line which defines the viscosity–temperature relationship of that particular oil over the contained temperature interval and, to a certain extent, to near extrapolations outside it.

While two points are sufficient to establish a straight line on a chart, two points may not necessarily define the *correct* straight line. Prudence indicates that to establish the viscosity–temperature relations expected of a particular hydrocarbon to be transported by a pipeline, viscosity measurements be made at least three, and preferably four, temperatures running from about five degrees below the lowest temperature expected to some five degrees above the highest expected temperature. If these points plot a straight line on a *Vis–Temp Chart* you may be confident in the intermediate values on the line. If, however, the points plot a curved line, the highest or lowest point is off the line, or plot a jagged line of some sort, you may have a non–Newtonian liquid, a violation of the cloud point or initial boiling point criteria, or errors in the measurements. In any of these cases the line on the *Vis–Temp Chart* cannot be trusted and further tests are necessary.

The charts are strictly applicable to hydrocarbons, and the *Standard* carries a note pointing out that erroneous results may be obtained with other kinds of liquids.

Vis–Temp Charts are available in several viscosity ranges from 0.3 to 20 000 000 cSt, temperatures from –70°C to +370°C, and in physical sizes running from 217 by 280 mm (8 1/2 by 11 inches) to 680 by 820 mm (26.75 by 32.25 inches). Not all ranges are available in all sizes.

The spread of temperatures on all of the *Charts* is so large that the operating temperature range of most pipelines, running from –10°C for arctic pipelines to +50°C for desert lines, is almost lost on the axis of abscissas. Four of the *Charts* have a temperature range of –70°C to +370°C, two run from –100°F to +700°F, and only Chart VII, *Kinematic Viscosity, Middle Range*, produced in the standard U.S. letter–page size of 8 1/2 by 11 inches with a range of –40°C to +140°C, has a fairly reasonable spread of temperatures for pipeline use. One of the solutions to this problem, especially useful when plotting blending charts (see later in this chapter), is to

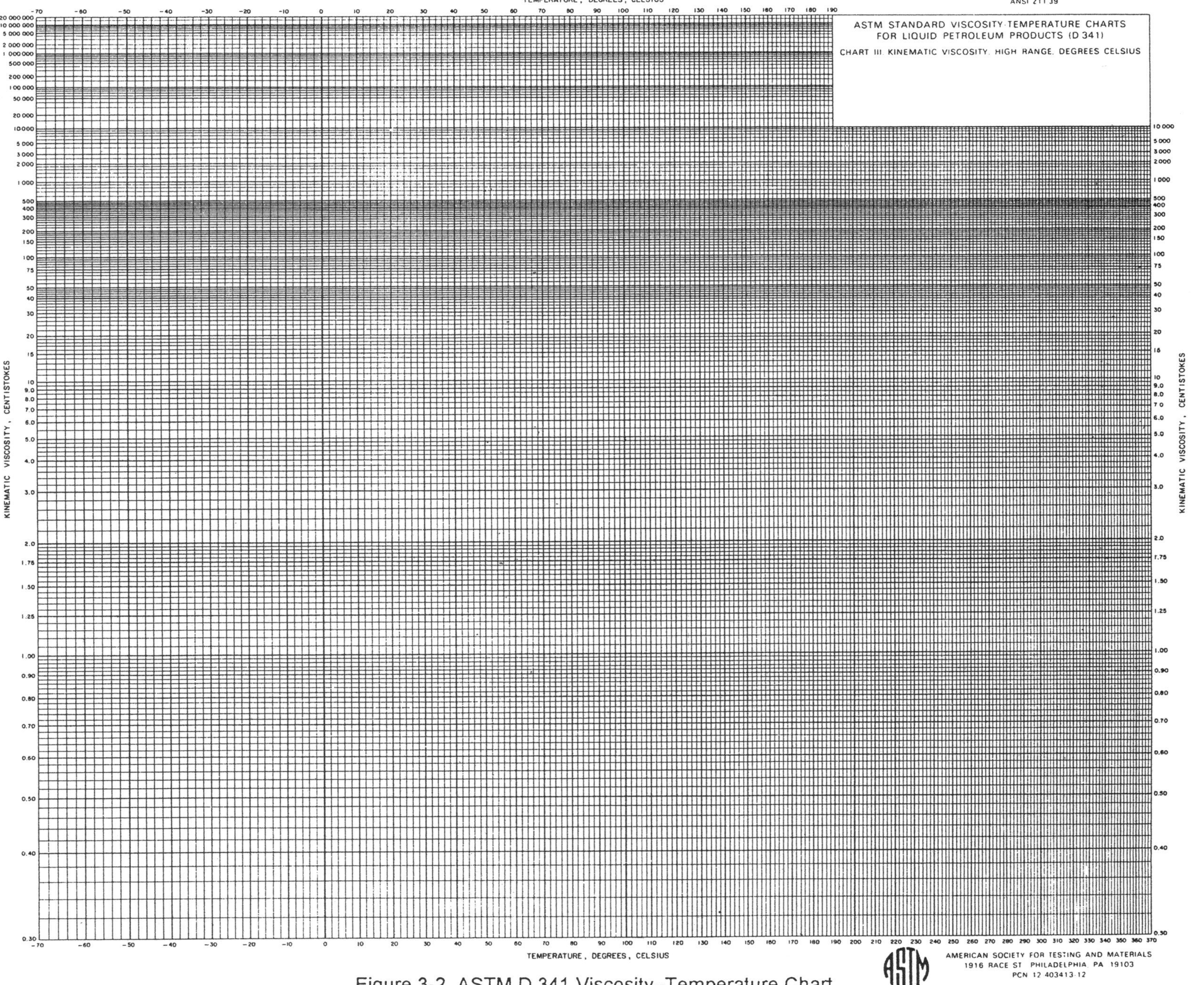

Figure 3-2. ASTM D 341 Viscosity–Temperature Chart

use a photocopier to expand the grid of the graph until the desired spread of temperatures occupies a reasonable physical spread on the plotting paper.

The Fahrenheit charts were originally scheduled to be deleted from the *Standard* after a period of time to allow users to convert from Fahrenheit to Celsius, so these may or may not be available. And, in any event, the 800°F spread of temperatures is much to large for pipeline use.

The old D 341–43 Fahrenheit charts (issued prior to publication of the new *Charts* in 1974), sometimes still available in technical bookstores or libraries, had a spread of temperatures much more useful to pipeliners.

The construction of the charts is described by Wright,[2] and the reference includes the equations used to develop the charts included in Appendix X1, *Mathematical Relationships*, of D 341.

The complete equation, involving (a) three logarithms, one of them a log–log function; (b) six exponentials, all involving a constant of six significant figures and a six figure multiplier for ν; (c) one temperature in °R or K; and (d) two constants, is, as indicated by the *Standard*, too complex for ordinary purposes. However, the equation, with all of its complexities, was designed for computer computation and programs nicely in Fortran, BASIC, and most spread–sheet languages so that, once programmed, the complexities disappear in the code.

The basic equation is

$$\log \log Z = A + B \log T \tag{3-7}$$

where $Z = (\nu + 0.7 + C - D + E - F + G - H)$
$\log$ = logarithm to base 10
ν = kinematic viscosity, cSt (or mm^2/s)
T = temperature, K or °R
A = constant
B = constant

The values of the other parameters, all exponential variables in ν, are as follows:

$$C = \exp(-1.14883 - 2.65868\,\nu) \tag{3-8}$$
$$D = \exp(-0.0038138 - 12.5645\,\nu) \tag{3-9}$$
$$E = \exp(5.46491 - 37.6289\,\nu) \tag{3-10}$$
$$F = \exp(13.0458 - 74.6851\,\nu) \tag{3-11}$$
$$G = \exp(37.4619 - 192.643\,\nu) \tag{3-12}$$
$$H = \exp(80.4945 - 400.468\,\nu) \tag{3-13}$$

The value of Z can be used in its full eight–parameter version to cover the full range of kinematic viscosities from 0.21 cSt to $2 \cdot 10^7$ cSt or, in progressively less complex formulations for Z, down to a formula without any of the six variables in ν that is valid for viscosities between 2.00 and $2 \cdot 10^7$ cSt. These formulas are as follows:

$Z = (\nu + 0.7)$	2.00 to $2 \cdot 10^7$ cSt
$Z = (\nu + 0.7 + C)$	1.65 to $2 \cdot 10^7$ cSt
$Z = (\nu + 0.7 + C - D)$	0.90 to $2 \cdot 10^7$ cSt
$Z = (\nu + 0.7 + C - D + E)$	0.30 to $2 \cdot 10^7$ cSt
$Z = (\nu + 0.7 + C - D + E - F + G)$	0.24 to $2 \cdot 10^7$ cSt
$Z = (\nu + 0.7 + C - D + E - F + G - H)$	0.21 to $2 \cdot 10^7$ cSt

The least complex formula, that using $Z = (\nu + 0.7)$, will cover all crude oils. The alternate formula using $Z = (\nu + 0.7 + C - D)$ will cover crude and all products down to the gasolines. Working with some gasolines, and with light LPG and LNG fractions, may require the more complex expressions for Z.

To work with the basic equation numerically instead of graphically, the constants A and B can be determined by rewriting Equation 3-7 and solving for A and B.

As a first approximation, the basic equation can be written as

$$\log \log(\nu + 0.7) = A - B \log T \tag{3-14}$$

where ν = kinematic viscosity, cSt
T = temperature, K or R
A, B = constants to be determined

Constants A and B can be determined by rewriting the equation in the following form and solving for the constants.

$$\nu = 10 \uparrow (10 \uparrow (A - B \log T)) - 0.7 \tag{3-15}$$

Solving the following equations for A and B allows calculation of the viscosity–temperature relationship for the subject oil.

$$A = \log(\log(\nu_1 + 0.7)) + B \log T_1 \tag{3-16}$$
$$B = (\log(\log(\nu_1 + 0.7) / (\log(\nu_2 + 0.7)))) / \log(T_2 / T_1) \tag{3-17}$$

This relationship fails for viscosities less than 2.00 cSt so LPG and LNG, gasolines, and most kerosines will not fit into the applicable limits of this rather simple relationship.

Appendix X1 has a set of equations due to Manning[3] which are applicable in cases where some of the data fall below the 2.0 cSt limit of the simple equations above. These can be used to calculate the temperature associated with a desired viscosity or, conversely, to calculate the viscosity associated with a given temperature. These equations are

$$\log\log Z = A - B\log T \tag{3-18}$$

$$Z = \nu + 0.7 + \exp\left(-1.47 - 1.84\,\nu - 0.51\,\nu^2\right) \tag{3-19}$$

$$\nu = [Z - 0.7] - \exp\begin{pmatrix} -0.7487 - 3.295[Z - 0.7] \\ +0.6119[Z - 0.7]^2 \\ -0.3193[Z - 0.7]^3 \end{pmatrix} \tag{3-20}$$

Inserting Equation 3-19 into Equation 3-18 permits solving for A and B. This form can also be used to calculate the temperature associated with a desired kinematic viscosity.

The kinematic viscosity associated with a given temperature can be calculated by solving for Z in the substituted Equation 3-18, and substituting this value of Z in Equation 3-20.

Often, and especially in computer programs for design or simulation of hot–oil pipelines, it is desirable to have a *predictor* routine which will produce a fairly accurate estimate of viscosity at one temperature given viscosity at another. If computer time is not important the best predictor would be Equation 3-15. Most computers powerful enough to run simulation programs with acceptable speed will have optimized logarithm and exponential functions in their floating–point package, and the added accuracy, though perhaps not necessary, will assure that the chaotic effect of chain computations is minimized insofar as the value of viscosity is concerned.

On the other hand, it is sometimes necessary to make an educated guess of the value of viscosity at some temperature and a 100 MIPS computer is not available. The following kinds of formulas are useful in these cases.

Horner and Fried[4] used a function of the form

$$\nu = 0.8 + e^{(CT)^n} \tag{3-21}$$

where $\nu =$ viscosity, cSt
$C, n =$ dimensionless constants

Zama[5] recommended the following:

$$\log \nu = \left(\log \nu_0 + 1.7\right) \times \left(100 + T_0\right) / \left(100 + T\right)^{0.52} - 1.7 \tag{3-22}$$

This equation, which Zama credits to Watson[6] computes ν at T given ν_0 and T_0 and of necessity assumes a certain slope to the viscosity–temperature curve.

Another, more rigorously derived formulation is due to Staender[7] who uses a three–constant function credited to von Vogel (source not known). This requires three sets of viscosity–temperature pairs to allow the determination of the constants. The actual–vs–calculated curves given by Staender show excellent correlation. Staender's function and method of calculating the constants is as follows.

The basic equation is

$$\nu = A\exp\left(B / (T - C)\right) \tag{3-23}$$

In logarithmic form, with ln symbolizing $\log_e$:

$$\ln \nu = \left(B / (T - C)\right) + \ln A \tag{3-24}$$

First calculate intermediate constant a:

$$a = \left(\ln(\nu_1 / \nu_2) / \left(\ln(\nu_2 / \nu_3)\right)\right) \tag{3-25}$$

Then solve for C:

$$C = \left[\frac{aT_1(T_3 - T_1) - T_3(T_2 - T_1)}{a(T_3 - T_2) - (T_2 - T_1)}\right] \tag{3-26}$$

Finally, solve for B and A:

$$B = \left[\frac{\ln\left(\dfrac{\nu_1}{\nu_2}\right)}{\dfrac{1}{(T_1 - C)}}\right] - \frac{1}{(T_2 - C)} \tag{3-27}$$

$$A = \frac{\nu_1}{\left[B / (\exp(T_1 - C)\right]} \tag{3-28}$$

With A, B, and C known, ν can be found explicitly for any T with a single calculation using Equation 3-23.

Hyman[8] gives a relationship which allows calculating the absolute viscosity μ at temperature T given two viscosity–temperature pairs. This function, which Hyman credits to Huang and Johnson,[9] is valid over fairly short temperature ranges.

$$\mu = a\exp\left(\frac{b}{T}\right) = \exp\left(c + \left(\frac{b}{T}\right)\right) \tag{3-29}$$

$$c = \ln \mu_1 - b / T_1 \tag{3-30}$$

$$b = \frac{\ln(\mu_2 / \mu_1)}{((1/T_2) - (1/T_2))} \tag{3-31}$$

Lastly, there is the single–point predictor relationship of Singh, Mutyala, and Puttagunta.[10] This equation is an interesting development in that in dozens of calculations involving the estimation of viscosity at a temperature other than that at which it was measured for the 20 crudes in the set reported by Amin and Maddox[11] the overall absolute deviation was only 0.82% and the maximum deviation for a single point out of the approximately 150 points reported was only 3.59%.

The basic equation is

$$[\log \nu + C]/[\log \nu_0 + C] = [T_0 / T]^S \tag{3-32}$$

where $\nu =$ kinematic viscosity, cSt
$T =$ absolute temperature, K
$C =$ 0.86960, when the log base is 10
$S =$ $0.28008 \times \log \nu_0 + 1.8616$ when
$T_0 =$ 310.93 K = 100° F = 37.78° C

Singh et al later modified this equation to a working equation as follows:

$$\log \nu = 10 \uparrow \left[B \times (310.93 / (t + 273.15))^S - C\right] \tag{3-33}$$

where $B = \log \nu_0 + C$
$C = 0.86960$
$S = 0.28008 \times B + 1.6180$
$= 0.28008 \times \log \nu_0 + 1.8616$
$T_0 = 310.93$ K = 37.78° C = 100° F

To predict the kinematic viscosity of a crude oil or its fractions for a wide range of temperatures, all that is required is to measure kinematic viscosity at 37.78°C = 100°F at atmospheric pressure. Parameters B and S can then be computed directly and the viscosity at any other temperature can be determined by substituting these parameters and ν_0 in Equation 3-33.

The following form allows direct calculation of the viscosity ν at any temperature t if the base viscosity ν_0 is measured at T_0= 273.15 K = 37.78°C =100°F.

$$\nu = 10 \uparrow \left[B \times (310.93 / (t + 273.15))^S\right] - C \tag{3-34}$$

For any other T_0 new values for B and S must be determined. This is an iterative procedure, but programs easily and closes quickly. Singh et al. have a short BASIC program in the reference.

All of the predictor kind of formulations except that of Singh et al. depend on having an a priori knowledge of the slope of the viscosity–temperature curve, and the Singh et al. relationship, for base temperatures other than for 273.15 K = 37.78°C = 100°F, requires a computer or calculator capable of making an iterative determination of the values for the variables B and S.

This same group published the results of later research that offers a simpler method without the *log log* function of their first work. Puttagunta, Miadonye, and Singh[12] proposed the following formula based on their work on heavy oils and bitumens in the Alberta, Canada, fields, where viscosity at 20°C runs from 500 cP (Lloydminster oil) to 1 200 000 cP (carbonate trend bitumen):

$$\log \mu = \frac{b}{\left(1 + \frac{T - 30}{303.15}\right)^s} + C \tag{3-35}$$

where $b = \log \mu_{30^\circ C,\ 0\ MPag} - C$
$C = -3.0020$
$s = 0.0066940b + 3.5364$

Given the viscosity μ in cP (or mPa–sec, which is the same unit) at 30°C and 0 MPag (or 101.3 kPa, or 14.69 psia, both being the same unit), the definitions of b, C, and s, Equation 3-51 will yield viscosity at any reasonable temperature T in Kelvin. Though developed in work aimed at predicting the viscosity of heavy oils and bitumens in situ, the correlation works very well for any well–behaved hydrocarbon; the average absolute deviation for 175 samples was 4.4% compared to measured values.

In summary, the ASTM method is very exact; if the input data are correct the output is going to be very close to reality. Staender's method does not cover nearly as wide range of vis–temp relations as ASTM but it is better than the others. Sometimes, however, an accurate answer is not needed, and a quick approximation from a function like that of Horner and Fried will yield a perfectly acceptable answer.

If you have the tools, the time, and a single viscosity–temperature point, the Singh et al. equations, for crude oils, would appear to be eminently satisfactory.

Experience provides the judgment required to choose the method to use.

Viscosity–Pressure Relations

The viscosity of a liquid increases with increased pressure. The absolute viscosity of a gas is constant with increasing pressure; but inasmuch as gases are highly compressible the kinetic viscosity of gases decreases with increasing pressure as ρ in the $\nu = \mu / \rho$ equivalency increases as pressure increases.

The theoretical equation for absolute viscosity in terms of pressure for a liquid is

$$\ln \mu = A + BP \tag{3-36}$$

This can be written to be explicit in μ as:

$$\mu = \exp(A + BP) \tag{3-37}$$

The best correlation I have been able to find for a viscosity–pressure predictor is that due to Kouzel.[13] His work commences with the data from the *ASME Viscosity Report*, Volumes I and II, 1953, which contains experimental data on over 40 hydrocarbons in terms of a correlation between

$$\left[(1/\mu)(d\mu/dP)_{P=0}\right]$$

and

$$\left[(1/\mu)(d\mu/dt)_{P=0}\right]$$

Both the pressure and temperature coefficients of viscosity are determined at zero gage pressure. The data are valid up to 425°F.

Kouzel sets these data in straight–line form as

$$10^3\left[\frac{1}{\mu}\frac{d\mu}{dP}\right]_{P=0} = 55 - (4.25\ \mathrm{E\text{-}03})\left[\frac{1}{\mu}\frac{d\mu}{dT}\right]_{P=0} \tag{3-38}$$

Substituting $2.3 \times \log \mu = d\mu / \mu$ in Equation 3-38 yields

$$\left[\frac{d \log \mu}{dP}\right]_{P=0} = 0.0239 - 4.2\left[\frac{d \log \mu}{dt}\right]_{P=0} \tag{3-39}$$

Hershey and Hopkins[14] recommended the following general form for the viscosity–pressure relationship.

$$\log \mu_p = \log \mu_a + bP \tag{3-40}$$

where μ_p = viscosity at pressure p, psig
μ_a = viscosity at a pressure of 1 atmosphere
P = 1000's of psig = 1000 × p

Kouzel found that the *ASME Viscosity Report* data plotted as *log* μ versus P was essentially straight–line in the range P = 0–5000 psig so that the slope b of the straight line established by the above equation could be written as

$$b = \left[\frac{d \log \mu}{dP}\right]_{P=1000} \tag{3-41}$$

Combining Equation 3-39 and Equation 3-40 with this explicit definition for slope b yields an equation which relates absolute viscosity at any pressure P to its absolute viscosity at a pressure of one atmosphere in terms of the pressure P and the viscosity–temperature function $(d \log \mu / dT)_{P=0}$ at one atmosphere pressure.

$$\log\left(\frac{\mu_p}{\mu_a}\right) = \frac{p}{1000} \times \left(0.0239 - 4.2 \times \left(\frac{d \log \mu}{dt}\right)_{P=0}\right) \tag{3-42}$$

To obtain the best accuracy it is necessary to use some care in determining the value of $(d \log \mu / dT)_{P=0}$. If two viscosity–temperature pairs are available the constants A and B in the Andrade formula $\mu = A \exp(B/t)$ can be determined and the value $(d \log \mu / dt)$ of Equation 3-41 can be calculated.

If only one viscosity–temperature pair is available, Kouzel recommends that $(d\log\mu/dT)_{P=0}$ be assumed equal to $0.0039\mu_a^{0.278}$. Kouzel says this is accurate for viscosities in the range from 0.5 to 200 cP. Substituting this value in the last equation yields

$$\log\left(\frac{\mu_p}{\mu_a}\right)=\frac{p}{1000}\left(0.0239+0.01638\mu_a^{0.278}\right) \qquad (3\text{-}43)$$

Or

$$\mu_p=\mu_a\times 10\uparrow\left(\frac{p}{1000}\right)\left(0.0239+0.01638\mu_a^{0.278}\right) \qquad (3\text{-}44)$$

One of Kouzel's examples will illustrate:

Estimate the viscosity of an oil at 4880 psig and 100°F. The viscosity at 1 atmosphere and 100°F is 2.32 cP.

$$\log\left(\frac{\mu_p}{2.32}\right)=\left(\frac{4880}{1000}\right)\left(0.0239+0.01638\times 2.32^{0.278}\right)$$

$$=0.218$$

and

$$\left(\frac{\mu_p}{2.32}\right)=10\uparrow(0.218)=1.652$$

thus

$$\mu_p=1.652\times 2.32=3.84\text{ cP}$$

Of more interest to pipeliners (after all, 4880 psig is not a *normal* pipeline operating pressure) might be the effect on a 20 cP burning oil in a products pipeline that might have a maximum discharge pressure of 2000 psig.

Pressure	Viscosity
0	20.00
500	21.47
1000	23.05
1500	24.74
2000	26.56

The viscosity at the 2000 psig discharge of the pump station is 6.56 cP, or nearly one–third, greater than the viscosity at the inlet of the subsequent downstream station, where a pressure of one or two atmospheres would be usual. This is enough to make a substantial change in Reynolds Number, even allowing for the increase in density of the burning oil at high pressures; and even though the friction factor in turbulent flow is an exponential function of Reynolds Number, there will be a measurably larger unit friction loss in the section of pipe leaving the station discharge than in that section entering the suction of the downstream station. This is the kind of effect that is usually not taken into account in the design of pipelines but must be considered in simulation studies.

There is a published nomogram prepared by George Mapstone, Geriston, South Africa (name of publication not known) based on Kouzel's work, but it is in terms of kinematic viscosity in cSt instead of absolute viscosity in cP. It is probably quantitatively correct, but it isn't correct because Kouzel's work, and his references, are based on absolute instead of kinematic viscosities.

Measurement of Viscosity

In the late 1830's a French scientist named Jean–Louis–Marie Poiseuille, who was studying the flow of blood in the veins of horses, experimentally arrived at an empirical relationship for the non–turbulent flow of Newtonian fluids in small diameter pipes. This empirical relationship was given sound theoretical standing by the work of a German hydraulic engineer named Gotthilf Heinrich Ludwig Hagen. Hagen published his work in 1839, Poiseuille published his in 1840, and the law they had discovered by two different approaches is generally called the *Hagen–Poiseuille Law* but, sometimes, just *Poiseuille's Law*.

The equation can be cast in many forms. The following is one of the most common.

$$V=\left[\left(\pi p r^4\right)/\left(8\ell\mu\right)\right]t \qquad (3\text{-}45)$$

where $V=$ volume of liquid flowing in time t
$p=$ pressure difference between inlet and outlet
$r=$ radius of the bore of the tube
$\ell=$ length of the tube
$\mu=$ absolute viscosity of the liquid

Since all the parameters except the viscosity can be measured, the viscosity can be calculated. Thus a long, small–bore tube carrying a Newtonian liquid in non–turbulent flow can be considered as a viscometer of sorts.

The U.S. National Bureau of Standards developed a formula which is more refined in that it makes a correction for kinetic energy and is explicit in absolute viscosity. The NBS formula is

$$\mu = \left(\pi d g r^4 t\right)\left(h - m v^2 / g\right) / \left(8Q(\ell + \lambda)\right) \qquad (3\text{-}46)$$

where μ = absolute viscosity
d = density in g/cm^3
r = radius of tube bore, cm
ℓ = length of tube, cm
Q = volume in cm^3 discharged in t sec
λ = a correction for length of the tube
h = average head in cm
m = coefficient of kinetic energy correction
g = acceleration due to gravity in cm/sec^2
v = mean velocity in cm/sec

Three standards issued by ASTM cover precision measurement of Newtonian kinematic viscosity for all reasonable purposes. These are:

ASTM D 446 Standard Specifications and Operating Instructions for Glass Capillary Kinematic Viscometers.

ASTM D 445 Standard Method of Test for Kinematic Viscosity of Transparent and Opaque Liquids (And the Calculation of Dynamic Viscosity).

ASTM D 2162 Standard Method of Basic Calibration of Master Viscometers and Viscosity Oil Standards.

ASTM D 445 is the standard for daily use, D 2162 describes standard calibration of viscometers, and D 446 describes how to make and use each of the acceptable viscometer designs referenced in the other two standards.

ASTM D 446 has three annexes which describe acceptable viscometers of three classes:

Annex 1: Modified Ostwald Viscometers for Transparent Liquids (four acceptable designs).

Annex 2: Suspended Level Viscometers for Transparent Liquids (five acceptable designs).

Annex 3: Reverse Flow Viscometers for Transparent and Opaque Liquids (three acceptable designs).

All of the acceptable designs depend on timing the flow of a measured quantity of test liquid as it falls by gravity through a capillary tube. The fall time, multiplied by the *viscometer constant* supplied by the manufacturer of the viscometer, yields kinematic viscosity in cSt. Viscometer constant is defined as:

$$C = \frac{\nu}{t} \qquad (3\text{-}47)$$

where C = viscometer constant
ν = kinematic viscosity, cSt, of standard fluid
t = flow time, seconds

The diameter of the capillary tubes in the viscometers is usually sized to produce a minimum fall time of 200 seconds for the appropriate range of viscosities, though 250 and even 300 second minimums are stipulated for certain tubes used to measure very low viscosities.

Each kind of viscometer is available in as many as twelve capillary sizes to accommodate viscosities from as low as 0.3 cSt for the Ubbelhode to as high as 100 000 cSt for the Ubbelhode, Cannon–Ubbelhode, Zeitfuchs Cross–Arm, and Lantz–Zeitfuchs designs.

Maximum flow time is usually taken as 1000 seconds. This is simply based on practical considerations: a 1000 second test is boring and, in the case of evaporating liquids, may spoil the test. If this flow time is exceeded, a viscometer with a larger bore capillary should be selected and the test rerun.

Kinematic viscosity expressed in cSt can be calculated from viscometer dimensions as follows:

$$cSt = \left(\frac{\pi d^4 h t}{128 V L}\right) - \frac{E}{t^2} \qquad (3\text{-}48)$$

where d = capillary diameter, cm
L = capillary length, cm
h = vertical distance between menisci, cm
V = timed volume passing through capillary, cm^3
E = kinetic energy correction, cSt / sec^2
t = flow time, seconds

This is another form of the *Hagen–Poiseuille Law* explicit in viscosity. If the efflux time is greater than 200 seconds and C is greater than 0.05, the kinetic energy correction is not significant and $cSt = Ct$ as described. If these conditions are not met, $cSt = Ct - E/t^2$; and while E is not a constant it may be approximated by the following equation:

$$E = 1.66V^{3/2}/L(Cd)^{1/2} \tag{3-49}$$

where V = volume of timing bulb, cm^3
L = capillary length, cm
d = capillary diameter, cm
C = viscometer constant

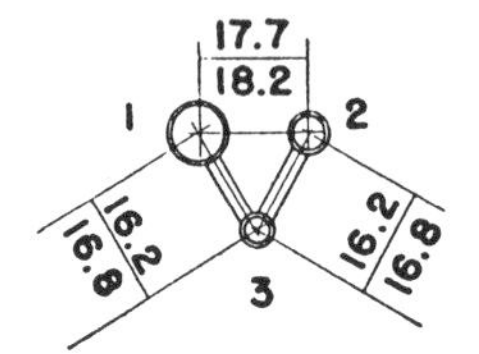

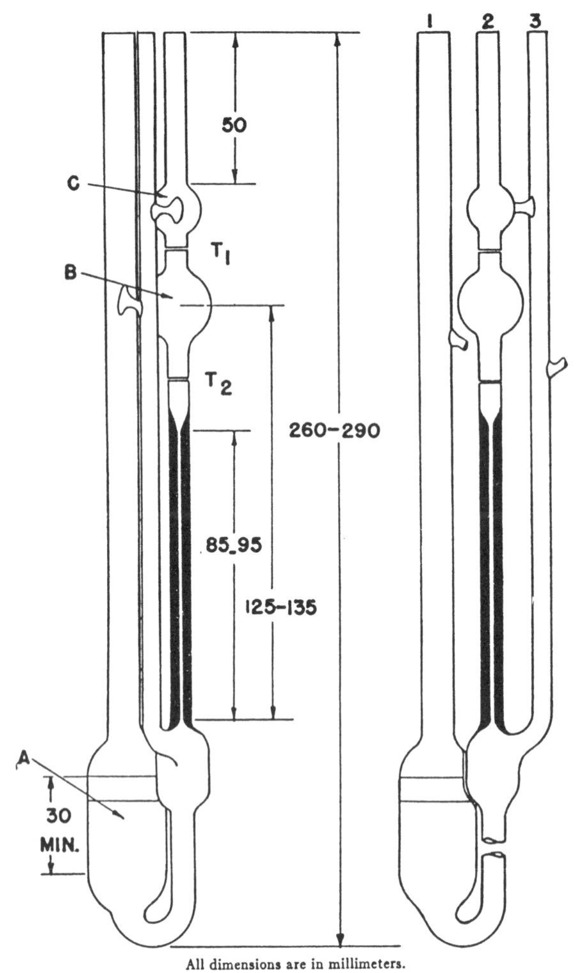

Figure 3-3. Ubblehode Viscometer.

Figure 3-3 is a line drawing of a Ubbelhode viscometer, which is one of the simplest and easiest to understand of the acceptable viscometers. To make a measurement, the viscometer is clamped in a vertical position, a sample introduced into bulb A, transferred by vacuum or pressure to bulb C, and allowed to flow from bulb C to bulb B. The time of passing of the upper meniscus between the timing marks E and F, multiplied by the viscosity constant of the viscometer, yields the kinematic viscosity in cSt.

Commercial Viscometry

The capillary type viscometers described in the ASTM standards previously cited herein are rather delicate for routine use, especially under oilfield conditions, and other, more physically robust, viscometers have been developed which, while not measuring cSt or cP directly, do have the desired characteristics of repeatability and simplicity of operation.

In the U.S. and Canada the most used of these kinds of viscometers are those of the *Saybolt* design. In the U.K. and commercially associated countries the *Redwood* viscometer is usually used; and Germany usually uses the *Engler* viscometer. There are many more.

The ASTM has decided, along with other changes having to do with the world-wide swing toward SI, to cancel ASTM D 88 which, for years, had been the world's standard for the design and use of Saybolt viscometers.

A typical Saybolt viscometer is shown in Figure 3-4. The measuring elements of the viscometer are the tube, shown in Figure 3-5, and the receiving flask, shown in Figure 3-6. The tube is filled with the sample, heated until the desired temperature is reached and stabilized, and then the cork sealing the nozzle is quickly pulled away. Viscosity is expressed as the number of seconds required for a 60 ml sample of test liquid to flow from the tube into the flask. If the tube is equipped with the *Universal* tip, the viscosity is reported in *Saybolt Universal Seconds* (usually *SUS* but, sometimes, *SSU*); if the *Furol* tip is used, it is reported in *Saybolt Furol Seconds* (usually *SFS* but, sometimes, *SSF*).

The efflux time for the Universal tip is about ten times that of the Furol tip. Thus, the Furol tip is used for heavier, more viscous, oils. The name Furol, itself, is derived from a combination of "fuels and road oils."

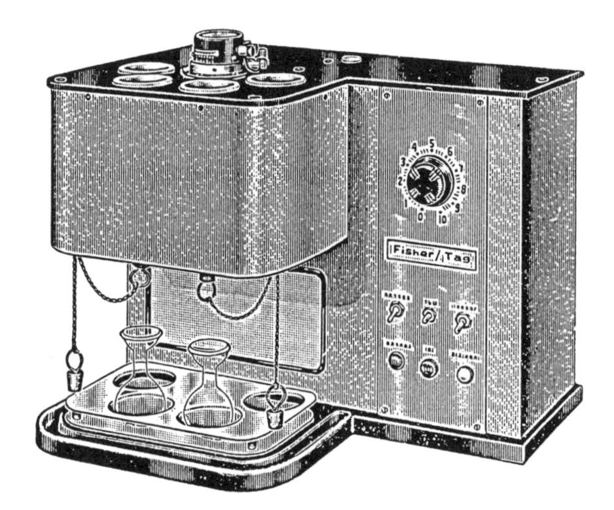

Figure 3-4. Saybolt Viscometer.

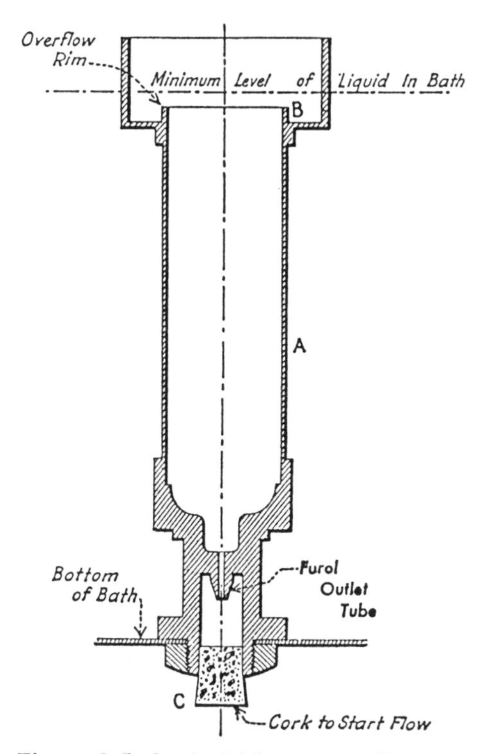

Figure 3-5. Saybolt Viscometer Tube.

The Universal tip is not reliable for efflux times less than 32 seconds, corresponding to a kinematic viscosity of about 1.81 cSt at 100°F, and thus cannot be used for gasolines and less viscous liquids. There was, at one time, a *Saybolt Thermal* tip used for measuring very low viscosities but it no longer has any official standing—if it ever did.

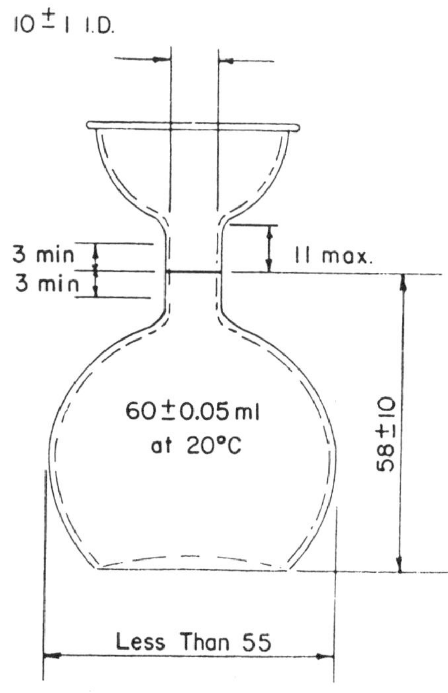

Figure 3-6. Saybolt Viscometer Flask.

Neither tip should be used for efflux times greater than 1000 seconds. If it is over 1000 seconds on the Universal tip, the Furol tip can be used. If it is over 1000 seconds on the Furol tip, recourse must be had to one of the capillary type viscometers or one of the power–driven viscometers described in the following chapter. There is no Saybolt test for viscosities corresponding to efflux times greater than 1000 seconds in the Furol tip.

There are many other kinds of viscometers, and new designs, some based on completely new principles, come on the market with regularity. Aronson and Nelson[15] contains descriptions of several dozens of viscometers manufactured by more than 40 companies. They classified viscometers as of four general types, each of which may have one or more of four different kinds of driving mechanism. Table 3-1 is taken from their classification diagram; and, while there are other ways to classify viscometers their way is as clear as any.

Viscometric Conversions

Saybolt Viscometers

Even though ASTM D 88 is gone, ASTM still publishes D 2161, *Standard Method for Conversion of Kinematic Viscosity to Saybolt Universal Viscosity or to*

Table 3-1

Viscometer Techniques

Kind of Measurement	Type	Manual Time	Manual Force	Automatic Time	Automatic Force
Absolute Viscosity- poise	Sliding Plate	•			
	Rotating				
	- cylinder	•	•		•
	- cone		•		
	- disc		•		
	Capillary				
	- positive displacement pump		•		•
	- fixed Δp	•	•		
Kinematic Viscosity- Stoke	Capillary				
	- fixed head		•		
	- liquid head	•			
	Rising Bubble	•			
	Consistometer	•			
Absolute Viscosity (with density correction)	Falling Ball	•			
	Falling Piston			•	
	Vibrating Reed			•	•
Gel Time	Rotating Probe	•			
	Windup Thread	•			
	Rising Disc	•	•		
	Vibrating Steel Rod	•	•		
	Thermal Incline	•			

Saybolt Furol Viscosity, and this is still the world–wide standard for these kinds of calculations. The standard has three sets of tables:

Table 1: Kinematic Viscosity to SUS, 100°F and 210°F.
Table 2: Conversion Factors, Kinematic Viscosity to SUS, 0°F >T>350°F.
Table 3: Kinematic Viscosity to SFS, 122°F and 210°F.

Table 1 covers kinematic viscosities from 1.77 cSt to 500.0 cSt, and Table 3 from 48 cSt to 1300 cSt. These are ranges sufficiently wide to cover most pipeline situations but there are two problems:

- The tables are for converting cSt into SUS and SFS, and thus are explicit in cSt, not SUS or SFS. In converting SUS into cSt, for instance, 206 SUS can be converted to 43.95, 44.00, 44.05, or 44.10 cSt. While this is no great difference, there is a certain mental discomfort in having such a range of choices.
- The tables are just that—*tables*; and it's difficult to enter a set of tables into a calculation except

when done with a computer when a look–up table can be established within the program.

Therefore, ASTM D 2161 provides equations for computer use, as follows:

$$SUS_{100F} = 4.6324cSt + \frac{1.0 + 0.03264cSt}{\begin{pmatrix} 3930.2 + 262.7cSt \\ +23.97cSt^2 + 1.646cSt^3 \end{pmatrix} \cdot 10^{-5}} \quad (3\text{-}50)$$

$$SUS_t = (1.0 + 0.000061(t - 100)) \times SUS_{100^\circ F} \quad (3\text{-}51)$$

The second equation can be solved explicitly for cSt, but the first must be solved by iterative methods because cSt appears as a variable in both the numerator and denominator of the fraction, but the function closes quickly.

Note Equation 3-51 is $1.0 + 0.000061(t - 100)$ times Equation 3-50; this multiplier is applicable to converting any SUS_{100F} viscosity measurement to SUS_t where t is a temperature in °F.

Another pair of equations has to do with SFS viscosities. These are

$$SFS_{122F} = 0.4717cSt + \frac{13924}{cSt^2 - 72.59cSt + 6816} \quad (3\text{-}52)$$

$$SFS_{210F} = 0.4792cSt + \frac{5610}{cSt^2 + 2130} \quad (3\text{-}53)$$

In all above equations, cSt is the kinematic viscosity in centistokes at any temperature t in °F.

There are four short–form equations which apply under certain conditions. These are

$$\begin{aligned} SUS_{100F} &= 4.6320 \times cSt, \ \text{cSt} \geq 500 \\ SUS_{210F} &= 4.6640 \times cSt, \ \text{cSt} \geq 500 \\ SFS_{122F} &= 0.4717 \times cSt, \ \text{cSt} \geq 1300 \\ SFS_{210F} &= 0.4792 \times cSt, \ \text{cSt} \geq 1300 \end{aligned} \quad (3\text{-}54)$$

Redwood Viscometers

The Redwood viscometer, a tube–and–flask viscometer similar to the Saybolt design, was once widely used in the U.K., but the U.K. Institute of Petroleum no longer publishes IP 70, which provided the standard method for measuring Redwood viscosity using the (British) National Physical Laboratory Redwood Viscometer, but the *IP Standards* still carry in Appendix B, Part II, *Conversion of Kinematic Viscosity to Redwood No. 1 Viscosity*, a one–page table equating kinematic viscosities from 4.00 to 100 cSt to *Redwood No. 1 seconds* at 70°F, 140°F, and 200°F. There are no equations—the relationships established by the table were drawn from experiments on various oils in Newtonian flow—but for viscosities in excess of 100 cSt a linear relationship is established as:

$$\text{Redwood No. 1 seconds}_t = A \times cSt_t \quad (3\text{-}55)$$

$$\begin{aligned} \text{where for } t &= 70^\circ F \quad A = 4.06 \\ t &= 100^\circ F \quad A = 4.10 \\ t &= 140^\circ F \quad A = 4.18 \end{aligned}$$

The IP Appendix notes:

> Since the values obtained by IP 70 for Redwood viscosity may be in error as much as one percent, the conversion table should not be used for converting *Redwood No. 1* viscosity to kinematic viscosity.

The IP Appendix also notes that if measurements of kinematic viscosity are made by ASTM D 445 (which also carries the designation IP 71) the *Redwood No. 1* values obtained using the table will be within one percent of what would be measured with the National Physical Laboratory Standard Reference Redwood Viscometer.

It should be noted that *Redwood No. 1 Seconds* are sometimes called *Redwood Standard Seconds*, and *Redwood No. 2 Seconds* are often called *Redwood Admiralty Seconds*. The efflux time of the No. 1 tube is approximately ten times that of the No. 2 tube, thus the same relationship as *Saybolt Universal Seconds* to *Saybolt Furol Seconds*.

Engler Viscometers

The Engler viscometer is sometimes used in the U.S. and U.K. to measure viscosity of heavy products, and is used in Germany and some other countries, and in certain specialized industries all over the world, for other kinds of measurements.

There are three relevant standards in English language having to do with the Engler viscometer. ASTM D 1665, *Test for Engler Specific Viscosity of Tar Products*, IP 212, *Viscosity–Bitumen Road Emulsion*, and BS 434, which is said to be technically the same as IP 212, carry descriptions of the Engler viscometer; and while ASTM D 1665 is obsolete, IP 212 and BS 434 were still in effect when this was written.

The basis of the Engler design is, once again, a measurement of the efflux time of a given volume of sample through the orifice in a sample tube.

In the case of the Engler viscometer, kinematic viscosity is measured by the time a 200–ml sample at 20°C takes to flow from the tube.

The result of the test, given in terms of the flow time of the sample divided by the flow time of an equal quantity of water, is called *degrees Engler*. The time required for the water sample to flow is called the *base time*, and must be between 50 and 52 seconds. IP 212 makes note that for very viscous samples—exceeding 25 *degrees Engler*, for instance—some non–Newtonian flow may occur.

Conversion Formulas

In addition to the described equivalency formulas relating viscosity expressed in *Saybolt SUS* and *SFS*, *Redwood No. 1 Seconds* and *No. 2 Seconds*, and *degrees Engler* viscosity to *kinematic viscosity*, there are many equivalency formulas for obtaining viscosity in the desired units from viscosity measured in other units. Some of these are fairly accurate; some of them are merely good estimates.

Undoubtedly the best of these are those developed by O'Donnell.[16] These are referenced in *ASTM Viscosity Tables for Kinematic Viscosity Conversions and Viscosity Index Calculations*, ASTM Special Publication 43 C, and have a quasi–authoritative standing insofar as conversions cSt–SUS–cSt are concerned; are fully authoritative for conversions cSt–SFS–cSt since the equations are those of ASTM D 2161 cited previously; and insofar as conversions *cSt–Redwood–cSt* and *cSt–Engler–cSt* are concerned, the conversion equations were, according to O'Donnell, "based on existing conversion tables which were in common use in the country of origin for each particular viscosity unit."

To convert from *Redwood No. 1* to *SUS*, for instance, convert *Redwood No. 1* to *cSt* and then convert *cSt* to *SUS*: O'Donnell's formulas are explicit in each direction.

Table 3-2 gives the equations written to obtain flow times given kinematic viscosity in cSt, and Table 3-3 gives the equations written to yield cSt from flow times. The inset at the bottom of Table 3-2 is the ASTM D2161 temperature conversion equation.

There are many approximate conversion formulas, and dozens of different published conversion tables, all of which establish a kind of measure between several different kinds of viscosity units. Tables 34 and 35 of the Pipe Friction Manual,[17] for example, contain data for converting SUS and cSt into 22 other kinds of viscometer measurements. The only one of these, other than Redwood and Engler, apt to be encountered by a pipeliner is *Barbey viscosity* which is measured by the Barbey viscometer sometimes used in France.

There are also many nomograms which can be used. One of the most widely known of these is that copyrighted by Texaco, Inc., in 1960 and since copied widely by others.

Short–form conversions are handy, however, in that they provide answers quickly without having to create a computation form to accommodate six–place constants, exponential variables, fractional powers, and the like. The equations shown in Table 3-4 for converting commercial viscometer measurements to *cSt*, and vice versa, are typical. The equations for converting *SUS*, *SFS* and *Redwood No. 1* are common and are in print in many versions; the equations for converting *Engler* and *Redwood No. 2* are from a private source I trust. The equations for SUS, SFS, and Redwood No. 1 check well with the ASTM and IP tables; the Redwood No. 2 formula fits within the definition that Redwood No. 2 = [Redwood No. 1/10] ± 3%; and the Engler conversion checks well with the Viscosity Conversion Table[18] from the *Handbook of Chemistry and Physics*.

Viscosity of Blends

There is a simple, general, scientifically sound equation for the viscosity of mixtures of pure liquids:

$$\mu_m = (\mu_1)^{x_1} \times (\mu_2)^{x_2} \times (\mu_3)^{x_3} \ldots \times (\mu_n)^{x_n} \tag{3-56}$$

where μ_m = absolute viscosity of the mixture
μ_n = absolute viscosity of component n
x_n = mol fraction of component n

Unfortunately, the values available as inputs to a blending problem usually do not provide the mol fractions of the components; and, in addition, most of the time the components of the blends are, themselves, blends.

There are ways to handle this problem which, more and more, is becoming a part of routine pipeline operations,

Table 3-2

Summary of Equations for Converting Kinematic Viscosity in Centistokes to Flow Times

$$T = DV + \left[\frac{1+E}{F+GV+HV^2+IV^3} \right]$$

$T =$ flow times in seconds or Engler degrees
$V =$ kinematic viscosity in centistokes
$D, E, F, G, H, I =$ constants given below

Eqn No	Used For	D	E	F	G	H	I	Range cSt
1	SUS at 100°F	4.6324	0.0	0.039911	0.000938	0.000280	0.00000274	>1.8
2	SUS at 210°F	4.6635	0.00671	0.039922	0.000938	0.000280	0.00000274	>1.8
3	SFS at 122°F	0.47170	0.0	0.4895	-0.005213	0.0000718	0.0	>48
4	SFS at 210°F	0.47916	0.0	0.3797	0.0	0.0001783	0.0	>48
5	Redwood No. 1 at 140°F	4.0984	0.0	0.038014	0.001919	0.0000278	0.00000521	>4
6	Redwood No. 2	0.40984	0.0	0.38014	0.01919	0.000278	0.0000521	>73
7	Engler degrees	0.13158	0.0	1.1326	0.01040	0.00656	0.0	>1.0

Note: For Saybolt Universal Seconds at other temperatures $SUS_t = cSt_t \times \left[1 + 0.000061 \times \left(t - 100^\circ F\right)\right] \times \left[SUS_{100^\circ F} / cSt_{100^\circ F}\right]$

Table 3-3

Summary of Equations for Converting Flow Times to Kinematic Viscosity in Centistokes

$$V = AT - \left[\frac{BT}{T^3 + C} \right]$$

$V =$ kinematic viscosity in centistokes
$T =$ flow times in seconds (or Engler degrees)
$A, B, C =$ constants given below

Eqn No	Used For	A	B	C	Range
11	SUS at 100°F	0.21587	11 069	37 003	SUS>32
12	SUS at 210°F	0.21443	11 219	37 755	SUS>32
13	SFS at 122°F	2.120	8 920	27 100	SFS>25
14	SFS at 210°F	20.87	2 460	8 760	SFS>25
15	Redwood No. 1 at 140°F	0.244	8 000	12 500	R No. 1>35
16	Redwood No. 2	2.44	3 410	9 550	R No. 2>31
17	Engler degrees	7.60	18.0	1.727	E>1

Table 3-4

Formulas For Approximate Conversion of Flow Times to Kinematic Viscosity in Centistokes

Form of Formula: $cSt = AT - B/T$

Type of Viscometer	Conditions of Measurement	Value of A	Value of B
Saybolt Universal	32>t>100	0.226	195
	t>100	0.220	135
Saybolt Furol	25>t>40	2.240	184
	t>40	2.160	60
Redwood No. 1	34>t>100	0.260	179
	t>100	0.247	50
Redwood No. 2	33>t>90	2.458	100
	t>90	2.440	0
Engler	1.35>t>3.20	8.000	8.64
	t>3.20	7.600	4

with the transportation of light fractions, such as LPG and NGL blends, as admixtures or *spikes* in crude oil mainstreams, and the preparation and shipment of crude–crude (or even crude–crude– crude) *cocktails*. The first kind of these blends is usually made up to provide an acceptable way of transporting the light fraction; the second is usually done to provide the most commercially attractive blend of hydrocarbons to the input of a dedicated refinery.

ASTM D 341 provides two ways to perform the calculation, one using graphical techniques and the other a completely analytical method suitable for programming into calculators or computers.

Graphical Method

The following is paraphrased from ASTM D 341:

> Assume we have the viscosity–temperature points of two oils at 40°F and 100°F. Plot the known data on the Viscosity–Temperature Chart and carefully draw straight lines through the two points for each crude. The lines should extend beyond the desired kinematic viscosity of the blend. Then draw a horizontal line through the desired blend viscosity passing through the lines of the component oils. Lay a centimeter scale along this line and carefully measure the distance between the lines for the two oils where they cross the line of the desired blend viscosity. Without moving the scale, on the same horizontal kinematic viscosity line read the distance from the low viscosity line to the temperature desired. Dividing the latter by the first measurement gives the volume fraction needed for the higher viscosity oil.

As an example, given a low viscosity oil which has a viscosity of 7.50 cSt at 100°C and 55.7 cSt at 40°C, and a high viscosity oil with a viscosity of 17.00 cSt at 100°C and 190.00 cSt at 40°C, find the volume fraction of the higher viscosity oil for a blend of 13.00 cSt at 100°C.

Figure 3-7 shows the solution from ASTM D 341-77. On the original ASTM plot the distance from the low viscosity oil line to the 100°C line was along the 13 cSt line was 2.26 cm. The distance from the low viscosity oil line to the high viscosity oil line along the 13.00 cSt line was 3.30 cm.

The volume fraction of the high viscosity oil equals 2.26/3.30 = 0.685.

While the illustrative example is fairly simple because the temperature of the desired blend is the same as one of the temperatures of measurement of the components, the method is solid; to obtain the volume fraction of the high viscosity oil to produce a blend viscosity of 30 cSt at 70°C requires the same kind of calculation.

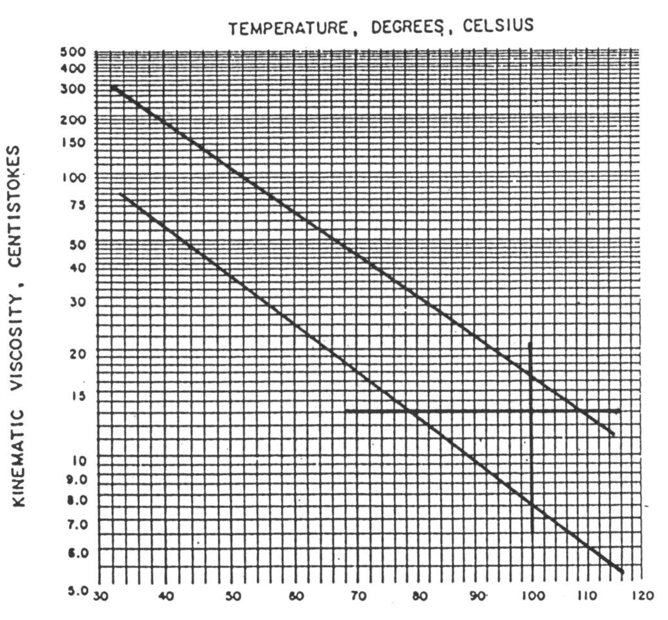

Figure 3-7. ASTM D 341-77 graphical blend problem.

The ragged appearance of Figure 3-7 is because it was necessary to use a copier to multiply the grid by 1.44 times, cut and paste a block to fit into the 21–pica width of a column in this book; and then reproduce it again. The version in the original ASTM D 341-77 is no better. In fact, the whole graphical calculation is difficult to make if the spread of temperatures is small or if the difference in viscosities is not very great; even using one of the large 26.75-inch by 32.25-inch charts the calculation area can easily shrink down to only an inch or two in each direction. The given example, on an 8 1/2-inch by 11-inch Chart III *Vis–Temp* form, has a base measurement line of only (about) 1.18 cm and the numerator line only (about) 0.86 cm and the result is 0.86/1.18 = (about) 0.73. Close, but not a satisfactory kind of calculation.

There are simpler solutions to these kinds of problems which do not require the very refined techniques of D 341. One of these, described in the original D 341-43, is shown on Figure 3-8. The problem above, which involves finding the viscosity of a blend at a temperature for which measurements of each of the components is available, can be solved by plotting one viscosity point on the 0°C line, the other on the 100°C line, and drawing a straight line between these points. Points on this line show the viscosities of blends of the components depending on the proportions of the two components.

Figure 3-8 plots a blend of 7.5 cSt and 17.0 cSt oil; a 13 cSt blend requires (about) 68 percent of the heavier component.

A second solution is to make use of the fact that D 341 is very precise, and sometimes precision is not needed. In this kind of case, we can realize that the $\log\log(\nu + 0.7)$ variable is not too far removed from $\log\log(\nu)$ and plot the Figure 3-8 solution on semi–log paper, as shown in Figure 3-9. This plot yields a 68% proportion of high viscosity oil as well as the other, accurate solutions.

The other simple solution is to program your calculator or computer to solve the blend problem analytically. See the following.

Analytical Method

The solution to ASTM demonstration problem can be obtained analytically with a quite simple algorithm based on ASTM D 341. This algorithm is the analytical equivalent of the mathematical manipulations of the graphical solution described above; and while it would be tedious to have to solve it with log–table–and–pencil techniques, what with all the $\log\log(x)$ calculations, it programs easily in almost any general programming language. The basic equations are

Volume fraction of high viscosity oil =

$$\left[\frac{(E-A)(C-D)}{(E-F)(A-C)}+1\right]^{-1} \text{ at } 40^\circ\text{C} \tag{3-57}$$

Volume fraction of low viscosity oil =

$$\left[\frac{(F-B)(C-D)}{(E-F)(B-D)}+1\right]^{-1} \text{ at } 100^\circ\text{C} \tag{3-58}$$

where $A = \log\log Z_{B(40)}$
$B = \log\log Z_{B(100)}$
$C = \log\log Z_{L(40)}$
$D = \log\log Z_{L(100)}$
$E = \log\log Z_{H(40)}$
$F = \log\log Z_{H(100)}$
$Z = (cSt + 0.70)$

and $B =$ blend
$L =$ low viscosity oil
$H =$ high viscosity oil
$40 = 40^\circ\text{C}$
$100 = 100^\circ\text{C}$

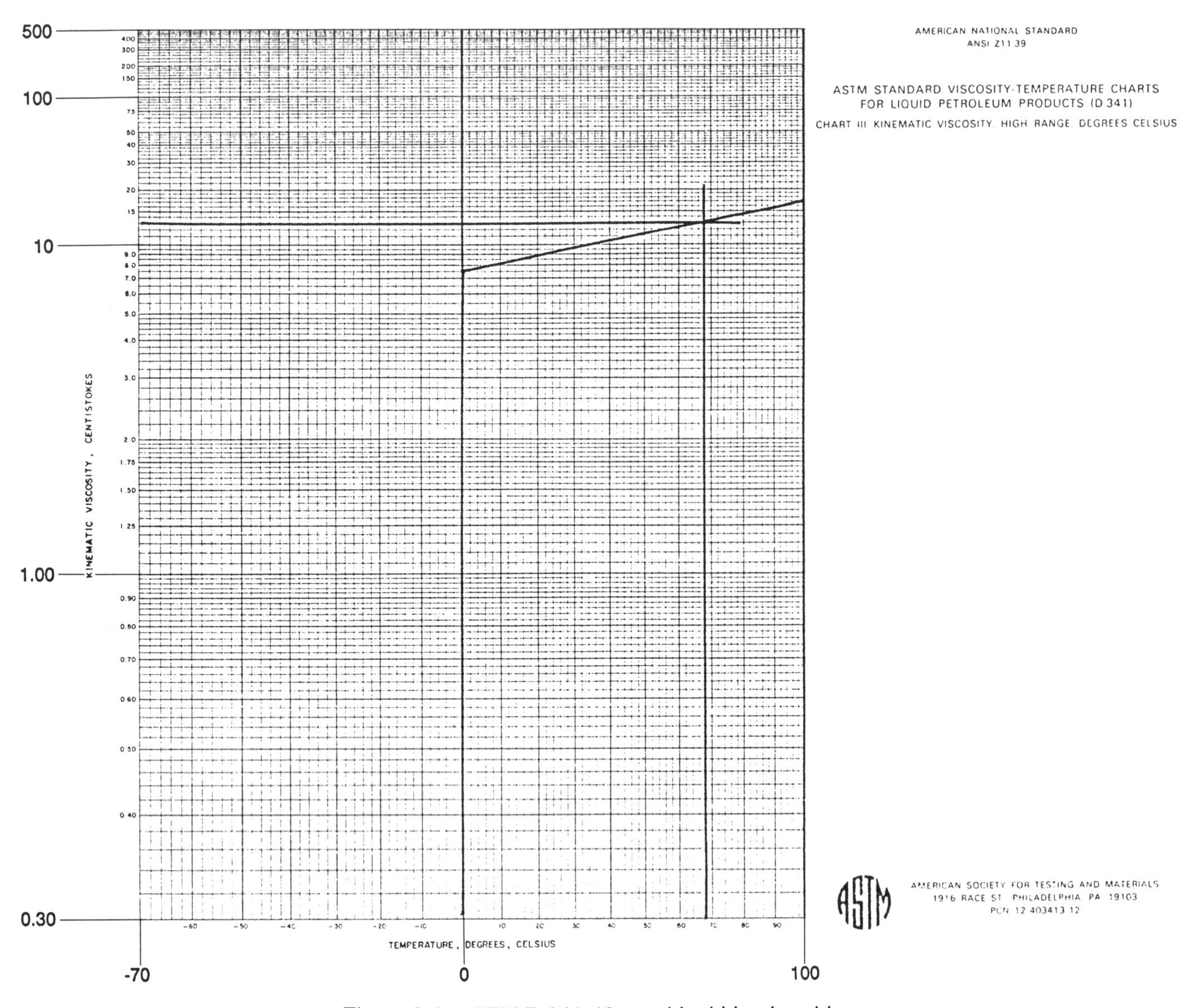

Figure 3-8. ASTM D 341-43 graphical blend problem.

We can solve analytically the problem solved graphically in the previous section as follows:

Volume fraction of high viscosity oil =

$$\left[\frac{(\log 17.70+\log\log 13.70)(\log\log 56.4-\log\log 8.20)}{(\log\log 190.7-\log\log 17.7)(\log\log 13.70-\log\log 8.20)}+1\right]^{-1}$$

$$\left[\frac{(0.09621-0.05565)(0.24336-0.03914)}{(0.35800-0.09621)(0.05565+0.03914)}+1\right]^{-1}$$

$$=0.68418$$

Note the answer is to five decimal places. If the output specification for my HP28S had been set for twelve places, the answer would have been 0.684024068011. Neither answer is particularly better than the 0.684 returned by the graphical solution and it, too, is probably too precise. A good answer would be "About 68 percent high viscosity oil will yield a blend having a viscosity of about 13 cSt at 100°F."

Inasmuch as the viscosity of the two components is not too far apart and the semi–log plot in Figure 3-9 worked pretty well, we might suspect that an analysis in terms of $Z=\nu$ instead of $Z=(\nu-0.7)$ might work out. And, we'd be right. See the following:

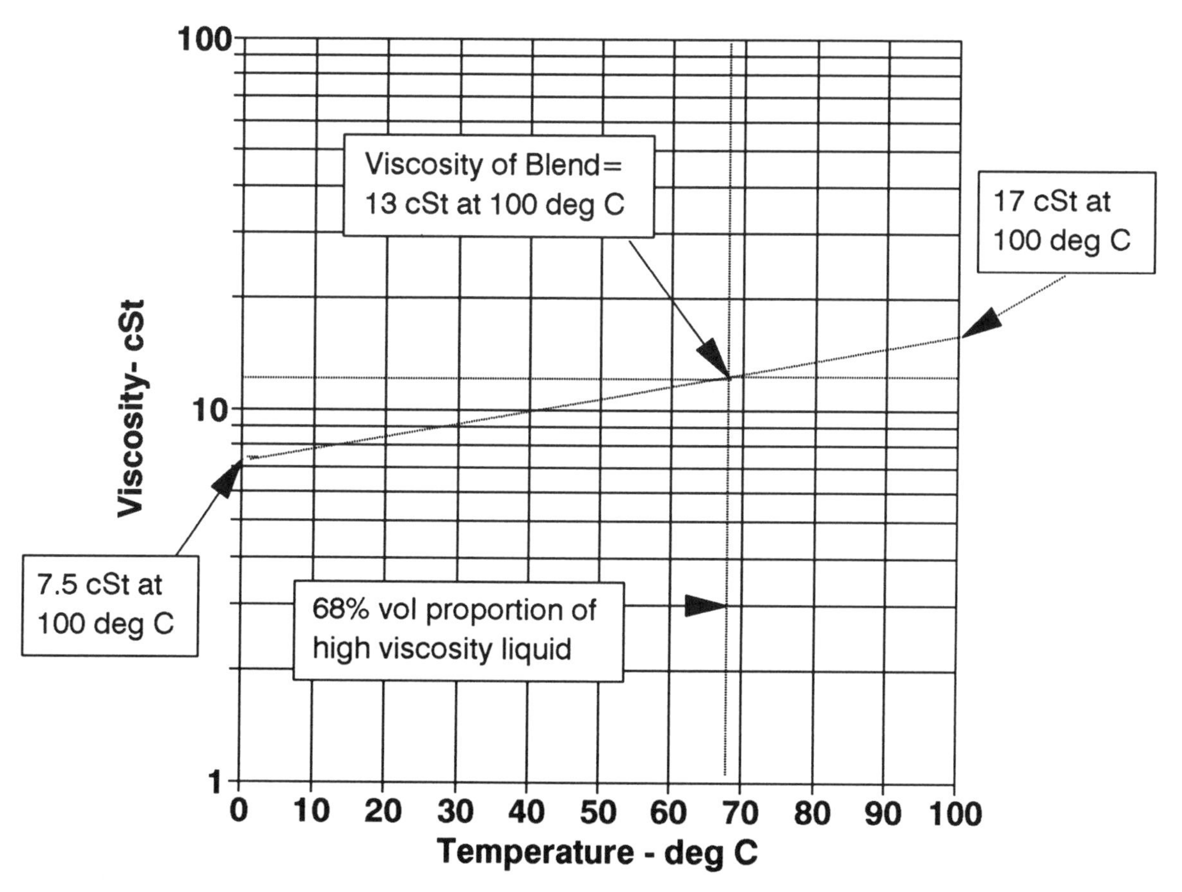

Figure 3-9. ASTM graphical blend problem on log-log coordinates.

Volume of high viscosity oil =

$$\left[\frac{(\text{loglog}\,17.00 - \text{loglog}\,13.00)(\log\log 55.7 - \log\log 7.50)}{(\log\log 190.00 - \log\log 17.00)(\log\log 13.00 - \log\log 7.50)}+1\right]^{-1}$$

$$=\left[\frac{(0.09006-0.04686)(0.24201+0.05796)}{(0.35770-0.09006)(0.04686+0.05796)}+1\right]^{-1}$$

$$=0.68403$$

Thus the $Z=\nu$ variation works about as well as the $Z=(\nu-0.7)$ solution; and while there is no real use programming in terms of $Z=\nu$ when it is so easy to program in terms of $Z=(\nu-0.7)$, it is comforting to know that the analytical solution backs up the semi–log graphical solution.

If nothing else is available, a piece of semi–log graph paper and a straight edge can solve a lot of engineering problems.

And always remember: Pipelining is a craft, not a science. Give judgment a chance.

Viscosity of Typical Hydrocarbons

There is no way to predict the viscosity of a petroleum liquid—except a pure hydrocarbon and then not entirely accurately—from other, perhaps more available or more easily measured, parameters. There are some measurable factors such as the *viscosity index*, the *viscosity–gravity constant*, and the *characterization factor* from which one can gain a feeling of the viscosity of a component, but the simple fact that such factors are available means that samples are available for measurement and the viscosity can, therefore, be measured.

Table 3-5 is a collection of viscosity–density data for some typical crude oils, and Table 3-6 is the same for some typical products. These have been gathered from many sources—all believed reliable—and can be used for conceptual designs and initial estimates of pipeline performance.

Table 3-5

Viscosities and Densities of Miscellaneous Crude Oils

		Kinematic Viscosity			
Origin	ρ 15°C	cSt	°C	cSt	°c
Hassi Messaoud (Algeria)	0.800	3.4	5	2.6	15
Qatar Export Blend (Qatar)	0.823	7.3	5	5.2	15
Edjeleh (Algeria)	0.845	8.0	5	5.5	15
Kirkuk (Iraq)	0.844	11.4	5	7.9	15
Agha Jari (Iran)	0.852	20.1	5	11.2	15
Zelten (Libya)	0.830	19.0	5	11.0	15
Gach Saran (Iran)	0.872	44.0	5	20.5	15
Forties (North Sea)	0.840	7.7	21	3.4	38
Mubarek (Sharjah)	0.840	3.2	38	2.8	50
Ninian (North Sea)	0.849	6.9	38	4.7	54
North Rumaila (Iraq)	0.853	10.6	21	6.4	49
North Slope (Alaska)	0.894	51.1	16	28.3	38
Berri (Saudi Arabia)	0.831	5.7	21	3.8	38
Arabian Heavy Blend (Saudi Arabia)	0.886	35.8	21	18.9	38
Arabian Light Blend (Saudi Arabia)	0.858	10.4	21	6.1	38
Iranian Heavy Blend (Iran)	0.872	9.8	38	7.5	54
Iranian Light Blend (Iran)	0.858	6.4	38	4.8	54
Taching (P. R. China)	0.860	45.3	25	29.4	38
Arjuna (Indonesia)	0.836	4.0	32	3.2	38
Attaka (Indonesia)	0.810	1.8	21	1.4	38
Basra (Iraq)	0.856	31.0	9	16.0	27
Boscan (Venezuela)	0.998	19 800	38	6 200	60
Brass River (Nigeria)	0.811	2.1	21	1.5	38
Brega (Libya)	0.823	5.6	21	3.5	38
Gulf of Suez Blend (Egypt)	0.868	46.6	20	7.8	54
Kuwait Export Blend (Kuwait)	0.868	34.0	5	18.7	15
Cactus Reforma (Mexico)	0.860	22.9	15	6.8	38
Argyle (North Sea)	0.848	21.2	20	4.7	50
Leduc (Canada)	0.829	10.0	10	2.8	38
Fosterton (Canada)	0.909	139.0	10	30.0	38
Tandjung (Indonesia)	0.825	130.0	40	4.5	60
Average Light Crude (USA)	0.788	3.7	10	2.0	38
Average Medium Crude No. 1 (USA)	0.825	8.6	10	4.2	38
Average Medium Crude No. 2 (USA)	0.847	17.7	10	7.4	38
Average Heavy Crude (USA))	0.862	29.0	10	10.3	38

Table 3-6

Viscosities and Densities of Miscellaneous Products

		Kinematic Viscosity			
Product	ρ 15°C	cSt	°C	cSt	°c
High viscosity fuel oil	1.044	460	70	88	98
Standard bunker fuel oil	1.014	320	50	39	70
Light fuel oil	0.966	155	40	45	60
Heavy diesel oil	0.893	8.5	40	4.9	60
Gas oil	0.887	7.9	40	4.5	70
Kerosine	0.824	3.3	10	1.8	40
Gasoline No. 1	0.751	0.96	10	0.65	40
Gasoline No. 2	0.702	0.66	10	0.47	40
Gasoline No. 3	0.682	0.48	10	0.36	40
Natural Gasoline No. 1	0.633	0.41	10	0.30	40
Natural Gasoline No. 2	0.611	0.38	10	0.27	40
n-Butane	0.584	0.29	10	0.22	40
Propane	0.508	0.22	10	0.18	40

For design purposes, or for development of operating programs for existing pipelines, there is no substitute for good measurements.

Working Equations

Calculators

Most of the working equations of Chapter 3 are summarized in Table 3-7, *Reference and Working Equations for Newtonian Viscosity*. To keep the table on one page, units of measurement have been eliminated for some equations, but these may be found in the text. In addition, some of the very complex equations, such as the complete version of the ASTM viscosity–temperature function, have not been included. These, also, may be looked up in the text.

Computers

There are five Computer Computation Templates included in this chapter. **CCT** No. 3-1 covers the ASTM D 341 method for finding the viscosity–temperature curve for a hydrocarbon, and **CCT** 3-2 and **CCT** 3-3 cover the short form and long form methods of finding the viscosity–pressure curve. **CCT** No. 3-4 covers Saybolt to cSt conversions and vice versa. **CCT** No. 3-5 gives the ASTM procedure for calculating the viscosity of a blend of hydrocarbons.

Table 3-7
Reference and Working Equations for Newtonian Viscosity

Constitutive Equation for Absolute Viscosity

$$\mu = \left(\frac{\tau}{\frac{dv}{dx}}\right) = \left(\frac{\tau}{\lambda}\right)$$

Defining Equation for Kinematic Viscosity

$$\nu = \frac{\mu}{\rho}$$

Viscosity–Temperature Relations — General

liquids $\log_e \mu = A + BT$

gases $\mu = CT^{1/2}$

Andrade's equation $\mu = A \exp B / T$

Viscosity–Temperature Relations — ASTM

general form: $\log\log Z = A + B\log T$

crude oil form: $\log\log(\nu + 0.7) = A - B\log T$

explicit ν $\nu = 10\uparrow(10\uparrow(A - B\log T)) - 0.7$

$$A = \log(\log(\nu_1 + 0.7)) + B\log T_1$$

$$B = \left(\log(\log(\nu_1 + 0.7)/(\log(\nu_2 + 0.7)))\right)/\log(T_2/T_1)$$

Viscosity–Pressure Relations

definition $\ln \mu = A + BP$

explicit definition $\mu = \exp(A + BP)$

Kouzel's Data

$$\log\left(\frac{\mu_p}{\mu_a}\right) = \frac{p}{1000}\left(0.0239 + 0.01638\mu_a^{0.278}\right)$$

explicit form $\mu_p = \mu_a \times 10\uparrow\left(\frac{p}{1000}\right) \times \left(0.0239 + 0.01638\mu_a^{0.278}\right)$

Measurement of Viscosity

Poiseuille's Law $V = \left[(\pi p r^4)/(8\ell\mu)\right]t$

NBS Formula $\mu = (\pi d g r^4 t)(h - mv^2/g)/(8Q(\ell + \lambda))$

Viscosity Conversion–Saybolt Viscometers

$$SUS_{100F} = 4.6324cSt + \frac{1.0 + 0.03264cSt}{\begin{pmatrix}3930.2 + 262.7cSt \\ +23.97cSt^2 + 1.646cSt^3\end{pmatrix}\cdot 10^{-5}}$$

$$SUS_t = (1.0 + 0.000061(t - 100)) \times SUS_{100^\circ F}$$

$$SFS_{122F} = 0.4717cSt + \frac{13924}{cSt^2 - 72.59cSt + 6816}$$

$$SFS_{210F} = 0.4792cSt + \frac{5610}{cSt^2 + 2130}$$

Viscosity Conversion–Redwood Viscometers

Redwood No. 1 seconds$_t = A \times cSt_t$

t=70°F A=4.06; t=100°F A=4.10; t=140°F A=4.18

Viscosity of Blends

General Function $\mu_m = (\mu_1)^{x_1} \times (\mu_2)^{x_2} \times (\mu_3)^{x_3} \ldots \times (\mu_n)^{x_n}$

μ_m = absolute viscosity of the mixture
μ_n = absolute viscosity of component n
x_n = mol fraction of component n

ASTM Procedure

Vol fraction high vis oil $= \left[\frac{(E-A)(C-D)}{(E-F)(A-C)} + 1\right]^{-1}$ at 40°C

Vol fraction low vis oil $= \left[\frac{(F-B)(C-D)}{(E-F)(B-D)} + 1\right]^{-1}$ at 100°C

$A = \log\log Z_{B(40)}$ $D = \log\log Z_{L(100)}$
$B = \log\log Z_{B(100)}$ $E = \log\log Z_{H(40)}$ $Z = (cSt + 0.7)$
$C = \log\log Z_{L(40)}$ $F = \log\log Z_{H(100)}$

References

1 Andrade, N. da C. *Nature*, Vol. 125, page 309, 1930.

2 Wright, W. A. "An Improved Viscosity–Temperature Chart for Hydrocarbons." *Journal of Materials*, Vol. 4, No. 1, 1969, pp. 19–27.

3 Manning, R. E. "Computational Aids for Kinematic Viscosity Conversions From 100 and 210°F to 40 and 100°C." *Journal of Testing and Evaluations*, Vol. 2, No. 6, November, 1974.

4 Horner, D. R. and Fried, P. "Computer Program for a Heated Crude Line." *Pipe Line Industry*, Nov. 1960.

5 Zama, W. A. "This Pipeline Simulation Program Recognizes Batch, Terrain, and Temperature." *Oil and Gas Journal*, Oct. 23, 1969.

6 Watson. *Industrial and Engineering Chemistry*, Vol. 28, p. 605, 1936.

7 Staender, W. "Method for the Computation of Non–Isothermal Head Loss and Optimum Dimensioning of Heated Pipelines Transporting Heavy Oil and an Exact Mathematical Treatment of the Attendant Heat Loss Problems."(Author's Note: Translated from the German–language title). *Rohre, Rohrleitungsbau, Rorhleitungstsransport*, February 1978.

8 Hyman, M. H. *Fundamentals of Engineering HPI Plants With the Digital Computer*, chapter 4, part 2. Petro/Chem Engineer. April 1969.

9 Huang and Johnson. *Petroleum Refiner*, Vol. 38, No. 5, 1959.

10 Singh, B., Mutyala, S., and Puttagunta, V. R. "Viscosity Range From One Test." *Hydrocarbon Processing*, pp. 39-41, September 1990.

11 Amin and Maddox. *Hydrocarbon Processing*, pp. 131-135, December 1980.

12 Puttagunta, V. R., Miadonye, A., and Singh, B. "Simple Concept Predicts Viscosity of Heavy Oil and Bitumen," *Oil and Gas Journal*, March 1, 1993.

13 Kouzel, B. "How Pressure Affects Liquid Viscosity." *Hydrocarbon Processing and Petroleum Refining*, Vol. 44, No. 3, March 1965.

14 Hershey and Hopkins. "Viscosity of Lubricants Under Pressure." *Proceedings of the ASME*, 1954.

15 Aronson, M. H. and Nelson, R. C. *Viscosity Measurement and Control.* Pittsburgh: Instruments Publishing Company, 1964.

16 O'Donnell, R. J. "Equations for Converting Different Viscosity Units." *Materials Research and Standards*, May 1969.

17 Pipe Friction Manual. Cleveland: *Hydraulic Institute*, 1961 (Fourth Printing 1975).

18 Hodgman, C. D., Weast, R. C., and Selby, S. M. (Eds.). *Handbook of Chemistry and Physics,* 41st Edition. Cleveland: C. R. Publishing Company, 1960.

Computer Computation Template

CCT No. 3-1: Viscosity–temperature relations of hydrocarbons. The method used is that of ASTM D 341, and is valid for the viscosity range 2.0–20 000 000 cSt. This will cover most crude oils. Given two vis–temp pairs, the method will find ν_t for any temperature for which the oil is Newtonian.

References: Chapter 3. Equations 3-7, 3-14, 3-15, 3-16, and 3-17.

Input

Factors $i := 0, 2 .. 20$

Input Data

$\nu_1 := 29.0$ cSt $\quad t_1 := 10$ deg C

$\nu_2 := 10.3$ cSt $\quad t_2 := 38$ deg C

Note: ν-t pairs are for average U.S. heavy crude. See Table 3-5.

Variables $T_1 := t_1 + 273$ K $\quad T_2 := t_2 + 273$ K $\quad t_i := 2 + 2 \cdot i$ deg C $\quad T_i := t_i + 273$K

Equations

$$B := \frac{\log\left(\log\left(\frac{\log(\nu_1 + 0.7)}{\log(\nu_2 + 0.7)}\right)\right)}{\log\left(\frac{T_2}{T_1}\right)}$$

$$A := \log(\log(\nu_1 + 0.7)) + B \cdot \log(T_1)$$

$$\nu_i := 10^{\left[10^{(A - B \cdot \log(T_i))}\right]} - 0.7$$

Results

$A = 9.175$

$B = 3.673$

i	t_i	T_i	ν_i
0	2	275	42.6
2	6	279	34.9
4	10	283	29.0
6	14	287	24.3
8	18	291	20.7
10	22	295	17.7
12	26	299	15.3
14	30	303	13.3
16	34	307	11.7
18	38	311	10.3
20	42	315	9.2

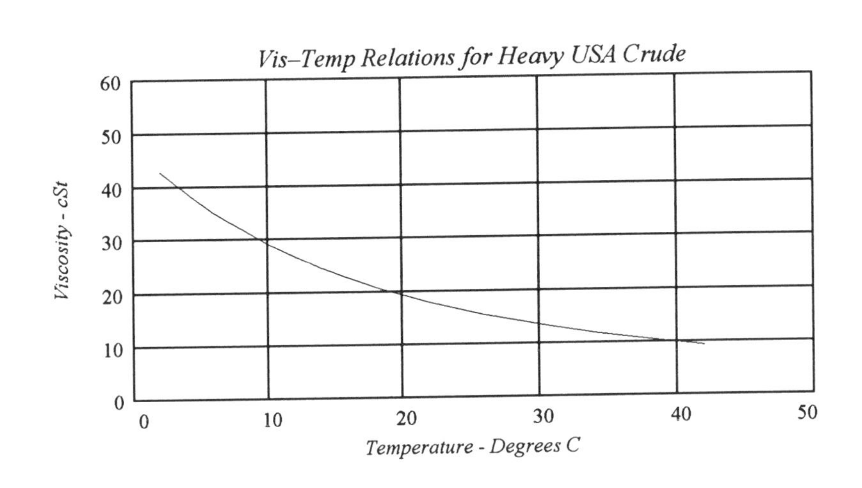

Computer Computation Template

CCT No. 3-2: Viscosity–pressure relations for hydrocarbons using the $d(\log\mu)/dT = 0.0039\mu_a^{0.278}$ assumption recommended by Kouzel. This is the short–form solution to the problem of finding the effect of high pressure on the absolute viscosity of a hydrocarbon. The long form solution, requiring finding the constants A and B (see **CCT** No. 3-01) is in **CCT** No. 3-03.

References: Chapter 3. Equations 3-43 and 3-44.

Input

Factors $i := 0, 1 .. 8$ $\quad p_a := 1000$ psig seed

Input Data $\mu_a := 2.32$ absolute viscosity, cP, at atmospheric pressure and 100 °F

Variables $p_i := i \cdot p_a$ psig

Equations

$$\mu_{p_i} := \mu_a \cdot 10^{\left(\frac{p_i}{1000}\right)\cdot\left(0.0239 + 0.01638 \cdot \mu_a^{0.278}\right)}$$

Results

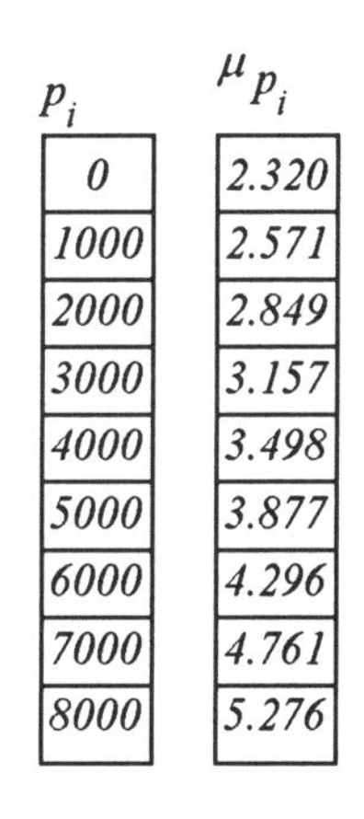

p_i	μ_{p_i}
0	2.320
1000	2.571
2000	2.849
3000	3.157
4000	3.498
5000	3.877
6000	4.296
7000	4.761
8000	5.276

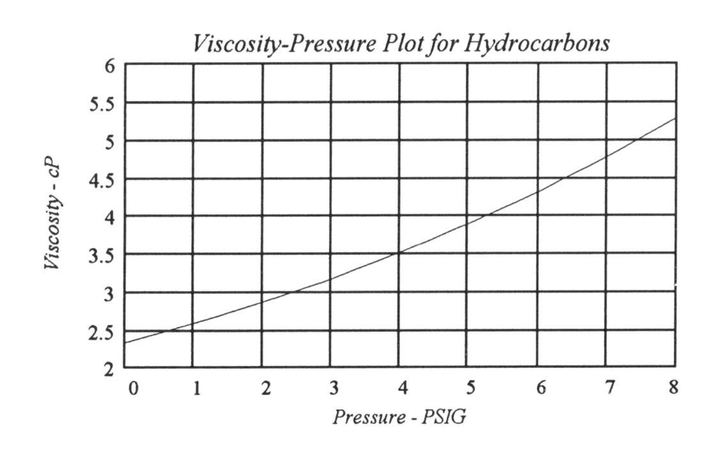

Computer Computation Template

CCT No. 3-3: Viscosity–pressure relations for hydrocarbons. This method is more reliable than the short–form method, which is usually acceptable for crude oils, of **CCT** No. 3-2. The *A* and *B* factors have to be determined from the procedures of **CCT** No. 3-1.

References: Chapter 3. Equations 3-37, 3-38, 3-39,3-40, 3-41 and 3-42. **CCT** No. 3-1.

Input

Factors $i := 0, 2 .. 20$ $\quad p_i := 400 \cdot i$

Input Data — Vis-temp pairs for selected oil used to establish *A* and *B* for Andrade equation.

$\nu_1 := 10.0$ cSt $\quad t_1 := 50$ deg F

$\nu_2 := 2.8$ cSt $\quad t_2 := 100.4$ deg F

Viscosity of selected oil at 100 °F $\quad \mu_a := 2.32$ *cP* $\quad t_a := 100$ deg F

Variables

$t_i := 2 + 2 \cdot i$ deg F $\quad T_i := t_i + 460$ deg R

$T_1 := t_1 + 460$ deg R $\quad T_2 := t_2 + 460$ deg R

Equations

(1) Calculate *A* and *B* using the method of **CCT** No. 3-1

$$B := \frac{\log\left(\dfrac{\log(\nu_1 + 0.7)}{\log(\nu_2 + 0.7)}\right)}{\log\left(\dfrac{T_2}{T_1}\right)} \qquad A := \log(\log(\nu_1 + 0.7)) + B \cdot \log(T_1) \qquad B = 6.766 \qquad A = 18.332$$

(2) Calculate μ as a function of *p*. μ=cP, *p*=psig, μ_α=cP at 100°F.

$$\mu := A \cdot \exp\left(\frac{B}{t_a}\right) \log\left[A \cdot \left(\frac{B}{t_a}\right)\right] \text{ by differentiation, yields } \frac{-1}{(t_a \cdot \ln(10))} \qquad N := \frac{-1}{(t_a \cdot \ln(10))} \qquad \mu_{p_i} := \mu_a \cdot 10^{\left[\left(\frac{p_i}{1000}\right) \cdot (0.0239 - 4.2 \cdot N)\right]}$$

Results

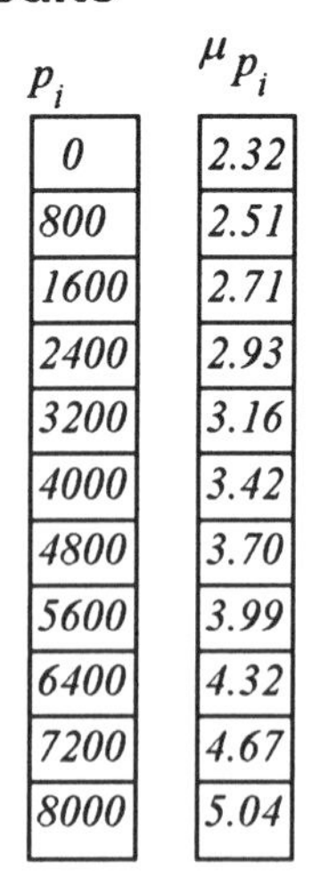

p_i	μ_{p_i}
0	2.32
800	2.51
1600	2.71
2400	2.93
3200	3.16
4000	3.42
4800	3.70
5600	3.99
6400	4.32
7200	4.67
8000	5.04

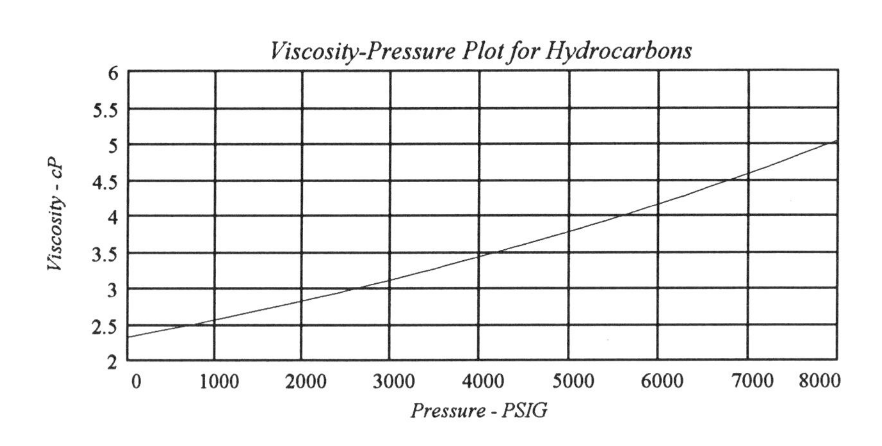

Computer Computation Template

CCT No. 3-4: Conversion of kinematic viscosities in cSt to/from Saybolt viscosisties in SUS and SFS using the methods of ASTM 2161. Part I uses the ASTM equations as written to obtain SUS and SFS viscosities given in cSt. Part II uses the MathCad *root* function to solve these equations iteratively to obtain cSt given SUS or SFS.

References: Chapter 3. Equations 3-50, 3-51, 3-52 and 3-53.

Input

$cSt := 20$ cSt $t := 60$ deg F

Part I: Equations and Results–cSt to SUS and SFS

$$SUS_{100F} := 4.6324 \cdot cSt + \frac{1.0 + 0.03264 \cdot cSt}{\left(3930.2 + 262.7 \cdot cSt + 23.97 \cdot cSt^2 + 1.646 \cdot cSt^3\right) \cdot 10^{-5}} \qquad SUS_{100F} = 97.8$$

$$SUS_t := (1.0 + 0.000061 \cdot (t - 100)) \cdot SUS_{100F} \qquad SUS_t = 97.6$$

$$SFS_{122F} := 0.4717 \cdot cSt + \frac{13294}{\left(cSt^2 - 72.59 \cdot cSt + 6816\right)} \qquad SFS_{122F} = 11.7$$

$$SFS_{210F} := 0.4792 \cdot cSt + \frac{5610}{cSt^2 + 2130} \qquad SFS_{210F} = 11.8$$

Part II: Equations and Results–SUS and SFS to cSt

Given SUS_{100F}, find cSt

$$root\left[\left[4.6324 \cdot cSt + \frac{1.0 + 0.03264 \cdot cSt}{\left(3930.2 + 262.7 \cdot cSt + 23.97 \cdot cSt^2 + 1.646 \cdot cSt^3\right) \cdot 10^{-5}}\right] - SUS_{100F}, cSt\right] = 20.0$$

Given SUS_t, find SUS_{100F}

$$root\left[(1.0 + 0.000061 \cdot (t - 100)) \cdot SUS_{100F} - SUS_t, SUS_{100F}\right] = 97.8$$

Given SFS_{122F}, find cSt

$$root\left[\left[0.4717 \cdot cSt + \frac{13294}{\left(cSt^2 - 72.59 \cdot cSt + 6816\right)}\right] - SFS_{122F}, cSt\right] = 20.0$$

Given SFS_{210F}, find cSt

$$root\left[\left(0.4792 \cdot cSt + \frac{5610}{cSt^2 + 2130}\right) - SFS_{210F}, cSt\right] = 20.0$$

Computer Computation Template

CCT No. 3-5: Viscosity of blends of hydrocarbons, using ASTM D 341. All viscosities are in cSt, and all temperatures are in °F.

References: Chapter 3. Equations 3-57 and 3-58.

Input

Conventions

Subscript **B** pertains to the *blend*, **L** to the *low viscosity* oil, **H** to the *high viscosity* oil, and **T** to °F. Thus cSt_L is the viscosity of the low viscosity oil, and cSt_{L40} is the viscosity at 40 °F. Input items required are vis–temp pairs for the components, and the desired blend viscosity at one of the pair temperatures.

Input Data

$cSt_{L40} := 55.7$ $\quad cSt_{L100} := 7.50$ $\quad cSt_{H40} := 190$ $\quad cSt_{H100} := 17.0$

Variables

$Z_{L40} := cSt_{L40} + 0.7$ $\quad Z_{L100} := cSt_{L100} + 0.7$ $\quad C := \log(\log(Z_{L40}))$ $\quad D := \log(\log(Z_{L100}))$

$Z_{H40} := cSt_{H40} + 0.7$ $\quad Z_{H100} := cSt_{H100} + 0.7$ $\quad E := \log(\log(Z_{H40}))$ $\quad F := \log(\log(Z_{H100}))$

Part I: Equations and Results–Percent of high viscosity oil for given blend viscosity at T_L

Input desired viscosity of blend at T_L $\quad cSt_{B40} := 70$ $\quad Z_{B40} := cSt_{B40} + 0.7$ $\quad A := \log(\log(Z_{B40}))$

At T_L, volume of high viscosity oil equals:

$$VolHiVisPct := 100 \cdot \left[\frac{(E - A)\cdot(C - D)}{(E - F)\cdot(A - C)} \right]^{-1}$$

$VolHiVisPct = 24.1$ percent

Part II: Equations and Results–Percent of high viscosity oil for given blend viscosity at T_H

Input desired viscosity of blend at T_H $\quad cSt_{B100} := 13.0$ $\quad Z_{B100} := cSt_{B100} + 0.7$ $\quad B := \log(\log(Z_{B100}))$

At T_H, volume of high viscosity oil equals:

$$VolHiVisPct := 100 \cdot \left[\frac{(F - B)\cdot(C - D)}{(E - F)\cdot(B - D)} + 1 \right]^{-1}$$

$VolHiVisPct = 68.4$ percent

4

Non-Newtonian Viscosity

The Other Kind of Viscosity

There is a relationship between shear stress and shear rate for any fluid. The general equation is

$$\gamma = f(\tau) \tag{4-1}$$

where $\tau =$ shearing stress
$\gamma =$ shearing rate

This is usually called the *constitutive equation* or, sometimes, the *rheological equation*.

For a large class of fluids, called *Newtonian* fluids, the ratio of shear stress to shear rate—the slope of the τ–γ curve—is a constant called the *absolute viscosity, dynamic viscosity,* or just *viscosity*. If this value is divided by density the result is *kinematic viscosity*. Chapter 3 is about these kinds of fluids.

The viscosity of Newtonian fluids is not dependent on the *thermal history* of the fluid, or on the *time duration* or the *time rate* of application of the shearing effort, but only on *temperature* and *pressure*. Water is the most Newtonian of all Newtonian fluids: Water at 60°F and 760 mm Hg will have the same viscosity after being boiled as before; the same after being frozen as before it froze; the same while drifting slowly in a forest stream as when flowing tens of feet per second in a fire hose; and it had the same viscosity yesterday as it has today, and will have the same viscosity tomorrow and all the days thereafter.

There are many other fluids that do not have these characteristics. They have no absolute viscosity, as such, but instead an *apparent viscosity* that may vary with thermal history, time duration of shear, time rate of shear; whether the fluid was previously sheared and if so when, how rapidly, and for how long, or all of these things. These are called *non–Newtonian* fluids.

Most pipeline problems deal with Newtonian fluids, but there are enough problems involving non–Newtonian fluids for us to take a good look at them in order to be able to handle the problems raised when they are encountered.

Classification of Matter

There are generally considered to be three *ordinary states* of matter: *gas, liquid*, and *solid*.

There are many formulas which are intended to define a certain state in terms of measurable parameters. These are called *equations of state*. The equations of state for *gases* are mostly derived from the kinetic theory of gases, are theoretically correct and quite accurate; those for *liquids* are mostly empirical, and may or may not be accurate; and there are, really, no equations of state for *solids* which would be of use in any ordinary pursuit.

Gases and *liquids* are *fluids*; and there is the technical, very theoretical, definition of *solids* as fluids with closely packed, essentially unmoving, atoms, being in effect a *super–cooled liquid*. This is not too far removed from reality, because when we look at shear stress versus shear rate diagrams for *fluids* we find the X–axis represents the *ideal liquid* whereas the Y–axis can be thought of as representing the *ideal solid*. And, some substances we think of as solids will, over long periods of time, actually flow: a lump of pitch left on the table top will flow over on the floor in a matter of days, and glass in the bottom of the windows of mediaeval cathedrals is thicker than that in the top because it has been flowing for centuries. The classic definition of the difference between liquids and solids—that liquids obey Newton's law of viscosity while solids obey Hooke's law of elasticity— fail in these cases. It also fails in the case of any viscoelastic fluid.

At one time it was popular to consider *vapor*, which is the gas–phase of a substance in contact with, and at temperature and pressure conditions at which it is in equilibrium with, the liquid–phase of the same substance, as a fourth ordinary state. However, except in the special case of water vapor (steam), vapors are more and more considered just a special case of a gas. Having said that, I note that Chapter 5 is entirely devoted to *vapor pressure*.

There has also been some consideration of dense phase pipelines[1], the term *dense phase* defining that state of a fluid which is in a single phase but exhibits properties which are between those of a liquid and a gas, as an alternative to chilled gas and LNG pipelines in arctic conditions. LNG weighs about 30 lb/cf at –250°F, chilled gas about 10 lb/cf in the range 0–15°F, and dense phase gas weighs about 20 lb/cf at –115°F. There have been no dense phase pipelines built as of this writing.

An *extra*–ordinary state of matter may be *plasma* —the state that exists in magneto–hydrodynamic and nuclear fusion installations—but that need not concern pipeliners for a long time.

We may define the salient physical characteristics of the three states, for our purposes, as follows:

- A gas is a fluid that will completely fill the container that confines it. If the gas is separated into parts, and the parts are then placed in contact in a different container, the gas will completely fill that container.
- A liquid fluid will fill that part of the container necessary to hold its volume and then, at rest, establish and maintain a plane upper surface. If the liquid is separated into parts, and the parts are then again placed in contact in a different container, it will occupy only that part of that container necessary to hold its volume and then again establish an upper surface.
- A solid has a shape that is independent of any container and it will maintain this shape until a force is applied to it. Any such force generates a stress which the solid tends to resist by deforming elastically and then, if the force is great enough, plastically. When the deforming force is removed, the solid will return to its original shape, if the deformation was entirely elastic, or to a somewhat different shape, if the deformation was partially plastic. If the force is large enough the solid will be crushed, pulled apart, or sheared, depending on the method of application of the force, and the parts, if again placed in contact, will not reform.

It is usual to say that fluids will not support shear, i.e., that a fluid has no *elasticity*; and while this is certainly true of gases, all liquids can be considered as having an *evanescent* kind of elasticity which means they will—if even for an infinitesimally short period of time—support shear. Some liquids exhibit a considerable elasticity in the ordinary sense in that they will attempt to revert to their original configuration when shearing stress is removed. *Viscoelastic* fluids are *elastic*; hence the name.

Solids, of course, support shear; and a solid, if stressed within its elastic limit, will quickly return to its original configuration when the stressing force is removed.

Classification of Fluids

Fluids may be classed in many ways. It has not been so many years when fluids were classed as *Newtonian*, being the class of fluids that exhibits a constant viscosity at constant temperature and pressure, or *non–Newtonian*, being the class of all other fluids.

This method is no longer sufficient to describe the different kinds of fluids which are encountered in practice and routinely handled with satisfactory results, and new and better methods are required. The classification of Govier and Aziz[2] is as good as any. Table 4-1 is based on their classification.

All fluids can be first classed as either *single–phase* or *multi–phase*.

Multi–phase fluids may be classed as *fine dispersions, coarse dispersions, macro–mixes,* or *stratified mixtures*. These names provide their own qualitative definitions of the fluid; quantitative definitions are not important for our purposes.

Fluids are then classed as being *true homogeneous*, *pseudo–homogeneous*, or *heterogeneous*.

A gas or a pure liquid is *true* homogeneous; a sample, however small, cannot be taken that will not be identical with the composition of any other sample of whatever size. A true homogeneous fluid is always in single–phase.

A gas–liquid or a liquid–solid multi–phase fluid may

Table 4-1
General Classification of Fluids

Single Phase	Multi–Phase				
	Fine Dispersion		Coarse Dispersion	Macro Mixed	Stratified
True Homogeneous	Pseudo–Homogeneous		Heterogeneous		
	Laminar Turbulent	Turbulent Only			
Flow Behavior as Single Phase					
			Flow Behavior as Multi–Phase		

have the heavier component so finely divided, so finely dispersed, and so thoroughly mixed in the carrier component that the mixture is *pseudo*–homogeneous: it is not *true* homogeneous, but for macro–effects it has the characteristics of a pure homogeneous fluid. Some pseudo–homogeneous mixtures exhibit these kind of characteristics only at high shear rates, such as those corresponding to turbulent flow in a pipeline, while others are of such dilution and dispersion that the characteristics of homogeneity are carried down to laminar flow at very low rates of shear and, in the limit, at rest.

A *heterogeneous* fluid is just what the name describes: a mixture that is not uniform.

A gas or a liquid flows as a single phase of matter; a water–coal slurry will flow as a multi–phase liquid–solid mixture. The slurry may be a fine dispersion, a coarse dispersion, or a combination of the two called a *two–phase vehicle* in which the water and the finer particles form a pseudo–homogeneous fine dispersion mixture that carries the larger particles of coal as a heterogeneous coarse dispersion. Such a slurry will flow as a pseudo–homogeneous fluid in turbulent flow and thus will behave, *while flowing*, as a single–phase fluid. See Wasp, et al.[3]

Single–phase fluids may be classified rheologically as in Table 4-2.

Of the *time–independent* fluids, *Newtonian* fluids are of greatest interest to pipeliners, and these have been covered in Chapter 3.

Pseudoplastic and *Bingham* plastic fluids are also of interest because both kinds of fluids are commonly seen in non–Newtonian pipelining, but *dilitant* fluids, which tend to set up as a solid under rapid rates of shear, are not encountered except in short lines in industrial facilities.

Table 4-2
Rheological Classification of Single–Phase Fluids

Purely Viscous Fluids		
	Time Independent	
		Newtonian
		Pseuodplastic
		Dilitant
		Bingham
	Time Dependent	
		Thixotropic
		Rheopectic
Viscoelastic Fluids		
	Many Forms	

Both of the *time–dependent* fluids are of interest, but in petroleum pipelines only *thixotropic* fluids, which tend to exhibit a lowered viscosity with increased agitation, are regularly encountered; many of the waxy crudes exhibit degrees of thixotropy. *Rheopectic* fluids, which tend toward a higher viscosity with increased agitation, are encountered only in industrial pipelines.

Until 1979, pipeliners had no interest in *viscoelastic* fluids, but with the advent that year of viscoelastic *drag reducing additives*—spectacularly introduced on the largest oil line in the United States: the Alyeska 48–inch crude line—pipeliners should at least understand the power–law representation of these kinds of fluids if only because the pumps and lines to move the additive from its storage to the point of injection on the mainline must be designed. The study of the use of drag reducing additives (usually called *DRA*) will be delayed until the proper place in this book.

Time Independent Fluids

Figure 4-1 shows the generalized viscosity situation for time–independent fluids from the idealized liquid to the idealized solid.

Values along the Y–axis typify the *ideal solid*. Shear stress in a solid is proportional to the *magnitude of the deformation*, which is usually called *strain*. For practical purposes, the certain time rate of application of the stress does not produce a difference in the strain from that produced by the same force acting over a different period of time.

Values along the X–axis typify the *ideal liquid*. This theoretical fluid has no elasticity, and *shear stress* is proportional to the *time rate of the deformation*, usually called *shear rate* or, sometimes, *shear strain rate*.

In terms of viscosity, ideal solids have a viscosity $\mu \rightarrow \infty$ while ideal liquids have a viscosity $\mu \rightarrow 0$.

Figure 4-2 is essentially the same information as that shown in Figure 4-1, except that it is in terms of *viscosity–vs–shear rate* rather than *shear stress–vs–shear rate*.

Newtonian fluids have absolute viscosity μ, which is the slope of the straight line shear stress–shear rate curve. It is a function only of temperature and pressure.

The *ideal plastic* also has a straight–line shear stress–shear rate curve, but only after a certain shear stress has been reached. This stress is called the *yield stress*, and the slope of the curve is μ_p which is called the *plastic viscosity*. The curve as shown is sometimes called that of a *Bingham plastic* or *yield–plastic*.

Drilling muds are thixotropic Bingham plastic fluids. Waxy crudes are usually Bingham plastics and may or may not be thixotropic; and there are some waxy crudes, once thoroughly stirred at a rapid rate—as in turbulent pipeline flow—seem never to return to their original non–Newtonian state. Never is not a technical word, but some crudes are known to have remained Newtonian for several months after having been thoroughly and rapidly sheared, and that is close enough to never for a crude oil which usually has an above–ground life of less than six months.

The constitutive equation for Bingham plastics is quite simple, considering that the two parameters required to

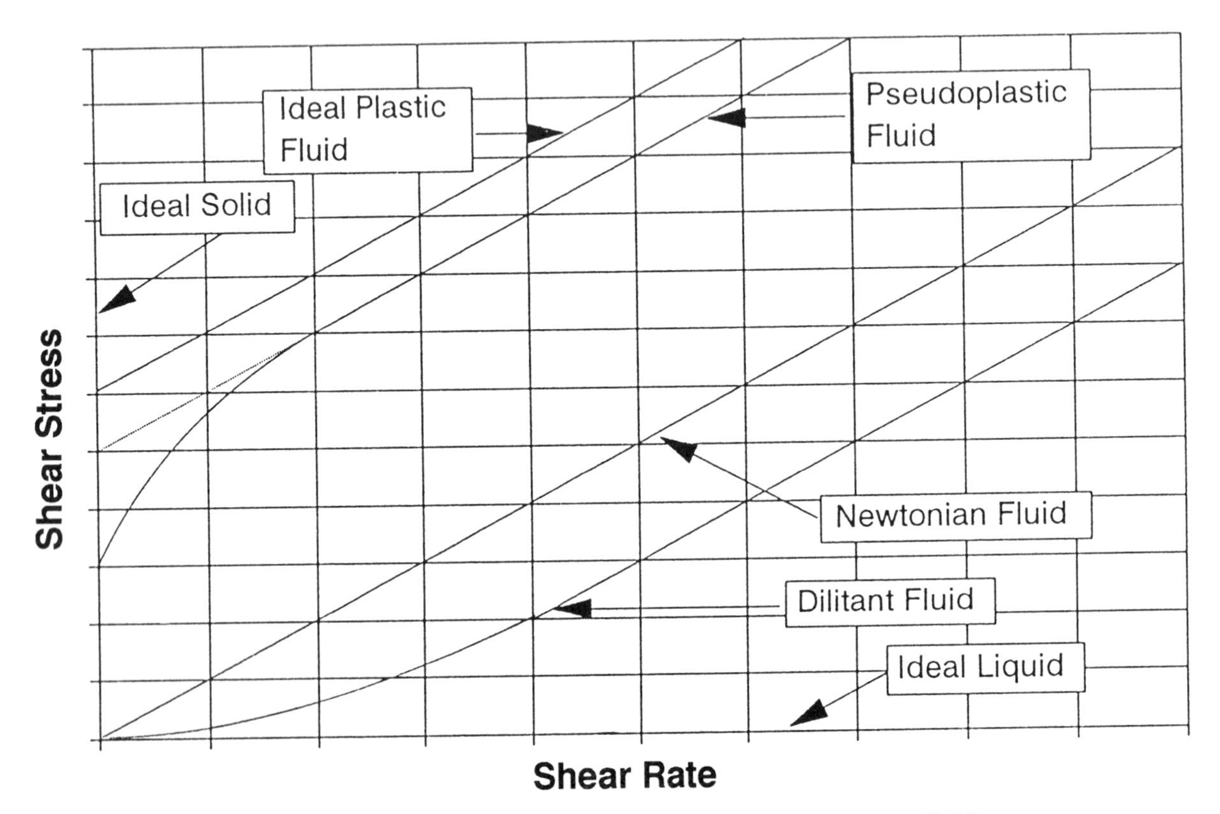

Figure 4-1. Generalized viscosity for time-independent fluids.

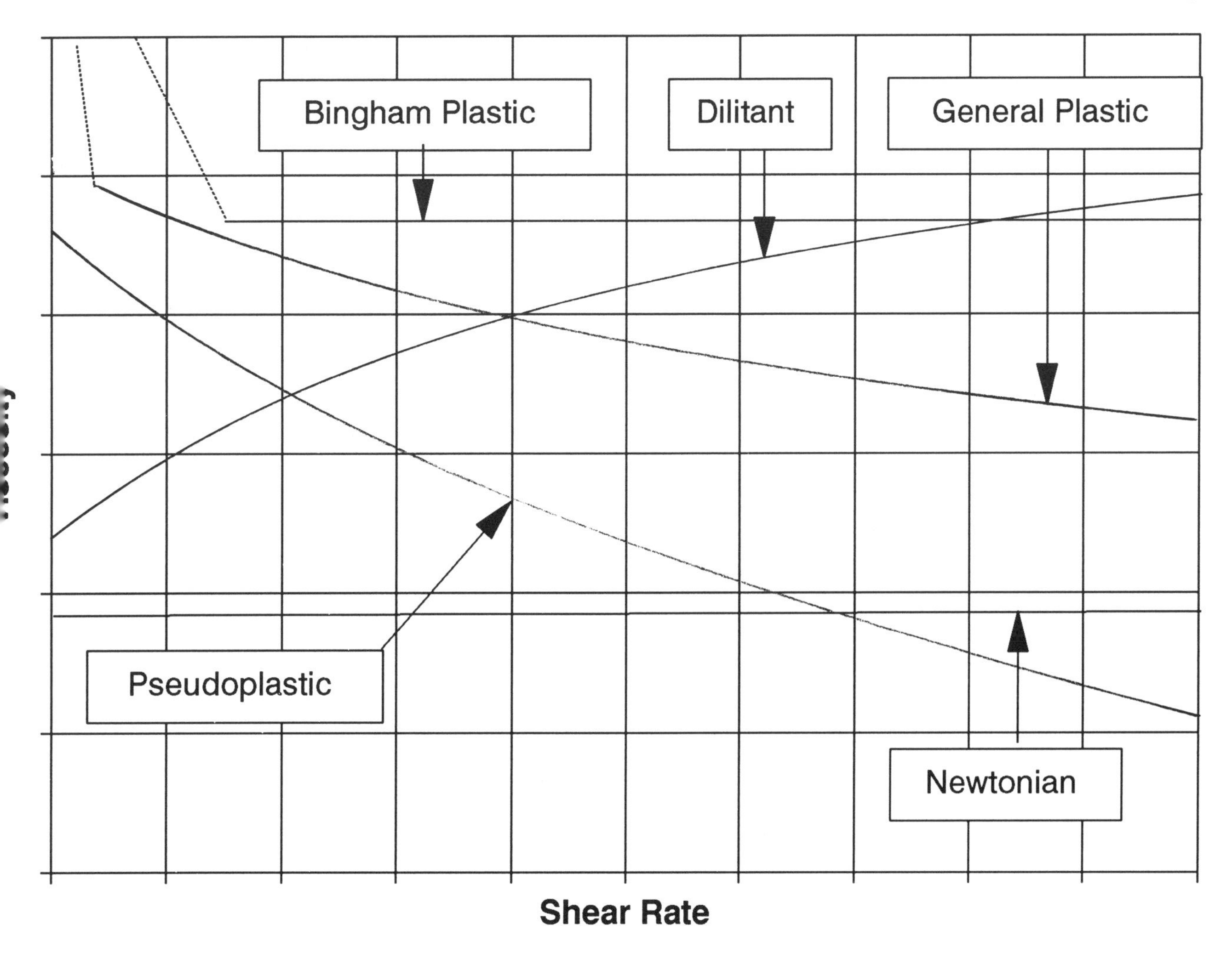

Figure 4-2. Viscosity -vs-shear rate for four non-Newtonian fluids.

yield τ in terms of γ are the *yield stress*, which is the intercept on the shear stress axis, and the *plastic viscosity*, which is the slope of the straight line portion of the relationship. Thus

$$\tau = \tau_y + \mu_p \gamma \tag{4-2}$$

where τ = shear stress
τ_y = yield shear stress
μ_p = plastic viscosity

Plastic viscosity is sometimes called the *coefficient of rigidity*.

Note that the constitutive equations for Newtonian fluids, and for Bingham plastic non–Newtonian fluids, are the only kinds of fluids which have a theoretically sound, simple constitutive equation.

All other fluids, *while under the conditions at which they are non–Newtonian*, have an apparent viscosity μ_a which is the slope of the shear stress–shear rate curve at a given time and shear rate. The reason for the qualifier is that some of these fluids commonly thought of as being non–Newtonian actually are Newtonian at very low rates of shear, very high rates of shear, or both.

The constitutive equations which describe, or attempt to describe, the viscosity of non–Newtonian fluids, require a minimum of two parameters and there are equations which use as many as four.

The *pseudoplastic* fluids have a continuously varying apparent viscosity which *decreases* with increasing shear rate toward some asymptotic value for μ_a where $\mu_a \rightarrow \mu$ where μ is, for all practical purposes, the absolute viscosity of the fluid. Pseudoplastics may or may not have a yield stress. Some crudes, paper–pulp suspensions in water, and mayonnaise are pseudoplastics.

Dilitant fluids are the opposite of pseudoplastic fluids; apparent viscosity *increases* with increasing shear rate, and some dilitant fluids will actually set up in solid form with a very rapid rate of application of shearing force. Some starch suspensions in water, and quicksand, are dilitant.

There have been many attempts to derive constitutive equations for pseudoplastic and dilitant fluids. The most commonly used for pseudoplastic fluids, and the one with the most faults, is the so-called *power law* developed by Ostwald which is

$$\tau = k\gamma^{n} \qquad (4\text{-}3)$$

where $\tau =$ shear stress
$\gamma =$ shear rate
$n < 1$
n and $k =$ constants for a particular fluid

The constant k is called the *consistency* of the fluid; the higher the value of k the more viscous the fluid. The constant n, called the *flow index*, is a measure of the degree of departure from the Newtonian fluid; the further n departs from unity the more pronounced the non–Newtonian properties.

If apparent viscosity is defined as $\mu_a = \tau / \gamma$, then substituting in Equation 4-3 yields

$$\mu_a = k\gamma^{(n-1)} \qquad (4\text{-}4)$$

Several things can be read into this equation if it is interpreted as written without qualification:

- When n is less than unity, as for pseudoplastic fluids, and thus $n{-}1$ is negative, as the shear rate γ approaches zero the value of the apparent viscosity μ_a approaches infinity. This does not happen for real fluids.
- The power law exponent n is constant; for real fluids it varies.
- The constant k is dependent on n; it is not an independent constant.

Nevertheless, the power law is a tool which makes possible relatively simple analyses of the flow of non–Newtonian fluids in pipes, and that is what is needed. For reference, following are some of the more complex equations, taken from Hughes and Brighton.[4]

$$\tau = A\sin\left(\frac{\gamma}{C}\right) \qquad \text{Prandtl Equation}$$

$$\tau = \frac{\gamma}{B} + C\sin\left(\frac{\tau}{A}\right) \qquad \text{Eyring Equation}$$

$$\tau = A\gamma + B\sinh^{-1}(C\gamma) \qquad \text{Powell–Eyring Equation}$$

$$\tau = \left(\frac{A\gamma}{B+\gamma}\right) + \mu_{\infty}\gamma \qquad \text{Williamson Equation}$$

In these equations A, B and C are constants (not the same in each equation).

Hughes et al. note that in instances where the power law does not yield reasonably correct results it may be better to resort to computer solutions, feeding into the computer the real, measured values of properties of the fluid, than to resort to the more complex approximations. This is true for more and more kinds of problems; the rigorous, closed general solutions sought in the past are being replaced by computer–aided numerical solutions to specific problems.

Figure 4-3 is a log–log plot of the power law for Newtonian, dilitant, and pseudoplastic fluids. Remember, the power law is an approximation; it does not make good *estimates*, but, given a point or two on the curve, it can make excellent *predictions*. Sometimes that is all we need.

Time Dependent Fluids

The *time–dependent* fluids of most interest to pipeliners are the *thixotropic* fluids; the *rheopectic* fluids are not of the kind usually, or even exceptionally, transported by pipeline.

The thixotropic fluids are not only sensitive to shear rate, but also exhibit a *decrease* in apparent viscosity with increasing time; many fluids ordinarily considered as *time–independent*, such as some Bingham plastic fluids, are also thixotropic.

Rheopectic fluids are also sensitive to shear rate, but exhibit an *increase* in apparent viscosity with increasing time.

Figure 4-4 illustrates the shear rate–shear stress history for a thixotropic fluid. The intercept on the shear stress axis is the yield stress; no movement occurs until this stress is reached. Thereafter, when the shear stress is applied at a constantly increasing rate the curve A–B–C will be traced out; and, if immediately on reaching C the rate of shear stress is decreased at a constantly decreasing rate the curve C–D–A will be traced back to the original

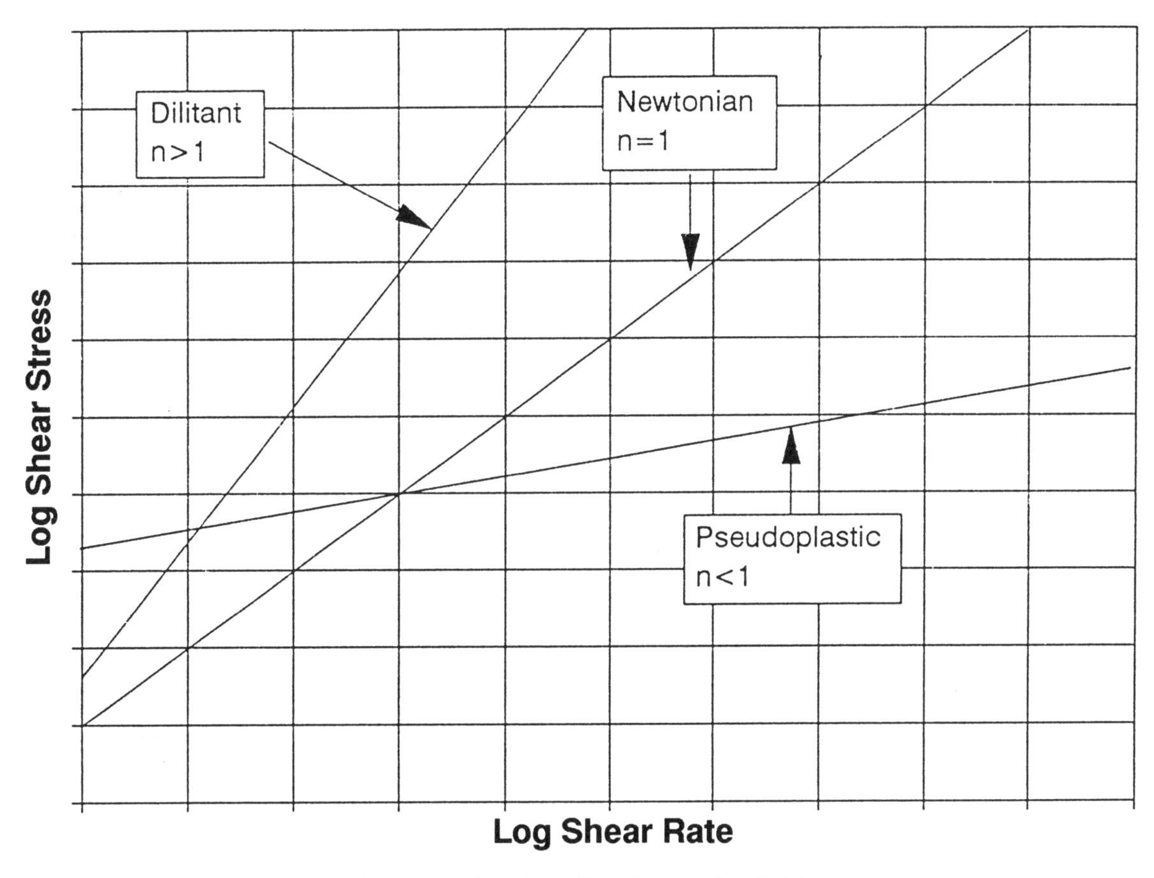

Figure 4-3. Log-log plot of power law fluids.

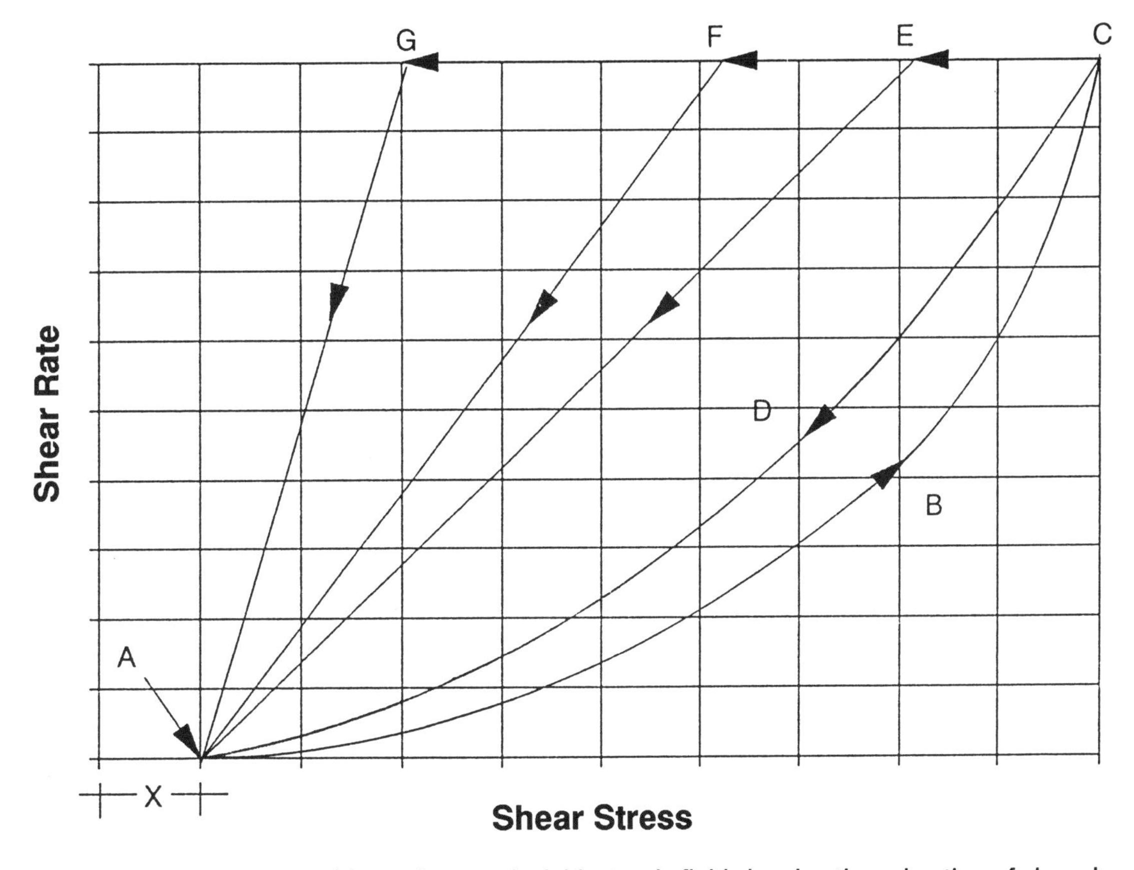

Figure 4-4. Shear stress-shear rate history for a typical thixotropic fluid showing time-duration of shear breakdown.

yield shear stress. If, however, after reaching C the rate of shearing is held constant, the shear stress will diminish with increasing time, going progressively back toward E, F, and G; and on reaching any of these points, there is the possibility, and in some cases the probability, that a straight–line shear stress–shear rate relationship will be developed as the shear rate is reduced. Thus, regression curves along C–E, E–F, and F–G develop over time at a constant shear rate; and curves along E–A, F–A and G–A are produced thereafter when the shear rate is reduced.

The figure traced out beginning at A and proceeding counter–clockwise around in a closed loop and back to A, is called the thixotropic *hysteresis loop*, comparable to the eddy–current hysteresis loop in the iron of a motor or transformer, and represents energy required to progress from A back to A. A hysteresis loop, by definition, is not a reversible process; and the energy expended in getting from A to C is not completely returned in getting back from C to A, without regard to the path traced in either direction. Hysteresis is derived from the Greek word *hysteros* meaning to lag behind, which describes the lag of the inbound track behind the outbound track in any hysteresis loop.

While Figure 4-4 describes the breakdown of a thixotropic fluid from the *time duration shear*, Figure 4-5 shows the breakdown of a fluid by the *increasing rate of shear*. The example is from experimental data developed by Green et al.[5] which illustrates the way a thixotropic fluid breaks down with increasing shear stress.

A thixotropic fluid with no yield stress component will draw a curve from A to B and then back through a *common point* if the increase–decrease in shear stress is carried out without interruption; or, if the shear stress was carried to C instead of just to B the curve A–C would be traced out with a return through the common point; and, finally, if the stress were carried from A to D the outward curve of the pseudoplastic fluid would be traced while the return trace would be that of a relaxing thixotropic fluid coming back through the common point. It is noted that the *common point* is a concept not usually shown in shear stress–shear rate diagrams for fluids, and most diagrams would show the return, relaxing curve returning to A, the starting point. The concept of viscosity breakdown by increasing shear is a common constant for all rheological studies; however, if you shear a thixotropic fluid long enough, or hard enough, it will become thinner, i.e., will have a lesser apparent viscosity.

A rheopectic fluid will also develop a hysteresis loop but, again, rheopectic fluids are not important here.

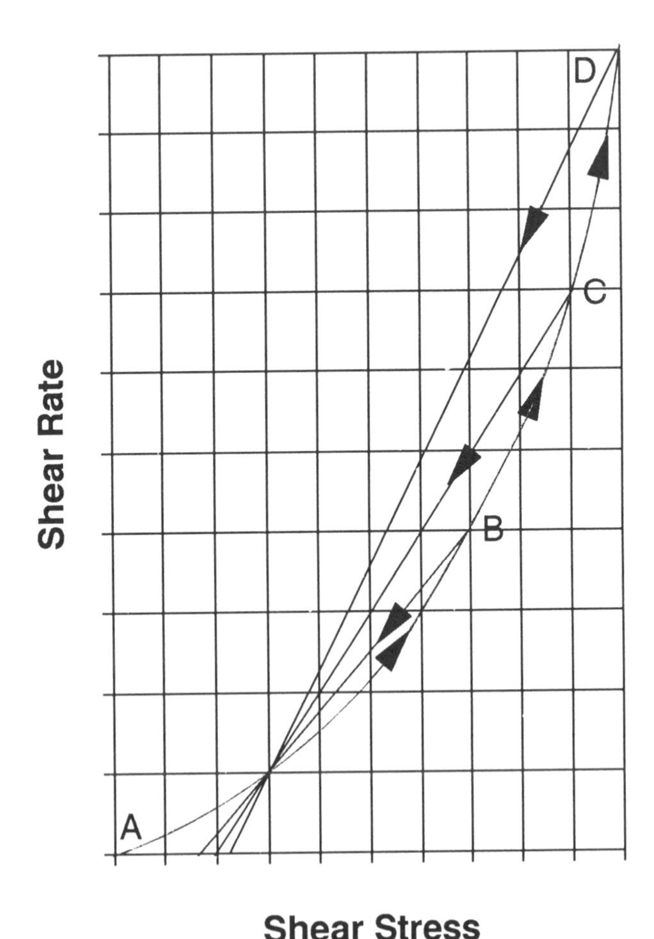

Figure 4-5. Thixotropic breakdown induced by rate of shear.

Viscoelastic Fluids

Viscoelastic fluids are the great unknowns of all fluids. They are little understood, though they have been studied for years; and only with the advent of the post World War II drive to develop polymer chemistry has there been any great amount of information available about them. But long–chain polymers—and tomato ketchup—are viscoelastic, and we had best learn what we can about them.

Certain viscoelastic fluids, mixed in tiny quantities with otherwise pretty ordinary kinds of carrier components, are *drag reducers*, and can reduce friction losses to a fraction of what they would be with the carrier alone. A fifty percent reduction in friction is possible with some pipeline fluids, seventy percent reductions are seen with down–hole oil well fracing fluids; and the city of Bristol, England, uses simple float–operated equipment to dump water soluble drag reducers into the drains of that ancient city so that the old brick–lined sewers can continue to handle the storm runoff of modern Bristol.

James Clerk Maxwell, one of the greatest of physicists, conceived of a liquid as being either viscous, elastic, or both, and arrived at the deceptively simple constitutive equation for a viscoelastic fluid that is named after him:

$$\bar{\gamma} = \frac{\tau}{\mu_0} + \frac{\bar{\tau}}{\lambda} \tag{4-5}$$

where $\bar{\gamma}$ = shear strain rate
μ_0 = viscosity at zero shear rate
τ = shear stress
$\bar{\tau}$ = shear stress rate
λ = modulus of rigidity
= modulus of elasticity in shear

The first of the two terms is Newton's law for viscous fluids; the second is Hooke's law for shearing stresses. If the second term approaches zero, the equation represents the action of a viscous fluid, viscosity becomes constant, and the material obeys Newton's law. At high viscosities, the first term approaches zero and the equation represents the action of an elastic solid in shear. After that it gets complicated.

Govier et al[2], pages 48–65, provide a fairly complete, if necessarily mathematically complex, further development of Equation 4-5, resulting in a constitutive equation for a viscoelastic fluid due to White and Metzner[6]:

$$\tau_{ij} = -2\mu_a d_{ij} + \theta_r \frac{\delta \tau_{ij}}{\delta t} \tag{4-6}$$

where d_{ij} = components of deformation rate tensor
$|d_{ij}|$ = S = shear rate, for simple shear
$\frac{\delta}{\delta t}$ = the Olroyd convected derivative
= the time rate of change of a quantity relative to a coordinate system moving with and deforming relative to the fluid
μ_a = apparent viscosity
θ_r = relaxation time of the substance

The first term can be changed by substituting a power law, or another empirical constitutive equation, for μ_a of a time–independent fluid, and if $\theta_r = 0$ or, if $\theta_r \to 0$, the equation backs off to that of a purely viscous fluid; the elastic properties of the fluid are all conveyed in the second term as would be expected from Equation 4-5.

It is in the determination of θ_r that prediction of the viscoelastic action of a fluid becomes complicated, and no further consideration will be given to it here because to do so would be to overpower the rest of the book. There is new information, both theoretical and practical, real–world data, becoming available daily. For reference, see Govier et al.,[2] Walters,[7] or papers presented at the International Conferences on Drag Reduction.[8]

And, as a practical matter, any viscoelastic fluid apt to be involved in pipelining will be purchased commercially and it is the problem of the vendor, not the pipeliner, to provide a usable constitutive equation for the fluid.

Viscometers for Non–Newtonian Fluids

Viscosity is measured by measuring the *time* required for a given volume of liquid to flow through a capillary or other restriction, or by measuring the *shear force* required to produce a given shear velocity gradient.

The *time–measuring* viscometers, described in detail in Chapter 3, measure *kinematic* viscosity, and are usually used to measure the viscosity of Newtonian fluids.

The *force–measuring* viscometers, described in the following, measure *absolute* viscosity. They can be used to measure the viscosity of any fluid and, especially, non–Newtonian fluids.

The sliding–plate analogy of Chapter 3 is theoretical in that it is impossible to realize the conditions imposed on the analysis. However, there are other methods of measuring viscosity by forcing the movement of one surface separated by a fluid from another nearby surface and measuring the force required to keep the second surface in place, that have been thoroughly analyzed and, while not theoretically correct, are sufficiently accurate for all but the most demanding applications. These are based on geometries similar to those shown in Figure 4-6.

Figure 4-6 [A] shows the concentric cylinder geometry of the Couette–type viscometer, named for Maurice Frédéric Couette. The outer cylinder is driven from the bottom by a shaft which can be rotated at whatever rate, the inner cylinder is suspended by a torsion wire, and the annulus is filled with the test fluid. When the outer cylinder is rotated the shear developed in the fluid in the annulus produces a torque on the inner cylinder which is measured by the deflection of the torsion wire. If dR is kept very small, and the rotation speed is such that only

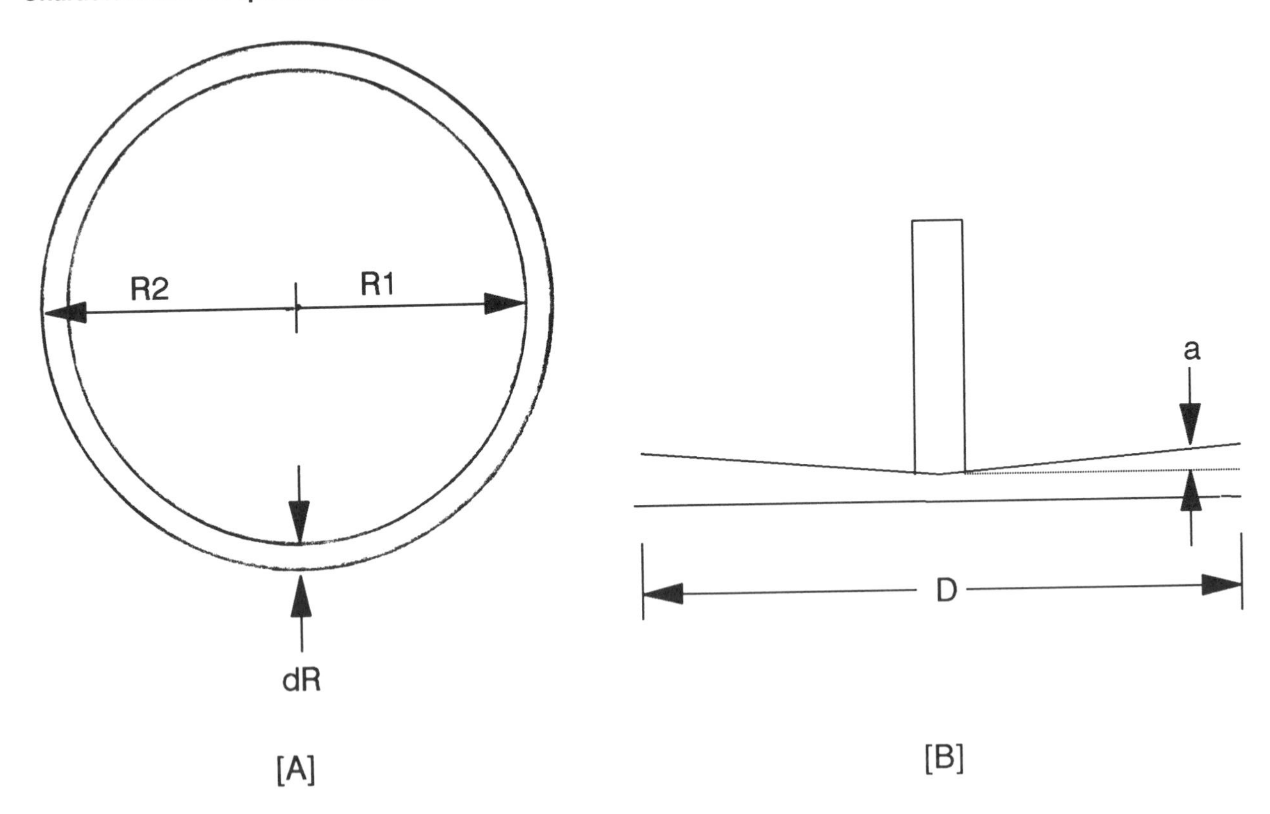

Figure 4-6. (A) Couette flow model. (B) Plate-and-cone model.

laminar conditions exist in the fluid, the viscosity can be calculated. First calculate the velocity gradient:

$$\frac{du}{dy} = \frac{2\pi R_2 N}{60(dR)} \tag{4-7}$$

where $\frac{2\pi R_2 N}{60} =$ fluid velocity at surface of R_2 for N in rpm

$dR =$ annulus clearance $<< R_2$

Then calculate the shear:

$$\tau = \frac{T_c}{2\pi R_1^2 (dR)} \tag{4-8}$$

where $T_c =$ torque in the torsion wire

Substituting τ from Equation 4-8 and *du/dy* from Equation 4-7 into Newton's equation yields the following value for μ.

$$\mu = \frac{15 T_c (dR)}{\pi^2 R_1^2 R_2 N h} \tag{4-9}$$

where $h =$ height of fluid level

Figure 4-6 [B] shows geometry for the plate–and–cone viscometer. The procedure is very similar: measure shear, measure torque, assure clearances are close, and calculate. It is important that the angle *a* is very small, 3° or less. Some viscometers work with angles less than 0.5°.

Note that the rate of shear can be varied from no–shear to very high shear rates in these kinds of viscometers.

And, note that while the coaxial cylinder and the plate–and–cone viscometers can be analytically calibrated there are any number of other configurations which can be calibrated against standard fluids of known viscosity. An especially valuable trait of forced–shear viscometers is that they can be built with large clearances to handle very viscous fluids, or very coarse mixtures, and still yield usable results.

One of the most widely used manual–force viscometers is the Brookfield design from Brookfield Laboratories. Figure 4-7 shows a Model DV-II+ which can be fitted with the spindles shown, with a plate–and–cone primary element, or any of several other recognized probes. The minimum viscosity reliably measured by the DV-II+ is 0.5 cP with the plate–and–cone element; the maximum is up to 104×10^6 cP for the Brookfield spindles, 136×10^6 cP for a coaxial cylinder element, 40 000 cP with a DIN element, or 264×10^6 for a T–Bar element. Accuracy is ±1% of full scale. Two ASTM standards[9,10] provide guidance as to operating the viscometer.

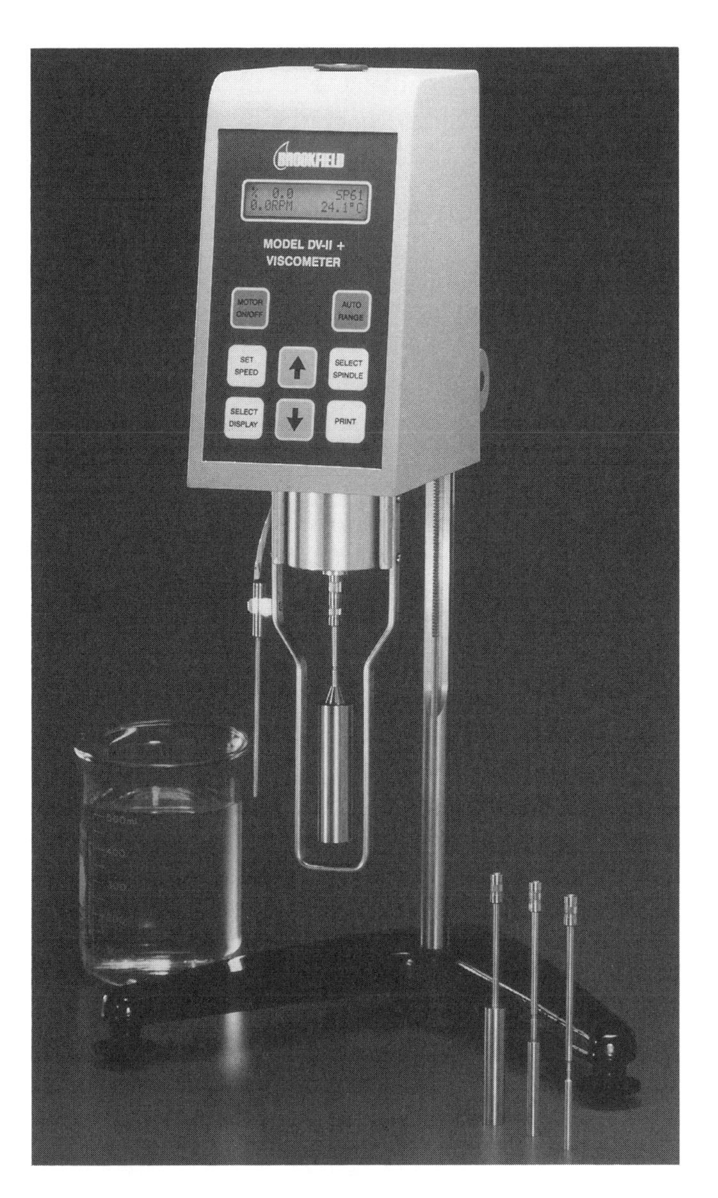

Figure 4-7. Brookfield Model DV-II+ viscometer. (Photo courtesy of Brookfield Laboratories.)

There are many designs of viscometers that measure shear rate, shear stress, or time, and thus viscosity. The *Stormer* design uses a falling weight rotating a free cylinder to induce an essentially constant shear stress in the tested material, and the time of fall is calibrated against known standards. The *Hallikainen* design uses two parallel plates—the standard viscosity analogy—as the primary element: a fixed glass plate is coated with the material, a moveable glass plate is squeezed against it, and the time required to move the plate a certain distance under a given force exerted by a falling weight is compared with known standards of viscosity.

A final note: Though the discussion up to this point has been confined mostly to the consideration of shear rate and shear stress, it must be remembered that temperature has an effect on the viscosity of non–Newtonian fluids just as it does on Newtonian fluids and, depending on the solid element of solid–liquid mixtures, pressure may have a larger effect on non–Newtonian fluids than on the more commonplace Newtonian fluids.

Non–Newtonian Petroleum Fluids

Petroleum liquids are *finely dispersed mixtures* or *multiple–component solutions*; there is little physical difference between the two.

Most petroleum liquids are handled at temperatures at which they are Newtonian; and at such temperatures they always have a measurable viscosity. However, almost any petroleum liquid will become non–Newtonian *above* a certain *high temperature,* where it commences to boil, or *below* a certain *low temperature,* where it commences to form wax crystals.

The first of these temperatures is called the *initial boiling point*, which is the temperature at which the first small fraction of distillate is produced by a liquid as its temperature is increased by heating. A more important parameter for pipeliners is the *flash point*, which is the lowest temperature at which a liquid, being heated, will ignite or *flash* when exposed to an open flame. The flash point is always lower than the initial boiling point.

The second temperature is the *cloud point*, which is the temperature at which wax crystals are first produced in a liquid as its temperature is decreased by cooling. A more important characteristic for pipeline purposes is the *pour point*, defined as the lowest temperature at which a liquid will flow under a low shearing force applied under certain specified cooling conditions. The pour point is always lower than the cloud point.

The trouble with these limits starts when one of them falls within the operating range of the pipeline; when the fluid starts to become non–Newtonian under temperature, pressure, and shear conditions expected in the pipeline.

Pipelines carrying NGL and LPG often carry liquids at temperatures above their initial boiling point, but these pipelines are designed so that the static pressure in the line is never allowed to drop below the vapor pressure of the product at pipeline temperatures and so distillation does not occur.

Hot oil pipelines have the same characteristic; the temperature at the output of the heating stations, and for some distance downstream, may be above the initial boiling point of the transported product, but line pressure is high enough to assure that no boiling occurs.

Other than these exceptions, there are no long–distance pipelines which carry liquids at temperatures such that distillation of the transported product occurs.

The cloud point is definitely the temperature at which non–Newtonian behavior commences, but since the shear stresses available and the shear rates produced in a pipeline are relatively high, it is usually best to work with the pour point instead while realizing that non–Newtonian behavior commences at low stresses and low shear rates at a temperature somewhat higher.

Figure 4-8 shows the ASTM D 97 pour point apparatus. The sample is heated to 115°F, the test jar is filled to the mark, set in the bath at a temperature well above the suspected pour point, and then, as the temperature is progressively lowered, the test jar is removed from the bath and very *slightly* tilted. The highest temperature at which the test jar can be held in a horizontal position for 5 seconds without movement is recorded, and the pour point is taken at 5°F above this temperature. Note that the shearing force on the bottom of the sample when held horizontally is only 30 mm of

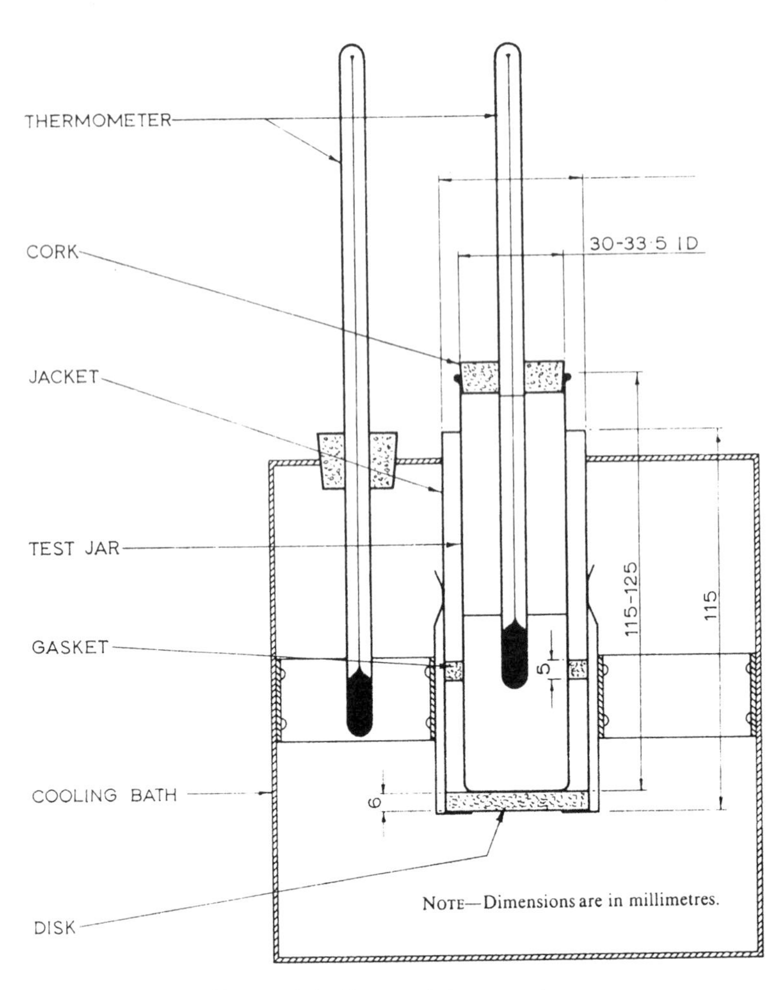

Figure 4-8. Pour point apparatus.

head. Obviously, the pour point is a delicate measure of the degree of stiffness or solidity of the fluid.

ASTM D 97 recognizes that thermal history effects can cloud the validity of pour point measurements. If it is known that the sample has been heated to a temperature above 115°F within the preceding 24 hours, the sample must be set aside for 24 hours at room temperature before starting the test. The result of this test is called the upper (maximum) pour point. The sample can be heated to 220°F and retested; the result of this test is called the lower (minimum) pour point. Both pour points are reported as the result of the tests.

The pour point test is not an especially reproducible procedure. For oils which do not have a sensitive thermal history, ASTM estimates the reproducibility in any one laboratory at 5°F and as between two laboratories about 10°F. My experience is not as good.

Also, the pour point test may not yield significant results for the pipeliner who is looking for an insight as to the reaction of the fluid under low shearing pressures. The ASTM recognizes this, and offers ASTM D 1659,[12] which is a test for the lowest temperature at which a sample will flow 2 mm in 1 minute through a 12.5 mm capillary under 15.2 cm Hg vacuum. The test involves parting the sample into six samples, and the maximum fluidity temperature is the maximum single fluidity temperature. It is, therefore, at least a conservative test; the *maximum*, not *minimum*, is reported.

If a really important pipeline is to be built, and a crude or some other product has to be handled under what are, or what are suspected to be, non–Newtonian conditions, laboratory–scale tests probably will not be sufficient, and bench–scale tests, which are small, model pipelines, may also fail. Then essentially full–scale, or what are often called *plant–scale,* test beds must be provided. The test loops built for the Alyeska 48–inch pipeline at Inuvik and for the BP–Hunt C-65 Libyan 34–inch at Kent are examples; if you are going to spend the annual income of a small kingdom on a single pipeline, one or two percent of that amount spent on an experimental pipeline may save that amount many times over. And, you can sleep at night.

The measured pour points for petroleum oils can be almost anywhere in the range of ordinary temperatures. Certain lubricants, and some aircraft jet fuels, are Newtonian at temperatures well below 0°F; and some crudes and heavy fuels have pour points above 100°F. Table 4-3 gives pour points for essentially the same list of crudes used to assemble Table 3-5. The lowest pour point is for Algerian Hassi Messaoud at –50°C; the highest is Indonesian Tandjung at +41°C. The range covered by the table is 91°C or nearly 194°F.

There is no table for pour points of products equivalent to Table 3-6 for viscosities. Pour points for products vary not only with the feedstock used to produce them but on the process used by the refiner. Nelson[13] lists No. 1 fuel oils with a pour point range from –65 to ±0°F; No. 2 fuels with a range of –60 +25°F; and No. 4 fuels with a range of –45 to +70°F, a range of 115°F or some 64°C.

It also is not usual to measure the pour point of light products. Instead, the *freeze point*, technically about the same thing as *cloud point,* is taken. The freeze point is measured on a sample as its temperature is *increasing*, while the cloud point is measured on a sample while the temperature is *decreasing*. Freeze point is an important parameter for jet fuels because of the low temperatures at the high altitudes where the jets fly.

Pipelining Non–Newtonian Fluids

Any petroleum fluid can be pumped at temperatures well below its pour point; the problem for the pipeliner is to determine how *much* lower, and *how*.

Defining the Problem

The literature of the oil industry is full of examples of non–Newtonian crudes, the problems they raise, and how they can be solved. Each of them is a separate case.

One of the very best analyses of the problems of pipelining waxy non–Newtonian oils is that by Ford, Ells, and Russell.[14] This group, all of British Petroleum, had the problem of trying to devise a way, or ways, of getting 55°F pour point Nigerian crudes and 75°F pour point Libyan crudes from their producing fields into tankers and then, after sea transport to U.K. and northern European ports, from the tankers to inland refineries. This series of four papers is really definitive in describing a major effort by a group of knowledgeable pipeliners to (successfully) solve some rather formidable problems with non–Newtonian crude oils. The method selected for solving their problems, heating the oil, was not original; their planning, and the methods they used, including use of a very sophisticated plant–scale test loop, were. The following are three of the basic truths from their work that should be written on gold tablets:

Table 4-3

Pour Points for Typical Crude Oils

Field	Country	Density 15°C	Pour Point °C
Hassi Messaoud	Algeria	0.800	–50
Qatar Export	Qatar	0.823	–24
Edjeleh	Algeria	0.845	– 9
Kirkuk	Iraq	0.844	–35
Agha Jari	Iran	0.852	–20
Zelten	Libya	0.830	+13
Gach Saran	Iran	0.872	–12
Forties	North Sea	0.840	± 0
Mubarek	Sharjah	0.840	–12
Ninian	North Sea	0.849	+ 7
North Rumaila	Iraq	0.853	–28
North Slope	Alaska, USA	0.894	–21
Berri	Saudi Arabia	0.831	+34
Saudi Arabian Heavy Blend	Saudi Arabia	0.886	–34
Saudi Arabian Light Blend	Saudi Arabia	0.858	–34
Iranian Heavy Blend	Iran	0.872	–21
Iranian Light Blend	Iran	0.858	–29
Arjuna	Indonesia	0.836	+27
Attaka	Indonesia	0.810	–34
Basrah	Iraq	0.856	+15
Boscan	Venezuela	0.998	+10
Brass River	Nigeria	0.811	–21
Brega	Libya	0.823	+ 7
Gulf of Suez Blend	Egypt	0.868	+ 5
Kuwait Export Blend	Kuwait	0.868	–30
Cactus Reforma	Mexico	0.860	–12
Argyll	North Sea	0.848	+12
Leduc	Canada	0.829	– 4
Fosterton	Canada	0.909	–15
Tandjung	Indonesia	0.825	+41
Ellenburger	Texas, USA	0.790	–32
East Texas	Texas, USA	0.834	+ 6
Julesburg	Colorado	0.825	– 4
Casper	Wyoming, USA	0.905	–29

- *Viscosity values*, effective for the full-scale pumping of waxy crudes at temperatures near or below their pour point, *can be predicted* with satisfactory accuracy, for both laminar and turbulent flow conditions, from laboratory tests.
- Prediction from laboratory work of full-scale *yield values* of waxy crudes at temperatures below their pour point *cannot be made with the same degree of accuracy,* and further investigations are required before this problem can be regarded as satisfactorily solved.
- The *sensitivity* of wax-bearing, high-pour crudes to *thermal treatment, rates of cooling*, and *mechanical shearing* is such that extreme care is required to establish the correct data for each individual crude.

Learning the Product

Certain oils can have a very complex thermal history. Measured values of viscosity, pour point, and especially the yield stress, can vary widely depending on how the oil had been treated in the period prior to the test. The following examples are from my personal files of non-Newtonian fluids I have had to handle.

- Redwater crude oil forms a gel at low temperatures which is difficult to break. Yet this crude, when quickly cooled to 32°F, exhibits a measured viscosity that is constant with a varying flow rate—thus Newtonian behavior—which is about half that of the varying viscosities measured on the same crude which has been cooled to 32°F for several weeks.
- East Texas crude which had been standing in storage tanks for some weeks had a measured pour point of 38°F. After being heated, and then quickly cooled and pour point re-measured, this procedure being repeated time after time, the pour point climbed two or three degrees on each successive test before stabilizing at 45°F. The so-called reheat-and-retest technique is a part of the *Navy Pour Point Test* which was in use during and immediately after World War II; I do not know if it is still being used.
- Holly Ridge crude, cooled in a laboratory vessel over 24 hours from room temperature to 32°F, formed a weak gel after a few more hours at 32°F. This same crude, loaded on barges in the lower Mississippi river in winter and cooled slowly over a matter of weeks as the barge was towed northward through progressively colder water for delivery to a port in the Ohio river, formed a very strong gel at 38°F. Small rocks dropped from a height of five or six feet would *splat*, and then rest on the disturbed, waxy surface of the crude.
- A sample of Rubelsanto crude taken at the choke manifold on the derrick floor of the drilling rig, cooled to atmospheric temperature over a period of several days while being transported to a testing laboratory, and then tested, exhibited a pour point of 50°F. Another sample, taken at the separator after the oil had been first run through a heater-treater at 130°F, showed a pour point of 70°F, nearly 20°F *higher*. A third sample, taken two months later from a still-standing stock tank at about 90°F had a pour point of 60°F.
- A heavy fuel oil (7.3 °API) had a viscosity at 130°F of 50 000 cSt. After being twice heated to 205°F and allowed to cool while being pumped down a pipeline it measured 52 000 cSt at 130°F—essentially no change.

A plant-scale test pipeline would have afforded me the opportunity of defining the problems raised by these fluids much better than by the equipment I had on hand. But none of the movements was of such size as to justify a plant-scale facility, and so I had to make-do with bench-scale testing.

Sometimes you just can't test, or, at least, *think* you can't test. When I decided—without testing of any kind—that it was safe to pig a pipeline, full of East Texas crude, in the winter season, the incoming pig traps filled with hard, congealed micro-crystalline wax, the strainer baskets exploded, and there was pure hell as things which should have been caught in the strainers went through the close-fitted mainline pumps. Thereafter pigging was carried out in a light West Texas stream run at least 24 hours after the tail end of the East Texas batch. The pipeline was my plant-scale test bed.

It is important to learn as much as possible about the thermal history of the oil, and what kind of shear it has recently been through, if you are considering handling the oil routinely below its pour point. There was considerable consternation at the outset of design of the Alyeska 48-inch pipeline because some of the early tests indicated that oil from the main formation was thixotropic in the operating temperature range of the pipeline. Subsequent tests, resulting in handling the oil in a certain way before introducing it into the pipeline and, for the most part, insulating the line in those areas where heat loss would be greatest, made possible the routine operation of this long artery under some very intimidating conditions.

Hackman et al.[15] report on tests undertaken for the U.S. Navy Supervisor of Salvage in which a typical No. 6 fuel oil with a pour point of 50°F flowed at 32°F with a head of one or two feet. The report didn't say how *much* it flowed.

Conversely, I have seen eight feet of Holly Ridge crude with a pour point of about 52°F set up solid in a barge at 38°F; and crude from the East Texas field—once the most prolific field in the U.S.—with a pour point of about 44°F, set solid for 30 feet at 28°F in an 80 000 bbl storage tank. The Holly Ridge crude was moved out by stirring it with a high pressure lance of another crude; the East Texas crude moved out when the weather changed.

Probably the difference in the Hackman experience and mine is that Hackman et al. were working with a refined oil which did not have a strong wax structure, whereas both Holly Ridge and East Texas crudes have a very strong wax matrix in their make up.

Using the high–pressure oil lance was a simple solution for the barge oil; the total depth was about 8 ½ feet and the agitation of the lance produced enough shear at such shear rates as to break the wax matrix down and let the crude flow to the suction nozzle under as little as 6 inches of head.

At the time—and maybe even now—there wasn't a good way to break out the 30 feet of East Texas crude set up in the mainline tank; if it wouldn't flow under 30 feet of head you just had to wait for spring. Then, when the temperature rose to a point where flow would start, the tank would drain rapidly and completely. A tank with a *spider*[16] in the bottom to help minimize tank bottoms can be used like the crude oil lance was used in the barge, but there are not too many tanks equipped with spiders.

Hot Oil Pipelines

There are *hot oil pipelines*, and there are *heated pipelines*; and a hot oil pipeline is not necessarily a heated pipeline.

A hot oil pipeline carries the oil at temperatures at which it is Newtonian, or near–Newtonian (which may or may not be a new definition). If the distance to be covered is greater than a mathematically pre–determined distance, the oil may have to be reheated one or more times, and some pipelines have both *pump–heater stations* and *heater stations*. The distance between heaters is a complex function of heat input and heat loss, which is discussed fully in the chapter on hot oil pipelines, and is limited in any event by the upper limit of temperature, usually taken as about 205°F, because to go much higher could raise the temperature of the surface of the pipe immediately outside the heater stations above the boiling point of water. This might not be a pipeline problem but it certainly is a public relations problem; very hot pipelines *steam* and this does upset the public.

Witness, however, Khedr, Nazar, and Weber,[17] who describe a 1.5 mile 6–inch vacuum residuum transfer line, running from a vacuum–distillation unit in one refinery to a delayed coker in another, designed for 650°F maximum, 230°F minimum. This line was designed primarily as an above–ground line with expansion loops; the underground portions were carefully insulated, supported, isolated, and protected from outside forces. The entire line was provided with an impedance heating system. Moral: If you *have* to move a product, you can probably do it, but you may have to move beyond the rules of ordinary pipelining.

A heated pipeline is what you might expect; the pipeline (and, usually, the source oil) is heated. This may be done by steam tracing, electrical resistance heatline tracing (inside or outside the pipe), skin–effect current tracing (insulated copper conductors encased in *heat tubes* welded axially to the pipe), or electric impedance heating (which heats the pipe steel itself).

Hot oil lines may be insulated. Insulation is expensive, but so is heat, and there is often an economic balance in favor of insulation, at least for the hottest part of the line. Also, insulation assists in keeping heat loss down in areas where ordinarily a large heat transfer out would occur, such as in a swamp or a river; heat loss from a hot, bare pipeline immersed in flowing water can be enormous.

One last thing: A hot oil line—sometimes even a heated line—is usually subject to freezing if flow is stopped long enough. The only safe way to handle this, other than thermally or chemically treating the oil prior to pumping, is to provide a displacement fluid and a safe set of displacement pumps. Some lines use a medium diesel fuel, and some use water. All of them have displacement pumps independent of the supply of third parties. If the line goes down for more than a predetermined period of time, the line is displaced. If water is used, there can be some nasty emulsion breaking problems, but that is better than having to abandon a frozen pipeline.

Pour Point Depressants

When the problems really get tough—when the crude and the weather work together to assure you will have a frozen pipeline if you don't do something about it—there are always pour point depressants. Holgate[18] describes treating a Libyan crude which had a natural pour point of 15–20°C with 1500–2000 ppm of pour point depressant to bring the pour point down to 2°C, well below the expected low pipeline temperature of 10°C.

Holgate's problem was with a 16–inch field trunk line, and 1500–2000 ppm of depressant was not too large a premium to pay to keep the three oil fields served by the pipeline on stream. Much more difficult was Trans–Alpine's problem described at a symposium at the *Institute of Petroleum* in 1971.[19] This 1 000 000 bpd 40–inch mainline from Trieste northward supplied Austria and south–east Germany with essentially all their crude needs and it, too, was faced with handling Libyan crudes at pipeline temperatures near 0°C. TAL didn't feel that using an additive in a million barrel per day stream would be an attractive economic solution to their problem, and instead decided to install tie–in points along the pipeline so that additional pumps, bringing additional shearing stress on line, could be used in case the line was shut down in cold weather and the crude could not be sheared and re–started with only the mainline pumps.

Several operators at the IP symposium reported on the use of pour point depressants. Price[20] described pour point depressants—sometimes called *flow improvers*, not to be confused with viscoelastic *drag reducer additives*, which are also sometimes called flow improvers—as chemicals which can change the wax crystal structure in the fluid. These chemicals work by modifying crystal habit, size, and adhesion to each other; and in many cases the needle–like structure of the wax is changed into a more spherical structure. Price noted that effective weight concentrations are in the range 0.002–0.05%. The first, which is in the order of 20 ppm, would be a reasonable commercial application; the second, being in the order of 500 ppm, is a little heavy. But Brod et al.[21] reported on a special movement through the Rotterdam–Rijn of an oil that "ordinarily would never be allowed near the pipeline at the time of year the test took place" (their quotes) that required a dosage of 0.12% by weight, or some 1200 ppm. Another test through the Pipeline de l'Isle de France with a neat African crude which had a 24°C pour point had a 0°C pour point after treating and no yield value at pipeline temperatures. The dosage was not given.

An especially difficult problem, handling high yield waxy crudes in underwater pipelines, was described by van Engelen et al.[22] This paper describes tests run on Shell's 6 mm bench–scale test pipeline to determine what was the best flow improver to use on the waxy Bombay High crude, which has a 27–30°C pour point, that had to be pumped ashore through waters that could be as cold as 20°C. Here, as usual with waxy crudes handled in turbulent flow, the problem was assuring the ability to restart the pipeline after an extended shutdown and cooling of the fluid. The flow improver selected, injected at 500 ppm, provided this assurance.

Heat Treating

Another interesting way to handle non–Newtonian waxy oils developed by Oil India to handle their Moran and Nahorkatiya crudes is described by Chandrasekharan et al[23]. These oils have a pour point in the range from 29°C to 34°C. The method developed after considerable research involves heating the oil to 90°C in gas–fired heaters; fairly rapid cooling to 65°C in heat exchangers with incoming crude on the cold side; and then slow, controlled cooling of stilled fluid in treaters, which are really just large, vertical, shell–and–tube heat exchangers, to the final temperature of 18°C. Final cooling was at a controlled rate no greater than 0.5°C/minute, the rate predetermined as that most favorable for producing a low viscosity, low yield value, crude product. The method is expensive and is not suited to short lines because the cost of the treatment facility, if measured per mile or per kilometer of movement, would be overpowering, but it is economic for Oil India in their particular pipeline situation. They note the treated crude is either a supercooled solution of the wax in the residual oil, thus a solution, or a suspension of wax crystals in the residual oil as the continuous component, thus being a pseudo–homogeneous fluid—no one seems to be sure what it is—but the wax matrix is seemingly permanently altered for this crude in any event. Chandrasekharan et al. provide a detailed analysis of the process and a flow diagram of the process plant, with its 36 treaters grouped in nine banks of four each, operating on a 270–minute cycle time and handling 45 000 bpd. Mukhopadhyay[24] describes the same process in less technical terms.

Diluents

There are lucky operators who have two crudes, one heavy, messy, and in short supply, and the other light, ordinary, and available in quantity; they can just mix the two.

Heavy fuel oils are regularly moved between refineries by blending down with naphtha and then separating the two at the receiving end of the line.

This kind of solution has to be a one–way movement; if not, a second line is required to return the blending stock to the source of the heavy stock and both capital and operating costs skyrocket.

Defroster Lines

Then there are the really imaginative solutions. Rupert et al.[25] describe a 150–km crude line transporting Lirik, Sumatra, crude, which has a pour point of 105–110°F and an extremely strong gel; a puddle of this oil is a liquid in sunlight, but at the coolest part of the night congeals to where it will support the weight of an average man. This 8–inch line has a 3–inch *defroster* line buried along with it. The defroster line receives water at 180°F at each pump station and pumps it each way at each station, the cooled water being discharged at a meeting point about halfway between pump stations.

It is interesting to note that Rupert et al. felt they had an alternate—a surface pipeline that would use the sun as the defroster, a perfectly feasible solution in tropical Indonesia. This was an economic solution but it had a problem—a rainstorm would cool the pipeline, congeal the oil, and shut it down. And, during the rainy season, it might be some time before the sun came out again and defrosted the pipeline.

Oil–Water Suspensions

Another interesting project was described by Lamb and Simpson.[26] This involved transport of 40 000 bpd of Tandjung, Sumatra crude oil with a 33% wax content and a 105°F pour point, some 238 km through the Indonesian tropics, where annual rainfall is over 100 inches, to a deepwater terminal and refinery at Balik Papan. A singular aspect of this crude is that it has a *specific heat* of approximately 1.0 kcal/°C, essentially the same as that of water, over the temperature range 30–43°C. In this range the specific heat of the *liquid* part of the crude, about 0.5 kcal/°C, is augmented by the heat of fusion of the wax crystals, and the *effective* specific heat is a variable with temperature depending on the quantity of wax involved in the melting or recrystallization.

Figure 4-9 shows the extreme variation of the viscosity of Tandjung crude with varying temperature. Ambient temperatures along the route run 25–30°C, and average ground temperature at pipeline depth is about 28–30°C. At 30°C the viscosity is in excess of 50 000 cP.

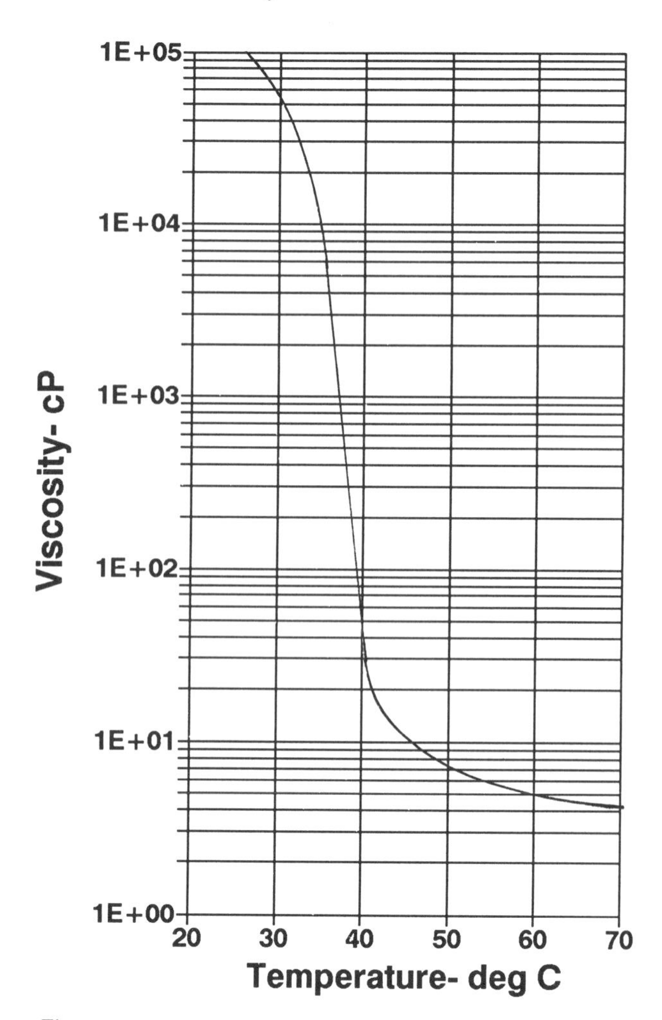

Figure 4-9. Viscosity-temperature curve for Tandjung crude.

A hot oil line was considered and ruled out because calculations showed that a restart pressure of 15 000 psi/mile—an impossible value—would be required for a 20–inch line. Pour point depressors were tried and found ineffective; diluents were an option but, considering the high wax content of the crude, would be required in large quantities and would necessitate a large, expensive return line; visbreaking, though technically feasible, was not economically attractive; and the only alternative left was mixing with water.

If oil is mixed with water it can be in an *oil–in–water suspension*, with water being the continuous phase; a *water–in–oil suspension*, with oil being the continuous phase; or in an *oil–and–water emulsion.*

There evidently was no thought given to the emulsion solution, though there are such pipelines operating in Indonesia. So the oil–in–water suspension solution was chosen.

Since there was then—and is now—little known about the theory of oil–and–water suspensions it was prudent to go into plant–scale testing, and a 0.5 mile 6–inch loop and a 1 mile 18–inch loop were constructed. These test beds were equipped with the necessary pumps, tanks, and instruments, and extensive tests were run. These proved the validity of the bench–scale work, and also proved that the migration of oil and water on a hill during a shutdown was not a problem. They also showed that inversion of the suspension, from an oil–in–water mix with water the continuous phase into a water–and–oil suspension with oil the continuous phase, would not occur.

Figure 4-10 shows the viscosity of the 30–70 water–oil suspension. At ground temperature, and at a pipeline shear rate of 10 sec^{-1}, which corresponds to a flowing velocity of about 2.4 fps, the reduction in viscosity is from approximately 100 000 cP to 400 cP, a truly phenomenal ratio.

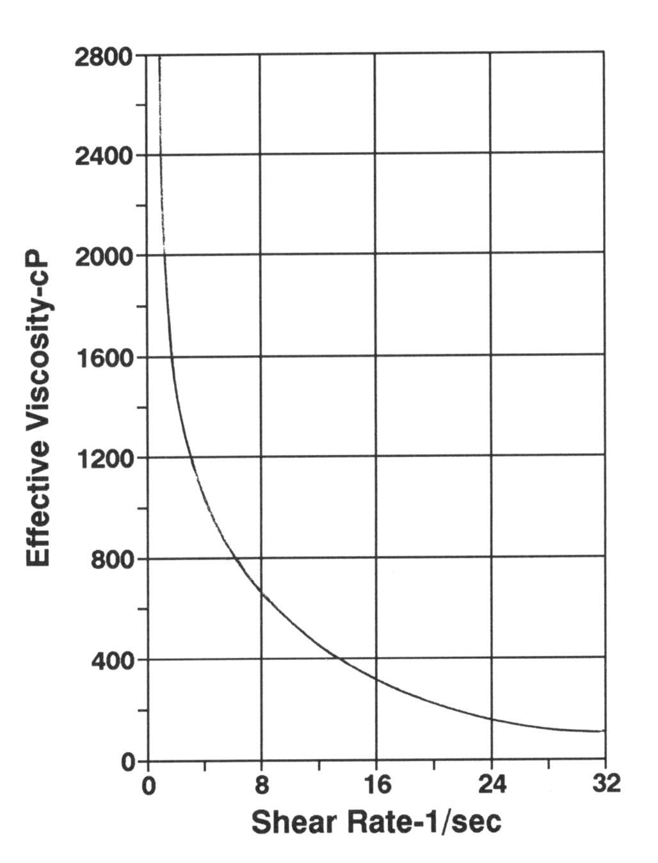

Figure 4-10. Viscosity-vs-shear rate for Tandjung suspension.

Lamb and Simpson concluded their paper with a list of three cautions along the lines of those of Ford et al.[14]

- The use of the suspension system is limited by the necessary combination of crude properties and local conditions (especially temperature) needed to provide an attractive operation.
- Each case should be thoroughly investigated in the laboratory, and carefully studied in the range of shear rates and temperatures to be employed in practice.
- The existence and the magnitude of the time dependence of flow properties should also be investigated in waxy oil–in–water suspension systems.

And it should be noted that before the mainline was built, a 0.5 mile 6–inch, and then a 1 mile 20–inch, test loop was built; and the studies described above were run on these facilities after laboratory and bench–scale tests were completed.

Oil–Water Emulsions

There has been little written about oil–emulsion pipelines, but there is work going on in this field. The Canadian Transoil group, using technology developed by BP and Venezuela's INTEVEP, demonstrated[27] they could mix 575 bpd of an emulsion of 65% tar sands bitumen and 35% water; pump 5900 bbls of the emulsion for 27 km without a change in the emulsion; and then, at the end of the line, store 6700 bbls of emulsion in tanks without any problems of emulsion stability or stratification. The BP–INTEVEP technology involves controlled mixing of bitumen with low concentrations of a commercial surfactant to produce oil–water emulsions.

Figure 4-11 is from my files, and shows tests on an oil–water emulsion of 35°API Lago, Venezuelan crude. The parameters K and n are those required to define the power law for the fluid (see Equation 4-3). I don't know the origin of this information, but considering where it was filed in my files, I believe it is accurate.

One emulsion pipeline of which I am aware involved an Indonesian crude that made the selection of the emulsion design almost mandatory; it went into the emulsion with very little agitation and shearing and, upon completing its

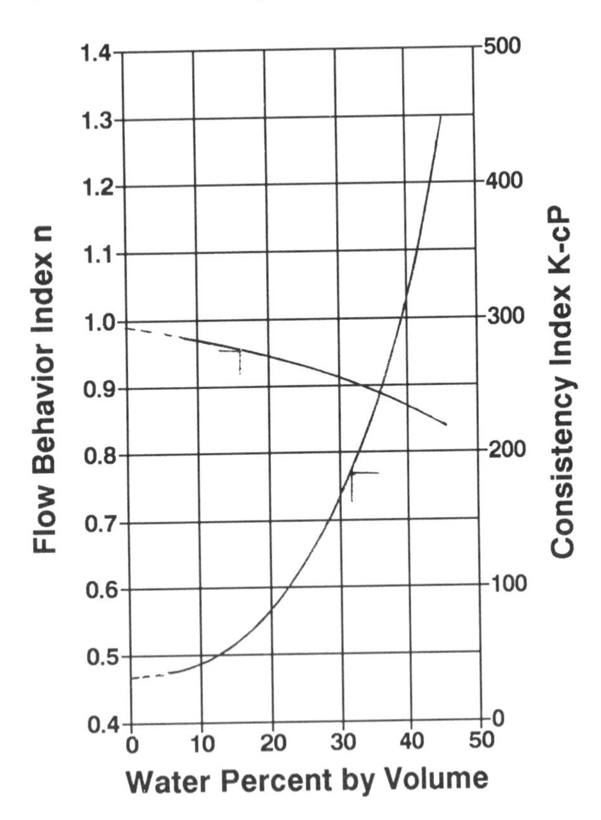

Figure 4-11. Power Law parameters for Lago crude emulsion.

journey through the pipeline, would break out completely in a matter of a few days if left sitting quietly in storage tanks. No heat, and no chemicals, were required; it was a self–breaking emulsion. I am told the water didn't even have to be treated before being released back into the environment, but I doubt that; it is too good to be true.

Last Words

Non–Newtonian petroleum oils, especially the very waxy crudes, are not to be taken lightly. There are two frozen pipelines beneath Bayonne Bay simply because the operators didn't believe they had to have a utility–free power source for their pumps in case the power went out for some days.

There are ways to handle these difficult fluids, and the examples above describe some of the better ones. Care in the investigatory and design process can assure that the product can be transported, though perhaps not in the way you might have thought when you first started on the problem.

And, always, given a non–Newtonian fluid and a large, expensive movement, don't forget at least bench–scale or, preferably, plant–scale testing: Time and money spent on testing before the line is built is much less costly than solving the problems raised by an improperly designed, constructed, or operated pipeline after it is in place.

References

1 Katz, D. L., and King, G. *Dense Phase Transmission of Natural Gas.* CGA 1973 National Technical Conference.

2 Govier, G. W., and Aziz, K. *The Flow of Complex Mixtures in Pipes.* New York: Van Nostrand Reinhold Ltd., 1972.

3 Wasp, E. J., Kenny, J. P., and Gandhi, R. L. *Solid–Liquid Flow Slurry Pipeline Transportation.* Clausthal, Germany: Trans Tech Publications, 1977.

4 Hughes, W. F., and Brighton, J. A. *Fluid Dynamics.* New York: McGraw-Hill Book Company, 1967.

5 Green, Henry and Weltman, Ruth N. "Equations of Thixotropic Breakdown for the Rotational Viscometer," *Industrial and Engineering Chemistry*, Vol. 18, No. 3, March 1946.

6 White, J. L., and Metzner, A. D. *Progress in International Research on Thermodynamic and Transport Properties*, pp. 748–762. New York: Academic Press, 1963.

7 Walters, K. *Rheometry*, London: Chapman and Hall, 1975.

8 Author's Note: *Conferences on Fundamentals and Applications of Polymer Drag Reduction in Liquid Flows* were held in Cambridge, U.K., in 1974 and 1977; in Bristol, U.K., in 1984; and in Zürich, Switzerland in 1989 under the sponsorship of the International Association for Hydraulic Research.

9 ASTM D 2669, *Standard Test Method for Apparent Viscosity of Petroleum Waxes Compounded With Additives (Hot Melts).* American Society for Testing Materials, Philadelphia.

10 ASTM D 2983. *Standard Test Method for Low–Temperature Viscosity of Automotive Fluid Lubricants Measured by Brookfield Viscometer.* American Society for Testing Materials, Philadelphia.

11 ASTM D 97. *Standard Test Method for Petroleum Oils.* American Society for Testing Materials, Philadelphia.

12 ASTM D 1659. *Standard Test Method for Maximum Fluidity Temperature of Residual Oil.* American Society for Testing Materials, Philadelphia.

13 Nelson, W. L. *Petroleum Refinery Engineering*, 4th ed. New York: McGraw-Hill Book Company, 1972.

14 Ford, P. E., Ells, J. W., and Russell, R. J. "Handling High Pour–Point Crudes in Pipelines," *Proceedings API Pipeline Conference*, 1965.

15 Hackman, D. J., Yates, J. B., and Tierney, J. M. *Summary Report on Piping Systems for Transferring High–Viscosity*

Oils, U. S. Navy Supervisor of Salvage Report No. 9-74, Columbus, Ohio: Battelle Columbus Laboratories, May 1974.

[16] Author's Note: A *spider* is a network of small pipes strung out in the bottom of a tank and connected to a manifold so that high pressure oil can be used to act as many small jets and move out the BS&W, and wax, that accumulates in the bottoms of crude tanks. The spider is so–called because it is something like a spider's web. I guess.

[17] Khedr, E. M., Nazar, S. M., and Weber, E. C. "Thermal Analysis Keys Design of Long Vac–Resid Line," *Oil and Gas Journal*, July 3, 1989.

[18] Holgate, J. P. "North Africa Line Brings Waxy Crude to Terminal." *Oil and Gas Journal*, August 10, 1987.

[19] "Here's Some New Technology for Handling Waxy Crudes," John Cranfield, Editor. *Pipe Line Industry*, January, 1971.

[20] Price, Dr. Roger. Paper at IP Symposium. See footnote [15].

[21] Brod, M., Deane, B. C., and Rossi, F. Paper at IIP Symposium. See footnote[15].

[22] van Engelen, G. P., Vos, B., Kaul, C. L., and Aranha, H. P. "Flow Improvers Aid Production of Bombay High Field." Offshore Technology Congress, 1979.

[23] Chandrasekharan, K. P., and Sikdar, P. "Here's How Waxy Indian Crude is Prepared for Transit." *Oil and Gas Journal*, October 1970.

[24] Mukhopadhyay, A. K. "Conditioning Process Solves Crude Transport Problem," *Pipe Line Industry*, June 1979.

[25] Rupert, J. M., and van Diemen de Jel, B. N. "Pipelining Crude With 110°F Pour Point," *Oil and Gas Journal*, June 10, 1957.

[26] Lamb, M. J., and Simpson, W. C. "Waxy Crude Moves in Water Suspension." *Oil and Gas Journal*, pp. 140–146, July 1, 1963.

[27] Heavy Oil–Water Emulsion Line Tests Successful. *Oil and Gas Journal*, p. 27, Dec. 21, 1987.

5

Vapor Pressure

The Vapor Phase

Aristotle believed—and therefore for some 2000 years all philosophers believed—that "nature abhors a vacuum." Then, in the middle of the 15th century, Torricelli invented the barometer and established a limit to nature's ability in this regard. That limit is called *vapor pressure.*

Torricelli's barometer was about 36 inches long and was filled with mercury; Blaise Pascal, who among other talents was one of the greatest of French mathematicians, undertook to confirm Torricelli's findings with a barometer over 34 feet long filled with red wine—which may or may not indicate that the French nation was running a surplus of *vin rouge odinaire* as early as the mid–1600's. It does have a lot to say about the density of mercury and red wine.

The mercury used by Torricelli has almost no vapor pressure at ordinary temperatures and thus, to this day, mercury barometers are the standard around the world because they are almost temperature–insensitive. The red wine of Pascal's barometer, however, had a high vapor pressure at atmospheric temperatures, and so the wasteful practice of filling barometers with red wine has, thankfully, disappeared.

If it can be understood that liquid mercury has a vapor pressure, however small, at ordinary temperatures, there should be no problem in understanding that the heaviest of petroleum fluids has a measurable vapor pressure down to very low temperatures. Light fluids may have vapor pressures of several *psia*; and NGL and LPG fractions may have vapor pressures in the hundreds of *psig*, at ordinary atmospheric temperatures.

Liquid petroleum pipelines are generally high–pressure undertakings. Pressures of 1500 psig are commonplace, and pressures in excess of 2000 psig are not unknown. Therefore, to the extent of long distance trunk pipelines, pipelining is a high pressure kind of endeavor; and the vapor pressures of crudes and, except for NGL and LPG, products transported is usually not of any great import.

But a considerable part of the petroleum transportation cycle—that apart from the trunklines—involves handling fluids at relatively low pressures. The surface of a fluid in a floating–roof tank is at atmospheric pressure, or 0 psig; and the surface of the same fluid in a pressure–vacuum vented cone–roof tank would be at a pressure only a few ounces more or less than 0 psig. The fluid at the nozzle level of the tank, or in a pipe connected to the nozzle, is at no more pressure than that due to the head of fluid in the tank—a maximum of 60 feet or so for tanks of normal configuration—while at the other end of the line where the flowing fluid en route from the tank arrives at the inlet of the tank booster pump, the tank head, and whatever elevation head adds to or subtracts from it, may be completely dissipated in friction, and the pressure of the fluid may be several psi *less* than atmospheric pressure; a *partial vacuum* may exist in the pipe. And, in the limiting case, that partial vacuum is limited by the vapor pressure of the fluid.

Whenever pressures are large compared with the vapor pressure of the fluid being handled, the vapor pressure is not a controlling factor in the design or operation of the facility. However, when vapor pressure is a significant proportion of the operating pressure, and especially when it is of the same order of magnitude as the operating pressure, to neglect the effects of vapor pressure on the design and operation of the facilities can be disastrous.

A high vapor–pressure fluid is more *difficult* to handle, and usually more *dangerous*, by which is meant it has a more pronounced tendency to burn or explode, than a low vapor–pressure product. There are certain exceptions to this rule, one of them being JP–4 jet fuel, but the rule generally holds.

The tendency of a petroleum fluid to ignite is measured by its *flash point*, the temperature that quantitatively describes the tendency of the fluid to ignite in the presence of an open flame; and there is a further characteristic, the *fire point*, which is the temperature at which a fluid will not only ignite but continue to burn. If the fluid is to be handled outside the pipeline, as in tanks, tank trucks and rail cars, and tank containers, it is important to have a knowledge of these characteristics of a fluid as well as to know its vapor pressure. Sometimes the legal definitions of *flammable*, *combustible*, or *hazardous* are tied to flash point, fire point, or both; and whether a fluid can be transported and stored in a normal manner, or in a strictly specified, defined (and usually much more costly) manner, depends on these values. The code of the *United Nations* (UN) and the *International Maritime Dangerous Goods Code* (IMDGC) agree that fluids with a closed–cup flash point greater than 61°C (141.8°F) are *non–hazardous*; those with a flash point between 0°C and 61°C are *semi–hazardous*; while those with a flash point below 0°C are *hazardous*. The manner, and thereby the cost, of transporting and storing each of these classifications of fluids is different. And, these classifications may not be acceptable to other entities; the U.S. Department of Transportation has a definition of flammable which is not in accordance with that of the IMDGC, and the FLAMMABLE placard has to be removed from a container after transport under DOT auspices before it is handed over to a shipping company for international transport by sea or risk problems at the port of discharge.

Note that *vapor pressure*, as well, sometimes enters into commercial problems; most tankship charter parties, for instance, limit the RVP (Reid Vapor Pressure) of products transported in open–tank ships.

Also note that fluids may be classified as other than non–hazardous for reasons not associated with vapor pressures, flash points, or fire points; corrosive fluids, for instance, or fluids of a high toxicity, can attract a hazardous classification. Acetylene can be handled commercially by dissolving it in some neutral liquid, and in this form it is quite safe, but in pipelines carrying *real* acetylene, C_2H_2, the danger is that the product will *dissociate* just because it wants to; and in most jurisdictions the design pressure of an acetylene pipeline is 50 times its proposed operating pressure.

Vapor Pressure and Flash Point

If a stilled fluid exists in equilibrium in a closed container, that part of the container not filled with fluid will be filled with vapor rising from the fluid and the entire vessel will be under a pressure called the *vapor pressure* of the fluid.

The single ASTM definition of vapor pressure contained in the *Glossary of Definitions* is

> Vapor pressure is the pressure exerted by the vapor of a solid or fluid when in equilibrium with the solid or fluid.

This is not one of the better ASTM definitions, but it does bring in the case of sublimating solids which is necessary to take into account substances which do not go through a liquid phase in passing from the solid phase to the gas (vapor) phase.

Tabor[1] gives a more complete description, and definition, of the process:

> Suppose a quantity of liquid is placed in a sealed container and the air above it is evacuated. Some of the liquid will evaporate, and some of this evaporated liquid will condense. Equilibrium will be reached when the rate of evaporation equals the rate of condensation. The pressure exerted by the vapor under these conditions is known as the *saturation vapor pressure* or, more usually, the *vapor pressure*. It increases as the temperature is raised. It will be almost identical in the presence of air or another gas, since the vapor and the alien gas exert their partial pressures independently.

Vapor pressure is difficult to measure accurately under conditions that approximate industrial situations. The theory, however, is straightforward.

For a pure liquid, vapor pressure may be described in terms of the latent heat of vaporization and the gas constant:

$$p = A\exp\left(\frac{-L}{RT}\right) \tag{5-1}$$

where $p =$ vapor pressure
$L =$ heat of vaporization per mole
$R =$ gas constant
$T =$ temperature, K or $°R$
$A =$ a constant

The change of vapor pressure with temperature is shown by the Clapeyron equation:

$$\frac{dp}{dT} = \left(\frac{L}{T}\right)\left(\frac{1}{V_x - V_y}\right) \tag{5-2}$$

where $V_x =$ volume occupied by 1 mole of vapor
$V_y =$ volume occupied by 1 mole of liquid

Inasmuch as $V_x >> V_y$, little accuracy is lost if V_y is disregarded, thus yielding an equation of the form:

$$\frac{dp}{dT} = \left(\frac{L}{T}\right)\left(\frac{1}{V_x}\right) \tag{5-3}$$

Since $pV_x = RT$ for any gas, if L is assumed constant then this equation can be written in terms of dp/p and integrated to yield one form of the Clausius–Clapeyron equation:

$$\log_e p = \left(\left(\frac{-L}{R}\right)\left(\frac{1}{T}\right)\right) + C \tag{5-4}$$

where $C =$ a constant of integration

This equation as written yields a straight–line plot of $\log_e p$ –vs– $1/T$, assuming L is constant over the range of T. This assumption is valid for pure liquids, and essentially valid for simple liquid mixtures. It is *not* a particularly valid assumption for most wide spectrum petroleum fractions or for crude oils.

This problem is solved in practice by what is called a *Cox Chart*, constructed by plotting the logarithm of the vapor pressure against temperature on a straight line, and then constructing a temperature scale to fit the straight line. This is tantamount to adding a constant to the temperature term in the Clausius–Clapeyron equation. One such modification is quoted by Nelson[2] as

$$\log P = -\left(\frac{A}{T + 382}\right) + B \tag{5-5}$$

where $P =$ vapor pressure, psia
$T =$ temperature, $°F$
A and $B =$ constants

Figure 5-1 is an example, originally published by Chicago Bridge and Iron Company,[3] of a Cox Chart for gasolines derived from a Cox Formula written as

$$P_T = P_{100°F} \times 10^Z \tag{5-6}$$

where $P_T =$ vapor pressure at $T°F$
$P_{100°F} =$ vapor pressure at $100° F$
$Z = 4 - \left(\frac{2240}{T + 460}\right)$

Vapor pressure also varies with externally applied pressure, and in a somewhat unexpected way—vapor pressure *increases* with an *increase* in externally applied pressure. This increase is, for most applications, almost negligible.

$$\Delta p = P_e\left(\frac{\rho_x}{\rho_y}\right) \tag{5-7}$$

where $\Delta p =$ increase in vapor pressure due to P_e
$P_e =$ externally applied pressure
$\rho_x =$ density of vapor phase of fluid
$\rho_y =$ density of liquid phase of fluid

Since the ratio of vapor density to fluid density is in the order of 1:1000 at atmospheric temperatures for most vapors, an increase in vapor pressure in the order of 0.001 units can be expected for each unit of externally applied pressure. This is certainly not a major variation, and can be neglected for all ordinary purposes.

Measurement of Vapor Pressure

The most common method of measuring vapor pressure in the petroleum industry is that of ASTM D 323,[4] called the *Reid Method,* which is applicable for measurements of volatile crudes, and of all volatile products except LPG; measurements of vapor pressures of LPG are usually carried out in accordance with ASTM D 1267,[5] called the *LP–Gas Method.*

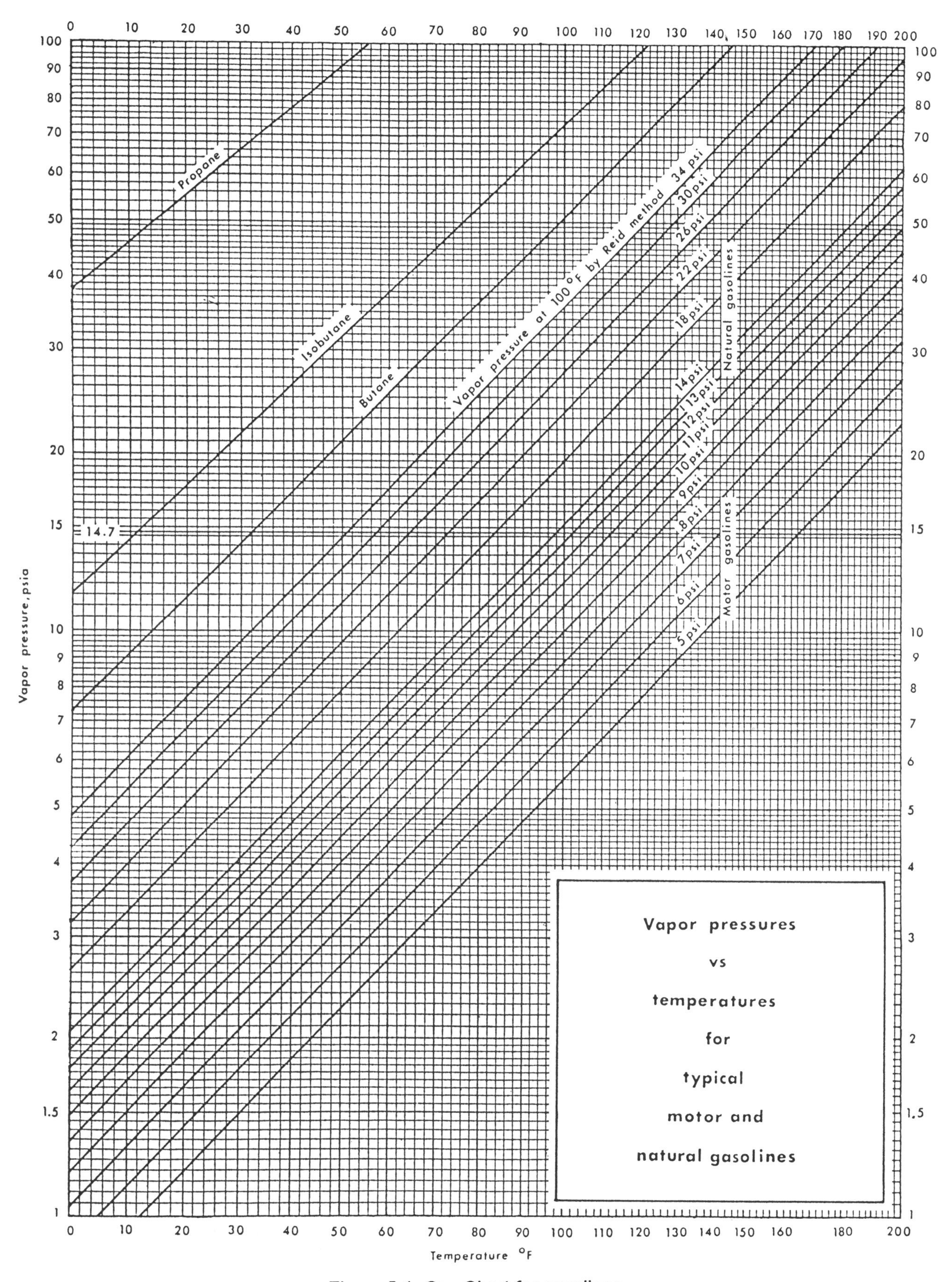

Figure 5-1. Cox Chart for gasolines.

The Reid Method yields a vapor pressure at 100°F called the *Reid Vapor Pressure*, abbreviated RVP. It differs slightly from the *true vapor pressure* at 100°F due to the small, but necessary, vaporization of the sample, and to the presence of water vapor and air in the sample space. Materials having a RVP of up to 26 psi can be accurately measured; higher pressures, up to 50 psia or so, can be measured but the accuracy is reduced. RVP is reported in terms of an absolute pressure, such as *psia*.

The LP–Gas method yields a vapor pressure at a given temperature between 100°F and 158°F. Materials having less than 5% by fluid volume of components having a boiling point higher than 0°C (32°F), and a total vapor pressure less than 225 psig, can be measured by this method. Note this test yields a result measured as a *gage* pressure, while the Reid method yields an *absolute* pressure.

The apparatus for the RVP test is shown in Figure 5-2. The ratio of the volume of the air chamber to that of the gasoline chamber (called the V/L ratio) is ideally 4:1, and must lie in the range 3.8:1 to 4.2:1; for aviation gasoline with a RVP of about 7 psia, the ratio must lie in the range 3.95:1–4.05:1. The one–opening chamber is used for fluids which have a vapor pressure of 26 psia or less; the two–opening chamber, which allows introduction of the sample into the chamber through a second, piped–solid, sample line, is for samples of higher pressures.

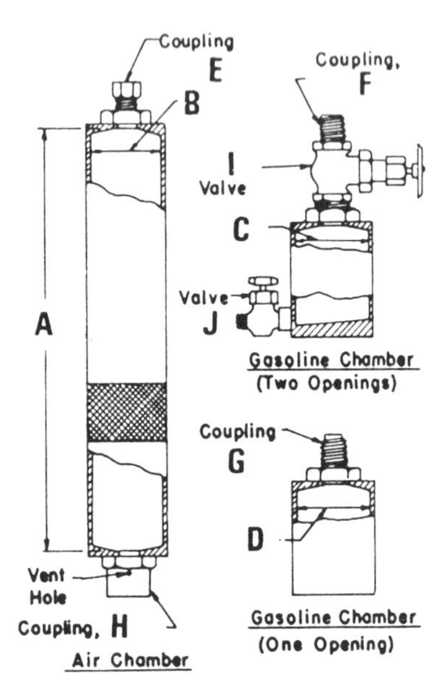

DIMENSIONS OF VAPOR PRESSURE BOMB

Key	Description	in.	mm
A	Air chamber, length	10 ± 1/8	254 ± 3
B, C, D	Air and gasoline chambers, ID	2 ± 1/8	51 ± 3
E	Coupling, ID min	3/16	
F, G	Coupling, OD	1/2	
H	Coupling, ID	1/2	
I	Valve	1/2	
J	Valve	1/4	

Figure 5-2. RVP bomb. (Courtesy of ASTM)

To use the apparatus, it is first cleaned and dried; the air chamber H is brought to some constant temperature, usually 100±0.2°F and a gage attached to coupling E; the gasoline chamber D is chilled to 32–40°F and then filled to overflowing with a sample chilled to the temperature of the chamber; the two chambers are then quickly connected, and the entire apparatus, save the pressure gage, is immersed in a 100°F constant–temperature bath. When the entire assembly reaches an equilibrium temperature of 100±0.2°F the gage is read, and readings are taken at not less than two–minute intervals until the reading is a constant. This is the *uncorrected RVP*. The pressure gage is checked against a standard, and the corrected value is recorded as the *manometer reading*. A correction is then applied for barometric pressure and initial air temperature (if other than 100°F; there is no air temperature correction to make if the initial temperature in the air chamber was 100°F), and the corrected manometer reading is recorded as the *Reid Vapor Pressure* (RVP).

Reid Vapor Pressure is by definition at 760 mm Hg and 100°F. For other temperatures and pressures a correction must be calculated as

$$C = \left[(p - p_t)(t - 100)/460 + t\right] - (p_{100} - p_t) \tag{5-8}$$

where C = correction, psi
t = air chamber temp at beginning of test, °F
p = barometric pressure, psi, at time of test
p_t = vapor pressure of water, psia, at t °F
p_{100} = vapor pressure of water, psia, at 100°F = 0.95

If the initial air temperature was below 100°F, the correction is subtracted from the manometer reading; if it was above 100°F, the correction is added. The vapor pressure of water can be obtained from almost any engineering handbook. The standard reference[6] is the ASME *Steam Tables*.

The equipment for D 1267 is similar to that for D 323 and, in fact, the D323 apparatus can be used for the D 1267 tests if it can withstand the requisite 1000 psig pressure test without permanent distortion. The D 323 apparatus corresponds to D 1267 apparatus with a *twenty percent chamber*.

The test is carried out essentially the same as it would have been performed with D 323 apparatus and a two–valve gasoline chamber, but the results are taken at the *temperature of measurement*, corrected for the calibration of the pressure gage, and recorded as the *corrected vapor pressure at T °F*. This value is then corrected for barometric pressure by the following.

$$\text{LPG gas vapor pressure} = (\text{corrected vapor pressure, psi}) - (760 - P_1) \times 0.01920 \quad (5\text{-}9)$$

The corrected value is called *LPG–gas vapor pressure at T °F*.

Again note that the temperature of measurement *must* be included to completely specify a *LPG–gas vapor pressure* measurement. This is *not required* for *RVP vapor pressure*; RVP measurements are taken only at 100°F, and this temperature is implicit in the definition.

Vapor pressure measurements are not very precise. Repeatability for the RVP test is 0.25 psi for 0–15 psig gages, 0.5 psig for 0–30 or 0–45 psig gages; and the corresponding reproducibility is 0.55 psi and 0.8 psi. Repeatability for LP–gas vapor pressure is 1.8 psi, and reproducibility is 2.8 psi. And, these relatively imprecise measurements must be considered in the light that the tests, themselves, are difficult to perform consistently and correctly, and that in the case of complex samples, such as crude oils, the tests may be carried out correctly yet still yield a vapor pressure that is far removed from the *true vapor pressure*. The effect of this aspect of the tests will be considered later on herein.

Vapor Pressure of Petroleum Fluids

Table 5-1 gives RVP values for several miscellaneous crude oils. These values have been taken at random from various publications. There are commercially available listings of the important parameters of crude analyses, including the RVP, available at low cost.

Table 5-2 is a similar table for typical product specifications; the source is given for each measurement.

These are for information and conceptual calculations, and not for design work. If you want a value you can trust, do your own work. And note, however, that your sample of a given crude may not be the same as someone else's sample; your testing lab may perform the test a little differently than someone else's lab; and your lab's equipment may be well kept while that of others may be pretty well knocked about; and that, if you really want to know what the RVP of a crude or a product is, supervise the taking of your own samples, and then pick a good laboratory that plays by all of the technical rules and has enough business to keep its equipment in good operating order. Otherwise, you may be stuck with a bad RVP, and that can cause endless trouble.

Measurement of Flash Point

There are several procedures in use around the world for measuring the *flash point* of volatile fluids. Each of these involves inserting a sample in a small vessel, called the *cup*; heating the cup and sample at a controlled rate; and determining the lowest temperature at which a flame (the *flash*) appears when a small igniting flame is inserted for an instant (about one second) in the vapor space above the sample. If the cup is enclosed, the test is called a *closed cup* test; if it is not, it is called an *open cup* test; no other description of the difference between these two broad classes of the test is required.

The closed–cup tests are usually used for measurement of flash point, and are more and more dominating this class of test. The open cup tests are more often used for go–no go testing, and for measurement of fire point.

The most commonly used apparatus are the *TAG*[7] and *Pensky–Martens*[8] closed–cup testers, the *Cleveland*[9] and *TAG*[10] open–cup testers, and the *Setaflash*[11] closed tester. The Pensky–Martens tester can be equipped with an open top for its cup and so can be used for open cup tests as well. In the U.K., an old test called *Flash Point by the Abel Apparatus*[12] is the legal go–no go test to determine if the sample has a closed–cup flash point below 73°F, the temperature stipulated in the *Petroleum (Consolidation) Act, 1928,* below which a fluid *shall not give off an inflammable vapor*. The Setaflash test, which uses the Setaflash apparatus (Setaflash is a registered trademark of Stanhope–Seta Limited), is a modern test which uses a series of very small samples—one for each trial temperature—and is excellent for go–no go testing, though true flash point testing can be, of course, another aim of the test.

As might be expected from the descriptions, the repeatability of these methods is not very good, and the reproducibility is worse.

- The Pensky–Martens and TAG closed–cup tests are repeatable to within 2°F for flash points below 140°F and to within 3°F for flash points

Table 5-1
Reid Vapor Pressures for Typical Crude Oils

Field	Country	Density 15°C	RVP psia
Hassi Messaoud	Algeria	0.800	11.0
Qatar Export	Qatar	0.823	7.6
Edjeleh	Algeria	0.845	4.0
Kirkuk	Iraq	0.844	4.5
Agha Jari	Iran	0.852	9.0
Zelten	Libya	0.830	4.8
Gach Saran	Iran	0.872	7.5
Forties	North Sea	0.840	5.1
Mubarek	Sharjah	0.840	4.8
Ninian	North Sea	0.849	5.4
North Rumaila	Iraq	0.853	7.0
North Slope	Alaska, USA	0.894	2.9
Berri	Saudi Arabia	0.831	5.5
Saudi Arabian Heavy Blend	Saudi Arabia	0.886	8.5
Saudi Arabian Light Blend	Saudi Arabia	0.858	4.2
Iranian Heavy Blend	Iran	0.872	6.6
Iranian Light Blend	Iran	0.858	6.5
Arjuna	Indonesia	0.836	7.1
Attaka	Indonesia	0.810	9.8
Basrah	Iraq	0.856	3.1
Boscan	Venezuela	0.998	1.0
Brass River	Nigeria	0.811	7.5
Brega	Libya	0.823	4.8
Safaniya	Saudi Arabia	0.894	7.0
Kuwait Export Blend	Kuwait	0.868	9.5
Cactus Reforma	Mexico	0.860	3.3
Ratawa	Neutral Zone	0.913	3.0
Redwater	Canada	0.850	7.9
Stettler	Canada	0.889	3.5
Lirik	Indonesia	0.855	1.0
Sprayberry	Texas	0.833	6.6
East Texas	Texas	0.834	6.0
Seeligson	Texas	0.823	2.2
Oklahoma Blend	Oklahoma	0.826	6.1

between 140 and 199°F. Reproducibility is 6°F for flash points below 55°F, 4°F for the range 55 to 139°F, and 6°F for flash points above 140°F and below 199°F.

- The Cleveland open–cup tests are repeatable only to within 15°F (for both flash and fire points), and can be reproduced only to 30°F for flash point and 25°F for fire point.
- The Setaflash tester, however, is repeatable and reproducible to within much smaller limits for low flash points, but both repeatability and reproducibility deteriorate at higher temperatures and for flash points above 150°F are not as reliable as the closed–cup tests. See the following.

Setaflash Apparatus
Repeatability and Reproducibility

Temperature °F	Repeatability °F	Reproducibility °F
20	0.5	1.4
70	0.5	2.9
93	1.3	4.9
150	2.0	7.5
200	2.6	9.9
260	3.3	12.4

The Setaflash is very accurate for low flash points, but the old Pensky–Martens closed–cup test, which is the most–used test in the petroleum industry, is almost as good in the range around 100–150°F, and the reproducibility is better. The Pensky–Martens closed–cup test is the standard used by the U.S. Department of Transportation (DOT), and the IOTTSG[13] bases all its classifications on closed–cup tests without specifying the precise type of equipment. And, in this regard, note that the *Institute of Petroleum* in IP 304[14] provides a method of test acceptable for use with any of the Abel (IP 170), Abel–Pensky (DIN 51755), Pensky–Martens (IP 34, ASTM D 93), or TAG (ASTM D 56) closed cups.

The Pensky–Martens closed–cup tester apparatus even has its own ASTM *Standard*[15] apart from the other standards describing its proper use. Figure 5-3 is from ASTM E 134, and describes the cup—the most important part of the apparatus—showing the thermometer position, the stirrer, and the flame exposure device. This device is shown in the measuring position and the shutter is, of course, open. When a measurement is not being taken, the shutter is closed and the flame exposure device lays horizontally, flame still burning, on top of the shutter plate.

Table 5-2

Limiting Reid Vapor Pressures and Flash Points Typical Product Specifications

Specification	Max RVP psia	Min Flash Point °F
ASTM D 396 Fuel Oil		
No. 1		100
No. 2		200
No. 4		130
No. 5 (light)		130
No. 5 (heavy)		130
No. 6		150
ASTM D 439 Automotive Gasoline		
Class A	9.0	
Class B	10.0	
Class C	11.5	
Class D	13.5	
Class E	15.0	
ASTM D 910 Aviation Gasoline		
Grade 80–87	7.0	
Grade 100–130	7.0	
Grade 115–145	7.0	
ASTM D 975 Diesel Fuel Oil		
No. 1–D		100
No. 2–D		125
No. 4–D		130
ASTM D 1655 Aviation Turbine Fuel		
Jet A		105
Jet B	3.0	
ASTM D 1835 L P Gas		
Commercial Propane	208	
Commercial Butane	70	
Commercial PB Mixture	<200	

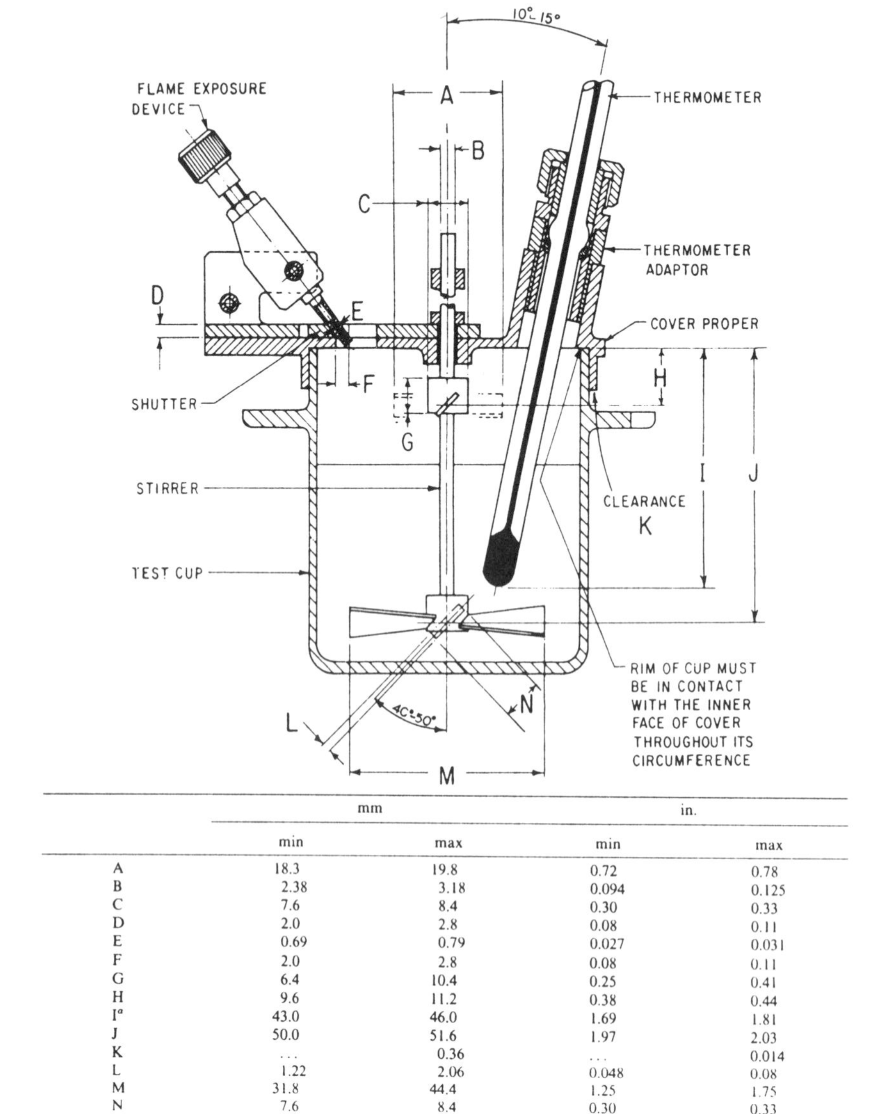

	mm		in.	
	min	max	min	max
A	18.3	19.8	0.72	0.78
B	2.38	3.18	0.094	0.125
C	7.6	8.4	0.30	0.33
D	2.0	2.8	0.08	0.11
E	0.69	0.79	0.027	0.031
F	2.0	2.8	0.08	0.11
G	6.4	10.4	0.25	0.41
H	9.6	11.2	0.38	0.44
I[a]	43.0	46.0	1.69	1.81
J	50.0	51.6	1.97	2.03
K	...	0.36	...	0.014
L	1.22	2.06	0.048	0.08
M	31.8	44.4	1.25	1.75
N	7.6	8.4	0.30	0.33

Figure 5-3. Pensky–Martens design of closed cup tester.

Flash Point of Petroleum Fluids

The flash point of crude oils is very rarely taken; the RVP is routinely measured as part of a crude analysis, and this is enough for most purposes; the uninformed public rarely gets its hands on crude oil and so doesn't need the assurance that the crude is apt to ignite at some low temperature, which it probably would.

Flash points are routinely taken on any fluid, including petroleum fluids, that is expected to reach the hands of the public. This process probably commenced in the *Oil for the Lamps of China* days when kerosene was routinely sold in tin cans which, if not carefully tended, emitted vapors which ignited from the flames of the lamps they filled. The lowest *D-number*, the number issued by ASTM *Committee D-2 on Petroleum Products and Lubricants*, in the 1983 *ASTM Annual Books of ASTM Standards* series, is D 56, *Flash Point by TAG Closed Tester*, which says something about the very senior nature of this test.

The exceptions to all of the above are the light refined fractions, the gasolines, liquefied gases and, sometimes, jet fuels. These kinds of products, *all of which will flash at almost any temperate atmospheric temperature*, are ranked by RVP instead of flash point.

Table 5-2 gives minimum flash points (and maximum RVP values) for several kinds of commercial refined products based on certain ASTM standard specifications.

Household kerosene is not included in the table, but it is covered by ASTM D 3699,[16] which specifies a minimum flash point of 100°F, still measured by ASTM D 56.

Matters of Safety

The break–point, commercially, between *hazardous* and *non–hazardous* fluids, seems to be a closed–cup flash of 61°C (141.8°F); fluids with flash points higher than this are usually, in commercial terms, classified as *non–hazardous*.

There are other measures that intrinsically involve the vapor phase that bear on the matter of safety. One of these is the *autoignition temperature* or, sometimes, the *autogenous ignition temperature*, which is the lowest temperature to which a solid, liquid, or gas is required to be raised to cause self–sustained combustion without initiation by spark or flame. For liquids, ASTM E 659[17] provides a procedure to determine the *hot flame autoignition temperature* (AIR), which is the temperature at which a yellow, red, or blue flame appears and temperatures in the test flask rise several hundred degrees; the *cool flame autoignition temperature* (CFT), which is the temperature at which a cool, blue flame, which sometimes can only be seen in a dark environment, appears, and temperatures in the test flask rise less than 100°C; and the *reaction threshold temperature* (RTT), which is the temperature at which very weak temperature increases, not accompanied by visible flames, occur.

Autoignition temperature is not important for most routine operations with petroleum fluids except in those instances where *pyrophoric* compounds of sulfur, oxygen and iron form on the roof members of fixed roof tanks, then loosen and fall, while glowing, to the floor of an essentially empty tank. No spark or flame is present, but a tank fire can occur.

Mixtures of air and petroleum vapors may be too *lean* to ignite, too *rich* to ignite, or just right to ignite and explode. The two limits are called the LEL (*Lower Explosive Limit*) and UEL (*Upper Explosive Limit*) or, sometimes, LFL and UFL, where the *F* instead of *E* stands for *flammable*.

An excellent treatment of the problem of calculating the flammability limits of mixtures is contained in Heffington and Gaines.[18] Their work provides ways to calculate the LEL and UEL and, the inverse, how to form a flammable mixture with a combination of a combustible and an inert gas.

They cite the empirical method called *Le Chatelier's Law*, referenced to two *Bureau of Mines* publications, which in equation form is as follows.

$$L = \frac{100\%}{\sum_{i=1}^{n} \frac{C_i}{L_i}}, \quad \sum_{i=1}^{n} C_i = 100\% \tag{5-10}$$

where C_i = vol% of *ith* combustible gas
L_i = LEL of *ith* combustible gas as vol%
n = total number of combustible gases present

The same form of equation can be used to calculate the UEL of a mixture by substituting the individual gas UEL values for L_i.

Following are LEL and UEL values for common combustible gases.

Gas	LEL Vol%	UEL Vol%
CH_4	5.0	15.0
C_2H_6	3.0	12.4
C_3H_8	2.1	9.5
C_4H_{10}	1.8	8.4
CO	12.5	74.0
H_2	4.0	75.0

In a tank containing a fluid with a very low vapor pressure (high flash point), the vapor space may be everywhere too *lean*; in tanks containing a high vapor pressure (low flash point), especially after setting still long enough for the vapors to cease building and reach equilibrium, the vapor space may be everywhere too *rich*; but in essentially all other cases there is at least a part of the vapor space which is above the LEL, and below the UEL, and is, therefore, explosive. The early JP–4 was this kind of fuel; under *almost* any condition, whether standing still or being filled or emptied; there was at least *part* of the vapor space that was explosive, and the number of explosions of fixed roof tanks of JP–4 is witness to this.

An excellent discussion of safety matters dealing with explosive mixtures is in Chapter II, *The Evolution and Dispersal of Hydrocarbon Gas on Tankers*, of reference [13]. Though written to be directly applicable to oil tankers, this reference is very applicable to all operations having to do with the storage of petroleum fluids. The reference also, in Chapter VI, carries a description of four *atmospheres* which can exist in a still, empty tank, these being

- *Atmosphere A*–an atmosphere which is not controlled; and thus can be above, below, or within the flammable range.
- *Atmosphere B*–an atmosphere made incapable of burning by the *deliberate reduction of the hydrocarbon content* to below the LEL. ...the reading given by an approved combustible gas indicator *should not exceed 50% of the LEL.*
- *Atmosphere C*–an atmosphere made incapable of burning by the introduction of an inert gas and the *resultant reduction of overall oxygen content*. In this connection the *oxygen content of the tank atmosphere should not exceed 8% by volume*.
- *Atmosphere D*–an atmosphere made incapable of burning by *deliberately maintaining the hydrocarbon content of the tank above the UEL*. For crudes and most liquid petroleum products, a *hydrocarbon content of at least 15% by volume* should be attained before work is done in the tank.

True Vapor Pressure

The point has been made earlier that the *correct* measurement of vapor pressure is a difficult, and tedious, procedure.

The RVP method of testing has its advantages: it is *simple*; it is *standardized*; and it is *accepted* world–wide as a measure of the volatility of one fluid as compared to that of another similar fluid. As a singular value, however —as the *vapor pressure* of a fluid—it sometimes just does not truly reflect the real world.

There are several reasons why the RVP method is not completely satisfactory for pipeline work.

Firstly—RVP is, by definition, taken at 100°F; and while this is a standardized temperature most pipeline problems are not based on this temperature and all RVP values must be translated to some other temperature.

Secondly—the method, while being simple, requires a certain dexterity in its performance, and the dexterity of testing personnel varies as among them.

Thirdly—the apparatus itself is notoriously subject to small damages; and while most of these would not be important if larger samples were being handled, tiny scratches in the mating surfaces of the connections between the upper and lower chamber, or in the threads of the screwed openings, or in any of the valves, can have a major effect of the measurement.

The greatest problem, however, lies in the assumption on which the validity of the measurement is based: that *the composition of the liquid remaining in the lower chamber after the upper chamber is filled with vapor is the same as it was at the start of the test.*

The RVP test will give excellent results on pure liquids, and the RVP may for all practical purposes be taken as the TVP, because after the partial vaporization of a pure liquid is completed that part which was not vaporized remains as it was before vaporization commenced, and the equilibrium between the liquid–phase and the vapor–phase is that of a pure vapor to a pure liquid of the same composition.

For most mixtures of hydrocarbons the assumption of constant composition falls down; and the composition of the vapor in the upper chamber may be markedly different from that of the liquid remaining in the lower chamber. For instance, a mixture of a fraction of one percent of propane, which has a vapor pressure of 184 psia at 100°F, with 99+% of octane, which at the same temperature has a vapor pressure of 0.54 psia, could conceivably be so constituted that at the end of the vaporization nearly all of the octane remained in the lower chamber while nearly all of the propane would be in the upper chamber, and the equilibrium so established would be for a mixture of propane gas, water vapor, and air, in the upper chamber, vis–à–vis essentially pure liquid octane, in the lower chamber. The assumption of constant composition would be completely negated, and the RVP value produced would be meaningless.

For narrow–range hydrocarbon mixtures the ratio of TVP to RVP can be estimated fairly accurately. Nelson[19] reported TVP and RVP values for a large set of samples of several kinds of petroleum fluids.

Table 5-3 is my recasting of one part of Nelson's data. Nelson gives no source for the natural gasoline data; the data for the refined gasolines is from an API oil–loss symposium.[20]

The natural gasolines have a spread of RVP from 24.0 maximum to 12.0 minimum, exactly 2:1. The average TVP/RVP ratio to three decimal places is 1.084, and the maximum deviations are +0.056, −0.054. These deviations, if applied to the average RVP in the group, would yield maximum deviations of +1.05, −1.01 psia,

Table 5-3

True Vapor Pressure and Reid Vapor Pressure Miscellaneous Volatile Products

Product	Vapor Pressure psia		Ratio TVP/RVP
	RVP	TVP	
Natural Gasolines			
No. 1	24.0	26.2	1.09
No. 2	22.5	24.2	1.08
No. 3	22.0	23.8	1.08
No. 4	20.4	22.2	1.09
No. 5	20.0	21.8	1.09
No. 6	19.7	20.3	1.03
No. 7	18.8	19.7	1.05
No. 8	18.4	20.1	1.09
No. 9	18.3	20.1	1.10
No. 10	18.0	19.5	1.08
No. 11	17.8	20.3	1.14
No. 12	16.0	17.5	1.09
No. 13	14.0	15.3	1.09
No. 14	12.0	12.9	1.07
Refined Gasolines			
No. 1	18.0	20.0	1.11
No. 2	16.0	17.8	1.11
No. 3	14.0	15.4	1.10
No. 4	12.0	12.9	1.07
No. 5	10.0	10.4	1.04
No. 6	8.0	8.3	1.04
No. 7	6.0	6.3	1.05
No. 8	5.0	5.2	1.04
No. 9	4.0	4.2	1.05
No. 10	3.0	3.1	1.03
No. 11	2.0	2.1	1.05

both of which are outside the range of acceptable values. However, if the two extremes are eliminated from the set, the average TVP/RVP ratio becomes 1.083—practically the same as before—but the maximum deviations become +0.017, –0.033; and these, applied to the average RVP value, yield deviations of +0.32, –0.62 psia. These deviations are outside the range of *repeatability*, but within the range of *reproducibility*, of ASTM D 323, and so it can be taken with reasonable assurance that the TVP of a natural gasoline is (about) 1.08 times its RVP.

The refined gasolines fall into two groups. The three gasolines with the highest RVP, two of which fall outside the range of ASTM D 439 for automotive gasolines, have a TVP/RVP ratio of about 1.10–1.11. The remainder, all but one of which fall in the ASTM Class C or lower RVP range, have a TVP/RVP ratio spread of 1.03–1.07. The average TVP/RVP ratio of this group is 1.046, with maximum deviations +0.024, –0.016. These deviations when applied to the average RVP of the group yield maximum deviations of +0.1 0.10 psia, within the repeatability limits of ASTM D323 for 5 to 16 psia fuels, and within the reproducibility limits of 0 to 16 psia fuels. Thus the TVP for (most) gasolines can be taken as about 1.05 times the RVP of the fuel.

Nelson's data on crude oils are recast in Table 5-4. Not shown are data from miscellaneous sources reported by Nelson which showed TVP/RVP ratios as high as 9.75 for Qatif, Saudi Arabian, crude and as low as 1.0 for a 40.2°API Mid–Continent (USA) crude. However, the data from these sources are highly random; and whenever there is a check analysis available from one of the API measurements on the same crude, the ratio of the miscellaneous–to–API measurement is sometimes so large as to cast doubt on the validity of the miscellaneous data.

Table 5-4

True Vapor Pressure and Reid Vapor Pressure Miscellaneous Crude Oils

Origin	Vapor Pressure psia		Ratio TVP/RVP
	RVP	TVP	
Kansas No 1	1.0	2.5	2.50
Kansas No. 2	2.0	3.6	1.80
Montana	3.0	5.0	1.67
Saudi Arabia	4.0	6.4	1.60
Bahrain	5.0	7.6	1.52
Texas No. 1	6.0	8.9	1.48
Texas No. 2	9.0	13.1	1.46
Oklahoma No. 1	8.0	11.8	1.47
Oklahoma No. 2	7.0	10.4	1.48
Oklahoma No. 3	10.0	14.3	1.43

The only real anomaly in the set of ten data points in Table 5–4 is the first one. If this is discarded the average TVP/RVP ratio is 1.546 and the maximum deviations are +0.254, –0.166 which, applied to the average RVP yield deviations of +1.52, –0.70 psia, considerably in excess of the repeatability or reproducibility of ASTM D 323. There is no simple TVP/RVP relation for crudes.

Nelson suggested the following as a kind of rule of thumb tabulation of typical ratios; the range of the values emphasizes the rule of thumb character of the averages; they are *not* engraved in stone.

Product	TVP/RVP Avg	TVP/RVP Range
Crudes		
RVP < 4 psia	4.52	1.67–10.0
RVP > 4 psia	1.72	1.00– 3.2
Refined hydrocarbons and bottled LPG	1.40	1.17– 1.6
Natural gasolines	1.08	1.03–1.14
Refined gasolines	1.07	1.03–1.45

It should be noted that there is a definite qualitative—if not yet a perceived quantitative—relationship between the *complexity* of the fluid and the TVP/RVP ratio. Thus, the refined gasolines have a TVP/RVP ratio of about 1.05; the natural gasolines, relatively narrow spectrum but still not truly refined, run about 1.08; while for the complex mixtures that are crude oil the ratio runs in excess of 1.7.

Measuring TVP is a difficult task requiring complex apparatus. API Bulletin 2512[21] describes the modified Chenicek–Whitman vapor pressure apparatus used by the API *Evaporation Loss Committee* in their definitive determinations that certainly fits the description; it is very complicated, and the procedure is very difficult. Though primarily used by the *Committee* for the determination of evaporation losses from tanks, TVP is a direct product of the measurement.

The real key to obtaining the TVP of a liquid is to have a knowledge of its distillation pattern. This is recognized in several standards:

- ASTM D 2878[22] says:
  Note 1—Most lubricants boil over a fairly wide temperature range, and this fact must be recognized in discussion of their vapor pressures. For example, the apparent vapor pressure over a range 0 to 0.1 percent evaporated may be as much as *100 times* (my italics) of that over the range 4.9 to 5.0 percent evaporated.
- ASTM D 2879[23] has a similar comment:
  Note 2—Most petroleum products boil over a fairly wide temperature range, and this fact must be recognized in discussion of their vapor pressure. Even an ideal mixture following Raoult's Law will show a progressive decrease in vapor pressure as the lighter component is removed, and this is *vastly accentuated* (the italics are mine) in complex mixtures...Such mixture may well exert a pressure in a closed vessel as much as 100 times that calculated from its average composition...

The *Evaporation Loss Committee*, in API Bulletin 2513,[24] correlates RVP and TVP –vs– temperature using what is called the *Slope of the ASTM Distillation Curve at Ten Percent Evaporated*. The slope is defined as

$$S = \frac{\left({}^{\circ}F \text{ at 15 percent} - {}^{\circ}F \text{ at 5 percent}\right)}{10} \tag{5-11}$$

In this equation S, the slope of the distillation curve at 10% evaporated, is approximated by taking the difference over the ten percent spread from 5% evaporated to 15% evaporated and dividing by ten. This may not be the precise slope of the distillation curve at 10°F, but its probably pretty close.

The ASTM distillation curve is established by the procedures and apparatus of ASTM D 86.[25] The slope of the curve at ten percent evaporated is a measure of the light fractions in the mixed liquid and thus is another measure of low temperature volatility. Since the results of a distillation are not always available, API 2513 suggests the following values for S:

Product	RVP	S
Light naphtha	9–14	3.5
Motor gasoline		3.0
Naphtha	2–8	2.5
Aviation gasoline		2.0

The Evaporation Loss Committee published a set of nomograms in API 2513—since repeated in several API publications and elsewhere—which relate RVP to TVP for varying temperature with ASTM ten-percent slope as a parameter. Figure 5-4 is one of these, taken from API 2513; it covers the entire range from 1.8 to 25.0 psia RVP. There are three other nomograms in the *Bulletin* that divide the RVP range into 1–7 psia, 5–15 psia, and 12–20 psia. These give a better precision than Figure 5-4 but, considering comments above on measured values of RVP, probably not much better accuracy.

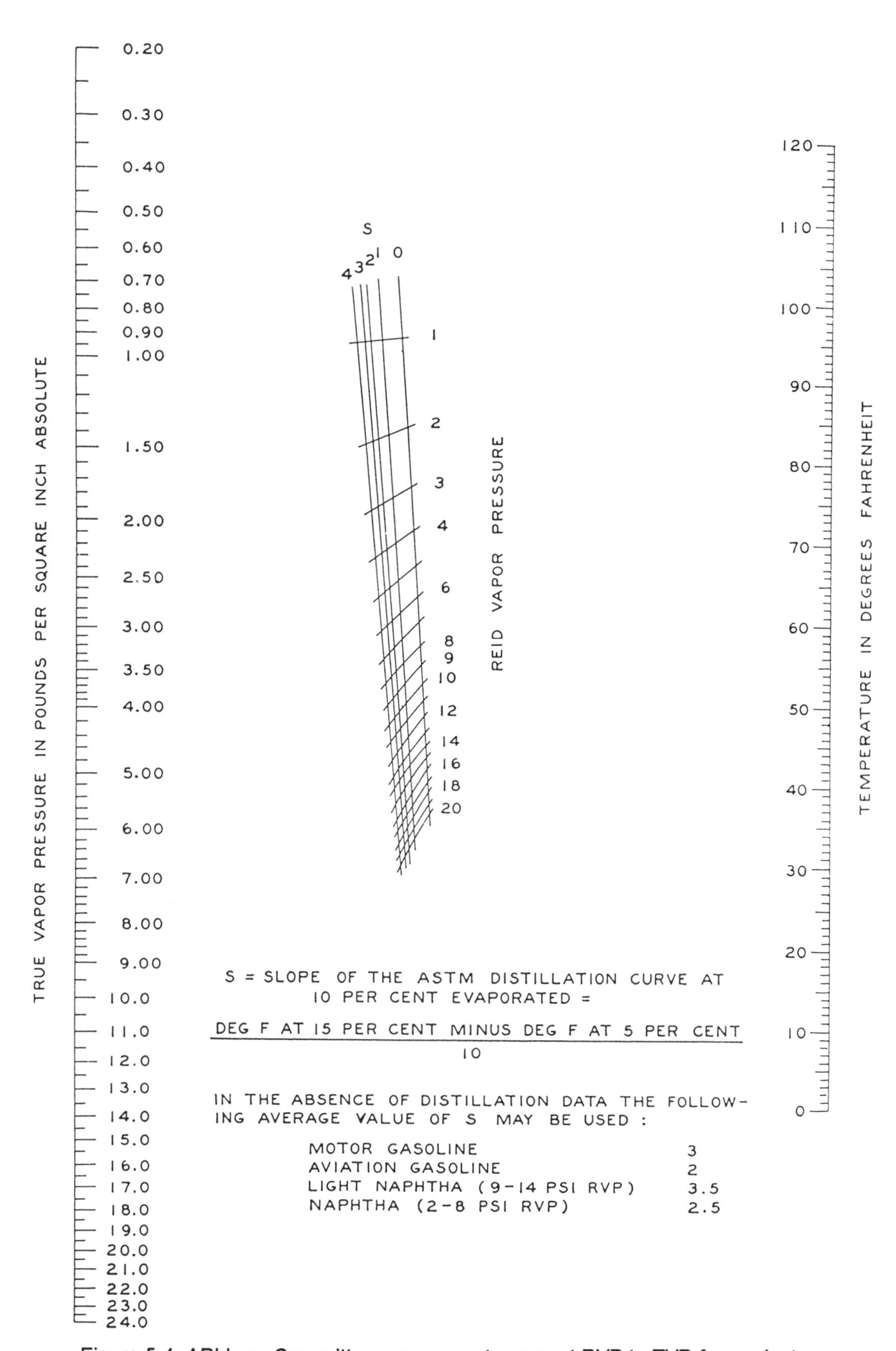

Figure 5-4. API Loss Committee nomogram to convert RVP to TVP for products.

Figure 5-5, also from Bulletin 2513, is for crude oil. The $TVP_{100°F}$–vs–RVP relation can be approximated by

$$TVP_{100°F} = 0.2858 + 0.8158 \times RVP + 0.0508 \times RVP^2 \qquad (5\text{-}12)$$

This equation is said to be valid within ±5% for RVP greater than 2.0 and less than 15.0 psia. Conversion to TVP at other temperatures may be calculated using equations already given.

Vapor Pressure of Blends

Theory

The thermodynamics of perfect gases has been well developed for many years, and Maxwell's kinetic theory of gases is over a century old. In addition, since the rapid rise of solid–state physics, especially in the electronics and ceramics field, there has been a recent emphasis on the theory of solids, and there is already a large body of literature on the subject. The study of liquids, however, has lagged the study of the other two fundamental states of matter, and it may therefore not be thought unusual that most theoretical treatments of the equation of state for liquids start with the assumption that a liquid is a melded solid or that it is a liquefied gas. However, what is of interest in the study of vapor pressure is the study of gas–liquid equilibrium; and, for pipeliners, only a small part of that large body of knowledge is of interest.

There are all kinds of relations between connected variables, usually called *laws*, that predict the effect of a change in an independent variable on the other variable or variables. These are usually qualified so that they apply only to *perfect* or *ideal* gases or liquids (or solids, but these are not of interest here). It is therefore of interest to look at the bases for these perfect or ideal fluids before proceeding to a consideration of real fluids.

Perfect gases are well defined for practical purposes by their *bulk properties*. The laws defining these properties are quite old.

Boyle's Law

For a given mass of gas at a fixed temperature the product of pressure and volume is a constant:

$$PV = A$$

Charles's Law

For a given mass of gas at constant pressure the volume increases with increasing temperature:

$$V = BT$$

Avogadro's Law

At the same temperature and pressure equal volumes of gases contain equal numbers of molecules; or, one mole weight of gas occupies the same volume as one mole weight of any other gas at the same temperature and pressure.

Avogadro showed that there are 2.73×10^{25} molecules per pound mole of any gas, and one pound mole of any gas occupies 379.4 cubic feet at 14.7 psia and 60°F.

Equation of State

The above three equations can be manipulated to yield the equation of state for perfect gases:

$$PV = nRT$$

where R = the gas constant
n = the number of moles

All of the above equations apply strictly only to perfect gases. Such gases have been defined in many ways, but the criteria of James Clerk Maxwell, who developed the *kinetic theory of gases*, still stand:

- The gas is composed of identical molecules.
- The molecules are infinitely small, and do not collide with each other.
- The molecules exert no forces on each other.
- The molecules are in random motion; and their collisions with the walls of the containing vessel are perfectly elastic.

There is, of course, no perfect gas; and while many gases, at atmospheric temperatures and pressures, exhibit properties which approximate those a perfect gas would have, at high temperatures and pressures all gases depart from the perfect gas laws. Petroleum gases begin to depart markedly from the perfect gas laws at about 70

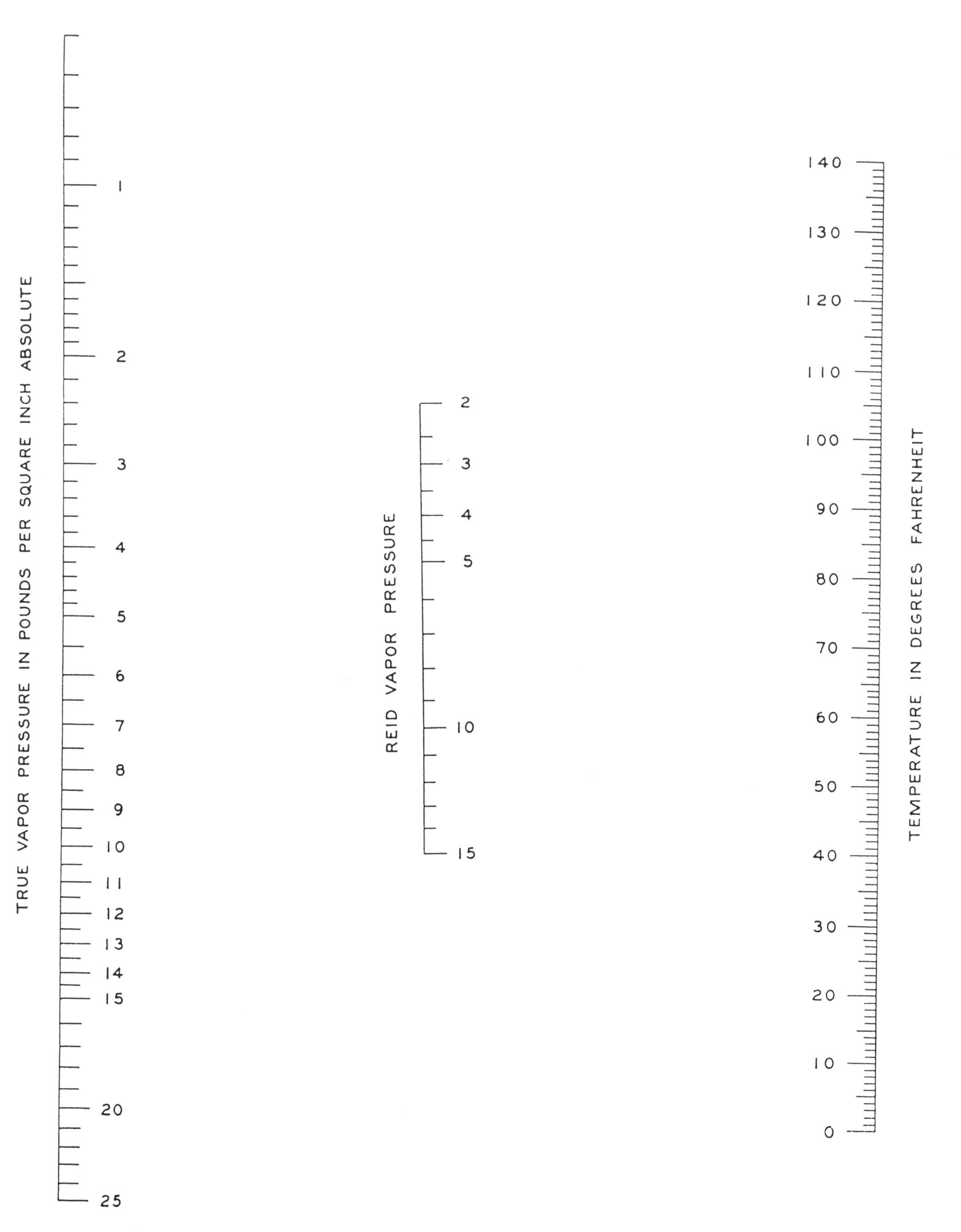

Figure 5-5. API Loss Committee nomogram to convert RVP to TVP for crude oils.

psia, so dealing with petroleum fractions no lighter than butane is not too complicated.

There are two further *laws* that are of interest in the study of vapor pressure. These are those due to Dalton and Amagat.

Dalton's Law

This is sometimes called the *Law of Partial Pressures*, and states that in a mixture of gases each component gas exerts a pressure equal to the pressure it would exert if it alone occupied the volume occupied by the mixture.

Amagat's Law

Amagat's Law, which is usually called the *Law of Additive Volumes*, says that the total volume occupied by a mixture of gases is equal to the sum of the volumes each of the component gases would occupy at the same temperature and pressure.

The first four of the principle basic laws were first discovered experimentally, though all of them can be derived theoretically from the kinetic theory of gases. The other two laws were derived from the first four.

Dalton's Law is useful in computing the pressure exerted by mixtures of petroleum vapors. In simplest mathematical form this law may be stated as follows:

Consider a mixture containing n_1 moles of gas_1, n_2 moles of gas_2, ...and n_j moles of component gas_j. The three components would have partial pressures respectively of

$$n_1(RT/V), n_2(RT/V), \ldots n_j(RT/V)$$

and the total pressure exerted on the walls of the containing vessel would be

$$p_t = p_1 + p_2 \ldots + p_j = n_1(RT/V) + n_2(RT/V) \ldots + n_j(RT/V)$$

Or, in general form:

$$p_t = \sum_{n=1}^{j} n_j(RT/V) \tag{5-13}$$

From Equation 5-13 can be derived the inverse form of Dalton's Law: The partial pressure exerted by a component gas is the product of the mole fraction of the component and the total pressure exerted by the mixture.

It is emphasized, again, that the relationships set down above apply only to perfect gases, or to real gases under pressure and temperature conditions such that their action closely approximates that of perfect gases. For real gases under higher pressures, or at higher temperatures, it is necessary to modify the equation of state to accommodate the peculiarities of real gases which, unfortunately, varies as pressure, temperature, and composition vary.

One way of doing this is to write the equation in terms of a power series in v or $1/v$

$$Z = \frac{Pv}{RT} = 1 + \frac{B}{v} + \frac{C}{v^2} + \frac{D}{v^3} + \ldots \tag{5-14}$$

where $B, C, D \ldots =$ virial coefficients
$Z =$ compressibility factor

Forgetting the power series, and taking K as a factor applicable to a specific gas at a specific pressure and temperature, the equation can be written

$$P = \frac{ZRT}{V} \tag{5-15}$$

The compressibility factor can be calculated in terms of reduced temperature T_r and pressure P_r, where $T_r = T/T_c$ and $P_r = P/P_c$, the subscript c indicating the value of T or P at the intersection of the critical temperature line and the critical pressure line on the P–T saturation curve for the fluid. Z is commonly read from graphs or charts, one of the most popular for hydrocarbon gases being the *Katz Chart*, after D. L. Katz who developed it.

Since having to go to the experimental data in Z–factor charts is, if nothing else, a nuisance, there have been many attempts to write an equation of state that is complete within itself. The most common, no longer much used, *generalized* equation of state for real gases is that due to van der Waal, which dates back to 1873. In pressure explicit form, this is

$$P = \frac{RT}{v - b} - \frac{a}{v^2} \tag{5-16}$$

b corrects for the volume occupied by the molecules
a/v^2 corrects for intermolecular attraction

One of the more elaborate empirical equations of state is the Beattie–Bridgeman equation with five constants that are obtained from experimental data for the gas:

$$P = \frac{RT}{v} + \frac{\beta}{v^2} + \frac{\gamma}{v^3} + \frac{\delta}{v^4} \quad (5\text{-}17)$$

where $\beta = B_0 RT - A_0 - cR/T^2$
$\gamma = -B_0 bRT + A_0 a - B_0 cR/T^2$
$\delta = B_0 bcR/T^2$

Dalton's Law of Partial Pressures expresses the relationship between the vapor pressure of the components of a gaseous mixture and the total vapor pressure exerted by the mixture; and there is a companion relationship, called *Raoult's Law*, which expresses the total vapor pressure exerted by the components of a liquid in terms of the characteristics of the component liquids. Raoult's Law is, however, applicable only to a perfect gas or an ideal mixture—a solution, in fact—which must be assigned certain properties. McCain[26] lists these properties as follows:

> "An ideal liquid solution is a solution for which mutual solubility results when the components are mixed, no chemical interaction occurs upon mixing, the molecular diameters of the components are the same, and the intermolecular forces of attraction and repulsion are the same between unlike as between the like molecules.
>
> "Such an ideal solution will on mixing be neither endothermic nor exothermic, and the volume of the solution will be the sum of the volumes of its components."

It can be shown that the evaporating (vaporizing) tendency of a component of a solution varies directly with the exposed area of that component on the surface of the solution, thus being proportional to the mole fraction of the component in the solution, and to the internal energy (vapor pressure) of the component. This is Raoult's Law:

$$p_j = p_{vj} x_j \quad (5\text{-}18)$$

where p_j = partial vapor pressure in the solution
p_{vj} = vapor pressure of the component at the temperature of the solution
x_j = mole fraction of the component

This can be written in terms of the total pressure p_t to yield

$$\frac{p_j}{p_t} = \left(\frac{p_{vj}}{p_t}\right) x_j \quad (5\text{-}19)$$

However, from Dalton's Law,

$$\frac{p_j}{p_t} = y_j \quad (5\text{-}20)$$

where y_j = mole fraction of the component in the gas or vapor

This equation relates the mole fraction of a component j in the gas (vapor) and liquid phases of a gas–liquid system in equilibrium at constant temperature. If there are two components in a solution:

$$y_1 = \left(\frac{p_{v1}}{p_t}\right) x_1$$
$$y_2 = \left(\frac{p_{v2}}{p_t}\right) x_2 \quad (5\text{-}21)$$

For perfect gases and liquids these relationships can be extended indefinitely:

$$\begin{aligned} p_t &= p_t(y_1 + y_2 \cdots y_j) \\ &= (p_{v1}x_1 + p_{v2}x_2 \cdots p_{vj}x_j) \\ &= p_{vt} \end{aligned} \quad (5\text{-}22)$$

Stated in words, the pressure exerted by the vapor phase is equal to the pressure of the liquid phase; the pressure exerted by the components of the vapor phase is proportional to the mole fraction of the components in such phase; and the vapor pressure exerted by the liquid phase is proportional to the products of the vapor pressures and the mole fractions of such components

If one set of mole fractions is known, the other set can be determined by making a material balance and solving the two equations simultaneously. One form of a material balance for n_T moles is

$$j_T n_T = x_j n_L + y_j n_G \tag{5-23}$$

where j_T = total moles of component j in the gas - liquid system
n_L = number of moles in the liquid phase
n_G = number of moles in the gas phase

All of this section of the theory of the vapor pressure of blends has dealt almost exclusively with *ideal* or *perfect* gases, or what is about the same thing, with real gases in the pressure–temperature conditions where they act as if they really were ideal or perfect.

For vapor pressure problems encountered in most liquid pipeline situations answers derived from the assumption of validity of the perfect gas laws are reasonable and useful. For higher pressures and temperatures p/p_v is not a constant; and one of the equations of state involving Z, or another complex representation of P–V–T relations, must be used to obtain useful answers.

Blends of Petroleum Fluids

Avogadro's Law states that equal volumes of gases at given temperature and pressure will contain equal numbers of molecules. One corollary of this is that equal volumes of gases at given temperatures and pressures will contain the same number of molecules, or moles, of a substance. This means that a volumetric analysis, wherein components are given as volume fractions of the total volume, can be carried out in the same way as if the fractions of the components were given in moles. For petroleum fluids, this is the save–all relationship insofar as the vapor pressure of mixtures is concerned because almost all blends of petroleum fluids are made up volumetrically.

As an example, suppose there are two equal volume containers X and Y into which are introduced equal volumes—less than the volume of X or Y—of fairly volatile liquids A and B; and suppose that after reaching equilibrium the pressure of A is 20 psia and that of B is 8 psia. T is equal, and because so little of A or B has been vaporized, V can be considered as being equal. Then suppose the two containers are connected and left until equilibrium is reestablished at the same T. The following can be derived.

I Final Vapor Pressure of Mixture

For A: $20\,psia \times \dfrac{X}{X+Y} = 20 \times 0.5 = 10\,psia$

For B: $8\,psia \times \dfrac{Y}{X+Y} = 8 \times 0.5 = 4\,psia$

For the (A + B) system $= 14\,psia$

II Final Composition of Vapor

For A: $\left(\dfrac{10}{14}\right) \times 100 = 71.43\%$

For B: $\left(\dfrac{4}{14}\right) \times 100 = 28.57\%$

Thus a liquid–liquid 50–50 mixture produced a 71–29 vapor mixture; and if the vapor were extracted and condensed it would contain 71 mole percent of A and 29 mole percent of B.

The example just given was for partial vapor pressure of vapors. A more common problem is to determine the vapor pressure of blends of liquids that do not vaporize because they are held at pressures above the vapor pressure of the blend. This problem requires measuring, or guessing at, the molecular weight of the component liquids.

Evans[27] illustrated the problem of working with two– and three–component blends of LPG and NGL with crude oils. Table 5-5 is Evans's tabulation of specific gravity, RVP, and molecular weight for three crudes, two LPGs, and one natural gasoline.

Table 5-5

Typical Parameters for Blending Calculations

Component	sp gr	RVP	Mol Wt
Crudes			
West Texas	0.835	6.0	230
Golden Trend	0.817	6.8	220
Scurry	0.818	8.0	220
Natural Gas Liquids			
Propane	0.508	190.0	44.1
iso-Butane	0.563	73.5	58.1
n-Butane	0.584	52.0	58.1
gasoline	0.628	26.0	73.4

Table 5-6 illustrates a sample problem included by Evans which involves blending 28 000 bpd of crude with 6000 bpd of natural gasoline and 6000 bpd of normal butane. The order of calculation is

1. bpd	given
2. vol%	bpd of component/bpd of mix
3. lb moles	8.33×spgr×vol% comp/mol wt comp
4. mole%	lb mol of comp/lb mol of mix
5. RVP	given
6. partial RVP	mol% of comp/RVP of comp
7. total RVP	sum of partial vapor pressures

The method can be easily programmed; and, if the same two or three products are involved in a routine daily blending problem—blending against a RVP limitation in a pipeline tariff or in a tanker charter party are good examples—graphs can be prepared that will allow quick estimates of the blending ratios of the products to meet the imposed limits.

Figure 5-6, for instance, is an example of a graph for two-component blending of propane, butane, or natural gasoline with 5 psia RVP crude; the dotted lines illustrate blending 60% crude with 40% butane to yield a 34 psia RVP mixture.

Figure 5-7 contains curves for three-component blends of crude, butane, and natural gasoline. The dotted lines illustrate mixing 30% NGL, 30% butane, and 40% crude to yield a blend with a RVP of 32+ psia.

Flash Point of Blends

Blends of Petroleum Fluids

Flash points can be predicted by the method of Butler, Cooke, Lukk and Jameson[28] as reported, and extended, by Lenoir.[29] The method is based on the fact that flash ignition occurs when a particular weight fraction of hydrocarbon is present in the vapor phase. This occurs for a single component when

$$MP_v = 15.19 \tag{5-24}$$

For mixtures substitute the following for MP_v.

$$\sum_{1}^{i} x_i M_i P_{vi} \tag{5-25}$$

where M = mol weight of component
P_v = vapor pressure of component
x_i = mol fraction of component in the liquid phase

The method gives good results with low or moderately volatile components, but for highly volatile components gives a flash point that is too low. The method is refined by Lenoir to use equilibrium ratios instead of vapor pressures and so modified will yield results generally within the limits of reproducibility of either the Pensky–Martens or TAG closed–cup testers.

Table 5-6
Calculation of Vapor Pressure of Three–Component Blend

Calculations	Average Crude	Natural Gasoline	Normal Butane	Total
Flow rates, bpd	28 000	6 000	6 000	40 000
Volume percent	70.00	15.00	15.00	100.00
Formula for lb–moles	8.33×0.835×70/210	8.33×0.628×15/73.4	8.33×0.584×15/58.1	
Lb–moles	2.319	1.069	1.256	4.644
Mole percent	49.9	23.0	27.0	100.0
RVP of component	5.0	26.0	52.0	
Partial vapor pressure	2.497	5.986	14.065	
Total vapor pressure				22.547

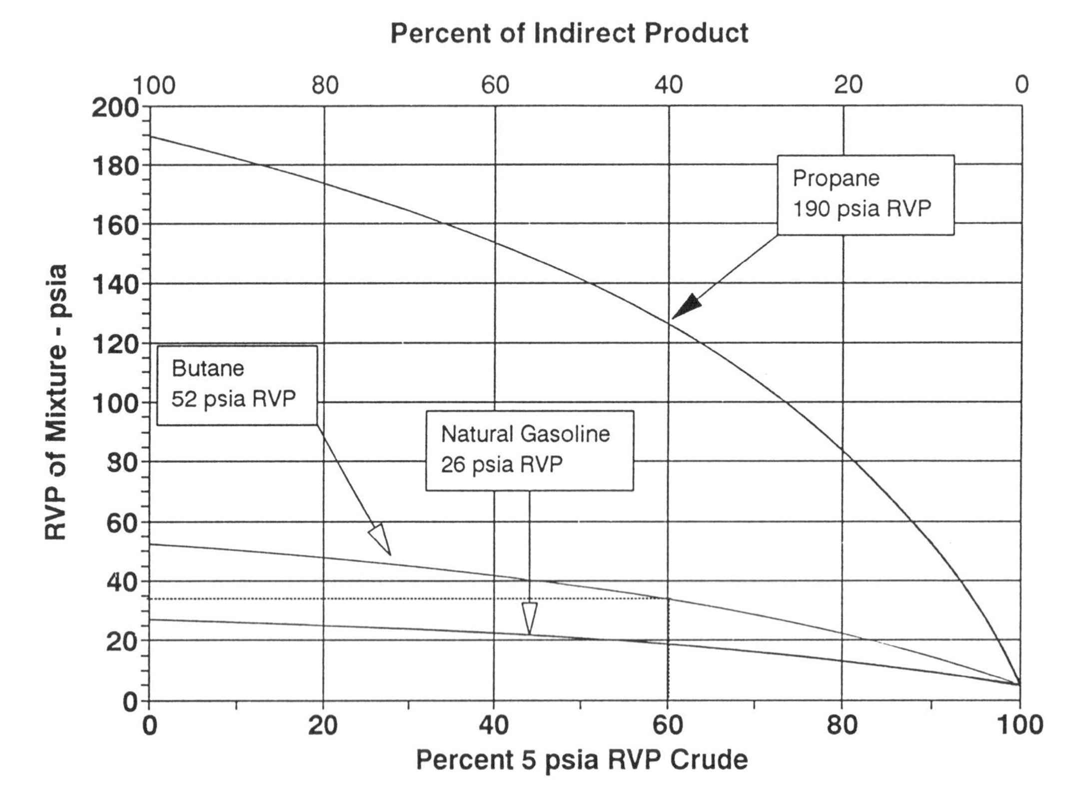

Figure 5-6. Blending chart for crude, butane, natural gasoline, and propane.

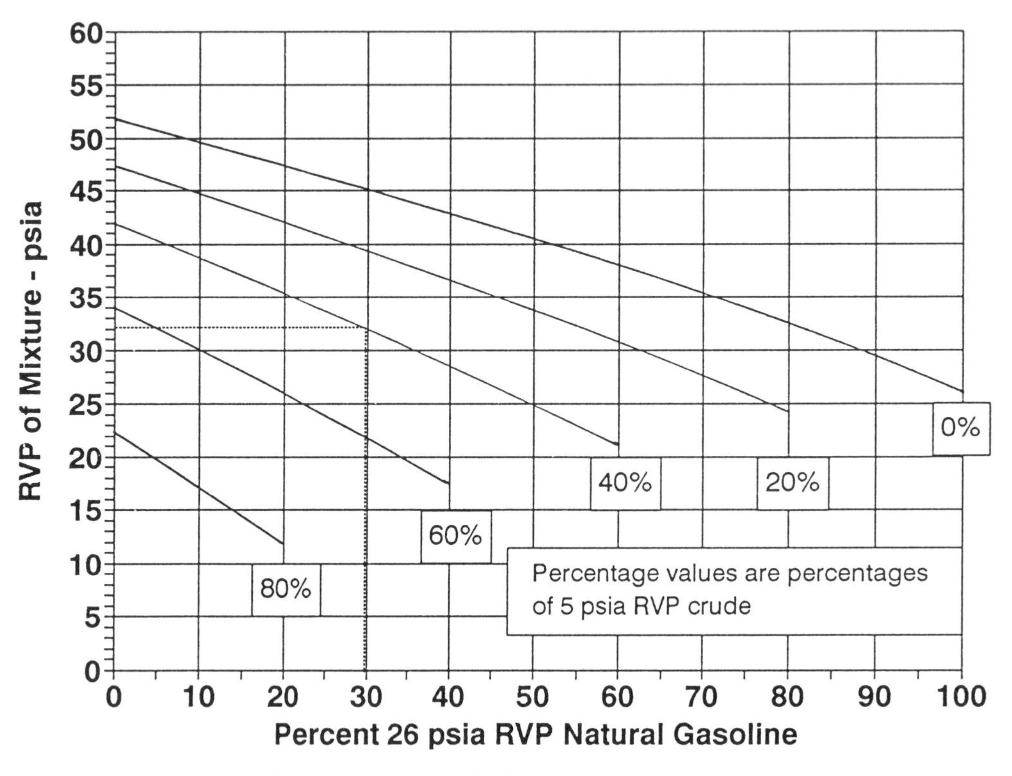

Figure 5-7. Graph for calculating three-component blends.

Ethyl Corporation[30] claimed a reasonable correlation using a formula of the form

$$T_b^{1/x} = V_1 T_1^{1/x} + V_2 T_2^{1/x} \ldots + V_n T_n^{1/x} \qquad (5\text{-}26)$$

where T_b = flash point of blend, °R
V_n = vol fraction of component n
T_n = flash point of component n
x = constant

The most likely value of x was found to be about –0.6, which yielded half of the predictions within ±4°F, which is the repeatability of the measurement. It was found that the value of x varied as among *refinery pools*, and therefore depended on feedstock as well as on refining processes; and that by customizing x, using past experience as a guide, it was possible to raise the accuracy of the predicted flash points from an average of 47%, using the general value of x, to 76%. The method is applicable to distillate fuel blends and similar.

There really are no good, simple, methods for predicting flash point. Nelson[31] gives two relationships between the flash point and T, the 0 to 10 percent boiling range (in °F) as;

For distilled fractions
$$\text{flash point} = 0.65 \times T - 100$$
For crude oils
$$\text{flash point} = 0.57 \times T - 110 \qquad (5\text{-}27)$$

Working Equations

Calculators

Table 5-7, Reference and Working Equations for Vapor Pressure, summarizes the two basic vapor pressure –vs– temperature equations, provides two equations having to do with Cox Charts—an approximate method of computing the effect of temperature on vapor pressure of hydrocarbons—and an approximate method for computing true vapor pressure from Reid vapor pressure at 100°F.

Table 5-7 also includes two approximation method of computing the vapor pressure and flash point of blends of hydrocarbons, as well as two formulas for predicting the flash point of hydrocarbons given the ASTM 10% boiling range.

Computers

CCT No. 5-1 computes the CB&I Cox Chart for hydrocarbons and, using *MathCad*, draws a curve of vapor pressure –vs– temperature. **CCT** 5-2 and 5-3 are rather complex templates that provide for computing the vapor pressure and flash point of blends of any reasonable number of components. **CCT** No. 5-4 is the $TVP_{100°F}$ –vs– RVP equation.

Two Personal Comments

Vapor pressure, and flash and fire points, are not very interesting to most pipeliners, but when you need the data you need a place to find it, and it is here and in the references in the end notes.

And as to needing it, I leave these personal experiences:

> I once had to junk 264 bid–package drawings of a large refined product storage project I had been called in to review, the majority of these being full–size, detailed, completed working drawings, because the system represented would not only not have worked well, it would not have worked at all. The designers—used to working with water—had not taken the relatively high vapor pressure of the products into account; and the combination of high vapor pressures, long pipe runs, and the topography of the site worked to produce a situation where no pump of any kind could have worked to boost the products from the tanks to the mainline pumps: The residual pressure at the suction of the tank booster pumps would have been a *negative absolute pressure*.

> I was called to visit a tank site immediately after a hardened (1 meter of reinforced concrete) cut–and–cover (2 meters minimum of earth cover) tank of JP–4 had exploded, taking the lives of 19 persons who were on the roof of the tank at the time. The concrete valve–shaft cover, about 3 meters square and 60 cm thick, was found some 400 meters away. I asked the guard what, in his opinion, had happened; there were no survivors to report. His reply was non–technical, but interesting; "I think somebody forgot he was standing on top of a tank of jet fuel." They truly were standing on a bomb.

Table 5-7

Reference and Working Equations for Vapor Pressure

Fundamental Equations

definition $$p = A \exp\left(\frac{-L}{RT}\right)$$

where p = vapor pressure
L = heat of vaporization per mole
R = gas constant
T = temperature, K or °R
A = a constant

Clausius–Clapeyron Equation

definition $$\log_e p = \left(\left(\frac{-L}{R}\right)\left(\frac{1}{T}\right)\right) + C$$

where C = a constant of integration

Cox Chart Equation–Nelson Version

definition $$\log P = -\left(\frac{A}{T+382}\right) + B$$

where P = vapor pressure, psia
T = temperature, °F
A and B = constants

Cox Chart Equation–CB&I Version

$$P_T = P_{100°F} \times 10^Z$$

where P_T = vapor pressure at T°F
$P_{100°F}$ = vapor pressure at 100° F
$$Z = 4 - \left(\frac{2240}{T+460}\right)$$

True Vapor Pressure-vs-Reid Vapor Pressure

$$TVP_{100°F} = 0.2858 + 0.8158 \times RVP + 0.0508 \times RVP^2$$

Vapor Pressure of Blends of Hydrocarbons

Evans Method

1	bpd	given
2	vol%	bpd comp/bpd mix
3	lb moles	8.33×spgr×vol% comp/mol wt comp
4	mole%	lb mol comp/lb mol mix
5	RVP	given
6	partial RVP	mol% comp/RVP comp
7	total RVP	sum of partial vapor pressures

Flash Point of Blends of Hydrocarbons

Ethyl Method

definition $$T_b^{1/x} = V_1 T_1^{1/x} + V_2 T_2^{1/x} \ldots + V_n T_n^{1/x}$$

where T_b = flash point of blend, °R
V_n = vol fraction of component n
T_n = flash point of component n
x = constant

Prediction of Flash Point of Hydrocarbons

Nelson Method

For distilled fractions
flash point $= 0.65 \times T - 100$
For crude oils
flash point $= 0.57 \times T - 110$

T=0 to 10 percent boiling range in °F

References

[1] Tabor, D. *Gases, Liquids, and Solids*. Hammond, Middlesex, England: Penguin Books, Ltd., 1969.

[2] Nelson, W. L. *Petroleum Refinery Engineering*, 3rd edition. New York: McGraw Hill Book Company, 1949.

[3] Technical Bulletin No. 20, *The Storage of Volatile Liquids*. Chicago: The Chicago Bridge and Iron Company, 1947.

[4] ASTM D 323, *Standard Test Method for Vapor Pressure of Petroleum Products (Reid Method)*. Philadelphia: American Society for Testing Materials.

[5] ASTM D 1267, *Standard Test Method for Vapor Pressure of Liquefied Petroleum Gas (LP–Gas Method)*. Philadelphia American Society for Testing Materials.

[6] *Thermodynamic and Transport Properties of Steam*. New York: American Society of Mechanical Engineers.

[7] ASTM D 56, *Standard Method of Test for Flash Point by TAG Closed Tester*. Philadelphia: American Society for Testing Materials.

[8] ASTM D 93, *Standard Test Methods for Flash Point by Pensky–Martin Closed Tester*. Philadelphia: American Society for Testing Materials.

[9] ASTM D 92, *Standard Test Method for Flash and Fire Points by Cleveland Open Cup*. Philadelphia: American Society for Testing Materials.

[10] ASTM D 1310, *Standard Test Method for Flash Point of Liquids by Tag Open–Cup Apparatus*. Philadelphia: American Society for Testing Materials.

[11] ASTM D 3828, *Standard Test Method for Flash Point by Setaflash Closed Tester*. Philadelphia: American Society for Testing Materials.

[12] IP 33, *Flash Point by the Abel Apparatus, Petroleum (Consolidation) Act 1928 Method*. London: Institute of Petroleum.

[13] Oil Companies International Marine Forum, *International Oil Tanker and Terminal Safety Guide*, 2nd edition. London: Applied Science Publishers, Ltd., 1974.

[14] IP 304, *Flash Test Using the Cup of Any Standard Closed Cup Apparatus*. London: Institute of Petroleum.

[15] ASTM E134, *Standard Specification for Pensky–Martens Closed Flash Tester*. Philadelphia: American Society for Testing Materials.

[16] ASTM D 3699, *Standard Specification for Kerosene*. Philadelphia: American Society for Testing Materials.

[17] ASTM E 659, *Standard Test Method for Autoignition Temperature of Liquid Chemicals*. Philadelphia: American Society for Testing Materials.

[18] Heffington, W. M., and Gaines, W. R. "Flammability Calculations for Gas Mixtures," *Oil and Gas Journal*, November 16, 1981.

[19] Nelson, W. L. "How Reid and True Vapor Pressures Vary," *Oil and Gas Journal*, 21 June 1954.

[20] "Symposium on Evaporation Loss," *Proceedings of the API*, Vol. 32, No. 1, 1952.

[21] API Bulletin 2512, *Tentative Methods of Measuring Evaporation Loss from Petroleum Tanks and Transportation Equipment*. Washington: American Petroleum Institute.

[22] ASTM D 2878, *Standard Method for Estimating Apparent Vapor Pressures and Molecular Weights of Lubricating Oils*. Philadelphia: American Society for Testing Materials.

[23] ASTM D 2879, *Standard Test Method for Vapor Pressure–Temperature Relationship and Initial Decomposition Temperature of Liquids by Isoteniscope*. Philadelphia: American Society for Testing Materials.

[24] API Bulletin 2513, *Evaporation Loss in the Petroleum Industry—Causes and Control*. Washington: American Petroleum Institute.

[25] ASTM D,86, *Standard Test Method for Distillation of Petroleum Products*. Philadelphia: American Society for Testing Materials.

[26] McCain, W. D. Jr., *The Properties of Petroleum Fluids*. Tulsa: The Petroleum Publishing Company, 1973.

[27] Evans, R. J. "Graphs Solve NGL–Crude Blending Problems," *Pipe Line Industry*, May 1958.

[28] Cooke, Lukk, and Jameson. "Prediction of Flash Points of Middle Distillates," *Industrial and Engineering Chemistry*, Vol. 48, 1946

[29] Lenoir, J. M. "Predicting Hydrocarbon Flash Points Accurately," *Hydrocarbon Processing*, January 1975.

[30] Ethyl Corporation, "Index Predicts Cloud, Pour and Flash Points in Distillate Fuel Blends," *Oil and Gas Journal*, November 9, 1970.

[31] Nelson, W. L. *Petroleum Refinery Engineering*, fourth edition. New York: McGraw–Hill Book Company, 1969.

Computer Computation Template

CCT No. 5-1: Convert vapor pressure at 100 °F to another temperature T °F. Based on Cox Chart equation for gasolines as published by Chicago Bridge and Iron.

References: Chapter 5. Equation 5-6. Figure 5-1.

Input

Factors $i := 0, 1 .. 10$

Input Data $P_{100F} := 34$ psia $VaporPressure_{100F} := P_{100F}$

Variables $T_i := 20 \cdot i$ $Z_i := 4 - \left(\frac{2240}{T_i + 460} \right)$

Equations

$$P_{T_i} := P_{100F} 10^{Z_i}$$

Results

T_i	Z_i	P_{T_i}
0	-0.870	4.6
20	-0.667	7.3
40	-0.480	11.3
60	-0.308	16.7
80	-0.148	24.2
100	0	34.0
120	0.138	46.7
140	0.267	62.8
160	0.387	82.9
180	0.500	107.5
200	0.606	137.3

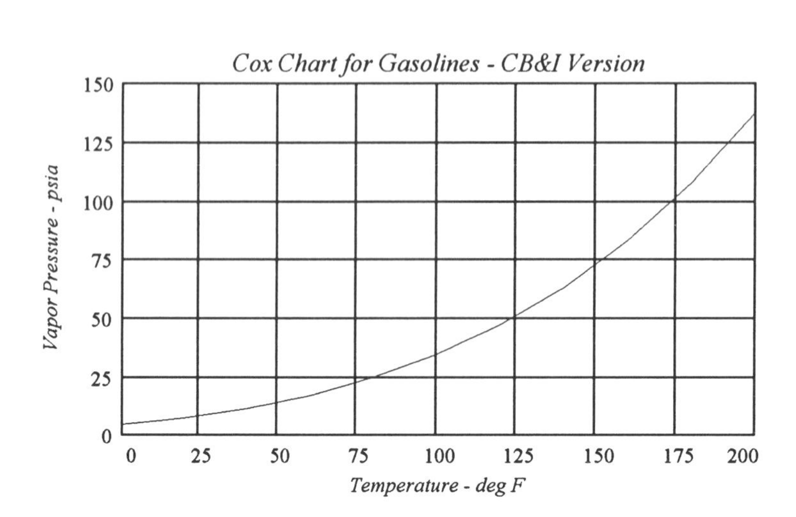

$VaporPressure_{100F} = 34$ psia

Computer Computation Template

CCT No. 5-2: Calculation of vapor pressure of blends using the Evans method. The example shown is for three components, but the procedure can be extended to any reasonable number of components; change the value of n and input data for all n components.

References: Chapter 5. Procedure under "Blends of Petroleum Fluids." Table 5-6.

Input

Define Scope Number of components: $n := 3$ $i := 1, 2..n$

Define Components

Volume of Component$_i$	RVP of Component$_i$	Mol Wt of Component $_i$	SpGr of Component$_i$
$Vol_1 := 28000$ bpd	$RVP_1 := 5.0$ psia	$MW_1 := 210$	$SpGr_1 := 0.835$
$Vol_2 := 6000$ bpd	$RVP_2 := 26.0$ psia	$MW_2 := 73.4$	$SpGr_2 := 0.628$
$Vol_3 := 6000$ bpd	$RVP_3 := 52.0$ psia	$MW_3 := 58.1$	$SpGr_3 := 0.584$

Equations

Variables

Vol total: $Vol_{Tot} := \sum_{i=1}^{n} Vol_i$

Lb Mol Component$_i$: $LbMol_i := \frac{\left[\left(\frac{Vol_i}{Vol_{Tot}}\right) \cdot 100\right] \cdot 8.33 \cdot SpGr_i}{MW_i}$

Lb Mol Total $LbMolTot := \sum_{i=1}^{n} LbMol_i$

Mol Pct Component$_i$: $MolPct_i := \frac{LbMol_i}{LbMolTot}$

Partial Pressure Component$_i$: $Pp_i := MolPct_i \cdot RVP_i$

Results

Vol_i	RVP_i	MW_i	$SpGr_i$	$LbMol_i$	$MolPct_i$	Pp_i
28000	5.0	210.0	0.835	2.319	0.499	2.497
6000	26.0	73.4	0.628	1.069	0.230	5.986
6000	52.0	58.1	0.584	1.256	0.270	14.065

RVP of Mixture: $RVP_{Tot} := \sum_{i=1}^{n} Pp_i$ $RVP_{Tot} = 22.547$ psia $Vol_{Tot} = 40000$ bpd $LbMolTot = 4.644$

Computer Computation Template

CCT No. 5-3: Calculation of flash point of blends using the Ethyl method. The example shown is for three components, but the procedure can be extended to any reasonable number of components; change the value of n and then input data for the n components. Note value for x is an average value; experimentally determined values may yield a better accuracy.

References: Chapter 5. Equation 5-26.

Input

Factors — Recommended value for x, lackinging experimental data $x := \frac{1}{-0.6}$

Define Scope — $n := 3$ $\quad i := 1, 2 .. n$

Define Components — Volume of Component$_i$ — Flash Point of Component$_i$

$Vol_1 := 10000$ bbl $\quad T_1 := 120$ °F

$Vol_2 := 3000$ bbl $\quad T_2 := 90$ °F

$Vol_3 := 3000$ bbl $\quad T_3 := 60$ °F

Variables $\quad Vol_{Tot} := \sum_{i=1}^{n} Vol_i \quad Vol_{Tot} = 16000$

Equations

$$d := \sum_{i=1}^{n} \frac{Vol_i}{Vol_{Tot}} \cdot \left(T_i\right)^x$$

Results

i	$\frac{Vol_i}{Vol_{Tot}}$	$\left(T_i\right)^x$	$\frac{Vol_i}{Vol_{Tot}} \cdot \left(T_i\right)^x$
1	0.625000	0.000343	0.000214
2	0.187500	0.000553	0.000104
3	0.187500	0.001087	0.000204

$d = 0.000522 \quad d^{\frac{1}{x}} = 93$ Flash point of blend °F

Computer Computation Template

CCT No. 5-4: Calculate TVP (true vapor pressure) at 100 °F from RVP (which, by definition, is always taken at 100 °F). Method is approximate, for 2 psia > RVP > 15 psia, which covers most crudes.

References: Chapter 5. Equation 5-12.

Input

Input Data $RVP := 15$ RVP input, psia

Equations

$$TVP_{100F} := 0.2858 + 0.8158 \cdot RVP + 0.0508 \cdot RVP^2$$

Results

$TVP_{100F} = 24.0$ psia

6

Specific Heat

Fundamental Thermodynamic Properties

There are five fundamental properties of substances considered from the standpoint of thermodynamics. These are *pressure, temperature, volume, entropy*, and *internal energy*. The almost universally acceptable symbols for these properties are P, T, V, S, and U.

The first three properties are instinctively understood; the other two are not. This is because while these are considered *fundamental* properties, they are actually *derived* properties. They are man–made combinations of P, V, and T with Q, *transferred heat*, and W, *mechanical work*, both characteristics of *processes* rather than of *substances*, that are convenient for thinking about and solving thermodynamics problems.

Entropy is often defined as the *measure of disorder* in a system. While this is true, the definition doesn't say much about entropy.

Entropy cannot be measured, but as it is always the change in entropy that is of interest and entropy can be defined in terms of change, this does not present a problem. The defining equation for entropy is

$$\Delta S = S_2 - S_1 = \int_1^2 \frac{dQ}{T}$$

where ΔS = the change in entropy
dQ = transferred heat
T = absolute temperature at which the heat is transferred

Entropy is a measure of energy. *Specific entropy*, which is the energy contained in a unit mass of a substance, would be expressed as ft-lb/lb or kcal/kg, for instance; a unit change in entropy can be written in terms of ft-lb/lb/°F, kcal/kg/°C, or any other consistent set of units expressing energy per unit of mass per degree of temperature.

For *all reversible* processes, the net change in entropy from the start to the end of the process is zero; for *any irreversible* process, the net change in entropy is always positive. Thus, the result of an irreversible process is an increase in entropy, an increase in the *disorder* of the system.

Internal energy can be described as the energy in a system that is *not* mechanical or chemical energy; and while, again, this tells what internal energy is *not* it does not say what it *is*.

But internal energy is just another man–made property of a substance that is useful in thermodynamics. Like entropy, it is defined in terms of a change:

$$\Delta U = U_2 - U_1 = Q - W$$

Or, in terms of differentials:

$$dU = dQ - dW$$

where dQ = a small amount of heat transferred
dW = a small amount of work done
dU = a small increase or decrease in internal energy

If it is assumed there are no chemical changes in a process so there is no change in *chemical energy*; and the changes in *mechanical kinetic energy* and *mechanical potential energy*, $wv^2/2g$ and wz, are accounted for elsewhere, then dU is the measure of change in *internal energy*.

Internal energy is therefore the sum of *intermolecular potential energy*, which is related to the forces between molecules; the *molecular kinetic energy*, related to the translational energy of individual molecules; and the *intramolecular energy*, the energy within the individual molecules associated with molecular and atomic forces. It is the energy *stored* in a substance, due to the activity and configuration of its molecules and to the vibrations of the atoms within the molecule.

The changes in internal energy are measured in terms of heat units per unit of temperature, and specific internal energy for a unit mass of a substance is defined in terms of Btu/lb/°F, or similar.

Note that ΔQ, heat gained or lost in a process, or ΔW, work done on or by the process during a process, depends not only on the initial and final states of the system but on the intermediate states as well, i.e., on the *path* of the process, while the entropy and internal energy gained or lost is not a function of the path; they are only a function of the initial and final conditions of the system. Entropy and internal energy are *point functions*.

There are also three other *derived properties* which are generally considered as fundamental properties. These are *enthalpy*, $H = U+PV$; the *Helmholtz* function, which is $A = U-TS$; and the *Gibbs'* function, which is defined as $Z = U+PV-TS = H-TS$. These latter two functions are sometimes called the Gibbs *psi*(Ψ) and *zeta*(Z) functions.

Secondary Properties

The eight fundamental properties described, if taken together with W, the *mechanical work* done by or on the system, and Q, the *heat energy transferred* in or out of the system, can be combined as properties, per se, or as first and second derivatives of properties, in so many ways there is truly an almost untold number of possible combinations making up new derived properties. It has been calculated that there are about 11×10^6 combinations which include first derivatives and some 9.5×10^{21} that include the second derivatives as well.

With such a huge set of possibilities it is not unusual that there is a tremendous number of possible *secondary properties* that can be deduced from these fundamental properties. If it is then considered that thermodynamics considers *reversible* and *irreversible cycles*, which may be *flow* or *non-flow*, and then may be further classified as being at *constant pressure, constant volume*, *constant entropy*, etc., it should not be thought unusual that classical thermodynamics is often described as a subject that places a heavy burden on the memory of the practitioner.

Pipeline hydraulics is a strictly limited form of the study and practice of thermodynamics, and *steady isothermal* or *steady adiabatic* flow of liquids in pipelines are specially limited fields.

However, there are always those unsteady and transient flow cases, and problems having to do with measurement of petroleum fluids and losses of such fluids, when we must have recourse to more generalized thermodynamic properties than those we usually use, and one of the most useful, and important, of the *secondary derivative* properties of matter that constantly shows up in such analyses is the *heat capacity* or, a slightly different but more often used concept, the *specific heat* of a substance.

This chapter looks at these two similar properties from the viewpoint of a pipeliner who is interested in the problems posed in the designing and operating of liquid petroleum pipelines; it is in no way intended as a treatise on the thermodynamics of fluids.

Definition of Heat Capacity and Specific Heat

The *heat capacity* of a substance is that quantity of heat required to increase the temperature of a given quantity of the substance by one degree of temperature. It can be expressed, for instance, as *kcal/°C*, *Btu/°F*, or, in SI, as *kJ/K*.

Molar (or *molal*) *heat capacity* is the quantity of heat required to raise the temperature of *one molecular weight* (one pound mole, for instance) of the substance by one degree (in this case, 1°F or 1°R).

More common expressions are *specific heat*, which is the quantity of heat required to raise the temperature of *one unit of mass* (one kg, for instance) of the substance by (in this case) 1°C or 1 K, and *molar* (or *molal*) *specific heat*, which is the specific heat stated in (say) Btu/lb/°F multiplied by the molecular weight of the substance. Specific heat is sometimes looked at as the ratio of the heat capacity of the substance to that of water at the same temperature, because the definitions of the calorie, kilocalorie, and Btu, define the amount of heat required to increase the temperature of one unit of mass of water by one degree of temperature.

Heat capacity is the more basic property but, because of common usage, specific heat is more often used in engineering calculations. *Specific heat* is usually, and will always be in this chapter, symbolized as the lower case c; *molar specific heats* will be symbolized by the capital C. Thus $C = Mc$, where M is the molecular weight of the substance.

The specific heat equation is

$$Q = \int_{T_1}^{T_2} c\,dT \tag{6-1}$$

The specific heat, of a substance is the slope of the Q–vs–T curve

$$c = \frac{dQ}{dT} \tag{6-2}$$

Note c is a function of T; the complicating factor is that c may be a very *complex* function of T.

There are two special values of c which have been measured over and over, and often published, for many substances. These are c_v, which is the *specific heat measured at constant volume*, and c_p, which is the *specific heat measured at constant pressure*. These two properties can be defined in terms of the specific heat equation, or, in terms of two of the derived properties:

$$C_v = \left(\frac{\partial u}{\partial T}\right)_v \tag{6-3}$$

$$C_p = \left(\frac{\partial h}{\partial T}\right)_p \tag{6-4}$$

Note here C_v is in terms of u, *internal energy*, and C_p is in terms of h, *enthalpy*. They can be expressed in many other ways.

The ratio c_p/c_v is also widely used and published. This ratio, usually symbolized in engineering texts as k, or in classical texts as γ, is, itself, an important property of a substance.

Theoretical Values for Specific Heat

It is interesting to again note, as in the preceding chapter on vapor pressure, that the study of gases and solids from a *theoretical* standpoint is considerably advanced over the study of liquids. Thus, while there is a large body of theoretical work on the thermodynamics of gases (and especially on *ideal* gases or *perfect* gases), and a considerable amount of work has been done on the thermodynamics of solids—especially down at the atomic level where the semi–conductor properties of solids are found—the body of theoretical work on thermodynamics of liquids (except as concerns water about to boil, or boiling) is quite small.

The kinetic theory of gases, early in the 19th century, showed that c_v and c_p (and therefore k) for perfect gases could be written in terms of ordinary fractions, R, called the *universal gas constant*, and J, the *mechanical equivalent of heat or Joule's Constant.* The following summarizes these relationships:

For monatomic gases

$$c_v = \left(\frac{3}{2}\right)\frac{R}{J} \quad c_p = \left(\frac{5}{2}\right)\frac{R}{J} \quad k = 1\frac{2}{3} \tag{6-5}$$

For diatomic gases

$$c_v = \left(\frac{5}{2}\right)\frac{R}{J} \quad c_p = \left(\frac{7}{2}\right)\frac{R}{J} \quad k = 1\frac{2}{5} \tag{6-6}$$

For polyatomic gases

$$c_v = \left(\frac{3}{1}\right)\frac{R}{J} \quad c_p = \left(\frac{4}{1}\right)\frac{R}{J} \quad k = 1\frac{1}{3} \tag{6-7}$$

These values are not usually precise enough for process calculations, but they are in the right order of magnitude and for light gases at low pressures they are essentially correct. Table 6-1 shows the relationships, especially the relationship for k, hold quite well for some gases.

There are actually five specific heats that enter into the study of reversible non–flow processes of perfect gases, and the specific heat may be defined for each process in terms of fundamental equations.

Constant pressure process

$$P = \text{constant} \quad c = c_p \tag{6-8}$$

Constant volume process

$$V = \text{constant} \quad c = c_v \tag{6-9}$$

Table 6-1
Specific Heats of Gases

Type of Gas	Gas	C_p	C_v	C_p–C_v	k
Monatomic	He	4.97	2.98	1.99	1.67
	A	4.97	2.98	1.99	1.67
Diatomic	H_2	6.87	4.88	1.99	1.41
	O_2	7.03	5.03	2.00	1.40
	N_2	6.95	4.96	1.99	1.40
	Cl_2	8.29	6.15	2.14	1.35
Polyatomic	CO_2	8.83	6.80	2.03	1.30
	SO_2	9.65	7.50	2.15	1.29
	C_2H_2	12.35	10.30	2.05	1.20
	NH_3	8.80	6.65	2.15	1.31

Constant temperature (isothermal) process

$$T = \text{constant} \quad c=\infty \tag{6-10}$$

Constant entropy (isentropic) process

$$S = \text{constant } c = 0 \tag{6-11}$$

Polytropic process

$$PV^n = \text{constant} \qquad c_n = c_v\left(\frac{n-k}{n-1}\right) \tag{6-12}$$

Some early work by Dulong and Petit published first in 1819 predicted that c_p for solids should be *about* 6 calories/gram–atom/°C; and this is *about* correct. Table 6-2, adapted from Tabor,[1] illustrates this principle. NaCl is 2 g–atom and CaF_2 and SiO_2 are 3 g–atom molecules. The ratio 6/g–atom holds fairly well.

Thus there are good bases for the theoretical prediction of the specific heats of gases, and fairly good bases for solids. There are no acceptable ways of deriving the theoretical specific heat of liquids.

The myriad relations of classical thermodynamics provide theoretical insights to many effects concerning specific heats that apply to *all* substances, and not just to gases and solids; they apply equally to *liquids*. One of these is very important, because it means that if enthalpy can be measured then c_p is immediately available.

Table 6-2
Specific Heats of Solids

Substance	c_p cal/g/°C	Atm or Mol weight	C_p cal/g-atom or g/mole/°C
Al	0.22	27	5.9
Cu	0.09	63.5	5.7
Ag	0.54	108	5.8
Pb	0.03	207	6.2
NcCl	0.21	57.5	12.0
CaF_2	0.22	78.0	17.1
SiO_2	0.28	60.0	16.8

$$\left(\frac{\partial H}{\partial T}\right)_P = c_p \tag{6-13}$$

Measure H –vs– T at constant P, express the resulting relationship in a power series, differentiate the series, and evaluate the result to calculate c_p.

Another relationship is

$$(c_p - c_v) = (-T)\left(\frac{\partial V}{\partial T}\right)_P^2 \left(\frac{\partial P}{\partial V}\right)_T \tag{6-14}$$

Thus the difference between c_p and c_v is the negative product of the *absolute temperature* by the *square of the volume coefficient of thermal expansion at constant pressure* by the *isothermal bulk modulus*. Or

$$C_p - C_v = \frac{\alpha^2 vT}{\beta_T} \tag{6-15}$$

where $\alpha =$ coefficient of expansion $\frac{1}{v}\left(\frac{\partial v}{\partial T}\right)_P$

$v =$ specific volume

$T =$ absolute temperature

$\beta_T =$ isothermal compressibility $-\frac{1}{v}\left(\frac{\partial v}{\partial P}\right)_T$

This equation, in either form, can be used to deduce two things:

- C_p will always be greater than C_v, and thus c_p will be greater than c_v, because the squared term will always be positive, the bulk modulus negative, and the negative product of the two will always be positive.
- The difference between c_p and c_v for liquids and, especially, for solids, will always be small or very small. This follows from the fact that the square of the volume coefficient of expansion will always be a very small number; and though the bulk modulus for liquids—and especially for solids—is a large number, the product of these two will be small. In fact, whenever specific heat of a liquid is given for ordinary temperatures and pressures, it is c_p that is given. It is difficult, if not impossible, to find published values for c_v for solids.

There are two equations due to Clapeyron that can be manipulated to yield another important parameter in terms of measurable quantities. These are

$$T\left(\frac{\partial P}{\partial T}\right)_V = J\left(\frac{\partial Q}{\partial V}\right)_T \tag{6-16}$$

$$T\left(\frac{\partial V}{\partial T}\right)_P = J\left(\frac{\partial Q}{\partial P}\right)_T \tag{6-17}$$

The second equation is derived from the first. Another generally applicable equation is

$$\left(\frac{\partial Q}{\partial T}\right)_X = \left(\frac{\partial Q}{\partial T}\right)_P + \left(\frac{\partial Q}{\partial P}\right)_T\left(\frac{\partial P}{\partial T}\right)_X \tag{6-18}$$

Here the first term on the right is c_p. The $(\partial Q/\partial P)_T$ term can be taken from Clapeyron's second equation, and is equal to $-(T/J)(\partial V/\partial T)_P$ which is $-(T/J)$ times the coefficient of volume expansion under constant pressure. The third term, $(\partial P/\partial T)_X$, is the change in pressure for a corresponding change in temperature, or the *inverse* of the change in temperature for a corresponding change in pressure, under Process X. Thus the specific heat c_X can be calculated in terms of tabulated parameters and the P -vs- T relationship for Process X.

One last useful equality.

$$c_P\left(\frac{\partial T}{\partial P}\right)_Q = \left(\frac{T}{J}\right)\left(\frac{\partial V}{\partial T}\right)_P$$

This allows the computation of the adiabatic increase in temperature of a liquid suddenly compressed, as in passing through a pump, in terms of T/J and the thermal coefficient of volume expansion, all of which are very readily obtainable values.

And it is useful to remember $(C_p - C_v) = R$. Since the gas constant in terms of cal/g-mol/°C, kcal/k-mol/°C, or Btu/lb-mol/°F, is 1.98719, or *about* 2, we have the justly famous

$$(C_p - C_v) = 2 \tag{6-19}$$

and

$$k = \frac{C_p}{(C_p - 2)} \tag{6-20}$$

which is the same as

$$k = \frac{mc_p}{(mc_p - 2)} \tag{6-21}$$

Table 6-3 is a collection of some of the values for R I have collected over the years. The conversions between the values aren't precise; and the temperatures at which the values apply have long since been lost in my files. They are pretty close, however, and interesting.

The most interesting is not on the table because it is useless, though noteworthy: R = 2.807 hp-sec/lb-mol/°F. I have always thought it was intriguing that

$$(C_p - C_v) = 2.807 \text{ hp-sec / lb-mol/}^\circ F$$

It should be remembered that while specific heat is relatively insensitive to changes in pressures over any ordinary range that, except for monatomic gases, it is highly variable with variable temperature; specific heat of monatomic gases is not temperature sensitive.

The general form of the relationship for gases is

$$\log c_p = a + bT \tag{6-22}$$

Table 6-3

Miscellaneous Universal Gas Constant Values

Pressure Units	Vol Units	Temp Units	Value	Gas Constant Units
bar	liter	K	83.14	bar-liter/kmol-K
bar	m³	K	0.08314	bar-m³/kmol-K
barye	cm³	K	8.31×10^7	barye-cm³/gmol-K
kg_f/cm^2	liter	K	84.78	kg_f/cm^2-liter/kmol-K
kg_f/cm^2	m³	K	0.08478	kg_f/cm^2-m³/kmol-K
kg_f/m^2	cm³	K	8.48×10^5	kg_f/cm^2-cm³/gmol-K
atmosphere	cm³	K	82.057	atm-cm³/gmol-K
atmosphere	liter	K	0.082057	atm-liter/gmol-K
atmosphere	m³	K	8.21×10^{-5}	atm-m³/gmol-K
atmosphere	in³	R	1260	atm-in³/lbmol-R
atmosphere	ft³	K	1.3145	atm-ft³/lbmol-K
atmosphere	ft³	R	0.7302	atm-ft³/lbmol-R
MPa	m³	K	0.008314	MPa-m³/kmol-K
kPa	m³	K	8.314	kPa-m³/kmol-K
psi	ft³	R	10.73	psi-ft³/lbmol-R
psi	in³	R	18510	psi-in³/lbmol-R
psf	ft³	R	1545	psf-ft³/lbmol-R
inHg	ft³	R	21.86	inHg-ft³/lbmol-R
inH_2O	ft³	R	295.9	inH_2O-ft³/lbmol-R
inH_2O	mole	R	0.7846	inH_2O-mol/lbmol-R
oz/in²	ft³	R	171.5	oz/in²-ft³/lbmol-R
inHg	liter	K	2.457	inHg-ltr/gmol-K
mmHg	liter	K	62.26	mmHg-ltr/gmol-K
Energy Units				
kcal		K	1.987	kcal/kmol-K
cal		K	1.987	cal/gmol-K
joule		K	8314.3	joule/kmol-K
BTU		R	1.987	BTU/lbmol-R
ft-lb		R	1544	ft-lb/lbmol-R
joule		R	1.987	joule/lbmol-R
kWh		K	0.001049	kWh/lbmol-K
hp-hr		R	0.000780	hp-hr/lbmol/-R

For real gases, multi–constant equations are usually used. The exponential form is

$$c_p = A + Be^{-C/T^n} \tag{6-23}$$

The other common form is derived by expanding the relationship in a series, as follows. The maximum value of n is usually 4.

$$c_p = a + bT + cT_2 \ldots + T^n \tag{6-24}$$

For liquids, the relationship *over small changes in temperature* and well removed from the critical pressure and temperature is

$$c_p = a + bT \tag{6-25}$$

This simple equation is a straight line of slope b; and given c_p at any two points c_p, T_1 and c_p, T_2, then c_p is known for all values between T_1 and T_2 and outside this range as long as the equation holds. For ordinary pipelining problems the equation is sufficiently precise so long as neither the temperature of fusion nor the temperature of vaporization is closely approached, or if the substance is not water—water has some peculiar characteristics, especially in the vicinity of 4°C where its density is at a maximum.

Measurement of Specific Heat

An academician of considerable standing once told me, "Nobody measures the specific heat of gases any more—they've all *been* measured." Which is probably true.

Practically all measurements of specific heats of gases were made at a time when the *cgs* system of units was in vogue in the classical scientific fraternity; specific heats reported in other units are more than likely a converted cgs value. Inasmuch as the definition of specific heat in one way or another includes the value of the calorie, a cgs unit, it is natural there would be some differences in the definition. Note the following:

Kind of Calorie	Value, Joules
20 deg calorie	4.1816
15 deg calorie	4.1855
Mean calorie	4.1897
TC calorie	4.1840
IT calorie	4.1868

The 20 and 15°C calories are measured around a Δt centered on the given temperature; the mean calorie is one–hundredth of the quantity of heat required to raise the temperature of a gram of water from 0 to 100°C; the thermochemical calorie is derived from chemical relationships; and the international steam table calorie is 1/860th of a watt–hour based on the 1934 *International Steam Tables*. The difference between the maximum and minimum standard value is less than 0.2 percent and is not important for pipeline calculations except those involving purchase or sale of heat. The difference does explain some of the minor differences in published figures, however.

The specific heat of gas mixtures is computed by taking an analysis of the gas and computing the contribution of the specific heat of each component, taken on a molal basis, to the specific heat of the whole. The specific heat of the gases may be taken from many sources, but one of the most respected is the ASTM *Physical Constants of the Hydrocarbons C_1–C_{10}*, 1916. Data from the *Project 44* of the API is also often cited. Table 6-4 listing specific heats for the paraffinic hydrocarbons through C_7 is from Project 44 data.

Given the weight analysis of a gas, the specific heat can be computed by the relationship

$$c_m = Xc_x + Yc_y + Zc_z \qquad (6\text{-}26)$$

where X, Y, Z = weight parts of gases x, y, z
c_x, c_y, c_z = specific heats of gases x, y, z

For a volumetric analysis, which is much more common, the relationship is

$$M_m c_m = XM_x c_x + YM_y c_y + ZM_z c_z \qquad (6\text{-}27)$$

where X, Y, Z = volume parts of gases x, y, z
M_x, M_y, M_z = mol wt of gases x, y, z
c_x, c_y, c_z = specific heat of gases x, y, z

Note that $M_x c_x$ are the molal specific heats of a gas that has a molecular weight M_x and a specific heat c_x. To derive the specific heat c_m it is necessary to calculate M_m, which is the sum of $XM_x + YM_y + ZM_z$ and divide it into $M_m c_m$.

For liquids there is no such easy way; and there is, especially, no easy way for petroleum hydrocarbons. One authority estimates there may be as many as ten thousand different hydrocarbon compounds in an ordinary crude; and to this can be added a thousand or so sulfur compounds, some inerts, a few contaminant metals, etc.

Table 6-4

Specific Heats of Natural Gas Components BTU/lb/°F At 60°F and 14.696 psia

Gas	Form	Mol Wt	c_p gas	c_v gas	c_v liquid
Paraffin Hydrocarbons					
methane	CH_4	16.043	0.523	0.403	
ethane	C_2H_6	30.070	0.410	0.344	0.926
propane	C_3H_8	44.097	0.388	0.343	0.592
i-butane	C_4H_{10}	58.124	0.387	0.353	0.570
n-butane	C_4H_{10}	58.124	0.387	0.353	0.564
i-pentane	C_5H_{12}	72.151	0.383	0.355	0.535
n-pentane	C_5H_{12}	72.151	0.388	0.361	0.544
n-hexane	C_6H_{14}	86.178	0.386	0.363	0.533
n-heptane	C_7H_{16}	100.205	0.388	0.368	0.528
Inert Gas Components					
carbon monoxide	CO	28.010	0.248		
carbon dioxide	CO_2	44.010	0.199		
hydrogen sulfide	H_2S	34.076	0.238		
molecular hydrogen	H_2	2.016	3.408		
molecular nitrogen	N_2	28.013	0.248		

Therefore, the specific heat of hydrocarbon liquids is usually either estimated, or predicted from a nearby measurement of a similar substance.

The standard method of measurement is that of ASTM D 2766[2] which is based on the procedure suggested earlier—measuring enthalpy of the substance at a series of different temperatures, writing these results in terms of a power series, and then differentiating the power series to yield c_p. The specific heat actually derived is a dimensionless number which is numerically equal to the thermal heat capacity of the substance. It should be mentioned that carrying out the test and subsequent calculations is equivalent to a refresher course in thermodynamics.

Calculation of Specific Heat of Petroleum Hydrocarbons

The specific heat of a petroleum hydrocarbon—and it should be remembered this is almost always reported in terms of c_p—is almost insensitive to changes in pressure in the ordinary sense, but it is an essentially linear function of temperature.

Specific heat of hydrocarbon *mixtures* is also a rather direct function of density, and it is also a somewhat less direct but definite function of what may be termed—for lack of a better word—the *kind* of molecules which go into the makeup of the mixture.

The standard way of computing specific heat of hydrocarbon liquids has been available since at latest the mid–1940's (I have a copy of the TEMA Standards for 1949 which uses the formula as a base for a graph of specific heat –vs– API gravity), but it was not accepted as a standard, per se, until ASTM issued D 2980[3] in 1970. The method is based on obtaining an estimated value of c_p in terms of API gravity, or specific gravity 60/60°F, and then making a correction, which may vary from 0.845 to 1.12 in the extreme or, realistically, from 0.97 to 1.03 for crude oils, by using a multiplier derived from the *Watson Characterization Factor*, sometimes called the *UOP Characterization Factor* or the *Nelson and Watson Characterization Factor*, defined as

$$K = \frac{\left(T_{ave\ molal\ B.P.,\ deg\ R}\right)^{1/3}}{spgr\ 60/60^\circ F} \tag{6-28}$$

The *average molal boiling point* (sometimes, as in ASTM D 2980, called the *mean* average boiling point) is obtained by deriving the volumetric average boiling point from a distillation according to ASTM D 86 and applying a correction based on the slope of the ASTM D 86 curve determined from the temperatures at 90% and 10% distilled.

The fundamental equation for specific heat measured in Btu/lb/°F as a function of specific gravity 60/60°F and temperature in °F is

$$c_p = 0.6811 - 0.308 \times spgr + \left(0.000815 - 0.000306 \times spgr\right) \times T \tag{6-29}$$

The multiplier to take into account the *kind* of hydrocarbon in terms of K is

$$(\text{multiplier}) = (0.055 \times K) + 0.35$$

For $K = 11.818$ the multiplier is 1.00. Graphs based on the above equation are generally drawn for $K = 11.8$ which is cited as "characteristic of Mid–Continent (U.S.) stocks."

The information required for the computation of K is not usually immediately available to a pipeliner, at least in the conceptual or design phases of a project, but those in Table 6-5 can be used as a guide. These are taken from a listing of the pertinent properties of over 200 crudes in Nelson.[4] There are also the original listings in the *Oil and Gas Journal*, later published in book form,[5] of analyses of hundreds of different crudes. These analyses include data on the ASTM distillation curves and the K-factor can be derived from such information by direct substitution in the formulas of ASTM D 2980. The two main parameters are

$$VABP = \text{volume average boiling point}$$
$$= (\text{sum of } 10, 30, 50, 70 \text{ and } 90 \text{ pct distilled temps})/5$$
$$\text{Slope} = (90 \text{ pct distilled temp} - 10 \text{ pct distilled temp})/80$$

ASTM D 2980 has a pair of graphs to assist in evaluating specific heat, but these are based on Equations 6-28 and 6-29 and there is no problem in solving these analytically if K is available or can be estimated.

The ASTM formula without the correction factor in K is probably sufficiently accurate for most pipeline use, noting that the accuracy of the result is less than the *probable* correction resulting from use of the K-factor multiplier, but there are other recognized formulas which are somewhat simpler and are widely used. One due to Cragoe[6] gives the *mean specific heat* between 0 °C and T °C as

$$c_{p,m} = \left(0.403 + 4.05 \times 10^{-4} \times T\right)\left(s^{-1/2}\right) \tag{6-30}$$

where $c_{p,m}$ = mean c_p between $0^\circ C$ and $T^\circ C$

s = specific gravity $60/60^\circ F$

$T = {}^\circ C$

It is reported that the formula gives results that are about two percent low for products of paraffin base crudes, and about two percent high for products of naphthenic base crudes.

A similar formula is due to Mills reported by Hobson and Pohl[7] as follows:

Table 6-5

Characterization Factors for Typical Crudes

Field	Country	SpGr 60/60°F	K-factor at 250°F
Qaiyarah	Iraq	0.970	12.10
Santa Maria	California	0.963	11.90
Tia Juana	Venezuela	0.959	11.65
Lloydminster	Canada	0.958	11.40
Kalamono	New Guinea	0.947	11.40
Lost Hills	California	0.944	11.60
Smackover	Arkansas	0.931	11.62
Cabimas	Venezuela	0.928	11.87
Tatums	Oklahoma	0.922	11.80
Kalimantan	Indonesia	0.916	11.40
Talco	Texas	0.908	12.10
Brookhaven	Mississippi	0.898	12.10
Poza Rica	Mexico	0.872	12.20
Kuwait Export	Kuwait	0.869	12.30
Santa Barbara	Venezuela	0.865	11.88
Bahrain Export	Bahrain	0.861	12.25
Air Dar	Saudi Arabia	0.855	12.25
West Texas Composite	Texas	0.854	11.96
Dammam	Saudi Arabia	0.850	12.03
Agha Jari	Iran	0.845	12.30
Abqaiq	Saudi Arabia	0.842	12.30
Kirkuk	Iraq	0.842	12.00
Sprayberry	Texas	0.834	11.78
East Texas	Texas	0.833	11.92
Qatif	Saudi Arabia	0.832	12.04
Louden	Illinois	0.831	11.94
Mid-Continent	Oklahoma	0.821	11.95
Santa Elena	Ecuador	0.820	12.00
Scurry	Texas	0.816	11.83
Santa Rosa	Venezuela	0.805	11.70
Cumarebo	Venezuela	0.790	12.10
Coldwater	Michigan	0.781	12.12
Camiri	Bolivia	0.760	12.35
Santa Anita	Bolivia	0.751	12.10

$$c_{p,T} = \left(0.403 + 8.00 \times 10^{-4} \times T\right)\left(s^{-1/2}\right) \tag{6-31}$$

The parameters have the same meaning as in the Cragoe formula except that s is in terms of density at 20°C.

Marks's Handbook[8] gives what is about the same formula as

$$c_p = \left(0.388 + 4.5 \times 10^{-4} \times T\right)\left(s^{-1/2}\right) \tag{6-32}$$

Note that the ASTM form of equation, without the K–factor correction, for an oil of 0.825 spgr 60/60°F would yield a specific heat at 68°F =20°C of 0.465; the Mills form yields a value of 0.461; and the Marks version gives the same value. For the same oil at 212°F =100°C the computed values are 0.546, 0.532, and 0.532. The differences between the ASTM and Cragoe forms, 0.87% at 68°F =20°C and 2.63% at 212°F =100°C , are within the *accuracy* of the ASTM method without the K-factor correction and so either approximation should be valid.

If the information necessary for the calculation of the *K*-factor is available the ASTM method should be used; there is no reason with today's calculators and computers to use anything but the best information available. If the *K*-factor is not available, the Cragoe–type formulas are within the reaches of engineering accuracy required of most pipeline designs. The only time real accuracy is required is when dealing with very high temperatures, or when operating at or near the fusion or evaporation temperatures of major constituents of the oil.

Working Equations

Calculators

Table 6-6, Reference and Working Equations for Specific Heat, summarizes the fundamental equations for specific heat, including the adiabatic ΔT due to ΔP that will come in handy later on when it appears in on–block pump efficiency tests. The c_p –vs– T relationships are covered briefly. Equation 6-26, specific heat of gas mixtures, given the proportions of the component gases by weight, is not included because of a lack of space on the table and, also, because the formula is little used. The calculation of the specific heat of gas mixtures, given the proportions of the component gases by volume, is covered, and an open–end **CCT** (see the following) is also provided. Finally, the ASTM/UOP K–Factor method of calculating specific heat of liquid hydrocarbons is provided, as well as the less precise, but satisfactory for most uses, formulas of Cragoe, Mills and Marks.

Computers

Three Computer Computation Templates are provided. **CCT** No. 6-1 is the above–mentioned gas analysis program, **CCT** No. 6-2 is the ASTM/UOP K-Factor equation, and **CCT** No. 6-3 contains all three of the Cragoe, Mills, Marks formulas.

Table 6-6

Reference and Working Equations for Specific Heat

Fundamental Equations

Specific Heat Equation $Q = \int_{T_1}^{T_2} c\,dT$

Definition of Specific Heat $c = \dfrac{dQ}{dT}$

Definitions of C_v and C_p

Specific heat at constant volume $C_v = \left(\dfrac{\partial u}{\partial T}\right)_v$

Specific heat at constantpressure $C_p = \left(\dfrac{\partial h}{\partial T}\right)_p$

C_v and C_p Relations

$$(C_p - C_v) = R \approx 2.0$$

$$(C_p - C_v) = f(\alpha,\beta,v,T) \;=\; (C_p - C_v) = \frac{\alpha^2 vT}{\beta_T}$$

$\alpha =$ coefficient of expansion $\dfrac{1}{v}\left(\dfrac{\partial v}{\partial T}\right)_P$

$v =$ specific volume

$T =$ absolute temperature

$\beta_T =$ isothermal compressibility $-\dfrac{1}{v}\left(\dfrac{\partial v}{\partial P}\right)_T$

$$k = \frac{C_p}{(C_p - 2)} \qquad k = \frac{mc_p}{(mc_p - 2)}$$

Adiabatic ΔT Increase Due ΔP

$$c_P\left(\frac{\partial T}{\partial P}\right)_Q = \left(\frac{T}{J}\right)\left(\frac{\partial V}{\partial T}\right)_P$$

Specific Heat –vs– Temperature Relations

General relationship for gases $\log c_p = a + bT$

General relationship for real gases $c_p = A + Be^{-C/T^n}$

Above expanded into a series $c_p = a + bT + cT_2 \ldots + T^n$

For Liquids, small changes in T $c_p = a + bT$

Specific Heat for Gases, Volumetric Basis

$$M_m c_m = XM_x c_x + YM_y c_y + ZM_z c_z$$

where $X, Y, Z =$ volume parts of gases x, y, z

$M_x, M_y, M_z =$ mol wt of gases x, y, z

$c_x, c_y, c_z =$ specific heat of gases x, y, z

Calculation of Specific Heat

ASTM Method
$$c_p = 0.6811 - 0.308 \times spgr + (0.000815 - 0.000306 \times spgr) \times T$$

$$(\text{multiplier}) = (0.055 \times K) + 0.35$$

$$\text{K= Characterization Factor} = \frac{(T_{ave\ molal\ B.P.,\ deg\ R})^{\frac{1}{3}}}{spgr\ 60/60^\circ F}$$

Multiplier for K=11.8, Mid-Continent Crudes, =1.0

Cragoe Formula $c_{p,m} = (0.403 + 4.05 \times 10^{-4} \times T)(s^{-1/2})$

where $c_{p,m} =$ mean c_p between $0^\circ C$ and $T^\circ C$

$s =$ specific gravity $60/60^\circ F$

$T = ^\circ C$

Mills Formula $c_{p,T} = (0.403 + 8.00 \times 10^{-4} \times T)(s^{-1/2})$

Marks Formula $c_p = (0.388 + 4.5 \times 10^{-4} \times T)(s^{-1/2})$

References

[1] Tabor, D. *Gases, Liquids and Solids*. Baltimore: Penguin Books, Inc., 1969.

[2] ASTM D 2766, *Standard Test Method for Specific Heat of Liquids and Solids.* American Society for Testing Materials, Philadelphia.

[3] ASTM D 2980. *Standard Method for Calculation of the Liquid Heat Capacity of Petroleum Distillate Fuels.* American Society for Testing Materials. Philadelphia.

[4] Nelson, W. L. *Petroleum Refinery Engineering,* 4th edition. New York: McGraw Hill Book Company, 1969.

[5] *Evaluation of the World's Important Crudes.* Tulsa: Petroleum Publishing Company, 1973 (and later editions).

[6] Cragoe, NBS Miscellaneous Publication No. 97, *Thermal Properties of Petroleum Products.* Washington: National Bureau of Standards, 1929.

[7] Hobson and Pohl, *Modern Petroleum Technology*, 4th edition. London: The Institute of Petroleum of Great Britain, 1975.

[8] *Mark's Standard Handbook for Mechanical Engineers,* 8th. edition. New York: McGraw–Hill Book Company, 1978.

Computer Computation Template

CCT No. 6-1: Specific heat of mixtures of hydrocarbon and inert gases based on a volumetric analysis of the gas. The procedure allows for any number of gases *n*, given the volume parts, molecular weights, and specific heats of the component gases.

References: Chapter 6. Equation 6-27. Figure 6-4.

Input

Factors

Number of components: $n := 6$ $\quad i := 1, 2 .. n$

Input Data

Gas No.	Name of Gas	Vol Part of Gas	Mol Weight of Gas	Mol Vol of Gas	Specific Heat of Gas at Constant Pressure Btu/lb/°F
1	carbon dioxide	$V_1 := 0.001$	$MW_1 := 44.01$	$MV_1 := V_1 \cdot MW_1$	$c_{p_1} := 0.199$
2	methane	$V_2 := 0.789$	$MW_2 := 16.043$	$MV_2 := V_2 \cdot MW_2$	$c_{p_2} := 0.523$
4	ethane	$V_3 := 0.060$	$MW_3 := 30.070$	$MV_3 := V_3 \cdot MW_3$	$c_{p_3} := 0.410$
4	propane	$V_4 := 0.034$	$MW_4 := 44.097$	$MV_4 := V_4 \cdot MW_4$	$c_{p_4} := 0.388$
5	n-butane	$V_5 := 0.010$	$MW_5 := 58.124$	$MV_5 := V_5 \cdot MW_5$	$c_{p_5} := 0.353$
6	nitrogen	$V_6 := 0.106$	$MW_6 := 28.010$	$MV_6 := V_6 \cdot MW_6$	$c_{p_6} := 0.248$

Equations

$$Vol_{Tot} := \sum_{i=1}^{n} V_i \qquad MW_{Tot} := \sum_{i=1}^{n} MW_i \qquad VMW_i := V_i \cdot MW_i \qquad \Sigma VMW := \sum_{i=1}^{n} V_i \cdot MW_i$$

$$c_{pTot} := \sum_{i=1}^{n} \frac{VMW_i}{\Sigma VMW} \cdot c_{p_i}$$

Results

$Vol_{Tot} = 1.000$ $\quad MW_{Tot} = 220.354$ $\quad \Sigma VMW = 19.556$

VMW_i	$\frac{VMW_i}{\Sigma VMW} \cdot c_{p_i}$
0.044	0.000
12.658	0.339
1.804	0.038
1.499	0.030
0.581	0.010
2.969	0.038

$c_{pTot} = 0.455$ Btu/lb/°F

Computer Computation Template

CCT No. 6-2: Specific heat of petroleum hydrocarbons by the ASTM D 2890 method. Includes use of the UOP–Watson Characterization Factor to refine results. Output is in terms of Btu/lb/°F. Accuracy should be about 4% for straight run fractions; somewhat less for more complex products and crude oils.

References: Chapter 6. Equations 6-28 and 6-29. Table 6-5.

Input

Factors $i := 0, 1 .. 10$ $T_i := 20 + 10 \cdot i$ °F

Input Data $SpGr := 0.825$ H_2O=1.000 $K_{UOP} := 11.818$ <==See Table 6-5 for typical K-Factors. A K-Factor of 11.818 yields a correction factor of 1.0.

Equations

$$c_{p_i} := [0.6811 - 0.308 \cdot SpGr + (0.000815 - 0.000306 \cdot SpGr) \cdot T_i]$$

$$K_{correct} := (0.055 \cdot K_{UOP}) + 0.35$$

$$c_{pcorrect_i} := c_{p_i} \cdot K_{correct}$$

Results

T_i	$c_{pcorrect_i}$
20	0.438
30	0.444
40	0.449
50	0.455
60	0.461
70	0.466
80	0.472
90	0.478
100	0.483
110	0.489
120	0.495

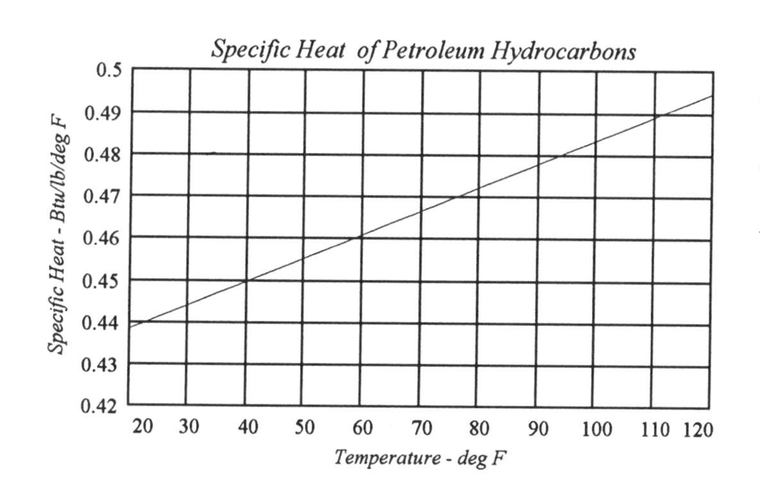

$SpGr = 0.825$

$K_{UOP} = 11.818$

$K_{correct} = 1.000$

Computer Computation Template

CCT No. 6-3: Calculation of specific heat of liquid petroleum hydrocarbons by the Cragoe, Mills, and Marks formulas. The results of calculations for all three formulas, using the input data, are shown in the **Results**, but only the calculations made with the Cragoe formula are plotted. This can be easily changed by changing the name of the variable on the Y–axis of the plot.

References: Chapter 6. Equations 6-30, 6-31, and 6-32.

Input

Factors $i := 0, 1 .. 10$ $\qquad T_{F_i} := 20 + 10 \cdot i$

Input Data $SpGr := 0.825$

Equations

$$c_{pmCragoe_i} := \frac{0.403 + 4.05 \cdot 10^{-4} \cdot T_{F_i}}{SpGr^{\frac{1}{2}}} \qquad c_{pmMills_i} := \frac{0.403 + 8.00 \cdot 10^{-4} \cdot T_{F_i}}{SpGr^{\frac{1}{2}}} \qquad c_{pmMarks_i} := \frac{0.388 + 4.5 \cdot 10^{-4} \cdot T_{F_i}}{SpGr^{\frac{1}{2}}}$$

Results

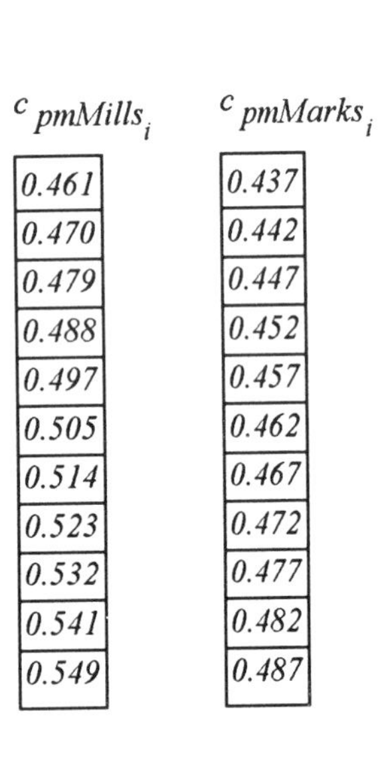

T_{F_i}	$c_{pmCragoe_i}$	$c_{pmMills_i}$	$c_{pmMarks_i}$
20	0.453	0.461	0.437
30	0.457	0.470	0.442
40	0.462	0.479	0.447
50	0.466	0.488	0.452
60	0.470	0.497	0.457
70	0.475	0.505	0.462
80	0.479	0.514	0.467
90	0.484	0.523	0.472
100	0.488	0.532	0.477
110	0.493	0.541	0.482
120	0.497	0.549	0.487

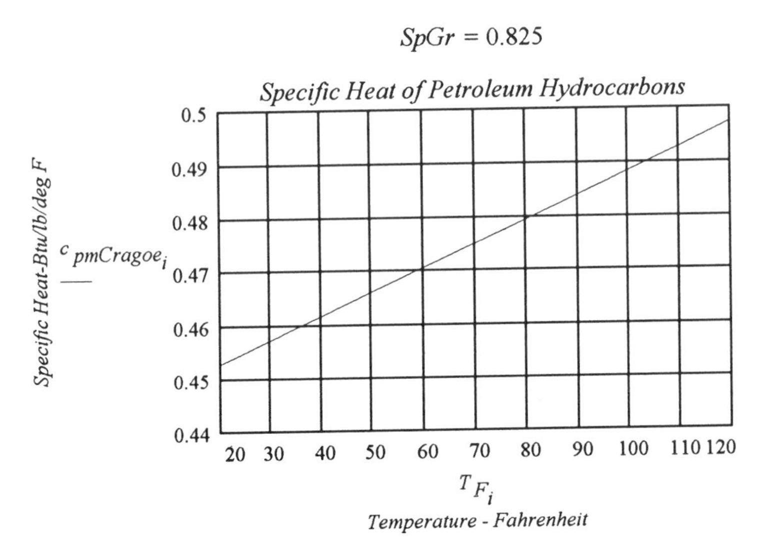

7

Heat of Combustion

Introduction

This complete—if short—chapter devoted to *heat of combustion*, a property of combustible substances not in any way connected with the flow of fluids in pipelines, is not included here by accident. There *is* a reason: money is involved.

The three most important costs in operating a large pipeline of ordinary configuration are, in order,

- Charges against capital
- Energy
- Labor

If we exclude those kinds of sources which make very small contributions to the available energy picture such as geothermal energy, solar energy, windmills, and the like, the energy costs will, in the end, derive from the heat of nuclear reactions, the kinetic energy of falling water, or the burning of hydrocarbons.

There is nothing that can be done about the cost of commercially purchased power derived from nuclear or hydropower sources, and little that can be done about the cost of commercial power generated by burning coal, oil, or natural gas. However, it is important to know how this power is generated, how to measure it, cost it, and how to place a value on *locally generated* power, i.e., gas engine, diesel engine, and steam or combustion turbine power, as compared to *imported* power, which generally means electricity, for the pump prime movers. To be able to do this you have to have a knowledge of and a ready reference on hydrocarbon fuels.

Theoretical Heat of Combustion

The theoretical heat content of a pure hydrocarbon can be calculated from the *ideal gas specific gravity*, *ideal gas calorific value*, and the *compressibility summation factor* for the substance.

When pure hydrocarbons burn two kinds of products are produced:

- oxides of carbon (CO and CO_2)
- oxides of hydrogen (water)

The first of these produces heat; the second absorbs it. If a pure hydrocarbon is burned and the heat generated is measured immediately after combustion is complete, the produced water will be in vapor form and the heating value obtained is called the *low heating value*; if the entire volume of the products of combustion is brought back to the temperature at which the combustion was initiated, the heat of vaporization of the water, equal to about 1055 Btu/lb = 2.395 MJ/kg = 586 kcal/kg at 20°C = 68°F, will be recovered, and the heating value obtained is called the *high heating value*. The low heating value is sometimes called the *net heating value*; the higher figure is then referred to as the *gross heating value*. Note that these differences apply strictly to hydrocarbons; when carbon monoxide, for instance, burns, the only oxide produced is CO_2, and there is no high heat–low heat problem.

Be careful in dealing with heating values. DEMA[1] says, with solid technical bases:

> Any heat engine can use only the low heat value of a fuel for producing power, and therefore, it is

> an accepted standard to use low heat value in all computations relative to engine performance.

Yet, in Marks,[2] an equally recognized authority, we read:

> In Germany, the low heat value of the fuel is used in calculating the efficiencies of internal combustion engines. In the United States, the high value is specified by the ASME Power Test Code.

The standard sources for the parameters mentioned above—*specific gravity*, *total* and *net calorific value*, and the *compressibility summation factor*—are API Research Project 44,[3] IGT Research Bulletin No. 32[4] and ASTM D3588.[5] The IGT and ASTM documents detail the arithmetic of calculating the heating value of a gas composed of a mixture of hydrocarbon and inert gases.

The reason for including three references to the same data is that if you are going to get involved in a heat balance, an efficiency study, or a gas purchase or sale contract, you might as well use data you can back up. The IGT and API references are hard to find; the ASTM reference is current. Note that Kemp,[6] who gives an excellent step-by-step description of calculations according to Research Bulletin No. 32, documents his literature search which resulted in *seventeen* Btu values for methane—the simplest of hydrocarbons—ranging from 1004.2 to 1014.50 per cubic foot (all corrected to 14.735 psia and 60°F), and only in three instances did any two agree; eleven were different. The variations for the heavier gases were even more disparate: ethane ran from 1757 to 1792, a two percent spread; propane from 2503 to 2590 and iso-butane from 3251 to 3363 ran over three percent. These aren't large percentages; but add to these a few measurement errors, such as bad static pressures, worn orifices—there are many of these kinds of things—and you can quickly come up with a substantial overall measurement error. I know of one measurement that went for years with a built-in error of nearly 11 percent.

There are minor differences in the heat values in some of the tables accompanying this chapter, and we don't apologize for them: the differences aren't *errors*, they are *inaccuracies*.

Gases

For pure, ideal gases, the heating value of a mixture of gases can be calculated by

$$H_T = X_1H_1 + X_2H_2 + X_3H_3 \ldots X_nH_n \tag{7-1}$$

where $X_1, X_2, \ldots X_n$ = mole fraction component gas
$H_1, H_2, \ldots H_n$ = heating value component gas
H_T = heating value composite gas

For accurate calculations, the supercompressibility of real gases must be taken into account. Research Paper No. 32 handles this as follows:

$$Z_T = 1 - \left(X_1\sqrt{b_1} + X_2\sqrt{b_2} + \ldots X_n\sqrt{b_n}\right)^2 \tag{7-2}$$

$$H_R = H_T/Z_T \tag{7-3}$$

where $X_1, X_2 \ldots X_n$ = mole fractions
$\sqrt{b_1}, \sqrt{b_2} \ldots \sqrt{b_n}$ = empirical summation factors

Thus H_T from the first equation is the *ideal* total heating value of a composite mixture (assuming the ideal heating values for the components are used), H_R is the *real* heating value of the mixture, and Z_T is the *compressibility factor* that relates the two.

Either the total heating value or the net heating value can be used for the values of H in the above calculations.

Table 7-1 gives values for specific gravity, total and net heating values, and the $\sqrt{b}$ factors for some important gases.

The calculation of heats of combustion of mixtures of gases using the mole fraction content of individual gases arises primarily because the modern method of gas analysis, that embodied in ASTM D1945,[7] continuous gas chromatography, yields results in mole percent.

Analyses in volume percent will work just as well. If H_i is the heating value of the *ith* component, and X_i is the volume percent of the *ith* component, then

$$H_{mix} = \sum_{i=1}^{n} X_iH_i \tag{7-4}$$

If the composition of the mixture is given in mass fractions Y_i, the volume fractions X_i can be calculated by

Table 7-1

Properties of Some Important Gases

Kind of Gas	Molecular Weight	Ideal Gas Specific Gravity*	Ideal Gas Total Heating Value† BTU/ft³	Ideal Gas Net Heating Value† BTU/ft³	Compressibility Summation Factor
Hydrocarbons					
Methane	16.043	0.5539	1012.0	911.2	0.0436
Ethane	30.070	1.0382	1772.9	1621.6	0.0917
Propane	44.097	1.5224	2523.0	2321.4	0.1342
Isobutane	58.124	2.0067	3260.1	3008.0	0.1744
n-Butane	58.124	2.0067	3269.6	3017.5	0.1825
Butanes, avg.	58.124	2.0067	3265	3013	0.178
Isopentane	72.151	2.4910	4009.4	3706.4	0.2276
n-Pentane	72.151	2.4910	4018.5	3716.0	0.2377
Pentanes, avg.	72.151	2.4910	4014	3711	0.233
Hexanes, avg.	86.178	2.9753	4758	4405	0.283
Benzene	78.108	2.6969	3750.6	3599.3	0.266
C_8 Aromatics, avg.	–	3.6654	5224	4972	0.36
Non–hydrocarbons					
Air	28.964	1.0000			0.0202
Carbon dioxide	44.010	1.5194			0.0640
Carbon monoxide	28.010	0.9671	321.3	321.3	0.0217
Helium	4.000	0.1382			−0.0170
Hydrogen	2.016	0.0696	324.9	274.5	
Hydrogen sulfide	34.076	1.1765	638.6	588.2	0.0985
Nitrogen	28.013	0.9672			0.0164
Oxygen	32.000	1.1048			0.0270

Notes: * Specific gravity based on ratio of molecular weight to that of air (28.9644).
† Combustion at 60°F; volume of a pound mole of ideal gas at 14.73 psia (101.56 kPa) = 378.601 ft³ (10.798 m³).

$$X_i = \frac{(Y_i / MW_i)}{\sum_{i=1}^{n}\left(\frac{Y_i}{MWi}\right)} \quad (7\text{-}5)$$

where MW_i = molecular weight of *ith* component

There are a few good reasons for measuring heat of combustion as Btu/lb or similar, and not the least is that the variation in heating value as among the usual hydrocarbons is not instinctively understood if it is always presented in terms of heat value/volume unit. Note this comparison of heat values for just two gases:

Compound	Btu/ft³	Btu/lb
Methane	1012	23 898
Propane	2523	21 768
Δ Btu	1511 Btu	2 220 Btu
Δ %	149 %	9.3 %

Liquids

There are no really good, *practical*, theoretical calculations for heating value of composite mixtures of liquids. This is primarily because we don't know what's in them. There are estimates that the number of hydrocarbons in a typical crude runs into the thousands, for instance. To even think about using Equation 7-1 to solve such a problem is completely out of reason even if the occurrence and composition of the component liquids were accurately known.

We can start out with a few factors for components of complex hydrocarbons and then reason from that point onward.

Component	Symbol	High Heat Btu/lb
Carbon to CO_2	C	14 096
Carbon to CO	C	3 960
CO to CO_2	CO	4 346
Sulfur to SO_2	S	3 984
Hydrogen	H	61 031
Pentane	C_5H_{12}	21 095
Hexane	C_6H_{14}	20 675
Octane	C_8H_{18}	20 529
n–Decane	$C_{10}H_{22}$	20 371

We immediately see the overpowering effect of hydrogen; however, it doesn't weigh very much.

Then we see the effect of burning separately–occurring non–hydrocarbon, components, such as carbon, carbon monoxide, carbon dioxide, and sulfur. These burn, but except for carbon, they don't liberate a lot of heat; they are not *inert*, however. Inert components, such as nitrogen, occupy space, don't burn, and soak up heat from the combustion reaction.

Lastly, we begin to see the constancy, if that's the word, of the heating values of pure liquid hydrocarbons —the pentane and heavier fractions; the heat value per pound only drops about 3.5% when the carbon element doubles from C_5 to C_{10}. Inasmuch as the molecular weight of pentane is about 72, and that of n–decane is about 144, we can expect hydrocarbons with a molecular weight of 200–300 or so to have a heating value of about 18 000–19 000, which they do, more or less.

There is a graph in Nelson[8] due to Hougens and Watson, *Chemical Process Engineering,* vol. 1, John Wiley & Sons, that relates gross heat of combustion to API gravity with the Nelson characterization factor, $K = \frac{\sqrt[3]{T_B}}{s}$, where T_B = average boiling point in °R and s = specific gravity 60/60°F, are parameters. Nelson says this graph is very accurate if the percentage of water, ash and sulfur are accounted for; I have not been able to deduce the mathematical basis for this statement.

Solids

About the only solid fuel the civilized world is interested in, aside from wood logs used to heat homes and animal chips used to heat water for tea, is coal. There is an entire body of literature on coal, and just about any aspect of coal has its own experts who have investigated it and written widely on their findings. However, as a pipeliner, all you are apt to be interested in is that coal is made up of carbon, hydrogen, oxygen, nitrogen, sulfur, and ash; and that there are ways of analyzing coal to determine the makeup of any particular kind of coal. And, coal has water in it; this may be *surface water* or *entrained water* depending, as you would expect, on where it is found.

If you know the analysis of dry coal its high heat can be found by the ASME Power Code formula

$$H_Q = 145.4C + \left(H_2 - \frac{O_2}{8}\right) + 40.5S \qquad (7\text{-}6)$$

Measuring Heat of Combustion

Liquids

The standard method of test for heat of combustion for liquid hydrocarbons is ASTM D 240.[9] This also is a bomb–calorimeter test and also yields gross heat at constant volume. If the percentage of hydrogen in the sample is known the net heat of combustion can be calculated as

$$H_n = 1.8H_g - 91.23 \times H \qquad (7\text{-}7)$$

where H_n = net heat of combustion, Btu/lb (1 Btu/lb = 2.326 kJ/kg)
H_g = gross heat of combustion, cal/g (1 cal/g = 4.1868 kJ/kg)
H = hydrogen in the sample, percent

If the percentage of hydrogen in aviation gasoline and turbine fuel samples is not known, the net heat of combustion may be calculated as

$$H_n = 4310 + (0.7195)(1.8)(H_g) \quad (7\text{-}8)$$

where H_n = net heat of combustion, Btu/lb
H_g = gross heat of combustion, cal/g

This equation is strictly empirical, being based on data from papers cited in the standard, but if nothing better is available, use it. A more accurate method, applicable to Avgas 100–130 and 115–145, JP–4 and JP–5 jet fuel, and kerosine, Jet A or A–1, is available in ASTM D1405.[10] This method requires knowledge of the aniline point (ASTM D 611, or IP/2) and the API gravity. The method yields repeatable results to within 5 Btu/lb and reproducible results within 15 Btu/lb. Since the heat value of JP–4 runs about 18 500 Btu/lb, the 15 Btu/lb limit represents an error of only a fraction of one percent, certainly an acceptable accuracy for any purpose except—maybe—fuel consumption tests on a new engine.

Gases

Once upon a time there was a bomb–calorimeter kind of test for gas fuels, ASTM D900,[11] but it has long been replaced by the continuous–calorimeter method, ASTM D1826[12], for gases in the range from 900 to 1200 Btu/ft³. The heating value reported is the total calorific value in *Btu per standard cubic foot* of gas obtained by complete combustion at constant pressure in air with the temperature of the gas, air, and products of combustion being at 60°F and all of the water formed by the combustion reaction being condensed to the liquid state. The standard cubic foot referenced is the so–called *wet gas* or *saturated* gas cubic foot, and at the standard pressure of 14.73 psia the partial pressure of the gas is 14.4739 psia, the 0.2561 partial pressure of water vapor at 60°F being subtracted from the 14.73 psia standard.

The U.S. National Bureau of Standards tested the continuous calorimeter method over a four year period and found that, given that the apparatus was operated and periodically calibrated in accordance with the standard, reproducible results within 0.5 percent could be expected.

Note the results are given at 14.73 psia, 60°F, saturated. ASTM D1826 has a method to translate a measurement in these terms to other bases, and Table 7-2 is a summary of such calculations for a wide range of special measurement bases.

Table 7-2

Factor to Apply to Heating Value Given in BTU per Cubic Foot at 14.73 psia, 60°F, Saturated, to Give Result in a Special Base of Measurement.

Note: V_t at 32°F = 0.0885 psi
V_t at 60°F = 0.2561 psi
V_t at 15°C = 0.2471 psi

Special Base Condition	Factor F
14.40 psia, 60°F, dry	0.9949
14.65 psia, 60°F, saturated	0.9945
14.65 psia, 60°F, dry	1.0122
14.70 psia, 60°F, dry	0.9979
14.70 psia, 60°F, saturated	0.9979
14.80 psia, 60°F, saturated	1.0048
14.90 psia, 60°F, saturated	1.0117
14.90 psia, 60°F, dry	1.0295
14.95 psia, 60°F, saturated	1.0152
15.025 psia, 60°F, dry	1.0381
15.20 psia, 60°F, saturated	1.0325
29.3 in Hg[A], 60°F, saturated	0.9766
30.0 in Hg[A], 60°F, saturated	1.0003
30.0 in Hg[A], 60°F, 15% saturated	1.0154
30.0 in Hg[A], 32°F, saturated	1.0695
30.0 in Hg[A], 60°F, dry	1.0180
One standard atmosphere[B], 60°F, dry	1.0154
One standard atmosphere[B], 60°F, saturated	0.9976
One standard atmosphere[B], 15°C, dry	1.0173
One standard atmosphere[B], 15°C, saturated	1.0002

Note: A Column of pure mercury at 32°F under standard gravity (32.174 ft/s²)
B 101325 Pa = 14.69595 psi.

Solids

The standard method of test for heat of combustion of coal is a ASTM D3286.[13] This is a bomb–calorimeter test yielding the *gross heat of combustion at constant volume*. There is little likelihood you will ever need this standard.

Estimating Heat of Combustion

The preceding two sections cover the theory and measurement of heat of combustion. Sometimes it is good enough to estimate—or predict—heating values. This section is intended to provide bases for doing that.

Hydrocarbon Liquids

One of the most interesting, practical, treatises on some important properties of hydrocarbons is very old, out of print, and nearly unobtainable except through library sources that will photocopy it for you. This is the justly famous *Thermal Properties of Petroleum Products*[14] compiled by C. S. Cragoe of the U.S. National Bureau of Standards in 1929. This little volume—it has less than 50 pages—has more useful information on the properties of petroleum hydrocarbons than any other publication of any size I have in my library. It is not precise, and most of its methods have been refined many times over the years, but you can obtain *useful* answers very quickly with its methods.

As to heat of combustion, Cragoe gives the following equations for Q_V, the gross (high) heat of combustion *at constant volume*:

$$Btu/lb = 22\,320 - 3\,780d^2 \tag{7-9}$$

$$Btu/gal = 186\,087d - 31\,515d^3 \tag{7-10}$$

$$cal/g = 12\,400 - 2\,100d^2 \tag{7-11}$$

$$cal/ml = 12\,400d - 2\,100d^3 \tag{7-12}$$

In these equations d is specific gravity 60/60°F. The equations are valid for $0.51 > d > 0.99$ for all ordinary temperatures and yield results accurate to within 1%.

For most practical applications of the combustion of petroleum products the process is carried out at *constant pressure* (atmospheric) and the water vapor is not condensed. The net (low) heat of combustion at *constant pressure* can be calculated with the above relations by correcting Q_V for gross (high) heating value, as follows:

$$Q_P = Q_V - \%H\left[(9 \times 585) - 220\right] \times 0.01 \tag{7-13}$$

In this equation, 9 is the number of grams of water formed from 1 gram of hydrogen, 585 is the latent heat of vaporization of water at 20°C (68°F), and 220 is a small correction to take into account the change in volume from the initial to the final product. An average value for the percentage of hydrogen in oils can be obtained by

$$\%H = 26 - 15d \tag{7-14}$$

This equation will yield hydrogen content within 1% for a wide range of petroleum fractions and oils.

All these equations are for clean, water free, oils. For commercial products, which may (and probably do) have appreciable quantities of water, sulfur, or ash, Cragoe gives the following equations for estimating values for dirty oils:

$$\overline{Q_V} = Q_V - \frac{Q_V}{100}(\%\,H_2O + \%\,ash + \%\,S) + X(\%\,S) \tag{7-15}$$

$$\overline{Q_P} = Q_P - \frac{Q_P}{100}(\%\,H_2O + \%\,ash + \%\,S) + X(\%\,S) - Y(\%\,H_2) \tag{7-16}$$

Values for X and Y are taken from the following.

Units	X	Y
cal/g	22.5	5.85
Btu/lb	40.5	10.53
Btu/gal	338d	87.8d

For a 25°API oil with 0.5% water, 0.1% ash, and 1.0% sulfur the gross heating value drops from 145 000 Btu/gal (clean) to 142,936 Btu/gal (as found), a reduction of 2064 Btu/gal or some 1.43%. Not a lot, unless you figure that every 100 days you'll burn an extra 1.43 days worth of fuel over that calculated from the clean oil.

There is a much later set of equations presented in a paper by Dick Foster–Pegg[15] which reviewed NBS No. 97 (see above), ASTM D1405 (see previous section), proposed ASTM 1963 (found unacceptable), and ASTM 1969 (also found unacceptable). Foster–Pegg's equations are:

$$HHV = 17\,454 + 77.3G - 0.52G^2 - 20S \tag{7-17}$$

$$LHV = 16\,660 + 66G - 0.52G^2 - 20S \tag{7-18}$$

where HHV = gross heating value, Btu / lb
LHV = net heating value, Btu / lb
S = weight percent sulfur
G = API gravity

These equations are said to be within the 55 Btu/lb repeatability and the 175 Btu/lb reproducibility of ASTM D240 and thus should be sufficiently accurate for any ordinary calculation.

Hydrocarbon Gases

There are similar estimating methods for gases. The August 1965 *Rules of Thumb*[16] gives the Headlee Formula:

$$\frac{Btu}{ft^3} = 1545spgr + 140 - 16.33(\%H_2) \tag{7-19}$$

The Btu value is total heating value, 30 in Hg, 60°F, saturated. Example: A gas with 0.670 specific gravity and 1.0% nitrogen content would have a heating value of (1545×0.670)+140–(16.33×1.0)=1159 Btu.

Yost[17] has a pair of formulas, one for Btu/ft³ at 14.73 psia and 60°F, and the other for dth/lb (1 *dekatherm* = 1 000 000 Btu), which have their origin in the AGA "Report on Fuel Gas Energy Metering" of 1968.

$$Btu/ft^3 = 1571.5(spgr) + 144 - (\text{vol inerts}) \tag{7-20}$$

$$dekatherm/lb_{mass} = 0.020350 + 0.001970/spgr - \left(\frac{mass\ inerts}{spgr}\right) \tag{7-21}$$

Note the first of these equations is about the same as the Headlee equation.

Caldwell,[18] in working with the same general form of equations, points out an interesting phenomenon:

> A plot of specific gravity -vs- heating value of pure compounds is non–linear. Strangely enough, however, any plot of hydrocarbon gas mixtures over any likely range of specific gravity or heating value will be linear.
>
> The logic lies in Boyle's law deviation of mixed hydrocarbon compounds; deviation cannot be found by the simple summation of mole fractions multiplied by their respective pure–compound heating values. This also indicates that on–stream fractionation equipment, such as process–type chromatographs, cannot be used to obtain calorific summations without some loss in accuracy.
>
> On the other hand, specific gravity, when taken as a ratio of molecular weight of air to that of a hydrocarbon gas mixture, will unvaryingly reveal heating value of a mixture. Accuracy cannot be surpassed by meticulous calculations or precise calorimetry. This point has been demonstrated many thousands of times throughout the range of any mixture likely to be encountered.

Caldwell reports the same equations as Equation 7-20 and 7-21 above, and cautions one must be sure to take into account the effects of the inerts, such as CO_2 and N_2.

Hydrocarbon Solids

You may think you will never need any information on coal–water mixtures until you have to figure a slurry pipeline. Then you'll need it. The following gives the bases.

Coal has water in it; how much water is always a question, but detailed methods of analysis are available. Given the analysis of a coal on a dry–coal basis, the properties of a coal–water slurry can be determined by multiplying each constituent by

$$K = \frac{100 - W_T}{100} \tag{7-22}$$

where W_T = percent water content

Table 7-3, taken from Shvartsburd,[19] shows the effect on the heating value of a particular coal for which the dry–coal analysis was available for making up slurries with 10, 20, 30 and 40 percent weight concentrations of water. The reference also includes a method to calculate the reduction in boiler efficiency over theoretical efficiency by taking into account lost heat in the dry gas, loss due to hydrogen combustion, loss due to fuel moisture and water and, what I like best about the analysis, the assumed losses due to *combustible refuse*, *radiation*, and *"unmeasured,"* which are all grouped into a nice, concise mathematical function: $\sum_q (the\ rest)$.

Table 7-3
Composition and High Calorific Value of Coal–Water Mixtures

Fuel Composition–Weight Percent							
H_2O	C	H_2	O_2	N_2	S	Ash	Q_H Btu/lb
Dry Coal							
0.0	75.7	5.4	11.1	1.5	0.7	5.6	13 524
Coal–Water							
10.0	68.1	4.9	10.0	1.4	0.6	5.0	12 172
20.0	60.6	4.3	8.9	1.2	0.5	4.5	10 819
30.0	53.0	3.8	7.8	1.0	0.5	3.9	9 467
40.0	45.4	3.2	6.7	0.9	0.4	3.4	8 114

Typical Values

Tables 7-1 and 7-2 give heating value for the basic hydrocarbon gases and some important non–hydrocarbon gases together with a method of correcting these data for Table 7-4[20] is a little different, in that it also gives values for some common commercial gases, including two natural gases that have a high ethane content, a low nitrogen content and, as a consequence, a high heating value. Some of the most prolific gas fields in the world are not so lucky; the famous Groningen field in The Netherlands, one of the world's largest, has some 15% nitrogen and as a consequence a low heating value. But there is a *lot* of it.

Table 7-5 is a selection from a set of factors that ran monthly for years inside the back cover of *Petroleum Press Service* under the title (as I recall) *Conversion*

Table 7-4
Heating Value of Gases

Type of Gas	Composition Per Cent by Volume										Heating Value Per 100 SCF		Ratio
	H	CO	CH_4	C_2H_6	C_3H_8	C_4H_{10}	C_nH_{2n}	CO_2	N_2	O_2	High	Low	Hi/Lo
Hydrogen	100.0										319.3	269.7	0.845
Carbon Monoxide		100.0									316.0	316.0	1.000
Methane			100.0								996.9	897.6	0.900
Ethane				100.0							1758.0	1608.0	0.915
Propane					100.0						2529.0	2327.0	0.920
n-Butane						100.0					3333.0	3076.0	0.923
Blast Furnace	2.0	27.0						13.0	58.0		91.7	90.7	0.989
Producer	10.5	22.0	2.6				0.4	5.7	58.8		135.0	127.0	0.941
Blue Water	51.8	43.4						3.5	6.9	0.6	302.0	276.5	0.916
Caraburetor Water	35.0	33.0	11.0				10.0	3.5	1.3		550.0	509.0	0.926
Oil	13.5	2.9	53.1				17.9	2.6	9.6	0.4	1020.0	938.0	0.919
Natural No 1			90.0	8.9					1.1		1055.0	951.0	0.901
Natural No 2			78.6	10.3				9.7	1.4		965.0	871.0	0.903
Propane-Air No 1					21.7				61.9		550.0	504.0	0.916
Propane-Air No 2					53.4				36.3	10.3	1350.0	1239.0	0.916
Butane-Air No 1						15.0			67.1	17.9	550.0	507.0	0.923
Butane-Air No 2						43.5			44.6	11.9	1450.0	1338.0	0.923

Note: Heating values are in BTU/cf at 30 inches Hg (14.73 psia), 60°F, saturated.

Table 7-5

Conversion Factors Used in The Oil Industry

Product Specific Gravity Ranges

Product	Specific Gravity	Bbls per Meton
Crude oils	0.80-0.97	8.0-6.6
Aviation gasolines	0.70-0.78	9.1-8.2
Motor gasolines	0.71-0.79	9.0-8.1
Kerosines	0.78-.084	8.2-7.6
Gas oils	0.82-0.90	7.8-7.1
Diesel oils	0.82-0.92	7.8-6.9
Lubricating oils	0.85-0.95	7.5-6.7
Fuel oils	0.92-0.99	6.9-6.5
Asphaltic bitumens	1.00-1.10	6.4-5.8

Heat Energy Content of Fuels

Product	MJ/kg	Btu/lb
Crude oils	42.6-45.4	18 300-19 500
Gasolines	47.7	20 500
Kerosines	46.1	19 800
Benzole	42.1	18 100
Ethanol	27.0	11 600
Gas oils	44.7	19 200
Bunker fuel oils	42.6	18 300
Bituminous coal	23.7-34.0	10 200-14 600
LNG	51.9	22 300

Energy and Power

Unit	Value
International table (IT) calorie	4.1868 joules
15°C calorie	4.1855 joules
Thermochemical calorie	4.184 joules
Kilocalorie (IT)	1 000 calories
	3.96832 Btu
	1.163 watt hours
	0.001 thermie
Kilowatt hour	3412.14 Btu
	859.845 kcal
	3.6 megajoules
	1.34102 hp-hrs
Therm	100 000 Btu
	105.506 megajoules
	29.3071 hp-hrs
	25.1996 thermies
Metric horsepower (CV or PS)	735.499 watts
	542.476 ft $pounds_F$/sec
	0.986320 British hp
British horsepower	745.700 watts
	550 ft $pounds_F$/sec
	1.01387 metric hp
Kilowatt	737.562 ft $pounds_F$/sec
	1.35962 metric hp
	1.34102 British hp

Factors Used in the Oil Trade; and, after PPS became the *Petroleum Economist* and SI appeared on the scene, ran as *Conversion Factors Used in the Oil Industry*. The original had several classic conversions, such as 20 000 bpd = 10^6 MMTA (which it does, if the crude is about 33°API) and 7.5 bbls = 1 LT (which is does, if the oil is about 34°API—close enough to 33°API not to make much difference), and many a pipeline tender or tanker cargo was calculated with just these two factors. The last copy of the *Conversion Factors* I have is rather rigid, with all factors being either exact or correct to six significant figures, unless marked otherwise, and so Table 7-5 has two sub-tables of approximations and one table of precise conversions. All useful.

Table 7-6 is another set of factors derived from Dierdorff[21] which were the bases of a nomograph for estimating fuel consumption and costs. In this age of hand calculators and desktop computers, nomographs are not as useful as they once were. Dierdorff's collection of fuel data are as valuable now as they were when they were first published.

Lastly is Table 7-7.[22] This probably doesn't belong in this chapter—maybe in a chapter on LNG transfer lines and such—but the data are so hard to come by I thought it only right that I include it in this chapter on fuels and energy. If you ever need data on LNG, it's in this table. Note LNG, which is pure methane, is taken at 1000 Btu/ft³ HHV. This isn't precise, as was pointed out earlier; and if you need better accuracy you can multiply or divide by 1000/nnnn, where nnnn is your value for the HHV of methane.

Working Equations

Calculators

On the left side of Table 7-8, Reference and Working Equations for Heat of Combustion, are the fundamental equations for heat of combustion for mixtures of gases.

Table 7-6

Fuels/Energy Conversion Factors Used in Energy Studies Miscellaneous Fuels

Kind of Fuel	Energy
Crude oil	5 850 000 Btu/bbl
	5 800 000 Btu/bbl
Products, weighted average	5 522 000 Btu/bbl
Products, including NGL	5 518 000 Btu/bbl
Distillate fuel oil	5 825 000 Btu/bbl
Residual fuel oil	6 287 000 Btu/bbl
NGL, average	4 620 000 Btu/bbl
	4 011 000 Btu/bbl
Butane	4 284 000 Btu/bbl
	4 339 000 Btu/bbl
Propane	3 843 000 Btu/bbl
	3 847 000 Btu/bbl
Natural gas	1035 Btu/cf
	1032 Btu/cf
Coal, average	12 600 Btu/lb
	13 094 Btu/lb
	12-14 000 Btu/lb
Coal, anthracite	12 700 Btu/lb
Coal, bituminous and lignite, avg	12 010 Btu/lb
Power generation fuel input	
1970	10 769 Btu/kWh
1975	10 000 Btu/kWh
1985	9 500 Btu/kWh
2000	8 000 Btu/kWh

Fuel Oils

Kind of Fuel Oil	HHV Btu/gal
No. 6 fuel oil	
2.7% S	152 000
1.0% S	145 500
0.3% S	143 800
No. 5 fuel oil	150 000
1.0% S	144 500
No. 4 fuel oil	144 000
0.4 S	143 000
No. 2 fuel oil	140 000
Crude oil	138 000
Kerosine	
high	135 000
low	134 000
Naphtha	
heavy	125 000
light	117 000

On the right side are formulas for estimating the heat of combustion of coal, given the hydrogen, carbon, and sulfur content; the net heating value of liquid fuels, given the HHV and the hydrogen content; and both the HHV and LHV of petroleum hydrocarbons. The equations on the left side are classic and theoretically correct; those on the right side are strictly empirical, though drawn from fundamental principles.

Computers

There are five Computer Computation templates in this chapter.

CCT No. 7-1 is the classic IGT Research Bulletin No. 32 method for computing the heating value of a mixture of hydrocarbon and other common (mostly inert) gases given a mole fraction analysis of the mixture. The template can be expanded to include the full form layout included in Bulletin No. 32 if the text above the **Input** line is cut and the **Input** line moved up. The template is arranged for *MathCad* computation to yield intermediate and final answers to the same precision as those of Bulletin No. 32.

CCT No. 7-2 is an example of heating value analysis based on a volumetric analysis of the gas mixture.

CCT No. 7-3 provides a template for converting a gravimetric analysis to volumetric terms for analysis by **CCT** 7-2.

CCT 7-4 and 7-5 cover Cragoe's and Foster–Pegg's methods of estimating the heating values of petroleum hydrocarbons given the density and, in the case of the Foster–Pegg method, the sulfur content. Either method is sufficiently accurate for most engineering purposes. For commercial purposes, don't estimate; run a sample of the liquid through an ASTM D 240 bomb calorimeter test.

References

1 *Standard Practices for Stationary Diesel and Gas Engines*, Diesel Engine Manufacturers Association, 6th ed., 1972.

2 Baumeister, Theodore, Avallone, Eugene A., and Baumeister III, Theodore, *Marks' Standard Handbook for Mechanical Engineers*. New York: McGraw–Hill Book Company. 8th ed., 1978.

3 *Selected Values of Properties of Hydrocarbons and Related Compounds*, API Research Project 44. American Petroleum Institute, Washington.

Table 7-7
Conversion Factors for LNG

Based on: LNG at –258.9°F (–161.6°C)
LNG density = 3.48 lb/gal
Gas volumes measured at 14.7 psia, 60°F
HHV natural gas = 1000 Btu/cf

	Meton Liquid	Cubic Ft Liquid	Cubic M Liquid	Bbl Liquid	Gal Liquid	Cubic Ft Gas	Cubic M Gas	Million Btu	Million kcal
1 Cubic Ft Liquid	0.01183	1	0.02831	0.1781	7.479	625.4	16.79	0.6254	0.1576
1 Meton Liquid	1	84.56	2.394	15.06	632.5	52 890	1420	52.89	13.33
1 Cubic M Liquid	0.4177	35.32	1	6.29	265.4	22 090	593.1	22.09	5.567
1 Bbl Liquid	0.0664	5.625	0.1590	1	42	3512	94.27	3.512	0.8850
1 Gal Liquid	0.00158	0.1337	0.00379	0.02381	1	83.62	2.245	0.08362	0.02107
1 Cubic Ft Gas x 10^6	18.91	1599	45.27	284.8	11 960	10^6	26 850	1000	252
1 Cubic M Gas x 10^6	704.4	59 560	1686	10 610	445 400	$35.3 \cdot 10^6$	10^6	35 320	8900
1 Million Btu	0.01891	1.599	0.04527	0.2848	11.96	1000	26.85	1	0.252
1 Million kcal	0.07502	6.345	0.1796	1.130	47.46	3968	112.4	3.968	1

Other approximate equivalents:
1 meton LNG = 1.12 metons residual fuel
1 meton LNG = 1.84 metons bituminous coal
1 bbl LNG = 0.590 Bbl residual fuel
1 bbl LNG = 0.134 short tons bituminous coal

Table 7-8
Reference and Working Equations for Heat of Combustion

Heating Value for Mixture of Ideal Gases

$$H_T = X_1H_1 + X_2H_2 + X_3H_3 \ldots X_nH_n$$

where $X_1, X_2, \ldots X_n$ = mole fraction component gas
$H_1, H_2, \ldots H_n$ = heating value component gas
H_T = heating value composite gas

Heating Value of Real Gases–Mole Fractions

$$Z_T = 1 - \left(X_1\sqrt{b_1} + X_2\sqrt{b_2} + \ldots X_n\sqrt{b_n}\right)^2$$

$$H_R = H_T / Z_T$$

where $X_1, X_2 \ldots X_n$ = mole fractions
$\sqrt{b_1}, \sqrt{b_2} \ldots \sqrt{b_n}$ = empirical summation factors

Heating Value of Real Gases–Volume Fractions

If H_i is the heating value of the *ith* component, and X_i is the volume percent of the *ith* component then

$$H_{mix} = \sum_{i=1}^{n} X_iH_i$$

If the composition of the mixture is given in mass fractions Y_i, the volume fractions X_i can be calculated by

$$X_i = \frac{(Y_i/W_i)}{\sum_{i=1}^{n}\left(\frac{Y_i}{Wi}\right)}$$

where W_i = molecular weight of *ith* component

Estimated Heating Value of Solids–Coal

ASME $$H_Q = 145.4C + \left(H_2 - \frac{O_2}{8}\right) + 40.5S$$

Net Heating Value of Liquids if H_g is Known

ASTM, if H_2 is known, $$H_n = 1.8H_g - 91.23 \times H$$

where H_n = net heat of combustion, Btu/lb
(1 Btu/lb = 2.326 kJ/kg)
H_g = gross heat of combustion, cal/g
(1 cal/g = 4.1868 kJ/kg)
H = hydrogen in the sample, percent

ASTM Jet Fuels $$H_n = 4310 + (0.7195)(1.8)(H_g)$$

where H_n = met heat of combustion, Btu/lb
H_g = gross heat of combustion, cal/g

Heating Value of Liquid Hydrocarbons

Cragoe Method

$$Btu/lb = 22{,}320 - 3{,}780d^2$$
$$Btu/gal = 186{,}087d - 31{,}515d^3$$
$$cal/g = 12{,}400 - 2{,}100d^2$$
$$cal/ml = 12{,}400d - 2{,}100d^3$$

where d = spgr 60/60°F

Foster–Pegg Method

$$HHV = 17454 + 77.3G - 0.52G^2 - 20S$$

$$LHV = 16660 + 66G - 0.52G^2 - 20S$$

where HHV = gross heating value, Btu / lb
LHV = net heating value, Btu / lb
S = weight percent sulphur
G = API gravity

[4] *Calculation of Heating Value and Specific Gravity of Fuel Gases*, Research Bulletin No. 32. Institute of Gas Technology, Chicago.

[5] ASTM D3588, *Standard Method of Calculating Calorific Value and Specific Gravity (Relative Density) of Gaseous Fuels.* American Society for Testing Materials, Philadelphia.

[6] Kemp, Jim, "Calculated BTU From Analysis," *ENERGY Pipelines and Systems*, August 1974.

[7] ASTM D1945, *Standard Method for Analysis of Natural Gas by Gas Chromatography.* American Society for Testing Materials, Philadelphia.

[8] Nelson, W. L., *Petroleum Refinery Engineering,* p. 200. 4th ed. New York: McGraw–Hill Book Company, 1969.

[9] ASTM D240, *Standard Method of Test for Heat of Combustion of Liquid Hydrocarbons by Bomb Calorimeter.* American Society for Testing Materials, Philadelphia.

[10] ASTM D1405, *Standard Method for Estimation of Net Heat of Combustion of Aviation Fuels.* American Society for Testing Materials, Philadelphia.

[11] ASTM 900, *Standard Method of Test for Calorific Value of Gaseous Fuels by the Water–Flow Calorimeter.* American Society for Testing Materials, Philadelphia.

[12] ASTM D1826, *Standard Test Method for Calorific Value of Gases in Natural Gas Range by Continuous Recording Calorimeter.* American Society for Testing Materials, Philadelphia.

[13] ASTM D3286, *Test Method for Gross Calorific Value of Solid Fuel by the Isothermal–Jacket Bomb Calorimeter.* American Society for Testing Materials, Philadelphia.

[14] Cragoe, C. S., *Thermal Properties of Petroleum Products,* Bureau of Standards Miscellaneous Publication No. 97. U.S. Department of Commerce, November 9, 1929.

[15] Foster–Pegg, Dick, (edited) "Users Being Short–Changed by Standard Heating Value Curves," *Gas Turbine World,* October–November 1972.

[16] Rules of Thumb, *Pipe Line Industry*, August 1965.

[17] Yost, Kenneth C., "BTU Correction for Accurate Gas Measurement," *Petroleum Engineer,* February 1975.

[18] Caldwell, Bruce L., "Thermal Content: Coming Basis of Natural Gas Measurement," *Oil and Gas Journal,* 15 July 1968 (from paper given at AGA Gas Transmission Conference, Cleveland, 27 May 1968).

[19] Shvartsburd, Vladimir, "Criteria Given for Finding Combustion Properties of Coal–Water Mixes," *Oil and Gas Journal,* November 14, 1983.

[20] Adapted from *A.G.A. Gas Measurement Manual*, 1st ed. New York: American Gas Association, 1963.

[21] Dierdorff, Lee H., "Nomograph Speeds Fuels and Energy Cost Conversions," *Pipeline and Gas Journal*, March 1974.

[22] Author's Note: This table, pretty much in the form presented here, was published in the *Oil and Gas Journal* in the April 17, 1972 issue. My copy is a photocopy of the original, and I have no idea who originated the table. It is the best I know covering nearly any conversion you can think of involving LNG.

Computer Computation Template

CCT No. 7-1 Heating value of mixtures of hydrocarbon and inert gases based on a mole fraction analysis of the gas (such as that returned by a gas chromatograph). The procedure is that recommended in IGT Research Bulletin No. 32, and includes the compressibility summation factor. The example is drawn from Bulletin No. 32, but any number of component gases can be accomodated. Always adjust *n* for the number of gases.

References: Chapter 7. Equations 7-2 and 7-3. Table 7-1.

Input

Input Data	Gas No.	Name of Gas	Ideal Gas SpGr G_i	Ideal Gas HHV H_i	Summation Factor $b_i^{1/2}$	Mole Fraction of Component X_i	Number of Components
	1	hydrogen	$G_1 := 0.06957$	$H_1 := 325.02$	$SRb_1 := 0$	$X_1 := 0.164$	$n := 13$
	2	nitrogen	$G_2 := 0.9668$	$H_2 := 0$	$SRb_2 := 0.0164$	$X_2 := 0.004$	
	3	carbon dioxide	$G_3 := 1.5188$	$H_3 := 0$	$SRb_3 := 0.0640$	$X_3 := 0.002$	
	4	methane	$G_4 := 0.5536$	$H_4 := 1012.32$	$SRb_4 := 0.0436$	$X_4 := 0.404$	
	5	ethane	$G_5 := 1.0377$	$H_5 := 1773.42$	$SRb_5 := 0.0917$	$X_5 := 0.053$	
	6	propane	$G_6 := 1.5217$	$H_6 := 2523.82$	$SRb_6 := 0.1342$	$X_6 := 0.006$	
	7	ethylene	$G_7 := 0.9681$	$H_7 := 1603.75$	$SRb_7 := 0.0775$	$X_7 := 0.274$	
	8	propene	$G_8 := 1.4522$	$H_8 := 2339.70$	$SRb_8 := 0.1269$	$X_8 := 0.061$	
	9	butenes	$G_9 := 1.9362$	$H_9 := 3079.00$	$SRb_9 := 0.1810$	$X_9 := 0.006$	
	10	acetylene	$G_{10} := 0.8985$	$H_{10} := 1476.55$	$SRb_{10} := 0.0833$	$X_{10} := 0.002$	
	11	butadiene	$G_{11} := 1.8667$	$H_{11} := 2888.52$	$SRb_{11} := 0.1580$	$X_{11} := 0.017$	
	12	cyclopentadiene	$G_{12} := 2.2812$	$H_{12} := 3362.40$	$SRb_{12} := 0.2240$	$X_{12} := 0.002$	
	13	benzene	$G_{13} := 2.6956$	$H_{13} := 3751.68$	$SRb_{13} := 0.2660$	$X_{13} := 0.005$	

Equations

$$\sum_{i=1}^{n} X_i = 1.000 \qquad \sum_{i=1}^{n} X_i \cdot G_i = 0.7231 \qquad \sum_{i=1}^{n} X_i \cdot H_i = 1249.6 \qquad \sum_{i=1}^{n} X_i \cdot SRb_i = 0.0582$$

$$z := \left[1 - \left(\sum_{i=1}^{n} X_i \cdot SRb_i\right)^2\right] + 0.0005 \cdot \left[2 \cdot 0.164 - (0.164)^2\right] = 0.99677 \qquad G_R := \frac{\sum_{i=1}^{n} X_i \cdot G_i}{z} \qquad G_R = 0.7255$$

Results

$$H_{RDry} := \frac{\sum_{i=1}^{n} X_i \cdot H_i}{z} \qquad H_{RDry} = 1253.6 \text{ Btu/SCF, 60 °F, 30" Hg, dry}$$

$$H_{RSat} := 1253.6 \cdot 0.9826 \qquad H_{RSat} = 1231.8 \text{ Btu/SCF, 60 °F, 30" Hg, saturated}$$

Computer Computation Template

CCT No. 7-2 Heating value of mixtures of hydrocarbon and inert gases based on a volumetric analysis of the gas. The procedure allows for any number of gases *n*, given the volume parts and heating values of the component gases.

References: Chapter 7. Equation 7-4. Table 7-1.

Input

Factors — Number of components: $n := 6$ $i := 1, 2 .. n$

Input Data

Gas No.	Name of Gas	Vol Part of Gas	HHV of Gas Btu/ft³ at 60°F
1	carbon dioxide	$V_1 := 0.001$	$HHV_1 := 0$
2	methane	$V_2 := 0.789$	$HHV_2 := 1012.0$
4	ethane	$V_3 := 0.060$	$HHV_3 := 1772.9$
4	propane	$V_4 := 0.034$	$HHV_4 := 2523.0$
5	n-butane	$V_5 := 0.010$	$HHV_5 := 3269.6$
6	nitrogen	$V_6 := 0.106$	$HHV_6 := 0$

Equations

$$Vol_{Tot} := \sum_{i=1}^{n} V_i \qquad HHV_{Tot} := \sum_{i=1}^{n} V_i \cdot HHV_i$$

Results

$Vol_{Tot} = 1.000$

$HHV_{Tot} = 1.023 \cdot 10^3$ Btu/ft³ at 60°F

V_i	$V_i \cdot HHV_i$
0.001	0
0.789	798.468
0.060	106.374
0.034	85.782
0.010	32.696
0.106	0

Computer Computation Template

CCT No. 7-3 Sometimes—not often any more—a gas analysis is given in terms of weight fractions (a gravimetric analysis) instead of volume fractions (volumetric analysis) or mole fractions. For these cases, convert the gravimetric analysis values to volumetric values as follows.

References: Chapter 7. Equation 7-5. Table 7-1.

Input

Factors — Number of components: $n := 7$ $\quad i := 1, 2..n$

Input Data

Gas No.	Name of Gas	Mass Fraction %	Molecular Weight
1	carbon monoxide	$W_1 := 4.90$	$MW_1 := 28$
2	oxygen	$W_2 := 18.90$	$MW_2 := 32$
3	nitrogen	$W_3 := 70.40$	$MW_3 := 28$
4	hydrogen	$W_4 := 0.87$	$MW_4 := 2.016$
5	methane	$W_5 := 1.61$	$MW_5 := 16.032$
6	carbon dioxide	$W_6 := 1.72$	$MW_6 := 44$
7	water	$W_7 := 1.60$	$MW_7 := 18.016$

Equations

$$W_{Tot} := \sum_{i=1}^{n} W_i \qquad W_{Tot} = 100.000 \qquad \sum_{i=1}^{n} \frac{W_i}{MW_i} = 3.940 \qquad Vol_i := \frac{\frac{W_i}{MW_i}}{\sum_{i=1}^{n} \frac{W_i}{MW_i}} \cdot 100 \qquad Vol_{Tot} := \sum_{i=1}^{n} \frac{\frac{W_i}{MW_i}}{\sum_{i=1}^{n} \frac{W_i}{MW_i}} \cdot 100$$

Results

$W_{Tot} = 100.000 \qquad Vol_{Tot} = 100.000$

i	W_i	$\frac{W_i}{MW_i}$	Vol_i	$\frac{Vol_i \cdot MW_i}{100}$
1	4.900	0.175	4.442	1.244
2	18.900	0.591	14.991	4.797
3	70.400	2.514	63.818	17.869
4	0.870	0.432	10.954	0.221
5	1.610	0.100	2.549	0.409
6	1.720	0.039	0.992	0.437
7	1.600	0.089	2.254	0.406

$$MW_{Mixture} := \sum_{i=1}^{n} \frac{Vol_i \cdot MW_i}{100}$$

$$MW_{Mixture} = 25.382$$

Computer Computation Template

CCT No. 7-4: Estimating the total (high) heating value at constant pressure of liquid hydrocarbons using the Cragoe method. The single–value results are those calculated for a single value of ρ, the plotted values are results plotted on a spread of d whch is a function of ρ .

References: Chapter 7. Equations 7-9, 7-10, 7011, and 7-12.

Input

Factors $i := 0, 1 .. 11$

Input Data $\rho := 0.700$ Ref $H_2O = 1.000$

Variables $d_i := \rho + i \cdot 0.020$

Equations

$$Btuperlb := 22320 - 3780 \cdot \rho^2$$

$$BtuperUSgal := 186087 \cdot \rho - 31515 \cdot \rho^3$$

$$Caloriespergram := 12400 - 2100 \cdot \rho^2$$

$$Caloriesperml := 12400 \cdot \rho - 2100 \cdot \rho^3$$

$$Btulb_i := 22320 - 3790 \cdot (d_i)^2$$

$$Btugal_i := 186087 \cdot d_i - 31515 \cdot (d_i)^3$$

$$Caloriespergram_i := 12400 - 2100 \cdot (d_i)^2$$

$$Caloriesperml_i := 12400 \cdot d_i - 2100 \cdot (d_i)^3$$

Results

$Btuperlb = 20468$

$BtuperUSgal = 119451$

$Caloriespergram = 11371$

$Caloriesperml = 7960$

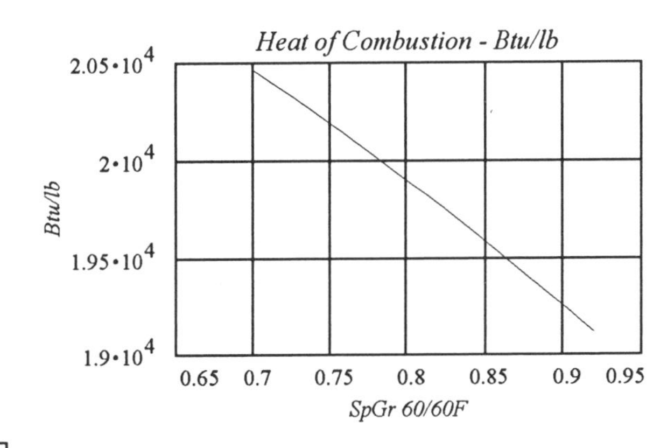

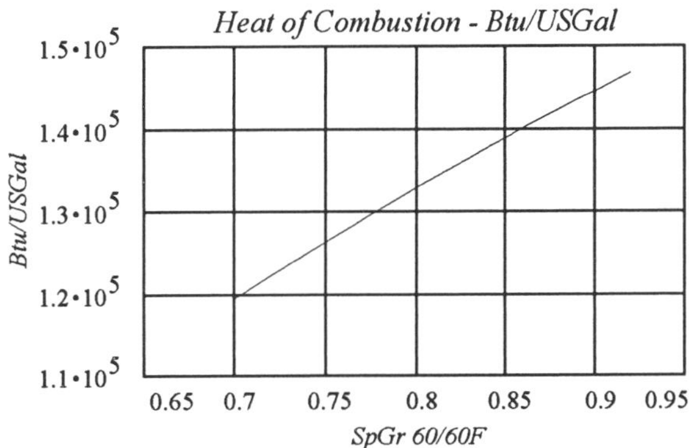

Computer Computation Template

CCT No. 7-5: Estimating the heating value of liquid hydrocarbons using the Foster–Pegg method. Note that the density function is in terms of °API. The single–value results are those calculated for a single value of *API*; the plotted values are results plotted on a spread of *d* which is a function of *API*.

References: Chapter 7. Equations 7-17 and 7-18.

Input

$i := 0, 1 .. 8$

Factors

Input Data $API := 10$ $S := 2.0$ Weight percent Sulfur

Variables $d_i := API + i \cdot 5$

Equations

$$HighHeatValue := 17454 + 77.3 \cdot API - 0.52 \cdot API^2 - 20 \cdot S$$

$$LowHeatValue := 16660 + 66 \cdot API - 0.52 \cdot API^2 - 20 \cdot S$$

$$HHV_i := 17454 + 77.3 \cdot d_i - 0.52 \cdot \left(d_i\right)^2 - 20 \cdot S$$

$$LHV_i := 16660 + 66 \cdot d_i - 0.52 \cdot \left(d_i\right)^2 - 20 \cdot S$$

Results

$HighHeatValue = 18135$ Btu / lb HHV

$LowHeatValue = 17228$ Btu / lb LHV

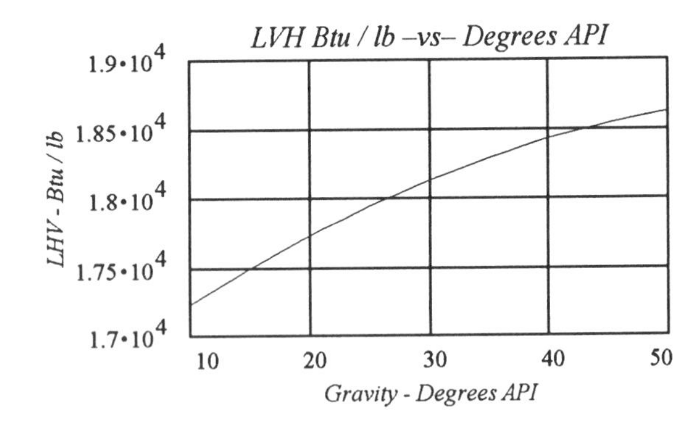

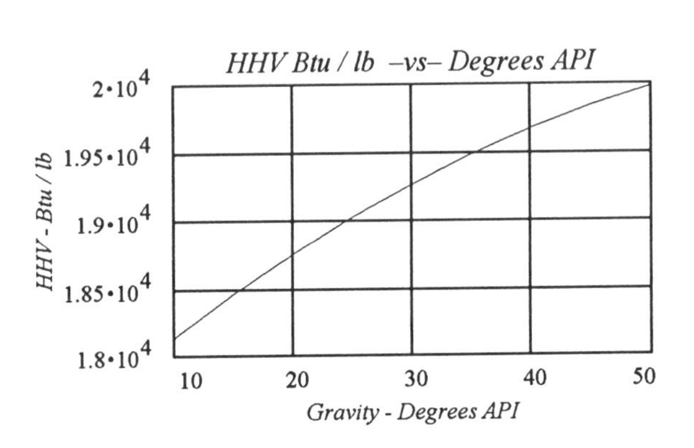

8
Adiabatic Bulk Modulus

Introduction

This has been the hardest chapter of the book, so far, to write. I rewrote it three times. You are looking at the fourth edition. It may be rewritten again if new data becomes available before the book goes to the printer. And, unless you get yourself into some severe pressure surge problems in a pretty large pipeline, you probably won't even need the information contained in it. But I have been in such situations, and I know the value of having this kind of information readily at hand. That's why this chapter is here.

The Bulk Modulus

Few pipeliners understand the difference between the bulk modulus given in tables of fluid properties, which is the *isothermal bulk modulus*, and the *adiabatic bulk modulus*, which is the bulk modulus pipeliners should be interested in any time a bulk modulus enters into a computation involving a moving fluid. Note the use of *fluid* instead of *liquid*; gases and, by extension, vapors, have a bulk modulus, too.

The Adiabatic Bulk Modulus

The compressibility and bulk modulus of liquids covered in Chapter 2 that have to do with measurement of volumes of liquids are the *isothermal compressibility* and the *isothermal bulk modulus*; they are *usually* measured on still–standing liquids at constant temperature, and are also *usually* measured at atmospheric pressure.

For years this kind of bulk modulus was about all that we had; nobody was measuring any other kind. The bad thing about this situation was that we were lacking a key ingredient in a part of pipeline hydraulics that has become more and more important as pipelines have become larger and larger and throughputs have become correspondingly greater—the calculation and control of pressure surges.

Fundamental in calculations involving pressure surges is the determination of the velocity of sound, or *celerity*, after the French word *célérité*, in the flowing medium; the velocity—*wavespeed*—of a transverse pressure wave moving up or down the length of a pipeline is exactly that of sound traveling through the same space.

Though the celerity in any fluid is modified by the size and material of the vessel containing it, it is essentially a function of bulk modulus and density, as follows:

$$c = \sqrt{\frac{dp}{d\rho}} = \sqrt{\frac{K_X}{\rho}} \tag{8-1}$$

These equations apply to *any fluid*. It applies as well to *any process X* if the value of K_X, the bulk modulus, is the modulus applicable to that process.

At standard conditions of 0°C and 760 mm Hg, the velocity of sound in open water is something less than 4800 ft / sec; in free air it is a little less than 1100 ft / sec. When fluids are confined in a pipe these velocities are modified. The speed of sound in a pipeline carrying crude oil is generally in the order of 1000 m or 1 km / sec.

In all of these situations, however, the passage of the wavefront is so quickly accomplished that, for practical purposes, there is no transfer of heat to or from the process—which is the *passage* of the wavefront—and the process is therefore an essentially *frictionless reversible adiabatic* or *isentropic* process. Thus the applicable bulk modulus for the calculation of celerity is the *adiabatic bulk modulus*, usually symbolized as K_S.

Early Developments

For years the pipeliner's references for compressibility and the bulk modulus were Jacobson[1] and Kerr[2] and the researches referenced by them. But essentially all these measurements were made on still liquids in closed containers, and the ratio of the change in pressure for a given change in volume, per unit volume, was the bulk modulus reported. The method allowed for varying the temperature—the bulk modulus could be determined at any temperature within reach of the apparatus—but the measurements were always made at constant temperature and the values so obtained were the *isothermal bulk modulus*, K_T.

For common substances—water is by far the best example—researchers measured and published the bulk modulus at different pressures and temperatures, but almost no table of values bothers to indicate, even in a footnote, that the value of bulk modulus is K_T; there are no other tables of any other kind of bulk modulus. Even Professor Bergeron,[3] in his deservedly famous treatise on the graphical method of solving pressure surge problems, took K for water as 2.07×10^8 kg / m²—about 294 000 psi—and that was all there was to it; no mention as to where the value came from, how it was measured, under what pressure and at what temperature, etc.

While those in the forefront of pressure surge theory and analysis have been aware since the times of Allievi, the father of the theory, that the isothermal bulk modulus is not the correct parameter to use in such analyses, three very important things worked together so that for years nobody worried too much about the fact that they were using the wrong number:

- Methods available for computing the effect of pressure surges didn't justify use of the correct, more accurate, modulus.
- The only fluid of interest was water, and the adiabatic bulk modulus of water isn't that far from the isothermal bulk modulus.
- The more accurate values weren't available, anyway.

Calculation of Bulk Modulus

For *any* fluid the bulk modulus may be defined as

$$K = -\frac{dp}{dV/V_1} = \frac{dp}{d\gamma/\gamma} = \frac{dp}{d\rho/\rho} \tag{8-2}$$

where K = bulk modulus, lbs/ft^2 or Pa
V_1 = volume of container, ft^3 or m^3
p = pressure, lbs/ft^2 or Pa
γ = specific weight, lbs/ft^3 or N/m^3
ρ = density, slugs/ft^3 or kg/m^3

The units shown are those for the FPS and SI systems, not necessarily those used in calculations. In the FPS system, for instance, pressures and bulk modulus are usually expressed in psi, not psf. Use of the units shown will assure coherency in the calculation, however.

If we assume the fluid is compressed in a cylinder, and further assume the cylinder and piston do not change in size in any way, we can write Equation 8-2 in a more easily understood way as

$$\left(\frac{F}{A}\right) = \left(\frac{dp}{dV/V_1}\right) \times \left(\frac{V}{V_1}\right) \tag{8-3}$$

The F/A term on the left is p, the *stress*; the V/V_1 term on the right is the *strain*; and the term in the center is K (Equation 8-2), the slope of the stress–strain curve. The stress–strain curve is not a straight line for any gas; for liquids it approximates a straight line over large changes in pressure. The value for the bulk modulus of a liquid is usually given as the value at atmospheric pressure, whatever the value of the temperature.

If the compression (or expansion) takes place at constant temperature (an *isothermal process*),

$$\frac{p}{\gamma} = \frac{p}{\rho} = \text{constant} \tag{8-4}$$

If it takes place in a frictionless process so no heat is exchanged, an *isentropic* or, what is nearly the same thing, an *adiabatic*, process results.

$$\frac{p}{\gamma^k} = \frac{p}{\rho^k} = \text{constant} \tag{8-5}$$

In both the above, γ is *specific weight*, ρ is *density*, p is *pressure*, and k is the *ratio of specific heats*. Thus Equation 8-2 can be written as

$$K = \frac{dp}{d\gamma/\gamma} = \frac{dp}{d\rho/\rho} \tag{8-6}$$

If Equation 8-6 is solved with Equations 8-4 and Equation 8-5 we find

$$\begin{aligned} K_T &= p \text{ for the isothermal process} \\ K_S &= kp \text{ for the isentropic process} \end{aligned} \tag{8-7}$$

Thus the difference between *isothermal bulk modulus* and the *isentropic* or *adiabatic bulk modulus* is the ratio c_p / c_v, the ratio of specific heats, sometimes called the *adiabatic exponent*; if you know the isothermal modulus and c_p / c_v, you know the adiabatic bulk modulus.

Classical Calculations

However, there are very few references in my library to bulk modulus determination, and this especially to the determination of the isothermal and adiabatic modulus of oils. I do have a reference from 1945, identified here as Cameron,[4] which was used as a basis for a private computation from Ludwig[5] in 1956, that guided me for years until the really seminal work by Wostl, Dresser, and Clark[6,7] in 1970 provided a way of measuring, and what is more important, a correlation that allows the estimation of, the adiabatic bulk modulus of petroleum oils with good accuracy.

Cameron was interested in the difference between the bulk modulus applicable to high–pressure hydraulic control systems, the isothermal bulk modulus, and that applicable to the direct injection of heavy oil fuel at very high pressures into the combustion chamber of diesel engines, which is the adiabatic bulk modulus. He started out first to make a complete literature search (probably a search only of British publications and those foreign publications available in the U.K.). There were only eleven, and one of these was a text book.

Cameron uses an equation of the form of Equation 6-15 for his computations:

$$C_p - C_v = \frac{TVA^2}{KJ} \tag{8-8}$$

which can be written in another form as

$$k = \frac{C_p}{C_v} = \frac{1}{1 - \dfrac{TVA^2}{\beta_T J C_p}} \tag{8-9}$$

where C_p = specific heat at constant pressure, cal/g
C_v = specific heat at constant volume, cal/g
T = absolute temperature, °C
V = volume per unit mass, cm^3
A = coefficient of expansion per °C
β_T = isothermal compressibility per g / cm^2
J = mechanical equivalent of heat
k = ratio of specific heats

Cameron has but one example, that of an oil having a C_p of 0.45 cal / g, a V of 1.11 cc / g for a density of 0.9 g / cc, A of 0.0007 / °C, β_T of 64.0 × 10^{-6} per kg / cm^2 (for a K_T of about 222 000 psi), at an assumed T of 25°C = 298°C absolute (today's terminology would be 298 K). This yielded k = 1.155 and β_S = 55.4 × 10^{-6} per kg/cm^2 (a K_S of about 257 000 psi).

Ludwig, who was interested in pressure surges on a 30–inch crude line—which was a very large pipeline in 1956, the date of the reference—wrote Equation 8-9 in terms of bulk modulus and density instead of compressibility and specific volume as follows:

$$k = \frac{C_p}{C_v} = \frac{1}{1 - \dfrac{\alpha^2 T K_T}{\rho C_p \times 778}} \tag{8-10}$$

Here the 778 ft–lb/Btu value for the mechanical equivalent of heat gives away the units of the equation which are, of course, in the FPS system.

Ludwig used an oil with a spgr of 0.85, corresponding to a density ρ of 53.04 lbs / ft^3; α, the coefficient of thermal expansion, of 0.445 × 10^{-3}/°F; bulk modulus K_T of 200 000 psi = 2.88 × 10^7 psf; C_p= 0.459 Btu/lb/°F; and T of 70°F = 530°R to yield k = 1.19 and an adiabatic bulk modulus, of 238 000 psi. The bulk modulus was an estimate of the modulus for a pressure of 500 psig.

Realizing that the values C_p and α for both Cameron and Ludwig were calculated from pre–WWII data, I remade both calculations using Equation 2-10 for α and Equation 6-29 for C_p. I arrived at $k = 1.18$ and K_S =

262 000 psi for the Cameron conditions and $k = 1.21$ and $K_S = 236\,000$ psi values for Ludwig's example. The old calculations weren't that bad.

Estimating Bulk Modulus

These kind of data—theoretical derivations of the adiabatic bulk modulus from sometimes dubious values of the isothermal bulk modulus—were all we had until the work of Wostl et al. (see References 6 and 7). This group measured the celerity of a pressure wave in oils under pressure under adiabatic conditions and from this value directly computed K_S as

$$K_S = V\left[\frac{\Delta P}{\Delta V}\right]_S = \frac{\rho c^2}{g} \tag{8-11}$$

where K_S = adiabatic bulk modulus, psi
ΔV = volume at pressure P_2 − volume at pressure P_1
ΔP = pressure P_2 − pressure P_1
ρ = density, lbs/in^3
g = acceleration due to gravity, in/sec^2
c = velocity of sound, in/sec

They could also measure K_T, the isothermal bulk modulus, with the same apparatus but felt that the accuracy of this measurement was not as good as that for K_S which was entirely different from any of the previous work.

The group compared their data on the isothermal bulk modulus of water and found their results for K_T at 68°F and 50 psig agreed well with the API 1101 value of 312 500 psi, which is an isothermal compressibility of 3.2×10^{-6}/psi, used as a *mean* value from 1 to 1000 psig, but their work at higher pressures, and at both higher and lower temperatures, differed substantially from the API 1101 values. Their data show that K_T for water is essentially linear with pressure but not with temperature. Curves are in Reference 7; their equations are not available.

For crude oil the basic equation for adiabatic bulk modulus is

$$\begin{aligned} K_S &= (1.286 \times 10^6) + (13.55 \times P) \\ &\quad - (4.122 \times 10^4 \times \sqrt{T}) \\ &\quad - (4.53 \times 10^3 \times A) - (10.59 \times A^2) \\ &\quad + (3.228 \times T \times A) \end{aligned} \tag{8-12}$$

where P = pressure, psig
A = API gravity at 60°F
T = temperature, °R = °F + 460

Figure 8-1 is a plot computed with Equation 8-12 for crudes from 10°API to 60°API at constant 1000 psig for temperatures from 40°F to 90°F at 10°F intervals.

The equation can be manipulated to illustrate some of the more important characteristics of K_S.

Figure 8-2 is a striking demonstration of the relative *insensitivity* of K_S to pressure; it increases from a little over 233 000 psi to a little over 260 000 psi—a ΔK_S of about 27 000 psi or about 12%—as pressure varies from 14.7 to 2014.7 psia, a change of nearly 14 000%. The change, calculated per the equation, is a straight line: $\Delta K_S = f(P) = 13.55 \times P$, where P is in psig.

Figure 8-3, on the other hand, shows the relative *sensitivity* of K_S to temperature. Here K_S varies from about 222 000 to 262 000 psi, a ΔK_S of 40 000 psi = 18%, for a 50° temperature change from 500 to 550°R which is a change of 10%. The variation is not quite linear, as can be seen from the equation.

Wostl et al. also derived a formula for calculating the isothermal bulk modulus K_T, and they go to considerable length to discourage its use for precise (i.e., commercial petroleum measurement) purposes, but for engineering purposes it is probably as accurate as anything we are apt to need in pipelining. The formula is

$$\begin{aligned} K_T &= 2.619 \times 10^6 + 9.203 \times P \\ &\quad - 1.417 \times 10^5 \times \sqrt{T} + 73.05 \times T^{1.5} \\ &\quad - 341 \times A^{1.5} \end{aligned} \tag{8-13}$$

Comparison of Calculated and Estimated Values

To compare the results of the above calculations, I took Cameron's and Ludwig's original results, the results I reached for the same problems using (what should be) better values for α and C_p, and the results of the calculations for the same two problems based on Equation 8-12 for 500 psig, and placed them in table form as follows:

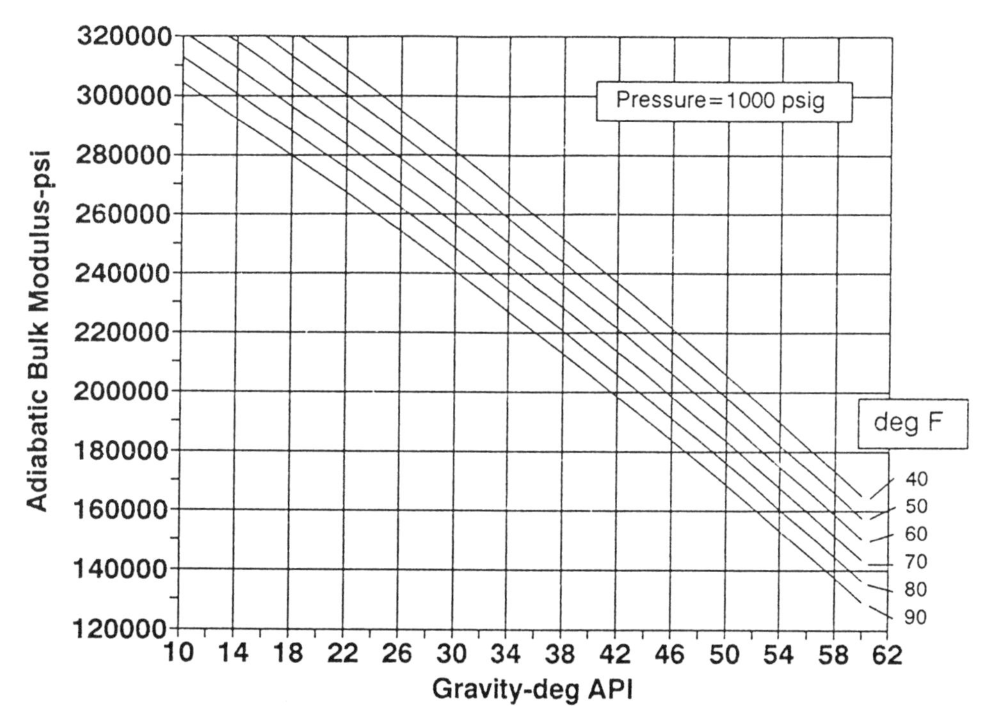

Figure 8-1. Adiabatic Bulk Modulus -vs- API Gravity for Crude Oils at Constant Pressure for Various Temperatures.

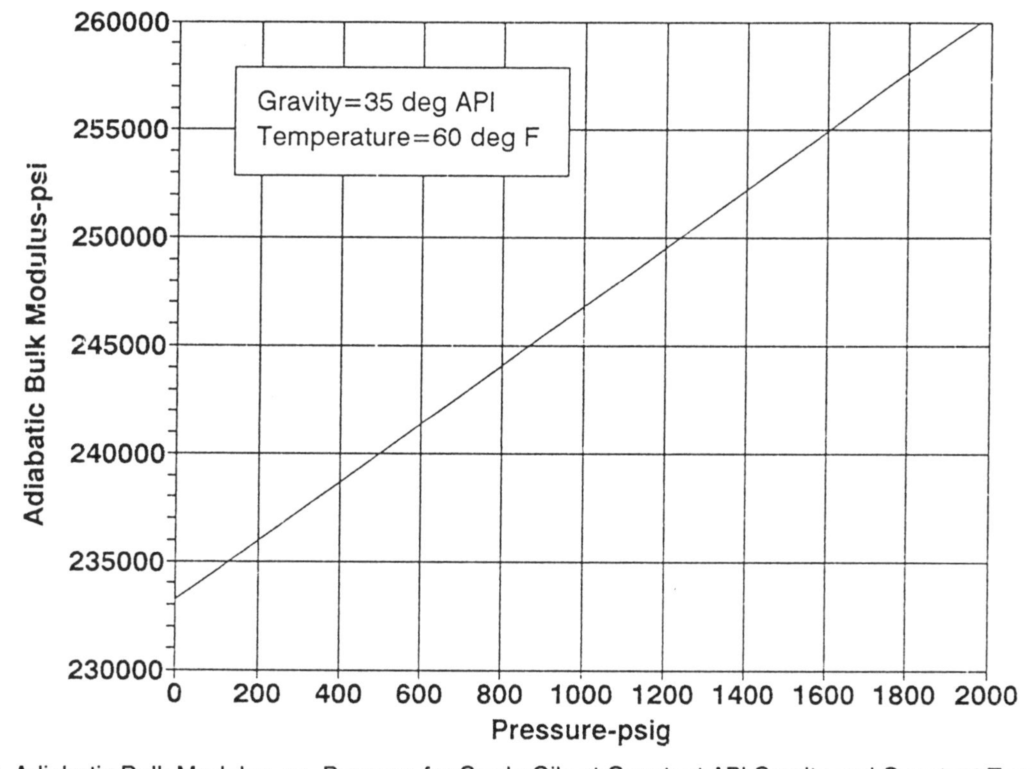

Figure 8-2. Adiabatic Bulk Modulus -vs- Pressure for Crude Oils at Constant API Gravity and Constant Temperature.

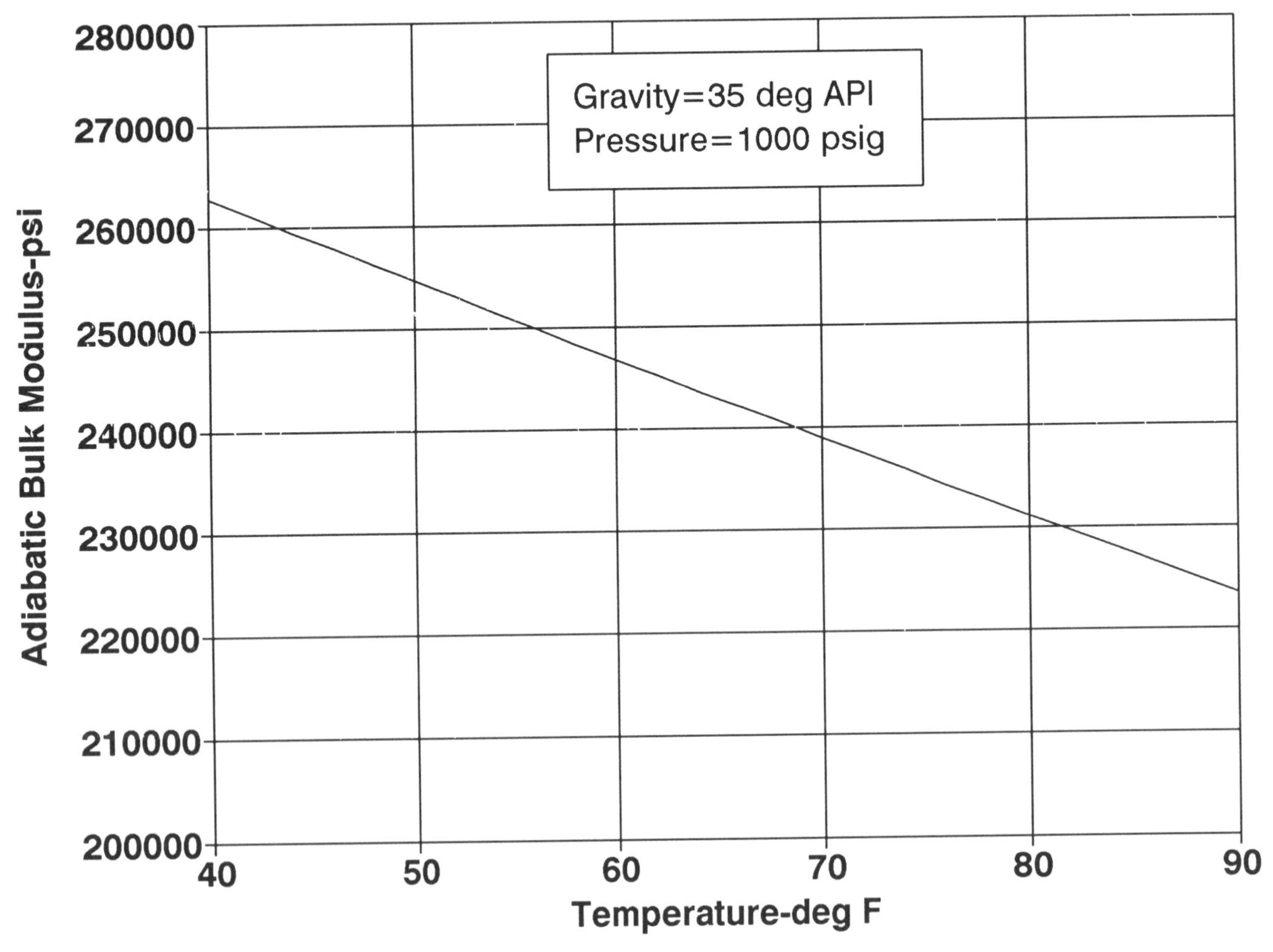

Figure 8-3. Adiabatic Bulk Modulus -vs- Temperature for Crude Oils at Constant API Gravity and Constant Pressure.

Calculation	Conditions	k	K_S
Original	Cameron	1.155	257 000
Modified	Cameron	1.18	262 000
Wostl	Cameron	n/a	259 000
Original	Ludwig	1.19	238 000
Modified	Ludwig	1.21	236 000
Wostl	Ludwig	n/a	232 000

Considering that both the original and modified calculations were based on dubious values of K_T the small spread is remarkable.

In any case, the Wostl equation, which was derived from tests on 22 liquids at 32°F, 68°F, and 120°F over a pressure range of 0–1800 psig, gives us a sound basis for calculating K_S for petroleum liquids directly without having to resort to calculations based on a value of K_T that was probably not made at the temperature or pressure of interest and, most probably, not on the same liquid.

Effective Bulk Modulus

Having decided that the proper bulk modulus to use is K_S instead of K_T, and then having available a method of computing K_S with considerable accuracy, it would be reasonable to believe that the bulk modulus problem has been solved. Unhappily, this is not the case.

The speed of sound, and thus the celerity of the pressure wave, in a pipe is a function not only of the fluid in the pipe but also on the characteristics of the pipe.

In the theoretical case of an infinitely rigid pipe the standard equation $c = \sqrt{K_S/\rho}$ holds directly; this is the speed of sound in an infinite universe of the fluid.

If the pipe is not infinitely rigid, the celerity is reduced because the effect of the expansion of the pipe concurrent with the compression of the fluid produces an *effective bulk modulus* that is always lower than K_S itself. The *general* equation is

$$c = \frac{\sqrt{K_S/\rho}}{\sqrt{1+(K_S/E)(D/t)m}} \quad (8\text{-}14)$$

where K_S = adiabatic bulk modulus of fluid
ρ = density of fluid
E = modulus of elasticity of pipe steel
D = pipe diameter
t = pipe wall thickness
m = coefficient for condition of pipe support

We may assume $m = 1$ without any loss in generality for purposes of this chapter.

The numerator of this equation is celerity in an unconfined fluid; the denominator is a number larger than unity which acts to reduce the value of celerity for the fluid confined in a pipe. If D and t are in consistent units, and K_S and E are in the same units, the denominator is a pure number. For a 30–inch pipe with a 0.250–inch wall thickness the pipe ratio is 30/0.25 = 120. If the modulus of elasticity of steel is 30×10^6 psi and we assume water has a K_S of 300 000 psi, and the elasticity ratio is $3 \times 10^5 / 3 \times 10^7 = 0.01$. The denominator thus becomes $\sqrt{1+(0.01 \times 120)} = 1.483$, and the celerity of the wave is $1/1.483 = 0.67$—almost exactly 2/3rds—that of the wave in the unconfined fluid.

The variation in c is the result of the variation of K_E, the *effective* bulk modulus, calculated as follows. Note that $\sqrt{K_E/\rho} = c$, which is Equation 8-14 in another form with K_E equal

$$K_E = \frac{K_S}{1+(K_S/E)(D/t)} \quad (8\text{-}15)$$

Figure 8-4 is a plot of K_E –vs– D/t for water with an assumed K_S of 300 000 psi and crude with an assumed K_S of 220 000 psi against the diameter/thickness ratio for steel pipes. Note that as the D/t ratio increases, or as t decreases for constant D, the effective bulk modulus decreases; the effect of the *stretching* of the pipe wall becomes a larger portion of the total volume increase in the $(dp/dV)(1/V)$ equation for K.

Figure 8-5 is another similar, somewhat surprising, presentation of the same data. Here, instead of having D/t plotted against K_E it is plotted against the celerity c in pipes containing water and oil. What is surprising is that the curves cross; at a D/t of 105 the calculated wavespeed is 3296 ft/sec, within 1 ft/sec, for both oil and water pipelines.

Aerated and Gassy Pipelines

Coefficients of Compressibility and Bulk Modulus For Gases

One would think that progressing from isothermal bulk modulus to adiabatic bulk modulus to effective adiabatic bulk modulus would be enough. Not quite.

Gases are *fluids*, and have many characteristics similar to those of liquids, the *other* fluid. The two equations under Equation 8-7

$$K_T = p \text{ for the isothermal process}$$
$$K_S = kp \text{ for the isentropic process} \quad (8\text{-}7)$$

apply to gases as well as liquids: the isothermal bulk modulus for an ideal gas at 200 psia is 200 psi; the adiabatic bulk modulus—assuming a k of 1.40—of the same gas at the same temperature is 280 psi. The comparable coefficients of isothermal and adiabatic compression are the reciprocals of these numbers: 1/200 = 0.005/psi, and 1/280 = 0.00357/psi.

For real gases the compressibility factor z comes into play. (Note that z is *not* a coefficient of compressibility; it is a correcting factor in the equation of state to make a computation involving a real gas more like that for an ideal gas.) A simple reading of this relationship is

$$c_g = \left(\frac{1}{p}\right) - \left(\frac{1}{z}\right)\left(\frac{\partial z}{\partial p}\right) \quad (8\text{-}16)$$

The factor z, and thus $1/z$, and the slope of the z -vs- p curve, can be picked off a chart of compressibility factors for the gas in question at the proper temperature.

The problem with gases, and why we are interested in them here, is that they become mixed with pipeline liquids; water becomes aerated, and crudes are not properly degassed, and the flowing fluid, while not in any way in two–phase flow, is not purely a flowing liquid, either.

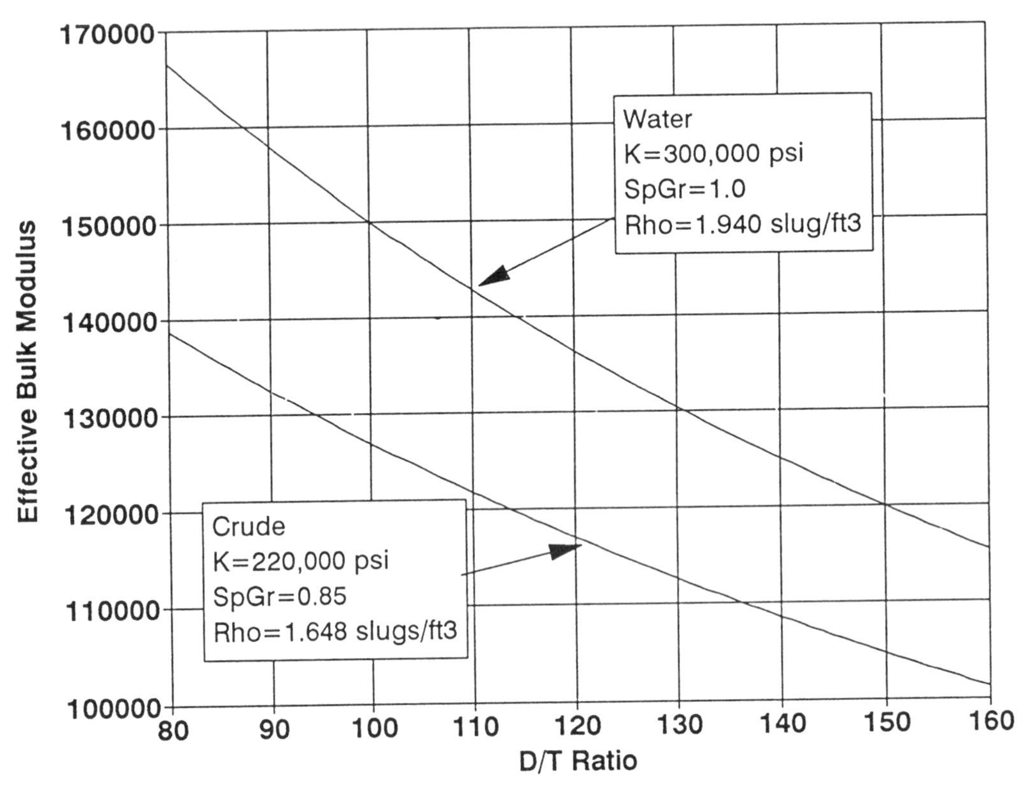

Figure 8-4. Graph Illustrating Effect of Pipe D/t Ratio on Effective Bulk Modulus.

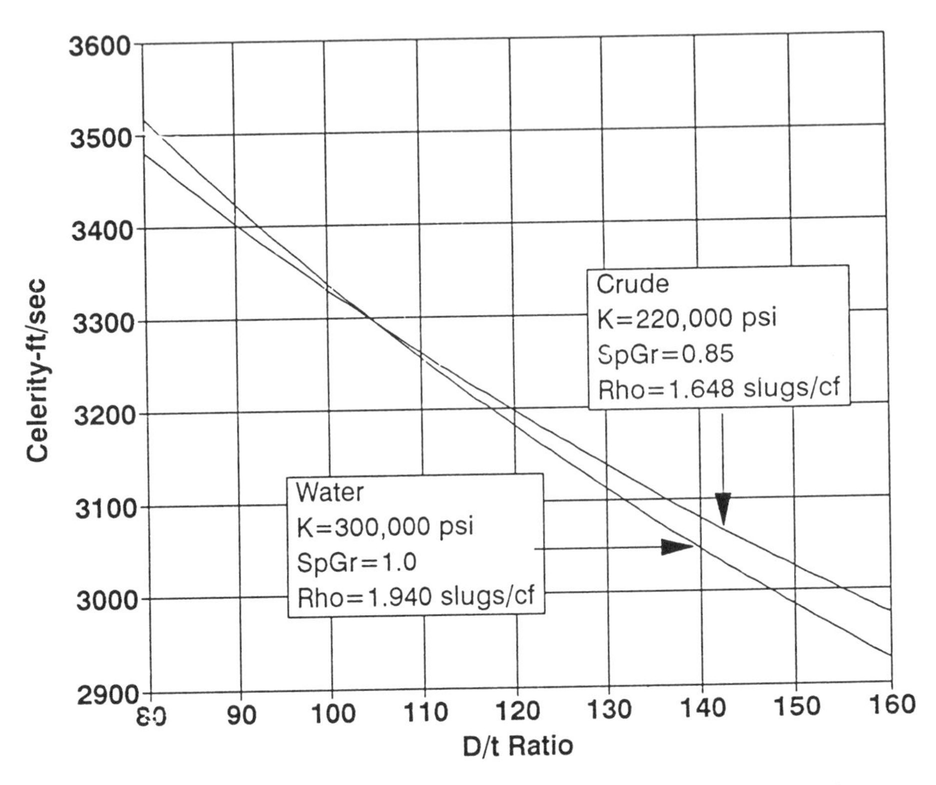

Figure 8-5. Graph Illustrating Effect of Pipe D/t Ratio on Wavespeed.

Aerated Pipelines

Aerated pipelines will be, generally, water lines; there is no good reason for a great deal of air being mixed with a petroleum oil.

But long–distance water pipelines are more and more being built like big–inch oil lines; and water lines of 56–inch and 60–inch diameter, operating at pressures in the hundreds of psig, have been built and are operating successfully. The pressure surge problem on these lines is not a simple one, however; there is always the problem of what effective bulk modulus is operative, and the distilled water lines of the Middle East, if not aerated, operate with the full adiabatic bulk modulus—essentially equal to the isothermal bulk modulus in all the reference tables—in play.

A very small amount of gas in a liquid flowing stream will have a major effect on its bulk modulus. If the amount of air is large enough, and if it is given the right set of *PVT* conditions, it may separate from the liquid carrier and become *entrapped*. However, entrapped air is another problem; it will not be dealt with here.

Streeter and Wylie[8] have a very simple explanation of aerated pipelines in terms of the effective bulk modulus, along the following lines.

We can assume that the total volume V is made up of V_{Liq} and V_{Gas}. A pressure change will then bring about a volume change which can be described as

$$\Delta V = \Delta V_{LIQ} + \Delta V_{GAS} \tag{8-17}$$

The bulk modulus of elasticity of the two components can be written as

$$K_{LIQ} = -\frac{\Delta p}{\Delta V_{LIQ} / V_{LIQ}} \tag{8-18}$$

$$K_{GAS} = \frac{\Delta p}{\Delta V_{GAS} / V_{GAS}} \tag{8-19}$$

Remembering that the bulk modulus for the fluid is

$$K = \frac{-\Delta p}{\Delta V / V} \tag{8-20}$$

We can substitute Equations 8-18 and 8-19 in Equation 8-20 and arrive at an *aerated K* for the fluid to use in the wavespeed equation, etc.:

$$K_A = \frac{K_{LIQ}}{1 + (V_{GAS} / V)(K_{LIQ} / K_{GAS} - 1)} \tag{8-21}$$

This *aerated bulk modulus* can be substituted in Equation 8-14 to calculate the wavespeed in the aerated pipeline. A few calculations will show that a very small amount of gas in the liquid will completely overpower the contribution to K_E of the pipe itself, and the K_E in play becomes the K_A of Equation 8-21. The effective density is

$$\rho_A = \rho_{GAS} \frac{V_{GAS}}{V} + \rho_{LIQ} \frac{V_{LIQ}}{V} \tag{8-22}$$

The wavespeed is therefore

$$c_A = \sqrt{\frac{K_A}{\rho_A}} \tag{8-23}$$

To illustrate the absolutely striking effect of a small amount of air in a water pipeline, consider the case of a water line operating at more–or–less atmospheric pressure with 1% entrained air.

The water, with a K of 300 000 psi = 43.2×10^6 psf and a ρ of $1.00 \times 62.4/32.17$ = 1.940 slugs / cf, would have a free wavespeed of 4719 ft / sec.

If we assume the air has a K of $144 \times 1.40 \times 14.67$ = 2958 psf and a ρ of 0.00238 slug / cf, it would support a free air wavespeed of 1115 ft / sec.

With this information in hand, we can calculate K_A as 2.98×10^5 psf, ρ_A as 1.92 slugs / cf, and a wavespeed of $\sqrt{K_A/\rho_A} = \sqrt{298\,000 / 1.92} = 394$ ft / sec.

Thus an air content of 1 percent caused a reduction in wavespeed from 4719 ft / sec to 394 ft / sec, a reduction of over 90%. Note the wavespeed in the pipe, 394 ft / sec, is *even less than the free air* wavespeed of 1115 ft / sec.

Figure 8-6 shows the dramatic effect of even smaller amounts of entrained air. It is obvious that entrained air has a major effect of wavespeed in a liquid pipeline.

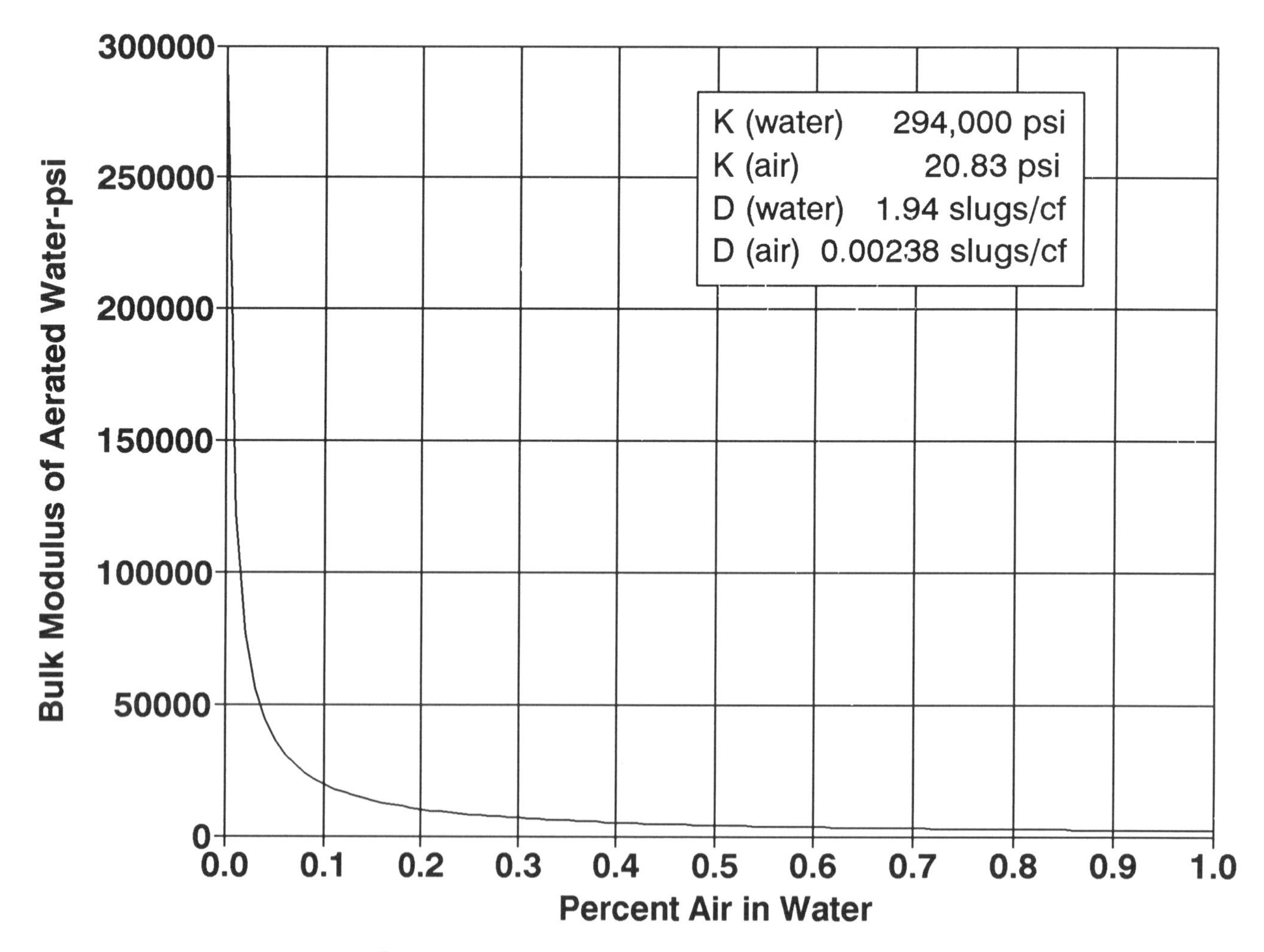

Figure 8-6. Effective Bulk Modulus of Aerated Water.

Gassy Pipelines

One would expect gassy pipelines to be carrying either

- A crude that has not been properly degassed in gas–oil separators in the point of production, or
- A blend of NGL or LPG fractions in a more stable crude or in a heavier refined product.

Such kinds of liquids would ordinarily not be worrisome in steady–state flow as long as pressure in the pipeline did not drop near the vapor pressure of the liquid, and this is not apt to happen at the usual operating pressures of trunk lines. For transfer lines, and especially for lines transferring incoming crude from tankships to onshore storage, which operate at lower pressures, either kind of liquid can cause a problem. I had a 36–inch 12–km line vapor-lock tight because a light North African crude came off the tanker with a vapor pressure (later found to be) much higher than normal with the result that gas pockets (read *embedded gas*) formed on the high points of the many two or three meter hill–and–dale undulations of the line and pressure available from the usual shore pumps couldn't break the series of manometer–like blockages.

However, crudes mixed with gassy LNG fluids, or almost any kind of carrier fluid mixed with propane or butane, can develop a K_A that will yield interesting results when trying to balance up a pipeline or compute its line pack: it is a pretty difficult calculation to compute the volume of oil in a segment of pipeline without knowing the K_E which, for practical purposes here, is the same as the K_A of aerated water lines. Be assured there is no reasonable way to accomplish this by laboratory analysis and measurement of the fractions and classic calculations such as those available for essentially pure liquids or gases.

The oil exploration and production industry has a term, the *formation volume factor of oil*, usually abbreviated as the B_O. It is defined as

$$B_O = \frac{\text{volume of oil + dissolved gas at reservoir } P \text{ and } T}{\text{volume of oil at stock tank } P \text{ and } T}$$

The B_O is usually greater than one, i.e., one reservoir barrel produces less than one STB (stock tank barrel). A B_O of 1.50—1.5 barrels at reservoir conditions produces 1 STB—is not unusual.

And so while it may be thought that the B_O kind of calculation—there *are* ways of making good estimates; see McCain[9]—might be useful one would not think that a B_O of 1.2, for instance, would exist in a pipeline. Until one remembers a 20% butane–propane spike in a medium crude oil and the whole mixture sitting at a tank temperature of 90°F for a couple of weeks. It *can* happen. The point is that the action of such liquids—or fluids, if you want to think of the LPG fractions as gases—cannot be accurately calculated by presently available means. What is acceptable in the E & P part of the oil industry, where you are really *guessing* how much oil is in place, anyway, is not acceptable in the pipeline industry where output had better equal input minus calculable losses or, what is nearly the same in this kind of case, calculable shrinkage, or there is a major problem somewhere in the system.

Measuring the Parameters in the Pipeline

There is, however, a way to find out how a pipeline will respond to surge conditions; *measure* it at one or two known conditions, and *project* the results to the other conditions of interest.

While this is especially applicable to aerated or gassy pipelines, each of which is very difficult to analyze—sometimes a good, sound, analysis is not reasonably possible for these kinds of systems—it is also applicable to, and is the best method of obtaining the data for further analysis of, any pipeline.

The basic wavespeed equation provides the basis for these measurements. Start the pipeline flowing, stop it, bottle up the resulting surge, and measure the transit time between known points to determine celerity; take a sample and measure ρ; and use the wavespeed equation $c = \sqrt{K_E/\rho_E}$ to calculate K_E.

Whatever the result is, it will be better than any calculation; the only problem is that this kind of operation cannot be done in the *design* stage of a project—it only works on an *existing* line.

There are many reports in the oil industry technical press over the years of wavespeed measurements, usually together with comparisons with calculated data. The following are a few examples:

- Wostl et al., previously referenced, reported on a comparison of calculated and measured K_S on a segment of the Colonial 36-inch products line flowing gasoline. They blocked surges into a 7.4 mile segment, measured the wavespeed and calculated K_S, and arrived at the following results:

Test No.	K_S Test	K_S Calc.	ΔK_S%
1	138 000	144 000	3.5
2	142 000	145 800	2.6

  The K_S calc values were made using the Wostl equation. The agreement is remarkable.

- Green[10] reported on three tests included in the *Final Report of Joint Surge Conference Covering Field Investigations Conducted During 1949*. For two of the tests the laboratory measurements and field (pipeline wavespeed) measurements were fairly close; for the other test the laboratory measurement was 254 000 psi compared to a pipeline value of 218 000 psi. The quite large discrepancy was attributed to "entrained gas or vapor during the test."
- Ludwig and Johnson[11] reported on another of the tests included in the *Joint Surge Conference* (see Green, above). They introduced the term *instantaneous bulk modulus*, as if all the kind of moduli mentioned herein were not enough, defined as

$$K_I = \frac{K_S}{1 - \left(\frac{p}{K_S}\right)\left(\frac{dK_S}{dp}\right)} \tag{8-24}$$

This modulus is intended to reflect the operative modulus when a pressure surge is generated starting from a very low pressure, in which case

Table 8-1
Reference and Working Equations
for
Adiabatic Bulk Modulus

General Definitions of Bulk Modulus

Classical Definition $K = -\dfrac{dp}{dV/V_1} = \dfrac{dp}{d\gamma/\gamma} = \dfrac{dp}{d\rho/\rho}$

where K = bulk modulus, lbs / ft^2 or Pa
V_1 = volume of container, ft^3 or m^3
p = pressure, lbs / ft^2 or Pa
γ = specific weight, lbs / ft^3 or N / m^3
ρ = density, slugs / ft^3 or kg / m^3

Common Definition $\left(\dfrac{F}{A}\right) = \left(\dfrac{dp}{dV/V_1}\right) \times \left(\dfrac{V}{V_1}\right)$

The F/A term on the left is p, the *stress*; the V/V_1 term on the right is the *strain*; and the term in the center is K, the slope of the stress–strain curve.

Definitions of Miscellaneous Kinds of Bulk Modulus

Process Definitions [γ = specific wt., ρ = density, p = pressure, and $k = c_p/c_v$]

Isothermal $\dfrac{p}{\gamma} = \dfrac{p}{\rho} = \text{constant}$

Isentropic (adiabatic) $\dfrac{p}{\gamma^k} = \dfrac{p}{\rho^k} = \text{constant}$

Form of definition $K = \dfrac{dp}{d\gamma/\gamma} = \dfrac{dp}{d\rho/\rho}$

Thus $K_T = p$ for the isothermal process
$K_S = kp$ for the isentropic process

Celerity–Wavespeed–General Expression

$$c = \sqrt{\frac{dp}{d\rho}} = \sqrt{\frac{K_X}{\rho}}$$

where K_x is bulk modulus for the process

Celerity of Pressure Wave in Pipes

Wavespeed in pipes $c = \dfrac{\sqrt{K_S/\rho}}{\sqrt{1 + (K_S/E)(D/t)m}}$

K_S = adiabatic bulk modulus of fluid
ρ = density of fluid
E = modulus of elasticity of pipe steel
D = pipe diameter
t = pipe wall thickness
m = coefficient for kind of pipe support

Wostl Equations For Calculation of Adiabatic and Isothermal Bulk Modulus of Hydrocarbons

$$K_S = V\left[\frac{\Delta P}{\Delta V}\right]_S = \frac{\rho c^2}{g}$$

$$K_S = (1.286 \times 10^6) + (13.55 \times P) - (4.122 \times 10^4 \times \sqrt{T}) - (4.53 \times 10^3 \times A) - (10.59 \times A^2) + (3.228 \times T \times A)$$

$$K_T = 2.619 \times 10^6 + 9.203 \times P - 1.417 \times 10^5 \times \sqrt{T} + 73.05 \times T^{1.5} - 341 \times A^{1.5}$$

where P = pressure, psig
A = API gravity at 60°F
T = temperature °R = °F + 460

the modulus calculated or measured in another way may not reflect the true state of the fluid during surge generation. Inasmuch as most pipelines will not generate an abrupt rise in inlet pressure to a pump or pump station, or in outlet pressure into open tankage, of more than a few hundred psi, this modulus isn't really important considering that the work of Wostl et. al shows the increase in the value of K_S is in the order of only $14 \times \Delta P$ (psig). Nevertheless, Ludwig et al. were able to take the laboratory and field data for two samples taken from field tests and arrive at computed wavespeeds of 4030 and 4140 ft / sec compared to measured speeds of 4100 and 4140; the agreement is, again, remarkable.

- Ludwig[12] makes some comments on the work done by the *Joint Surge* group which are interesting: "We received laboratory results that could be recomputed over the range for 177 000 psi for Wood–Leduc (Canadian) crude at 80°F to 260 000 psi for McCamey (Texas) crude at 40°F. The interpretation of the laboratory results is a little tricky, because they report on an *average* basis starting from 0 psig *where there is almost always some dissolved gas*. (The italics are mine.) In other words, laboratory results are usually too low for a line that is already under pressure."

I note here that essentially all bulk modulus determinations you are apt to run into in the usual technical literature suffer from this kind of indeterminate imprecision. Only the Wostl sound velocity measurements of K_S, which were always made under constant pressure, are valid point values for K_S.

While these data are included under a subject having to do with aerated or gassy pipelines, it should be obvious that field measurement of wavespeed in the *pipeline of interest* flowing the *fluid in question* is by far the best way to determine the value of K to use; every other kind of determination is second class by comparison.

Working Equations

Calculators

Table 8-1, Reference and Working Equations for Bulk Modulus, contains the definition equations, the free fluid and fluid–in–pipe wavespeed equations, and the equations developed by Wostl et al. for estimating the values of K_S and K_T for petroleum hydrocarbons.

The classical equation for K_S in terms of α, ρ, T, c_p, J, and K_T (Equation 8-10) is not listed as a working equation. It is perfectly valid and can be used, of course, but depends on having an acceptable value for K_T which is usually difficult to obtain and almost impossible to obtain if a K_T under pressure is wanted.

Computers

Seven Computer Computation Templates are included in this chapter.

CCTs No. 8-1 and 8-2 compute point values for K_S by the classic method and both K_S and K_T using the Wostl equations. **CCT**s No. 8-3, 8-4, and 8-5 compute and graph the relationships between the two bulk modulus values and fluid density, pressure, and temperature as predicted by the Wostl equations.

CCT No. 8-6 computes and graphs the value of K_E as a function the D/t ratio of the confining pipe, and **CCT** No. 8-7 computes and graphs a similar kind of function: the relationship between the celerity c and the pipe D/t.

If implementing these algorithms in other than a canned program, note that **CCT**s Nos. 8-3, 8-4, and 8-5 use a vector (a one–dimensioned matrix) as the independent variable, and **CCT**s No. 8-6 and 8-7 use a matrix with two dimensions to hold both the independent variables (the free adiabatic bulk modulus K_S and the pipe ratio D/t). The templates as written use the *MathCad* matrix feature.

References

1 Jacobson, E. W., et al, "Compressibility of Liquid Hydrocarbons," *Proceedings of the American Petroleum Institute,* Vol. 4, No. 25, 1945.

2 Kerr, S. Logan, Kessler, Lewis H., and Gamet, Merrill B., "New Method of Bulk Modulus Determination," Joint Petroleum and Hydraulic Session, *The American Society of Mechanical Engineers*, St. Louis, 1950.

3 Bergeron, L., *Du Coup de Beliér en Hydraulique, au Coup de Foudre en Électricité,* Paris: Dunod, 1950.

4 Cameron, A., "The Isothermal and Adiabatic Compressibilities of Oil," published in November 1945, in Vol. 31, No. 263, of a British (name cannot be determined from my copy) engineering society.

5 Ludwig, Milton. Private communication, 1956.

[6] Wostl, et al., *Review of Scientific Instrumentation*, Vol. 37, No. 12, 1966, p. 166S.
[7] Wostl, W. J., Dresser, T., and Clark, B. G., "Velocimeter Measures Bulk Moduli," *Oil and Gas Journal*, December 7, 1970.
[8] Streeter, Victor L. and Wylie, Benjamin. E., *Hydraulic Transients.* New York: McGraw-Hill Book Company, 1967.
[9] McCain, William D., Jr., *The Properties of Petroleum Fluids.* Tulsa: The Petroleum Publishing Company, 1973.
[10] Green, John E., "Pressure Surge Tests on Oil Pipe Lines," *Pipeline Hydraulics*, Tulsa: Petroleum Publishing Company, 1956.
[11] Ludwig, Milton, and Johnson, Sidney P., *Prediction of Surge Pressures in Long Oil–Transmission Lines.* Proceedings of the API, Division of Transportation, Vol. 36[5], 1950.
[12] Ludwig, Milton, *Surge Pressures on Trans Mountain Pipe Line.* (private communication). 1956.

Computer Computation Template

CCT No. 8-1: This is the classical method for computing adiabatic bulk modulus. The procedure follows the technically correct, but practically difficult, method of factoring the adiabatic bulk modulus off the isothermal bulk modulus which, itself, is a very difficult measurement, and in addition, enters into the calculation of the factor *k*.

References: Chapter 8. Equations 2-10, 6-29, and 8-10.

Input

Input Data

$SpGr_{6060} := 0.8500$ $H_2O = 1.000$ $F := 70.0$ deg F $K_T := 195000$ psi

$SpGr_{7060} := 0.8450$ $H_2O = 1.000$

$UOPK := 12.03$ Saudi Arabia Dammam crude (for example)

Variables

$\rho := 62.37 \cdot SpGr_{7060}$ lbs per ft^3 $T := F + 460$ deg R $K := 144 \cdot K_T$ psf

$\rho = 52.703$ lbs per ft^3 $T = 530$ deg R $K = 28080000$ psf

Equation 2-10

$$\alpha := \frac{341}{(1000 \cdot SpGr_{6060})^2} \qquad \alpha = 4.720 \cdot 10^{-4}$$

Equation 6-29

$$c_p := 0.6811 - 0.308 \cdot SpGr_{7060} + (0.000815 - 0.000306 \cdot SpGr_{7060}) \cdot F \qquad c_p = 0.460$$

Multiplier for Equation 6-29

$m := 0.055 \cdot UOPK + 0.35$ $m = 1.012$ $c_{pcorrected} := m \cdot c_p$ $c_{pcorrected} = 0.465$

Equations

$$k := \frac{1}{1 - \dfrac{\alpha^2 \cdot T \cdot K}{\rho \cdot c_{pcorrected} \cdot 778}} \qquad K_S := k \cdot K_T$$

Results

$C_p/C_v = k$ $k = 1.210$ Adiabatic bulk modulus = $K_S = 236027$ psi

Computer Computation Template

CCT No. 8-2: This template calculates both the adiabatic and isothermal bulk modulus using the equations of Wostl, Dresser, and Clark. The computed value of the adiabatic bulk modulus is sufficiently accurate for all pipeline engineering purposes. Wostl et al. do not recommend the use of the isothermal bulk modulus for commercial purposes though it, too, is sufficiently accurate for most all engineering calculations.

CCTs Nos. 8-3, 8-4, and 8-5 calculate and plot the effect of specific (API) gravity, temperature, and pressure on the bulk modulus.

References: Chapter 8. Equations 8-12 and 8-13.

Input

Input Data $SpGr_{6060} := 0.876$ Ref $H_2O = 1.000$ $F := 70.0$ deg F $P := 500$ psi

Variables $API := \frac{141.5}{SpGr_{6060}} - 131.5$ $T := F + 460$ deg R

$API = 30.0$ $T = 530$ deg R

Equations

$$K_S := \left(1.286\cdot 10^6\right) + (13.55\cdot P) - \left(4.122\cdot 10^4\cdot\sqrt{T}\right) - \left(4.53\cdot 10^3\cdot API\right) - \left(10.59\cdot API^2\right) + (3.228\cdot T\cdot API)$$

$$K_T := \left(2.619\cdot 10^6\right) + (9.253\cdot P) - \left(1.418\cdot 10^5\cdot\sqrt{T}\right) + \left(73.05\cdot T^{1.5}\right) - 341\cdot API^{1.5}$$

Results

Adiabatic Bulk Modulus = $K_S = 249611$ psi Isothermal Bulk Modulus = $K_T = 194351$ psi

Computer Computation Template

CCT No. 8-03: This template calculates the adiabatic and isothermal bulk modulus using the equations of Wostl, Dresser, and Clark. The graph shows the effect of density on these moduli. The seed value for *API* is 5° so that with $0>i>11$ the range covered is 10–60 °*API,* which covers most pipeline hydrocarbons. Temperature is set at 60 °F and pressure at 500 psig, but either can be changed as desired.

References: Chapter 8. Equations 8-12 and 8-13.

Input

Factors	$i := 1, 2..11$	$API_{Seed} := 5$ deg API $API_i := 5 + i \cdot API_{Seed}$
Input Data	$F := 60$ deg F	$P := 500$ psig
Variables	$T := F + 460$ deg R	$T = 520$ deg R

Equations

$$K_{S_i} := \left(1.286 \cdot 10^6\right) + (13.55 \cdot P) - \left(4.122 \cdot 10^4 \cdot \sqrt{T}\right) - \left(4.53 \cdot 10^3 \cdot API_i\right) - \left[10.59 \cdot \left(API_i\right)^2\right] + \left(3.228 \cdot T \cdot API_i\right)$$

$$K_{T_i} := \left(2.619 \cdot 10^6\right) + (9.253 \cdot P) - \left(1.418 \cdot 10^5 \cdot \sqrt{T}\right) + \left(73.05 \cdot T^{1.5}\right) - 341 \cdot \left(API_i\right)^{1.5}$$

Results

API_i	K_{S_i}	K_{T_i}
10	323241	245520
15	307660	236493
20	291550	225803
25	274910	213678
30	257740	200271
35	240041	185695
40	221813	170036
45	203055	153366
50	183767	135741
55	163950	117212
60	143604	97821

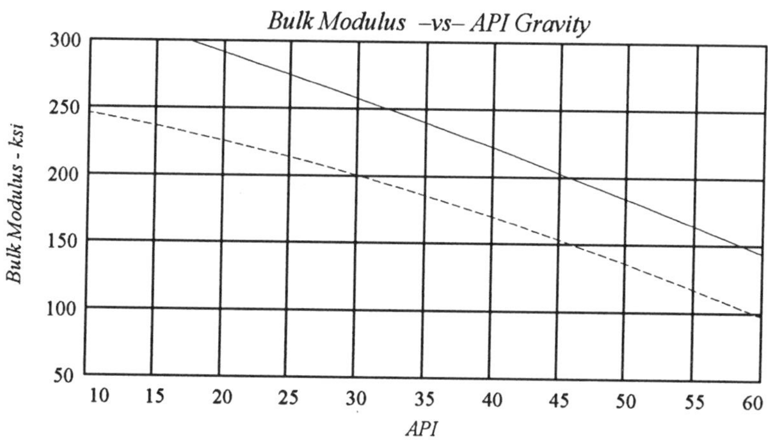

Computer Computation Template

CCT No. 8-4: This template calculates the adiabatic and isothermal bulk modulus using the equations of Wostl, Dresser, and Clark. The graph shows the effect of changing pressure on the bulk modulus. The value for density can be entered as *SpGr*, in which case the program will calculate *API*, or two lines down, as *API* in the *API=* variable line. Temperature could be entered as °R using the same procedure, but °F is the usual measurement. The seed value for pressure is 200 psig, so that with 0>*i*>11 the pressure range covered is 0 – 2000 psig. This can be changed as desired.

References: Chapter 8. Equations 8-12 and 8-13.

Input

Factors $i := 0, 1 .. 10$ $P_{Seed} := 200$ psig $P_i := i \cdot P_{Seed}$

Input Data $SpGr_{6060} := 0.850$ $H_2O = 1.000$ $F := 70.0$ deg F

Variables $API := \frac{141.5}{SpGr_{6060}} - 131.5$ $T := F + 460$ deg R

$API = 35.0$ $T = 530$ deg R

Equations

$$K_{S_i} := \left(1.286 \cdot 10^6\right) + \left(13.55 \cdot P_i\right) - \left(4.122 \cdot 10^4 \cdot \sqrt{T}\right) - \left(4.53 \cdot 10^3 \cdot API\right) - \left(10.59 \cdot API^2\right) + (3.228 \cdot T \cdot API)$$

$$K_{T_i} := \left(2.619 \cdot 10^6\right) + \left(9.253 \cdot P_i\right) - \left(1.418 \cdot 10^5 \cdot \sqrt{T}\right) + \left(73.05 \cdot T^{1.5}\right) - 341 \cdot API^{1.5}$$

Results

P_i	K_{S_i}	K_{T_i}
0	225506	175320
200	228216	177171
400	230926	179021
600	233636	180872
800	236346	182723
1000	239056	184573
1200	241766	186424
1400	244476	188274
1600	247186	190125
1800	249896	191976
2000	252606	193826

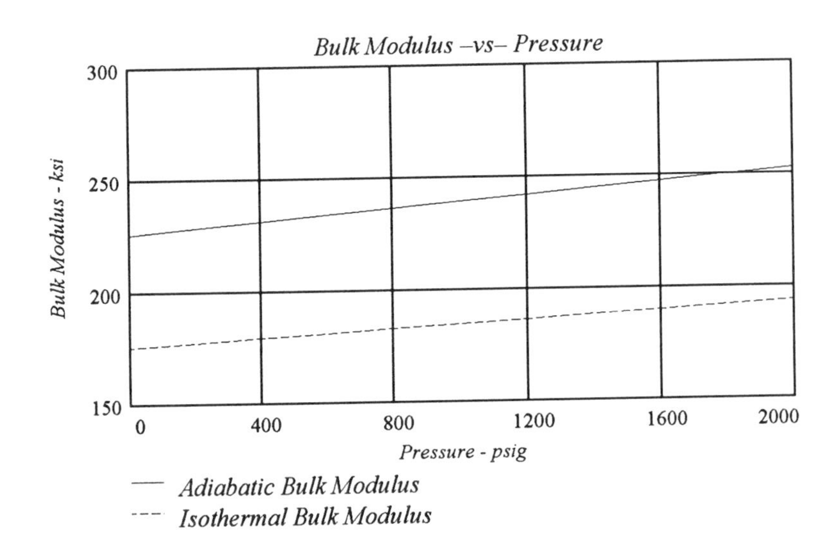

Computer Computation Template

CCT No. 8-5: This template calculates the adiabatic and isothermal bulk modulus using the equations of Wostl, Dresser, and Clark. The graph shows the effect of changing temperature on the bulk modulus. The value for density can be entered as *SpGr*, in which case the program will calculate *API*, or as *API* in the *API*= variable line. The seed value for temperature is 40°F, so that with 0>*i*>10 x 5 the maximum temperature is 90°F which covers most pipelines. This can be changed if needed. Pressure is set at 500 psig but can be changed as desired.

References: Chapter 8. Equations 8-12 and 8-13.

Input

Factors $i := 0, 1..10$ $\quad F_{Seed} := 40$ deg F $\quad F_i := F_{Seed} + i \cdot 5$ $\quad T_i := F_i + 460$ deg R

Input Data $SpGr_{6060} := 0.850$ $\quad H_2O = 1.000$ $\quad P := 500$ psig

Variables $API := \dfrac{141.5}{SpGr_{6060}} - 131.5$ $\quad API = 35.0$

Equations

$$K_{S_i} := \left(1.286 \cdot 10^6\right) + (13.55 \cdot P) - \left(4.122 \cdot 10^4 \cdot \sqrt{T_i}\right) - \left(4.53 \cdot 10^3 \cdot API\right) - \left(10.59 \cdot API^2\right) + \left(3.228 \cdot T_i \cdot API\right)$$

$$K_{T_i} := \left(2.619 \cdot 10^6\right) + (9.203 \cdot P) - \left(1.417 \cdot 10^5 \cdot \sqrt{T_i}\right) + \left[73.05 \cdot \left(T_i\right)^{1.5}\right] - 341 \cdot API^{1.5}$$

Results

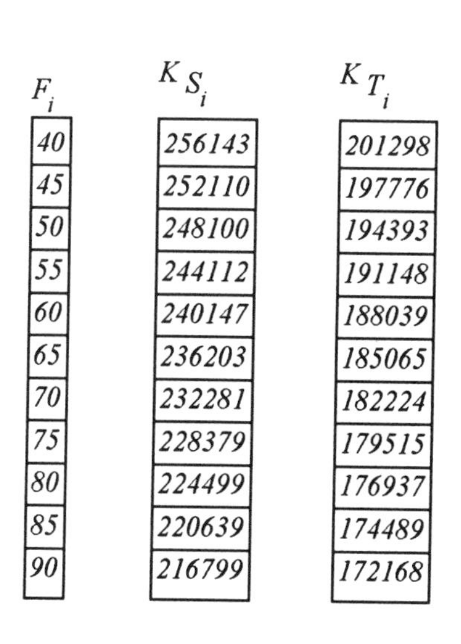

F_i	K_{S_i}	K_{T_i}
40	256143	201298
45	252110	197776
50	248100	194393
55	244112	191148
60	240147	188039
65	236203	185065
70	232281	182224
75	228379	179515
80	224499	176937
85	220639	174489
90	216799	172168

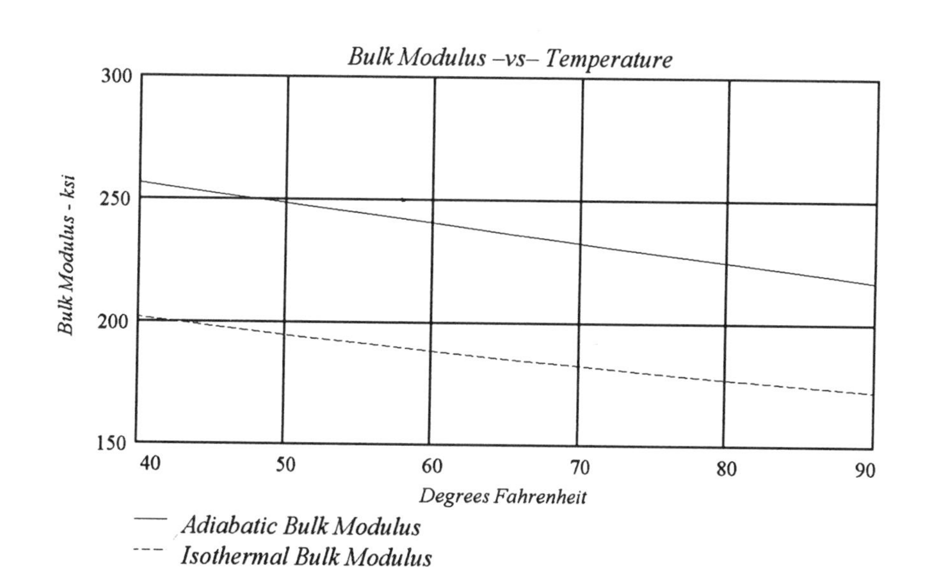

Computer Computation Template

CCT No. 8-6: This template calculates the effective adiabatic bulk modulus K_E, given the calculated or measured free bulk modulus K_S, in terms of the ratio of the free bulk modulus to the modulus of elasticity of the pipe material (K_S/E), and the geometry of the pipe (D/t).

References: Chapter 8. Equation 8-15.

Input

Factors $i := 0, 1 .. 10$ $j := 0, 1 .. 5$ $D_{seed} := 18$ inch $K_{Sseed} := 180000$ psi

Input Data $t := 0.375$ inch $E_S := 30000000$ psi

Variables $E := 144 \cdot E_S$ psf $D_{S_j} := D_{seed} + 6 \cdot j$ inch $K_{S_i} := 144 \cdot (K_{Sseed} + 10000 \cdot i)$ psf

Equations

$$K_{E_{i,j}} := \frac{1}{144} \cdot \left[\frac{K_{S_i}}{1 + \left(\frac{K_{S_i}}{E} \cdot \frac{D_{S_j}}{t} \right)} \right]$$ Effective Adiabatic Modulus, psi

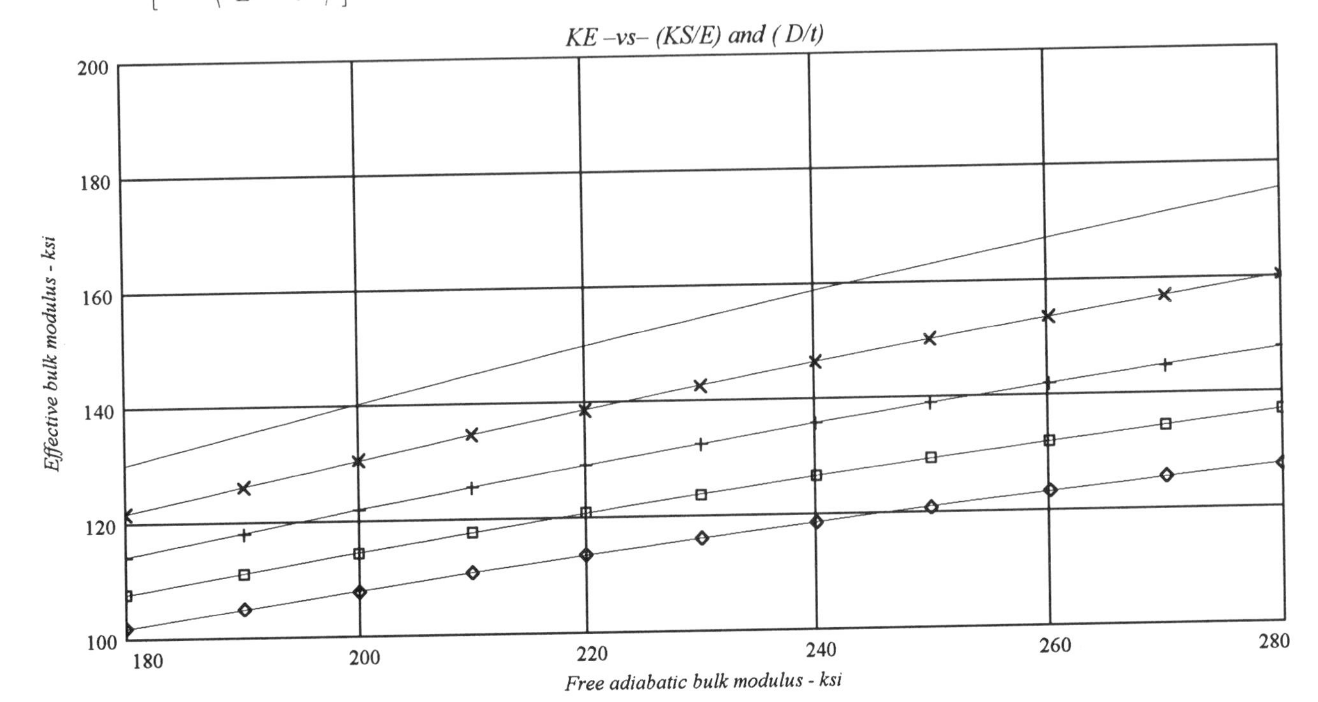

Computer Computation Template

CCT No. 8-7: This template calculates the wave speed—*the celerity*—of *sound*, and thus of a *pressure wave*, in a pipeline as a function of the free bulk modulus K_S, in terms of the ratio of the free bulk modulus to the modulus of elasticity of the pipe material, (K_S/E_S), and the (D/t)geometry of the pipe. E_S = 30 000 000 psi (for steel).

References: Chapter 8. Equation 8-14.

Input

Factors	$i := 0, 1..10$	$j := 0, 1..4$	$K_{Sseed} := 180000$ psi	$D_{seed} := 18$ inch
Input Data	$E_s := 30000000$ psi	$t := 0.375$ inch	$\rho_s := 0.850$ $H_2O = 1.000$	
Variables	$\rho := 1.938 \cdot \rho_s$	$\rho = 1.647$ psf		
	$E := 144 \cdot E_s$ psf	$D_{s_j} := D_{seed} + 6 \cdot j$	$K_{S_i} := 144 \cdot (K_{Sseed} + 10000 \cdot i)$	

Equations

$$c_{i,j} := \frac{1}{1000} \cdot \frac{\sqrt{\frac{K_{S_i}}{\rho}}}{\sqrt{1 + \left(\frac{K_{S_i}}{E} \cdot \frac{D_{s_j}}{t}\right)}} \quad \text{kilofeet/sec}$$

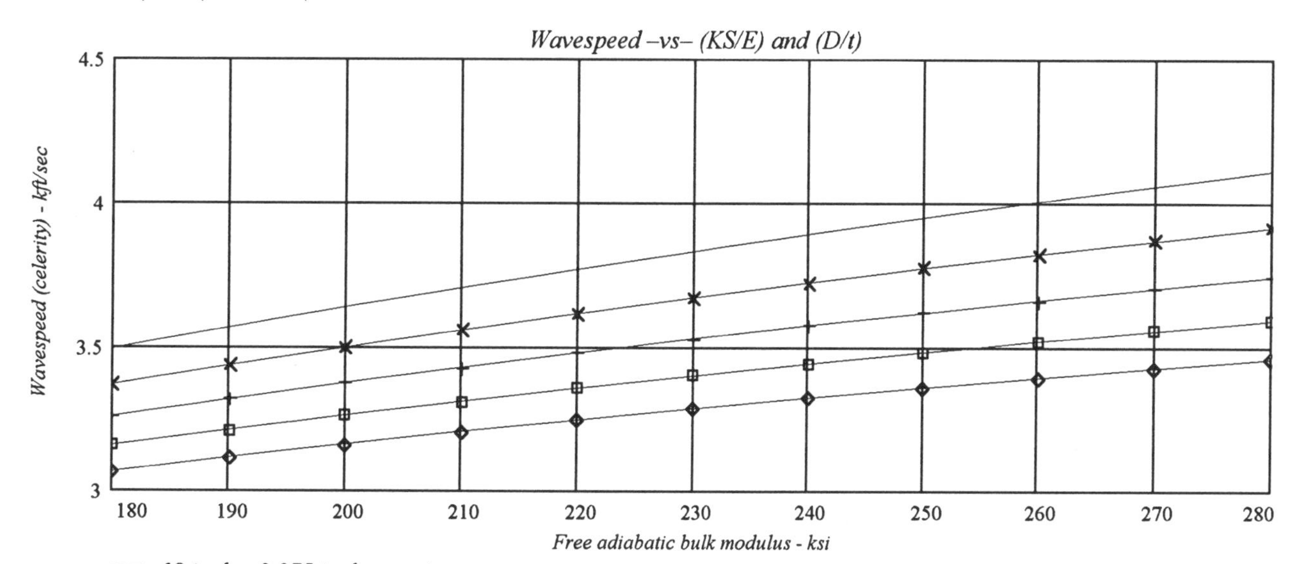

18 inch x 0.375 inch w.t. pipe
24 inch x 0.375 inch w.t. pipe
30 inch x 0.375 inch w.t. pipe
36 inch x 0.375 inch w.t. pipe
42 inch x 0.375 inch w.t. pipe

9

Miscellaneous Characteristics of Petroleum Fluids

Introduction

There are other characteristics of fluids that could in some way, some day, be of interest to a practicing pipeliner. While the characteristics covered in Chapters 2–8 are usually the most important, this chapter considers some of the less important characteristics of petroleum fluids—primarily liquids—that can become important in certain kinds of work.

Surface Tension

Surface tension is a physical characteristic of *liquids*. Surface tension can be measured and reported as *static* surface tension or *dynamic* surface tension. Normally we are mostly interested in the surface tension of *hydrocarbon liquids*, and then only if it is an important parameter in a problem of some kind.

For instance, a crude oil that has been produced at substantial pressure, but has not yet been passed through the low or medium pressure element of a gas–oil separator, is a very strange fluid: it is *full of gas*; it contains a very complex mixture of *liquid petroleum fractions* and, most probably, *salt water*. By definition it is at the *bubble point*, so that any relaxation of pressure produces a *vapor phase* in contact with the *liquid hydrocarbons phase* while, if the fluid is still, the *water phase* may try to fall out of the fluid completely.

If pressure is relaxed on this kind of fluid in an open chamber—such as that of a gas–oil separator—it either *will* foam or it *will not* foam.

If it *does not foam*, for purposes here the surface tension is not important. If it does foam, *the foam may be unstable*, and in this case it will quickly break down, disappear, and cause no further trouble. This case is not of interest to us at this time, either.

But if *the foam is stable*, it may continue to expand as new fluid is introduced into the chamber, to eventually fill the chamber, and then move on out into the system through one or more of the outlets. In this case the separator will not perform its function, the input fluid will pass essentially unchanged to the outlet, and the entire system will cease to function as it should.

If the input fluid appears in an outlet line, and then is again subjected to a further relaxation of pressure, the fluid will *flash*, i.e., it will change into a two–phase fluid containing petroleum gases, as a vapor, along with liquid petroleum components. To add to the problem, the liquid water phase—the water that does not emulsify with the liquid hydrocarbons—can appear, and all the calculations made about piping and pipelines carrying a properly degassed and dewatered crude become invalid; there is, really, no way to calculate what a system will do under this kind of circumstance. This is one of the reasons the study of two–phase flow in petroleum pipelines is such a difficult and inherently inaccurate calculation.

Foaming is first associated with the kind of crude, meaning primarily density and viscosity, and the gas–oil ratio; a heavy crude with a low gas–oil ratio will not foam, for instance. However, a heavy crude with a high gas–oil ratio may foam, and heavy crudes are more apt to form stable foams than light crudes.

Temperature and, of course, pressure, water content; and particulate matter—whether from the formation or from drilling mud or completion fluids—also have a hand in producing foam.

What all of the above factors have in common is that they are not controllable; the rate of production and setting of the production manifolding can have an effect on some of these factors but they cannot actually control them.

But there is a further factor that enters into the foaming equation—if that is what it is—the *surface tension* of the fluid, and this can be controlled to a certain extent though such control may be costly. This characteristic property is also important any time where *wetting*, *emulsification*, and *droplet formation* are encountered; and the last of these is of considerable importance in designing and operating multi–phase pipelines.

Surface Tension Calculations

Surface tension is usually defined as

$$\sigma = \frac{\text{work necessary for an increase in surface area}}{\text{amount of increase in surface area}}$$

It is expressed in terms of lb/ft in the FPS system or in N/m in SI. Most commonly, however, except in the most recent technical literature, surface tension is expressed in dynes/cm. Surface tension is dependent on the fluid in contact with the liquid, and most tables of values for surface tension are based on liquid in contact with air. Water, for instance, is usually taken to have a surface tension at atmospheric pressure and 68°F of 0.073 N/m = 73 dynes/cm = 0.0050 lb/ft.

The surface tension of most liquids—mercury, with a surface tension of 0.51 N/m = 0.035 lb/ft at 60°F and atmospheric pressure is a singular exception—is less than that of water. Vennard and Street[1] give a value for 0.880 spgr crude oil of 0.002 lb/ft = 0.03 N/m at 20°C = 68°F. Nelson[2] has a tabulation taken from Katz and Saltman[3] that gives surface tension in dynes/cm for nine pure hydrocarbons, and six hydrocarbon mixtures defined by their molecular weight, for eight temperatures running from −150°F to +400°F. Of these values, only those for the mixtures are of interest here and, of these, only the values in the temperature range that could be encountered in pipelining, i.e., from 0°F, the ground temperature in an arctic climate, to 200°F, which could be the temperature at the surface of a fluid rising from a deep well. These are summarized in Table 9-1.

The relation between surface tension and temperature is usually given as

$$\sigma_2 = \sigma_1 \left(\frac{T_c - T_2}{T_c - T_1} \right)^{1.2} \tag{9-1}$$

where σ_1 = surface tension at T_1 °R

σ_2 = surface tension at T_2 °R

T_c = critical temperature °R

Zanker[4] gives a nomogram to solve this equation but it is easily handled by any calculator.

Table 9-1
Surface Tension of Hydrocarbon Mixtures
Dynes per Centimeter

Hydrocarbon Mixture	0 °F	100 °F	200 °F
140 mol wt	27.6	22.5	17.4
160 mol wt	28.7	23.7	18.9
180 mol wt	29.6	24.8	20.0
200 mol wt	30.4	25.8	21.2
220 mol wt	31.0	26.6	22.1
240 mol wt	31.7	27.3	22.9

Surface Tension of Petroleum Fluids

Surface tension is a very difficult measurement, and often cut–and–try methods of design are required because a usable value of surface tension is not readily available. However, for the oil industry, ASTM now has ASTM D 3825 which can produce measurements of both *static* and *dynamic* surface tension on either pure liquids or on mixed fluids in which one or more components migrate to the surface. This is the case for gassy crude, or for *any* crude under proper temperature and pressure conditions.

Besides tables of values of surface tension, there are ways of estimating the surface tension of a hydrocarbon when an experimental value is not available.

The GPSA *Engineering Data Book*[6] given an equation

$$\sigma^{1/4} = \left(\frac{P}{M} \right) (d_L - d_V) \tag{9-2}$$

where σ = surface tension, dynes /cm

P = parachor

M = molecular weight

d_L = liquid density, g/cm^3

d_V = vapor density, g/cm^3

This form of equation was apparently first suggested by Macleod.[7] Sudgen[8] suggested and named the P term *parachor*, an additive function of the atoms or groups of atoms in the molecule. Parachor is almost independent of temperature. The parachor of a mixture is the sum of the pure component parachors multiplied by their mole fractions in the mixture.

Govier and Aziz[9] include a table in their Appendix A that lists the parachor factor for several different kinds of liquids. The GPSA *Engineering Data Book* lists *Group Contribution Values* for the calculation of parachors, and also gives a formula for computing the parachors of *hydrocarbons* that is derived from the work of Baker and Swerdloff:[10]

$$P = 40 + 2.38(\text{mol wt of liquid}) \tag{9-3}$$

Equation 9-3, together with Equation 9-2, can be used to make an engineering estimate of the value of σ for any fluid when its density and molecular weight are known or can be estimated. Equation 9-1 can be used to translate σ calculated or measured at one temperature to another.

A relatively recent development offers a method of predicting the dynamic surface tension—as distinguished from static surface tension, which is the surface tension measured on stilled liquids—in terms of the Watson characterization factor $K = (1.8 \times MABP)^{1/3}/spgr$ where $MABP$ is the molal–average boiling point; see Equation 6-28. This correlation, by Gomez,[11] is

$$\sigma = \left[\frac{681.3}{K}\right] \times \left[1 - \frac{T}{\left(13.488K^{1.7654} \times spgr^{2.1250}\right)}\right]^{1.2056} \tag{9-4}$$

As an example, the surface tension of kerosine with a specific gravity of 0.799 and a *Watson K–factor*[12] of 11.8 is 27.5 dynes/cm at 300 K.

Several of the two–phase gas–liquid flow correlations use σ as a parameter, sometimes calling it *interfacial tension* rather than surface tension, but true interfacial tension is defined by

$$\sigma_I = |\sigma_1 - \sigma_2| \tag{9-5}$$

where σ_I = interfacial tension
σ_1 and σ_2 = surface tension of components 1 and 2 against a common gas

Surfactants

If a high surface tension is interfering with a process and none of the usual tampering with the PVT relations works, you always can have recourse to commercially available additives that will produce favorable changes in surface tension of the fluid. These are not inexpensive, but are often well worth their cost in treating produced fluid from a foamy well. They are also sometimes used to enhance the solubility of drag reducer additives, and in this case they can be very economically effective. These agents are usually silicones or heavy alcohols.

Thermal Conductivity

Thermal conductivity is a parameter that falls out of the fundamental law of heat conduction, sometimes called Fourier's Law

$$\frac{dQ}{dt} = -kA\left(\frac{dT}{dx}\right) \tag{9-6}$$

This equation describes the time rate of heat transfer dQ/dt across the area A, dT/dx is the *temperature gradient* over the thickness x, and k is a constant of proportionality called the *thermal conductivity*.

For our purposes we can assume *thermal equilibrium* has been reached in a *homogeneous medium*, and dT/dx can therefore be assumed to be constant. In this case we can integrate the equation with respect to t and the usual engineering form of Fourier's Law appears

$$Q = \frac{kA(T_a - T_b)t}{x} \tag{9-7}$$

where Q = kcal
T_a, T_b = °C
A = m^2
t = seconds
x = meters
k = kcal / m · s · °C

This is more–or–less Ohm's Law for heat in MKS units. The Q/t term is analogous to electric current I; $(T_a - T_b)/x$, the temperature gradient, to the driving electric potential V; and k to electric conduction, the reciprocal of resistance R. The major difference between the flow of heat and the flow of electric current is that while k varies only by about 10^3 from the best to the poorest thermal conductors, the ratio of electric conductance may vary over a range of 10^{30}.

The dimensions of k are $MLT^{-3}\theta^{-1}$. The factors of k in the MKS system, probably the easiest to understand, are $kcal/s \cdot m^2 \cdot {}^\circ C/m$, as given above.

In the FPS and SI systems the factors of k become somewhat involved.

In the FPS system, with A in ft^2 and x in inches, k has units of $\frac{Btu \cdot in}{ft^2 \cdot hr \cdot {}^\circ F}$. In SI it becomes $\frac{W}{m \cdot K}$.

As to conversions

$$1\ kcal/m \cdot s \cdot {}^\circ C = 29\ 029\ Btu \cdot in/ft^2 \cdot hr \cdot {}^\circ F$$
$$= 4184\ W/m \cdot K$$

Several things to note about k:

- In the FPS system, the temperature gradient is often given in terms of °F/ft instead of °F/in. This makes a factor of 12 come into the value of k; a gradient of °X/1 inch is 12 times greater than a gradient of °X/12 inches.
- In *most* FPS applications, the period of time is the *hour* instead of the *second*; it should be obvious that 60×60 = 3600 times as much heat can be transferred in an hour as compared to a second.
- Many older references are in terms of the cgs system. The conversion from MKS to cgs is quite simple: multiply by 10, because 1 m = 100 cm while 1 kcal = 1000 cal.
- There are almost as many sets of units defining k as there are defining the *Universal Gas Constant*. One of the most useful references for handling problems involving transfer of heat in the pipelining industry is the *TEMA Standards*. The 6th Edition, the last in my library, lists many conversions to and from $Btu/hr \cdot ft^2$ for °F/inch and °F/foot including, among others, $cal/sec \cdot cm^2 \cdot ({}^\circ C/cm)$, $kcal/hr \cdot m^2 \cdot ({}^\circ C/cm)$ and $watts/cm^2 \cdot ({}^\circ C/cm)$.

Thermal Conductivity Calculations

If the medium is not homogeneous, the temperature difference will still be $(T_a - T_b)$ but it will be divided over the several layers in proportion to their thickness and thermal conductivities. Thus, for three layers

$$Q = \frac{A(T_a - T_b)t}{\frac{x_1}{k_1} + \frac{x_2}{k_2} + \frac{x_3}{k_3}} \tag{9-8}$$

If the temperature difference is not applied precisely and directly to the exterior faces of the medium there is a further parameter to consider: the *surface conductance*. If the temperature differential is being applied from a body of still air to a still surface there may be a surface conductance of as little as $1.5\ Btu/hr \cdot ft^2 \cdot {}^\circ F$ or, if a modest wind is blowing, as much as $6.0\ Btu/hr \cdot ft^2 \cdot {}^\circ F$. This effect can be added to Equation 9-8 to allow the calculation of the heat transfer through a multi–layer body which has a surface conductance to be overcome on each external surface.

$$Q = \frac{A(T_a - T_b)t}{\frac{1}{f_1} + \frac{x_1}{k_1} + \frac{x_2}{k_2} + \frac{x_3}{k_3} + \frac{1}{f_2}} \tag{9-9}$$

Here f_1 and f_2 are the two *surface conductances*, and the other variables are as previously defined.

This equation, with the numerator replaced by 1 (unity) defines U, the *transmittance*, or *overall heat transfer coefficient*. Thus

$$U = \frac{1}{\frac{1}{f_1} + \frac{x_1}{k_1} + \frac{x_2}{k_2} + \frac{x_3}{k_3} + \frac{1}{f_2}} \tag{9-10}$$

An effect of surface conductance different from thermal conductivity is that while the latter parameter, for any substance, will remain relatively constant, the former may vary widely with varying conditions and with time. The usual method of handling these variations is to provide for two, or sometimes even more, sub–parameters. Thus *TEMA*, in setting up a working equation for U, provides for h_o and h_i, the *film coefficients* of fluid outside and inside the exchanger tubes, plus r_0 and r_i, the *fouling resistance* outside and inside the tubes, plus r_w, the tube wall resistance itself.

The fouling resistance may appear, and increase, as the result of deposits forming on the surface of the body to which heat is being transferred to or from; by wear and tear, over time, to the surface; or by corrosion of the surface. The effect of none of these can be calculated; they must be estimated or, better, predicted. Then, too, there are ways of removing fouling when it does appear, though in some cases the cost of removal may exceed the cost of letting the fouling resistance run. And, finally, especially in the case of corrosion, fouling can be reduced by additives which act to reduce the formation of deposits or corrosion.

Thermal Conductivity of Petroleum Fluids

The measurement of thermal conductivity is a fairly simple matter, technically, but it is difficult and tedious, physically. The measurement procedure most applicable to hydrocarbon fluids, and these only to a vapor pressure of 50 psia, is ASTM D 2717.[14] The issue that I have reference to is D 2717–1978, the original issue of this standard; and while *repeatability* and *reproducibility* standards had not been developed at the time repeatability appears to be about 10 percent of the mean of two results by the same operator. It is most certainly not a precise measurement, nor need it be. The calculations involving thermal conductance tend to be those requiring only ordinary, reasonable results.

Cragoe[15] gives a simple formula for calculating thermal conductivity of petroleum liquids:

$$k = \left(\frac{0.813}{d}\right)\left(1 - 0.0003(t-32)\right) \tag{9-11}$$

where $k =$ thermal conductivity in $\dfrac{\text{Btu}}{\text{hr} \cdot \text{sq ft} \cdot \left({}^\circ\text{F}/\text{inch}\right)}$

$d =$ spgr $60/60^\circ$F

$t =$ temperature, °F

Cragoe says this equation, when compared to values measures at atmospheric pressure on 18 petroleum oils by seven different observers, was shown to be accurate to within 10 percent. The calculated values are probably too low for high pressures. He gives (his) Table 10, which was calculated by the Equation 9-11, compares these calculations with the experimental data, and then gives experimental data on bituminous substances. Consistent with the remark earlier herein, these old, but reliable, data, are for k in terms of

$$k = \frac{\text{cal} \cdot \text{cm}}{\text{sec} \cdot \text{cm}^2 \cdot {}^\circ\text{C}} \text{ and } \frac{\text{Btu} \cdot \text{inch}}{\text{hr} \cdot \text{ft}^2 \cdot {}^\circ\text{F}}$$

The *TEMA Standards* are an excellent source of tables of values of thermal conductivity, including a graphical presentation of values of k in terms of $\text{Btu}/\text{hr} \cdot \text{ft}^2 \cdot {}^\circ\text{F}$ of the parafinnic hydrocarbons from methane through octane and for mixtures of petroleum fluids from 10°API to 50°API, all at temperatures from –300°F to +400°F.

The *Standards* also contains procedures for estimating the effect of pressure on thermal conductivity of both liquids and gases in terms of the reduced pressure P_r and reduced temperature T_r where $P_r = P / P_c$ and $T_r = T / T_c$.

Note that the thermal conductivity of a mixture is *not* the weighted average of the thermal conductivities of the individual components. Zanker[16] makes the point that binary mixtures of pure liquids always have a point on the k -vs- % mixed curve that is lower than the arithmetic weighted average of the k values for the pure liquids. Zanker gives a complex formula he credits to Bates et al.[17] for computing the thermal conductivity of mixtures in terms of a characterization factor called ϕ and *sinh*, the hyperbolic sine. The method is explicit if ϕ is known from prior work but can be solved by iterative techniques if one value of the thermal conductivity of the mixture is known.

$$k_H \sinh \phi = k_H \sinh(m_H \phi) + k_L \sinh(m_L \phi) \tag{9-12}$$

where $k =$ thermal conductivity, any units

$m =$ weight fraction of component

$H, L, M =$ subscripts indicating *higher*, *lower*, and *mixed* components

$\phi =$ characterization factor for the two components

Zanker includes a nomograph to solve the equation.

Electrical Parameters

The electrical characteristics of petroleum fluids of most interest are *resistivity* or its inverse, *conductivity*, and the *dielectric constant*. These should be simple, easy–to–measure parameters. They aren't.

Conductivity

First we must understand that $R = \frac{V}{i}$ while $\rho = \frac{E}{j}$.

The first is Ohm's Law for the *macroscopic properties* of matter. We can measure these values with ordinary measuring instruments: V is in *volts*, i is in *amperes*, and R is in *ohms*. Resistance R is a singular property of a defined item that can be described in terms of its dimensions and the substance of which it is constituted.

The second is Ohm's Law for *microscopic properties* of a substance. We can measure these parameters only if we assure the sample of material to which we refer is *isotropic*, which means it has the same electrical properties everywhere, and is of a form, or shape, we can geometrically, or mathematically, manipulate. Thus, for instance, given a cylindrical conductor of cross section A and length ℓ carrying a steady current i. If we apply a potential difference V between the ends of the cylinder, the current density at all points in the cylinder will be the same, and thus $E = \frac{V}{\ell}$ and $j = \frac{i}{A}$. The resistivity ρ, denominated in *ohms*, may be written $\rho = \frac{E}{j} = \frac{V/\ell}{i/A}$. But $\frac{V}{i}$ is the resistance R and so $R = \rho\frac{\ell}{A}$. The reciprocal of the resistance is the conductance, usually σ, denominated in *mhos*.

In sum, V, i, and R are *macroscopic properties* that apply to a particular body or region; E, j and ρ are *microscopic properties*: they have values at every point in the body. The macroscopic properties are related by $V = iR$; the microscopic properties by $E = j\rho$.

We might be able to *calculate* the *resistance* of a *tank of crude*, but it wouldn't be of much interest. We can *measure* the *resistivity*, or its reciprocal, *conductance*, of the *crude* itself, and that can be of considerable interest.

Measuring Conductivity

It would appear that measurement of the conductance, usually called σ, of a petroleum hydrocarbon would be no different than measuring σ for any other substance: set up a sample of simple geometric configuration, apply a voltage across its end surfaces, measure the current flow, and divide. But some strange effects begin to appear in these kinds of measurements of hydrocarbon resistivities; the resistance computed from measurements made in the first seconds of a test may be as much as 10 times less than that computed from measurements made days, hours, or even minutes, later. Klinkenberg[18] considered several reasons for this, and eventually decided—in my opinion, *proved*—that this phenomena is due to the very low ionic content of these kinds of liquids; and that such liquids experience a depletion of the electrolyte—an *electrical clean–up*—so that after being subjected to a potential stress for a (relatively) long period of time, all the ions available for carrying an electrical current are *used up*.

Since the work of Klinkenberg, doubt has been cast on all prior measurements of resistivity or conductance of petroleum hydrocarbons; just the simple setting up and making of trial runs on a sample can effect the measured results by a factor of 5 or 10.

ASTM D 3114[19] provides a procedure for measuring the *rest electrical conductivity* of an oil, defined as the conductivity of uncharged hydrocarbon in the absence of ionic depletion or polarization–which is, in actual effect, conductivity at the initial instant of a DC measurement. The generation and dissipation of static electrical charges in oils has been shown to depend largely on this parameter. Output is in terms of *conductivity units*, a name given to 1 picomho/meter which, in SI, is 1 pS/m.

ASTM D 2624[20] provides a procedure for routine tests of aviation turbine and other distillate fuels to assure that the *static dissipator*—an additive that tends to increase the conductivity of the oil—is having the desired effect. Two kinds of instruments are described. One covers the *Portable Meter Method*, where the one–shot instruments themselves are carried to the point of measurement, and the other describes the *Continuous In–Line Conductivity Monitor Method*, which provides for measurements on a controlled, continuous flow of sample liquid through the cell (and thus a continuously renewed sample), assuring the measurements are comparable to the *rest conductivity* measurements.

Conductivity and Static Electricity

The relative resistivity, or conductivity, is important in the generation and dissipation of static electricity in oils. *Pure hydrocarbons* produce only small amounts of static electricity; but any time there is *relative motion of liquid hydrocarbon and a second phase*—which may be solid, gas, or another liquid—*electrical charge separation can take place at the interface between the two*. If the liquid hydrocarbon is the continuous phase, and if its electrical conductivity is sufficiently low, very high charge densities develop and surface discharges—intense enough to ignite hydrocarbon vapors—can initiate a fire and, sometimes, an explosion.

Rogers and Schleckser[21] cited 19 references having to do with this problem—many of them involving JP–4 jet fuel–and attempted to put some of the more technical research into engineering terms. This paper is based on a body of prior work reported by Rogers, McDermott, and Munday,[22] and extends that work to engineering application data.

They note there is a very wide spread in the tendency of hydrocarbons to generate and contain static electricity. They referenced straight–run distillates, including JP–4, with conductivities in the range of 1×10^{-15} to 5×10^{-13} mho/cm, up to some heating oils, diesel fuels, and untreated kerosines which may have a conductivity as high as 5×10^{-11} mho/cm.

Qualitatively, Rogers et al. note that it is, after all, movement of the liquid phase *relative to something* that generates static electricity. For instance

- The rate of production of charge in pipes increases with flowrate. In turbulent flow, charge increases roughly at the 1.75 power of velocity.
- In filters charge production is essentially a linear function of velocity.
- The effect of temperature is variable. With some samples, increasing temperature increases charge production; with others, it decreases it.
- Dispersed gases and finely divided solids generally increase charge production.
- The effect of small amounts of water is variable; sometimes it increases production of charge somewhat, sometimes it reduces charge production markedly.
- The rate of charge production varies widely with the nature of the charge–separating surface: (1) Rough, rusted pipes produce more charges than clean ones. (2) Filtering media are potent charge generators; in a typical fueling facility the filter–separator can produce more charges than all the other elements in the system. (3) An inactive fuel, on being pumped through a filter containing adsorbed contaminants, can become active and develop a high charge density.

The *potential gradient* in ordinary tanks containing hydrocarbons can reach values over 10 000 kV/ft.

The *breakdown gradient* for mixtures of air and a volatile hydrocarbon is in the order of 600 to 1000 kV/ft. Note, however, that aircraft fueling studies have shown that electric fields of *about one–sixth the breakdown field* can produce an *incendive* spark (a spark hot enough to ignite vapor).

If the intensity of an electric field *in air* is greater than about 3000 kV/m—about 1000 kV/ft—a spark results, i.e., it *breaks down*.

The major methods for reducing the electric charge in liquid hydrocarbons are

- Relaxation—Let the fluid rest and the charge disappears.
- Antistatic Additives—Increase the conductivity of the fluid.
- Neutralization—Use counter voltages to neutralize charge.
- Radiation—No practical use in the pipeline industry.

The relaxation principle is simplest if time is available; let the fluid rest, and charges will eventually leak away or neutralize. The only other practical method is the use of anti–static additives; if the conductivity of the liquid can be increased to about 1×10^{-11} *mho/cm*, the probability of a dangerous static buildup is minimal. The counter–voltage and radiation procedures are not, to my own knowledge, in present–day use in the oil industry.

There is another way to reduce static: *don't generate it in the first place*. For crude oil this usually involves holding the velocity in a pipe discharging to an open surface below a limiting velocity; or, what is simpler, don't let such a discharge take place at all: Keep the discharge opening always submerged. This is considered in more detail in **Static Electricity and Safety** later in this chapter.

Dielectric Constant

Dielectric constant, sometimes called *specific capacity* or *relative permittivity*, is that property of a dielectric which determines the electrostatic energy stored per unit volume for unit potential gradient. The dielectric constant of a uniform material is defined by Coulomb's equation

$$F = \frac{QQ'}{\varepsilon r^2} \qquad (9\text{-}13)$$

where Q and Q' = two charges
F = force between Q and Q'
r = separation between Q and Q'
ε = dielectric constant

The dielectric constant is a pure numeric, usually given in terms of the *relative permittivity* of a vacuum, which is unity. The *permittivity* of a vacuum, ε_0, is usually taken as 8.85×10^{-14} farad/cm = 8.85×10^{-12} farad/meter, and so the permittivity for any substance x is $\varepsilon_x \times \varepsilon_0$.

Water has a very high dielectric constant—about 80. It varies with temperature in accordance with the formula $\varepsilon = 80 - 0.4(t - 20)$, where t is in °C.

Oils, on the other hand, have a dielectric constant that is quite low, in the range 2.2–4.7. Hydrocarbon oils in the density range 850–900 kg/m^3 have a dielectric constant of about 2.2. Table 9-2, taken from Beyers,[23] gives ranges of dielectric constant for typical pipeline and production liquids, including the range 80.37 max, 16.00 min, for formation water, depending on its salinity.

Table 9-2
Dielectric Constants of Some Typical Pipeline and Production Liquids

Liquid	Dielectric Constant	
	Min	Max
Gasoline–premium	1.977	2.051
Gasoline–regular	1.960	2.015
Kerosine	2.080	2.110
Fuel Oil No. 1	2.170	
Fuel Oil No. 2	2.260	2.300
Diesel fuel	2.110	2.160
West Texas Sour crude	2.280	2.340
West Texas Semi–sweet crude	2.201	2.280
Texas Scurry County crude	2.180	2.230
West Texas Ellenberger crude	2.170	2.220
Kansas Arbuckle crude	2.260	
Water–distilled	80.37	at 20°C
Water-formation max/min	80.37	16.00

From Handbook of LACT Operations, *Pipe Line Industry, April 1963.*

The dielectric constants of gases at normal pressures and temperatures are substantially the same as that of a vacuum—unity.

Thus the order of magnitude of the dielectric constants of the three phases most apt to be contained in a mixture of petroleum oils are

- 80 for water
- 2 for oil
- 1 for air or gases

This is a very wide spread and, when looking back at Equation 9-13, it is easy to see how charges built up in one phase may be separated and accumulate in another. The largest potentials develop across the materials with the lowest dielectric constant, so water droplets carrying a charge drop them off in the gas or liquid oil phase and high voltages can develop in or on the surface of those phases.

Static Electricity and Safety

Rogers et al. firmly established that static electricity is produced in liquid hydrocarbons when there is relative motion of the hydrocarbon and a second phase, be that phase solid, liquid, or gas.

Appendix F, *International Oil Tanker and Terminal Safety Guide*,[24] has a very fine description of the way static electricity is *generated*, and how it may be reduced or controlled in the oil industry, based on the work of Rogers and others.

It starts out with an enumeration of the ways *charge separation* can occur:

- Flow of oil through pipes or fine filters
- Flow of oil/water mixtures through pipes
- Settling of solids or water in oil
- Spraying, splashing, or agitation of oil
- Ejection of particles or droplets from a nozzle

Electric charges which have been *separated* attempt to *recombine* and so neutralize each other; but if one of the separated materials has low conductivity, recombination is difficult and the material *accumulates* charges on or in it. If left alone, the charges will eventually recombine or, in effect, neutralize each other. A measure of the time the charges are retained is given by the *relaxation time* of the material. The relaxation time is related to conductivity and the dielectric constant by $\tau = \varepsilon / \sigma$.

If a material has high conductivity the recombination can counteract the accumulation and very little static electricity appears; alternatively, if the material has a very low conductivity, large static charge accumulations can occur.

For an oil of relatively high conductivity, i.e., with a conductivity greater than 100 picomho/m, and relaxation time less than 0.18 seconds, the recombination effect can outweigh the separation effect and so no great amount of charge appears at any given moment. Thus, *refined black oils*, which contain residual matter, and *crude oils*, both of which are in the range of 10 000–100 000 picomho/m, *are not* static accumulators whereas *clean oils* (distillates), all of which have low conductivities, *are* static accumulators.

This does not mean that black oils and crudes can be handled with complete safety; however, these oils can *generate* charges which can then *accumulate* on insulated conductors, in static accumulators being pumped into or are already in the tanks, and, especially, in mists and sprays, even of crude, where the air insulates the charged droplets.

Appendix F[24] contains a rather complex, even though simplified, method of calculating the maximum safe flow rate through the critical pipe section feeding into a tanker manifold or an individual tank.

The current j_∞ carried by a dry petroleum liquid after flowing in a pipeline of infinite length is

$$j_\infty = \alpha R^2 V^2 \tag{9-14}$$

where $\alpha = 15\times10^{-6}$, a constant
$R =$ radius of pipeline, m
$V =$ linear velocity, m / s

For a pipeline of length L, meters, j_L is

$$j_L = j_\infty\left(1-e^{-L/V\tau}\right) \tag{9-15}$$

where $\tau =$ relaxation time, seconds.

Rogers et al. point out that the electrostatic charge in a semi–conducting medium tends to decay or *relax* in accordance with the following equation:

$$Q = Q_0 e^{-t\sigma/\varepsilon\varepsilon_0} \tag{9-16}$$

where $Q =$ charge density after elapsed time t
$Q_0 =$ initial charge density
$e =$ base of natural logarithms
$t =$ elapsed time, seconds
$\sigma =$ liquid conductivity, mho/cm
$\varepsilon =$ liquid dielectric constant
$\varepsilon_0 = 8.85\times10^{-14}$ farad/ cm

The equilibrium charge Q, coulombs, that accumulates in a tank during filling is

$$Q = j_L\tau\left(1-e^{-t/\tau}\right) \tag{9-17}$$

where $t =$ filling time

This quantity of charge in a rectangular tank of length a, breadth b, height c, and filled to ullage p, all in meters, produces a maximum electric field in the center of the surface of the liquid given by Equation 9-18 below.

$$E_{max} = \frac{16\, j_L\tau(\cosh\beta d - 1)\cosh\beta p}{\Sigma_0\pi^2 abd(\Sigma_V\cosh\beta p\sinh\beta d + \Sigma_L\sinh\beta p\cosh\beta d)}$$

where $\beta = \pi\sqrt{\frac{1}{a^2}+\frac{1}{b^2}}$

$d = c - p$

$\tau = \frac{\Sigma_L\Sigma_0}{\sigma}$

In the above, $\Sigma_V, \Sigma_L, \Sigma_0$ are dielectric constants of the vapor, liquid, and vacuum phases, and σ is the liquid conductivity.

For practical purposes, the time of filling of a tank or a tanker is much greater than the relaxation time. In addition, a pipeline of length L greater than $3V\tau$ can be assumed to be of infinite length. Equation 9-18 then simplifies to Equation 9-19 below.

$$E_{max} = \frac{16\alpha R^2V^2\tau(\cosh\beta d - 1)\cosh\beta p}{\Sigma_0\pi^2 abd(\Sigma_V\cosh\beta p\sinh\beta d + \Sigma_L\sinh\beta p\cosh\beta d)}$$

This equation can be used to calculate the maximum velocity of flow V to produce no more than a specified E_{max}. The physical parameters that affect the value of the field are the tank dimensions, ullage, pipe diameter, and liquid conductivity. As to liquid conductivity, it appears that 0.2 picomho/m is the lowest measured *on the jetty* so, to be absolutely safe, a value of 0.1 picomho/m is recommended.

For a given Q, E_{max} will be greater in a small tank than in a large one; will be greater for a small ullage than for a larger one; will be less, for a given flow rate, for a large diameter pipeline than for a smaller one; and will always be greater for lower conductivity liquids than for those of higher conductivity.

And remember that a high value of Q can be generated in any kind of splashing or spraying oil, and piping design and operating procedures should be established such that these actions are minimized or, preferably, eliminated.

Above all, *keep the velocity of any kind of oil entering a new or newly cleaned and empty tank, or any tank with the entrance nozzle at or above the level of the liquid, as low as is reasonably possible until the nozzle is well covered.* These kinds of situations can generate and store enormous values of charge, and thus can, in the right circumstances, develop enormous potential differences along the surface of the fluid with subsequent sparking, ignition, and explosion.

Think how many times you have heard of a new tank firing up on the initial fill.

Appendix F[24] includes several useful tabulations and graphs, but their Figure F-2, based on a loading pipeline of theoretically infinite length, with tank dimensions of $6.1 \times 4.9 \times 12.5$ m = (about) $20 \times 16 \times 41$ ft, ullage at the outset of 0.3 m = (about) 1 ft, a $\rho = 0.1$ picomho/m, and for an $E_{max} = 500$ kV/m, is indicative of the care that must be taken with some hydrocarbons. We have redrawn it here as Figure 9-1.

Note that the old rule of thumb, *Keep the entrance velocity below 1 meter/second*, is safe. To comply with this rule, when a higher velocity would be safe, however, just costs time—and money.

The entirety of Reference 24 is unique and valuable; it should be in every pipeliner's library.

Trace Metals

With the continuing refinement of chemical analysis techniques it would appear that there is a little bit of

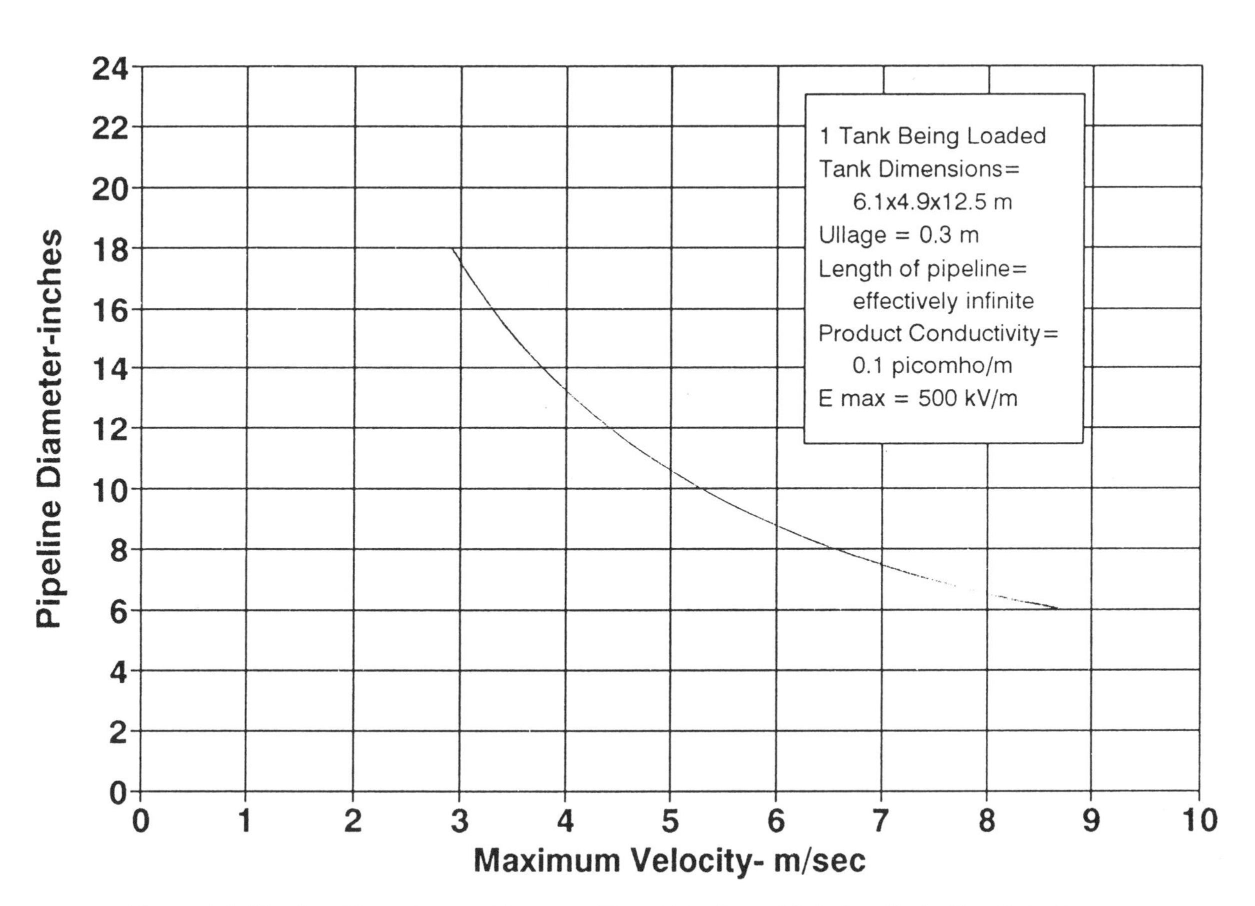

Figure 9-1. Pipeline Diameter -vs- Maximum Flow Velocity to Minimize Static Electricity Danger.

everything in everything; if there isn't one part per billion of a contaminant, there may be one part per trillion. And so forth. Valkovic'[25] is about as good a reference as I have found for these kinds of things.

Four trace metals are really important in petroleum fluids, one in natural gases and the others in liquids.

Mercury

Mercury is a sleeper; it was not even considered a contaminant until the aluminum cold box tubing of an LNG plant amalgamated with mercury in the incoming gas stream, destroying the protective oxide integrity of the piping and opening it up to corrosion, ruining the cold box, causing a months–long shut down of the plant while a new cold box with stainless steel tubing was installed, and generating a years–long legal battle about "who is responsible."

Really, nobody was responsible. The problem was that, firstly, nobody looked for mercury in the gas stream and, secondly, they probably wouldn't have found it, anyway—in the tiny quantities in which was present in the gas—with the techniques available at the time. But large plants handle millions and billions of pounds of gas; 1 ppb of mercury is 1 lb of mercury in 1 billion pounds of gas; and 1 pound of mercury can amalgamate with, passivate, and thereby open to corrosion and ruin, a lot of aluminum.

Kloosterman[26] reported that untreated gas from the Groningen field—one of the world's largest fields—in The Netherlands (this is *not* the field that delivered the gas to the LNG plant) has a mercury content of 180 $\mu g/m^3$, the treated gas contains 12 $\mu g/m^3$, and even the gas delivered to the consumer has a content of 5 $\mu g/m^3$. If the Dutch gas *had* been fed into a 10×10^9 m^3/year LNG plant (about 1000 MMBTU/day or 1 000 000 MCFD) the total intake of mercury would have been $10\times 10^9\times5\times10^{-6}$ g/year = 50 metons/year. Not only would this cause all kinds of trouble with any LNG plant aluminum but—though the carryover or solubility of mercury in LNG is not known to me—probably as well in any aluminum tube heat exchangers used to regasify the LNG. Kloosterman remarks that mercury may be more common in Dutch minerals than would be expected; he says Dutch coal contains between 45 and 400 μg of mercury per kg of coal.

Leeper[27] covers the problem of small amounts of mercury loosed on aluminum surfaces quite completely. He also points out the reason aluminum is not normally attacked by water or free oxygen is its tightly adhering oxide film, *which is not present when aluminum is amalgamated with mercury in an anaerobic atmosphere* (Leeper's italics). He sites one case where the mercury content was only about 12 $\mu g / m^3$, yet there was severe failure in the spiral wound exchangers in a LNG plant.

The point here is that the most sensitive test for mercury known to Valkovic' has a sensitivity of 10 ng/g, or roughly 10 ppb. If you had a gas with 5 ng/g of mercury in it, every 10^9 grams of gas you pumped would have 5 grams of Hg available to be attracted to an unprotected aluminum surface, and you wouldn't know you had the problem until you had pumped a few thousand billion grams of gas and something failed somewhere in the process.

There are ways of handling mercury, once you know you have it; the problem is *knowing* you are going to have it before you do.

Vanadium, Sodium, and Potassium

Vanadium is not something you worry about in a crude or fuel oil until you have to burn it at high temperature in a turbine. Then it tears up the nozzles and blades in the power set and you have a major failure on your hands.

Though not important to pipeliners, a vanadium and nickel content of more than 150 ppm in residual feedstock can also result in enormous metal deposits on the catalyst followed by serious deactivation of the catalyst.[28] Use of continuous vanadium analyzers allows some refiners to process high vanadium crudes without detrimental metal buildup on the catalyst,[29] but high vanadium feedstock has a low market value compared to other *cleaner* stocks.

Crudes can have vanadium contents in the range 0.003 ppm for Table Mesa, New Mexico crude to 1400 ppm for the heavy crude of Boscan, Venezuela. In most cases, high nickel content accompanies high vanadium content, and these heavy metals tend to end up in residual fuels. Nelson[30] cites a Sag River, Alaska crude, that has 129 ppm vanadium and nickel combined; the 7.3°API residual fuel from this crude contained 610 ppm of the metals.

Turbine fuels, whether crude oils or a refined distillate or residual fuel, containing vanadium along with sulfur and alkali metal compounds (i.e., *sodium* and *potassium chlorides*), pose very severe corrosion problems at normal turbine operating temperatures because of the formation during combustion and the deposition of sodium and potassium sulfates, sodium and potassium vanadates, and vanadium pentoxide. These compounds are semi–molten and corrosive at temperatures as low as 1150°F, and the corrosion rate increases exponentially with temperature. The life of the nozzles and blades is shortened, and the deposition of these compounds, and other ash materials, within the turbine significantly reduces turbine efficiency.

Vanadium pentoxide corrosion can be inhibited by the introduction of a magnesium sulfate additive (Epson salts), but this increases the rate of deposition and fouling within the turbine.

There is no additive known to me that will inhibit corrosion in the hot end of a turbine from sodium and potassium salts. However, fuels can be washed and centrifuged, or washed and passed through an electric desalter, and either method can reduce the concentration of alkali metal compounds to an acceptably low level.

The first method involves washing the oil with a clean water and centrifuging the mixture to remove the water and, of course, the salts: the salts are *not* in the oil phase.

The second method involves washing and then *forcing* the natural settling–out tendency of the water phase by setting up an electric field to induce coalescence of the water droplets and thus produce a more rapid settlement of the water containing the salts. There is an excellent article[31] on this procedure, with good examples, pointing out that usually the wash–and–centrifuge method is less expensive for low flow rates and the wash–and–electrically desalt method is less expensive at high flow rates. The dividing line between *low* and *high* rates will obviously depend on the economic environment at the place and time the problem appears.

At firing temperatures above 1500°F, corrosion from the alkali sulfates is probably more severe than that from the vanadium pentoxide.

If a vanadium–containing fuel is washed with water, centrifuged, and then treated with magnesium sulfate, new kinds of deposits appear. One test[32] showed a composition as follows:

Component	Weight Percent
$MgSO_4$	85–87
Na_2SO_4	1–4
V_2O_5	1–5
$CaSO_4$ and metallic oxides	Balance

These deposits are dry, powdery, very hygroscopic, and relatively innocuous. They tend to be self–removing; a two–hour shut down will generally result in 80% removal due to cracking produced by thermal gradients, swelling and crumbling produced by the absorption of moisture from the air, and from blow–through on subsequent re–starting.

These kinds of deposits will, however, reduce efficiency and, thereby, power output. Johnson[30], estimates that the power loss for large turbines is in the order of 0.5%/100 operating–hrs.

Organic Salts

There is another class of salts that can cause trouble in the high temperatures of a refinery though not detrimental at pipeline temperatures—the *organic salts*. Unlike the metal salts, these do not occur in nature but are generated in the oil field by cleaning and treating fluids, which often contain one or more of the halogens, in contact with crude oil. Craig[33] pointed out that as little as 5 ppm of organic chlorides can cause corrosion; and that while some refineries, with prior notification, can handle up to 7 ppm organic chlorides, others cannot take a concentration greater than about 1 ppm.

The only solution to the problem is to find the oils which are being contaminated with organic chlorides and eliminate the contamination at the source, because unlike inorganic salts, organic salts cannot be removed from the crude by any ordinary process. Preventing contamination requires first determining that it exists, and Craig found a simple distillation test could obtain the naphtha fraction, react the organic chlorides in the naphtha with a sodium agent and so convert the organic chlorides to sodium chloride, and then measure the amount of sodium chloride. He also points out that a gas chromatograph can be used to give a go–no go reading, and ppm of chlorides can be obtained with an accessory calculation.

Common contaminants are tetrachloroethylene and carbon tetrachloride; there are many other similar compounds used in the oil field.

Sulfur

Sulfur can exist in crudes and refined products in many ways, and essentially all crudes, and most products, contain some percentage of some sulfur compound. An 18–year ongoing study of sulfur in crude oils conducted by the *Bartlesville Energy Research Center*[34] discovered and classified 176 different sulfur compounds in only four different crude oils.

High Sulfur Crudes and Sour Crudes

The Bureau of Mines of the U.S. Department of the Interior has published[35,36] two large lists covering both U.S. and non–U.S. crudes listing the sulfur content of each. The problem here is that these very extensive lists—there are several thousand crudes listed—do not distinguish between *sour* crudes and *high sulfur* crudes. A crude may have a *high sulfur content* and not be a *sour crude*, which is a crude containing H_2S. And while a high sulfur crude may be a problem to a refiner, it is not a problem to a pipeliner unless it ends up, in some transformed, corrosive form, in a refined product to be carried in a products pipeline.

Nelson[37] has arbitrarily defined a *sour* crude as a crude that contains more than 0.0006 wt%, about 0.05 ft³/bbl, of dissolved H_2S. This amount is considered dangerous; an amount 10 times larger can lead to active corrosion. Reference 35 contains what Dr. Nelson—who was the world's leading engineer in the petroleum refining industry for years—believes was the only complete list *at the time* of sour crudes for which specific data on H_2S content had been published. Several crudes run to as much as 0.035 wt%; the most sour, Iraq's Kirkuk, which contains 0.153 wt% H_2S, is today never introduced into commercial channels in an unstabilized form. Stabilized Kirkuk has a negligible H_2S content.

H_2S is present in many natural gases; and, in some of them, the concentration of this gas in with the inert and hydrocarbon gases is deadly: there have been many cases of oil–field deaths resulting from carelessness with H_2S.

H_2S is, of course, the *rotten egg* gas. Very few have ever smelled rotten eggs, but almost any person has, at one time or another, smelled H_2S. And this is disarming; everyone *knows* that you can smell H_2S so it couldn't be dangerous—you would *smell* it first. This is probably true, but at high concentrations H_2S has a paralyzing effect on the olfactory nerves: you may smell it with breaths 1 through 5, think it is going away for breaths 6 through 10, and start to die with breath 11 because you can no longer smell the rotten eggs and you think it has gone away. And, you may, over time, become hypersusceptible to H_2S; the body does not build up a tolerance to H_2S, but goes the other way. The toxic effect is cumulative; the more one is exposed to H_2S the more susceptible one becomes.

The concentration of H_2S in gas is usually measured in terms of grains/100 ft³, which is a bastard measurement, per ASTM D 2385 (see below). The standard conversion is 1 grain/ft³ = 2.2884 gram/m³. Wehmeyer[38] gives the following relationships that can be used for conversions:

1 grain/100 ft³ = 15.8 volume ppm
1 mole percent = 628 grains/100 ft³
H_2S partial pressure of 0.001 atm
  = 18.5 grains/100 ft³ at 50 psia
  = 0.76 grains/100 ft³ at 1200 psia
H_2S partial pressure of 0.05 psia
  = 63 grains/100 ft³ at 50 psia
  = 0.76 grains/100 ft³ at 1200 psia

From a toxicity standpoint, almost any concentration of H_2S is suspect; Texas[39] legally defines any area with a H_2S concentration in excess of 100 volume ppm as a hazardous area. Their formula for computing the location of the 100 ppm *radius of exposure* is

$$R = \left[(1.589)(\text{mole fraction } H_2S)(Q)\right]^{0.6258} \tag{9-20}$$

where R = radius of exposure, ft
Q = volume available for escape, ft³/day

There is another formula, identical to Equation 9-20 except that the 1.589 multiplier is reduced to 0.4546, that legally describes the 500 ppm radius.

These formulas, while originally intended to describe toxic areas around wells producing crude or condensate carrying H_2S, should suffice for any kind of operation that regularly introduces H_2S into the atmosphere.

High sulfur crudes may have many times more total sulfur than a sour crude; and crudes with total sulfur contents over 3.0 wt% are not unusual. One analysis of an Oxnard, California crude[40] gives a total sulfur content for this 7.2°API oil of 7.47 wt%.

One relationship that holds around the world, however, is that total sulfur content is higher in high density crudes than in low density crudes; this relationship, though spotted with exceptions, is consistent within the groups.

Measurement of Sulfur in Petroleum Fluids

Measurement of *total sulfur* in *oils* can be carried out by several methods. ASTM D 129,[41] first issued in 1922, is the oldest of the group. Then comes ASTM D 1266,[42] first issued in 1969; IP 63[43] (which, with its companion, ASTM D 1551, has been withdrawn but is sometimes still used); and ASTM D 1552,[44] first issued in 1979. Any method is acceptable for some measurements on some kinds of samples. *Total sulfur* in *LPG* fractions can be determined by ASTM D 2764.[45]

ASTM D 3227[46] measures *mercaptan* sulfur in gasolines and distillates *after* the H_2S has been removed; ASTM D 2385[47] measures mercaptan sulfur *and* H_2S in *natural gas*.

Some Thoughts on Sulfur

Total sulfur can, or can not be, a problem. If the sulfur is in crude at ordinary temperatures, it certainly is no problem. But it can be converted into more corrosive sulfur compounds, such as SO_2, in the refining process, appear in water or as a dissolved gas in a distillate, and become a problem. Use of inhibitors, or careful selection of materials, can handle these kinds of problems. *But they won't kill anybody.*

H_2S *is dangerous. It will kill anybody who breathes too much of it.* And it kills other things; go out at midday and look in the basin formed by a tank's fire walls at the dead birds and small animals that were poisoned during the night from H_2S that escaped from the tank vents and settled in the still air of night within the firewalls. I once worked a field where a bird—which breathes hundreds of times per minute—could fly over the vent of a 55 000 bbl cone roof tank being filled and receive enough H_2S to kill it before it could clear the roof. The roof eventually was covered with dozens of dead birds. There were also, over the years, a few dead humans as well.

There are commercially available H_2S detectors, H_2S alarms, and H_2S protective clothing, fresh air masks, etc., for humans. Use these things, and be careful.

BS&W

The infamous *BS* of the English–language oil industry is *basic sediment*. It is usually paired with *W*, which stands for *water*. Thus, BS&W incorporates, in its definition if not always in real fact, all of the particulate matter and all of the liquid–phase water in a petroleum hydrocarbon.

Measurement of BS&W for routine purposes is a simple centrifuge procedure; a sample is collected, diluted with a clean solvent, and centrifuged in a hand– or electric–powered centrifuge until the particulates and water cease to accumulate in the narrow end tube of a specially shaped sample bottle.

Routine measurement of BS&W is covered by ASTM D 96 (API Chapter 10.4),[48] which governs tests made in the oil field where the crude is gathered, and by ASTM D 4007 (API Chapter 10.3),[49] which governs laboratory testing of BS&W. There are other tests for determination of water by distillation and of sediment by extraction, and there are some centrifuge tests designed especially for products, but these two standards are widely used and are the tests almost always referenced in any discussion of BS&W.

There are almost no published qualitative reports on basic sediment. The only one of any scope and detail is quite old, dating from the time when metering of a crude stream for commercial custody transfer was a new and difficult technique, and solids in the crude were grinding away the close tolerances required in the positive displacement meters of the day.

Green et al.[50] examined 31 samples from 5 crude streams and passed each sample through 60, 150, 250, 325 and 400 mesh screens. The openings in the screens were, respectively, 0.0092, 0.0041, 0.0024, 0.0017 and 0.0015 inches.

They found the suspended solids commonly found in crude are silica or quartz sand, silt, iron oxides and sulfides, metallic particles, crystalline salt, carbon, and sulfur. The size of the particles ranged from colloidal—in effect, dust—to about 0.80 in. A sample collected over a 60–day period on a run of 700 000 bbls of one crude showed 37% was magnetic and 63% was non–magnetic.

They found the method of collection of samples had a considerable bearing on the amount and kind of sediment collected; and this finding resulted in a years–long search for the best way of *catching a sample* of a moving stream. This is a proper problem for later consideration, but not in this chapter.

The first method used to measure BS&W indirectly involved the use of a capacitance probe, and measuring the dielectric constant of the fluid containing the BS&W. Beyers[23] points out that capacitance monitors of BS&W measure mostly the water content, on the basis that the

BS contains a large amount of water and the sediment is encapsulated in little drops of water.

Thompson and Nicksic[51] pointed out studies had shown that capacitance probes may be in error by a considerable amount—they quote a sample measured as having 6.7% BS&W that showed clean on the centrifuge test. Their investigations showed that two factors are responsible for the error: (1) it had been incorrectly assumed that the dielectric constant of a given oil is constant with time, and (2) the observed dielectric constant of a given crude is dependent on the frequency that is used to make the measurement. While the first factor presents a calibration problem, the proof that the *dispersion* of the dielectric constant, which is the change in the constant with frequency, is common among crude oils, is important in bringing capacitance probe measurements up to the standards of centrifuge measurements. They attribute the dispersion to colloidal clays in the crude, and found that by increasing the measurement frequency from 10 mHz to 100 mHz the accuracy of the BS&W measurement improved greatly.

Hanzevack, E L., Martin, J. L., and Milliken, J. B.[52] made extensive tests to determine the cause of errors in measuring BS&W and found the primary problem was in obtaining a properly mixed sample, i.e., one that included a representative proportion of the water phase as well as adsorbed water. They concluded that hand shaking of a sample, using an electric–motor driven laboratory shaker, or a homogenizer were not satisfactory. Tests using *dried* samples of crude mixed with 0.50% water measured out from 0.10 to 0.30% with hand shaking and a lab shaker, from 0.20 to 0.50% with a homogenizer, and from 0.45 to 0.50%—quite accurate measurements—with an inline static mixer, which is the only method they recommend. This paper is highly recommended reading if you have a problem getting consistently correct measurements of BS&W.

A later paper on this matter by Stewart[53] showed that excellent results could be obtained by using capacitance probes to measure the capacitance of a *wet* and a *dry* sample of the same crude; the difference between the two (because sediment particles *look* like water, apparently because each particle is enclosed in water) represents BS&W content. Stewart claims the method, which uses a centrifugal action to throw the BS&W out of the wet stream to make it dry, is accurate within the ±0.025% water accuracy of the regular API/ASTM centrifuge bottle.

Working Equations

Calculators

Table 9-3, Reference and Working Equations for Surface Tension and Thermal Conductivity, includes several equations important for each of these two disparate characteristics. The empirical formulas for estimating values of surface tension and thermal conductivity of hydrocarbons are simple, and easy to use.

Table 9-4, Reference and Working Equations for Static Electricity in Petroleum Hydrocarbons contains, on the left side, the basic Ohm's and Coulomb's laws, and on the right side the empirical equations taken from the *International Oil Tanker and Terminal Safety Guide.*[24] The last two equations given, those for E_{max}, are complex and should be solved with a computer where the code can be written once, checked thoroughly, and from then on used as a recipe.

Computers

CCT N0. 9-1 and 9-2 are the GPSA and Cragoe formulas for surface tension. **CCT** No. 9-3 is Cragoe's method for estimating thermal conductivity of petroleum of hydrocarbons. **CCT** No. 9-4 and 9-5 solve the current flow and E_{max} equations from the *Tanker and Terminal Safety Code*. **CCT** No. 9-6 computes the Texas Railroad Commission's equation for the 100 ppm safe radius for H_2S contained in an escaping gas stream.

None of these **CCT**s require canned solvers, and can be coded in any procedural computer language with a good set of floating point routines. Note that the hyperbolic functions are required for **CCT** No. 9-5.

Table 9-3
Reference and Working Equations for Surface Tension and Thermal Conductivity

Surface Tension

Definition

$$\sigma = \frac{\text{work necessary for an increase in surface area}}{\text{amount of increase in surface area}}$$

σ as $f(T)$

$$\sigma_2 = \sigma_1 \left(\frac{T_c - T_2}{T_c - T_1} \right)^{1.2}$$

where σ_1 = surface tension at T_1 °R
σ_2 = surface tension at T_2 °R
T_c = critical temperature °R

GPSA Formula

$$\sigma^{1/4} = \left(\frac{P}{M} \right) (d_L - d_V)$$

where σ = surface tension, dynes/cm
P = parachor
M = molecular weight
d_L = liquid density, g/cm^3
d_V = vapor density, g/cm^3

Baker and Swerdloff parachor formula

$$P = 40 + 2.38(\text{mol wt of liquid})$$

Gomez dynamic surface tension predictor

$$\sigma = \left[\frac{681.3}{K} \right] \times \left[1 - \frac{T}{\left(13.488 K^{1.7654} \times spgr^{2.1250} \right)} \right]^{1.2056}$$

where T = temperature, K
K = UOP K – factor
$spgr = spgr_{60/60}$ ref $H_2O = 1.000$

Thermal Conductivity

Definition

Usual form for Fourier's Law $Q = \frac{kA(T_a - T_b)t}{x}$

where Q = kcal
T_a, T_b = °C
A = m^2
t = seconds
x = meters
k = kcal / m·s·°C

Q for three layers x_1, k_1, x_2, k_2, and x_3, k_3

$$Q = \frac{A(T_a - T_b)t}{\frac{x_1}{k_1} + \frac{x_2}{k_2} + \frac{x_3}{k_3}}$$

Q as above w/fouling inside f_1 and outside f_2

$$Q = \frac{A(T_a - T_b)t}{\frac{1}{f_1} + \frac{x_1}{k_1} + \frac{x_2}{k_2} + \frac{x_3}{k_3} + \frac{1}{f_2}}$$

Thermal Conductivity of Hydrocarbons

Cragoe Method

$$k = \left(\frac{0.813}{d} \right) (1 - 0.0003(t - 32))$$

where k = thermal conductivity in $\frac{\text{Btu}}{\text{hr} \cdot \text{sq ft} \cdot (°\text{F / inch})}$
d = spgr 60 / 60°F
t = temperature, °F

Table 9-4

Reference and Working Equations for Static Electricity in Petroleum Hydrocarbons

Ohm's Laws for current flow

Ohm's Law for macroscopic properties

$$\text{resistance} = R = \frac{V}{i} \text{ ohms}$$

$$\text{conductivity} = G = \frac{i}{V} \text{ mhos}$$

where $V =$ volts

$i =$ amperes

Ohm's Law for microscopic properties

$$\text{resistivity} = \rho = \frac{E}{j} = \frac{V/\ell}{i/A} \text{ ohm - meters}$$

$$\text{conductivity} = \sigma = \frac{1}{\rho} = \frac{i/A}{V/\ell} \text{ mho per meter}$$

where $E = \frac{V}{\ell}$

$j = \frac{i}{A}$

Coulomb's Law

Coulomb's Law for macroscopic properties

$$F = \frac{QQ'}{\varepsilon r^2}$$

where Q and $Q' =$ two charges

$F =$ force between Q and Q'

$r =$ separation between Q and Q'

$\varepsilon =$ dielectric constant

Static Electricity in Hydrocarbons

Current In Infinite Pipeline $\quad j_\infty = \alpha R^2 V^2$

where $\alpha = 15 \times 10^{-6}$, a constant

$R =$ radius of pipeline, m

$V =$ linear velocity, m / s

Current in line of length $L \quad j_L = j_\infty\left(1 - e^{-L/V\tau}\right)$

where $\tau =$ relaxation time, seconds

$= \frac{\varepsilon}{\sigma}$

$L =$ length, meters

Maximum electric field in the center of the surface of a liquid in a rectangular tank, kV/m

$$E_{\max} = \frac{16 j_L \tau(\cosh\beta d - 1)\cosh\beta p}{\Sigma_0 \pi^2 abd(\Sigma_V \cosh\beta p \sinh\beta d + \Sigma_L \sinh\beta p \cosh\beta d)}$$

tank length = a

tank breadth = b

tank height = c

tank ullage = p

measured in = meters

where $\beta = \pi\sqrt{\frac{1}{a^2} + \frac{1}{b^2}}$

$d = c - p$

$\tau = \frac{\Sigma_L \Sigma_0}{\sigma}$

Formula to compute maximum tank filling velocity V to keep E_{max} to a safe value

$$E_{\max} = \frac{16 \alpha R^2 V^2 \tau(\cosh\beta d - 1)\cosh\beta p}{\Sigma_0 \pi^2 abd(\Sigma_V \cosh\beta p \sinh\beta d + \Sigma_L \sinh\beta p \cosh\beta d)}$$

Note: $E_{\max}$ should be held to 500 kV/meter

References

[1] Vennard, John. K., and Street, Robert L., *Elementary Fluid Mechanics,* 6th ed. New York: John Wiley & Sons, 1982.

[2] Nelson, W. L., *Petroleum Refinery Engineering,* 4th ed. New York: McGraw–Hill Book Company, 1969.

[3] Katz and Saltman, *Industrial Engineering Chemistry,* Vol. 31, p. 91, 1939.

[4] Zanker, Adam, "Estimate Surface Tension for Changes for Liquids," *Hydrocarbon Processing,* (date not known).

[5] ASTM D 3825, *Standard Test Method for Dynamic Surface Tension by the Fast–Bubble Technique,* American Society for Testing Materials, Philadelphia.

[6] *Engineering Data Book.* Tulsa: Gas Processors Suppliers Association, 9th ed., 1972.

[7] Macleod, D. B., *Transactions of the Faraday Society,* Vol. 19, p. 38, 1923.

[8] Sudgen, S., "The Variation of Surface Tension. VI: The Variation of Surface Tension With Temperature and Some Related Functions," *Journal of the Chemical Society,* Vol. 125, p. 32, 1924.

[9] Govier, G. W., and Aziz, K., *The Flow of Complex Mixtures in Pipes,* New York: Van Nostrand Rheinhold Company, 1972.

[10] Baker, O., and Swerdloff, W., "Calculations of Surface Tension-3: Calculations of Surface Tension Parachor Values," *Oil and Gas Journal,* 05 December 1955.

[11] Gomez, Jose Vicente, "Method Predicts Surface Tension of Petroleum Fractions," *Oil and Gas Journal,* 07 December 1987.

[12] Author's Note: If you don't have the K-factor, or the numbers to calculate it, use a K of 11.8. This seems to be characteristic of most U.S. medium quality crude oils.

[13] *Standards of the Tubular Exchanger Manufacturers Association.* New York: Tubular Exchanger Manufacturers Association, 6th. ed., 1978.

[14] ASTM D 2717, *Standard Test Method for Thermal Conductivity of Liquids,* American Society for Testing Materials, Philadelphia.

[15] Cragoe, C. S., *Thermal Properties of Petroleum Products,* National Bureau of Standards Miscellaneous Publication No. 97, U. S. Department of Commerce, Washington, D. C., 1929.

[16] Zanker, A., "Get Thermal Conductivity of Liquid Mixes Quickly," *Hydrocarbon Processing, March 1981.*

[17] Bates, et al. *Industrial and Engineering Chemistry Annual Edition,* Vol. 10, p. 314, 1938.

[18] Klinkenberg, A., "Electrical Conductivity of Low Dielectric Constant Liquids by DC Measurement," *Journal of the Institute of Petroleum,* Vol. 53, No. 517, January 1967.

[19] ASTM D 3114, *Standard Test Method for D-C Electrical Conductivity of Hydrocarbon Fuels,* American Society for Testing Materials, Philadelphia.

[20] ASTM D 2624, *Standard Test Method for Electrical Conductivity of Aviation and Distillate Fuels Containing a Static Dissipator,* American Society for Testing Materials, Philadelphia.

[21] Rogers, D. T., and Schleckser, C. E., "Engineering and Theoretical Studies of Static Electricity in Fuels," *Fifth World Petroleum Congress,* New York, 1959.

[22] Rogers, D. T., McDermott, J. P., and Munday, J. C., "Theoretical and Experimental Observations of Static Electricity in Petroleum Products," *Proceedings of the American Petroleum Institute,* Vol. 37, Section III, 1957.

[23] Byers, Donald P., "Handbook of LACT Operations, Part 3–BS&W Monitor and its Calibration." *Pipe Line Industry,* April 1963.

[24] Oil Companies International Marine Forum, *International Oil Tanker and Terminal Guide,* 2nd. ed. London: Applied Science Publishers, Ltd., 1974.

[25] Valkovic′,Vlado, *Trace Elements in Petroleum,* Tulsa: The Petroleum Publishing Company, 1978.

[26] Kloosterman, A. H., private communication, 1974.

[27] Leeper, J. E., "Mercury—LNG's Problem," *Hydrocarbon Processing,* November 1980.

[28] Nelson, W. L., "How Much Metals in Crude Oils?" *The Oil and Gas Journal,* 07 August 1972.

[29] Suboch, W. P., "Vanadium Analyzer Useful Tool," *The Oil and Gas Journal,* 28 February 1966.

[30] Nelson, W. L., "How Much Metals in Crude Oils?" *The Oil and Gas Journal,* 07 August 1972.

[31] (Author unknown), "Florida Power Cuts Crude Oil Sodium to Less Than 1.5 ppm," *Gas Turbine World,* December 1973.

[32] Harris Johnson, private communication.

[33] Craig, John. E., "Pipeline Program Combats Organic Chloride Contamination," *The Oil and Gas Journal,* 13 October 1986.

[34] Rall, H. T., Thompson, C. J., Coleman, H. J., and Hopkins, R. L., "Sulphur Compounds in Crude Oils," *Report No. 659,* U.S. Bureau of Mines, 1972.

[35] Hughes and Blake, "Bibliography of Reports Containing Analyses of Sulphur in Crude Oils," *I.C. Report No. 7470,* U.S. Bureau of Mines, 1948.

[36] Carrales, M. Jr., and Martin, R. W., "Sulphur Content of Crude Oils," I. C. Report No. 8676, U. S. Bureau of Mines, 1975.

[37] Nelson, W. L., "Sulfur Content of Oils Throughout the World," *The Oil and Gas Journal,* 13 February 1967.

[38] Wehmeyer, C. L., "Design of Sour Crude Separation Facilities," *Petroleum Engineer,* August 1975.

[39] Railroad Commission of Texas, Oil and Gas Docket No. 20-65,354, *Special Order Amending Rule 36 of the*

General Conservation Rules of Statewide Application, State of Texas, Having Reference to Oil and Gas Operation in Hydrogen Sulfide Areas, 15 March 1976.

[40] Nelson, W. L., "Sulphur in Crude Oils Around the World," *The Oil and Gas Journal*, 11 November 1968.

[41] ASTM D 129, *Standard Test Method for Sulphur in Petroleum Products (General Bomb Method)*, American Society for Testing Materials, Philadelphia.

[42] ASTM D 1266, *Standard Test Method for Sulphur in Petroleum Products (Lamp Method)*, American Society for Testing Materials, Philadelphia.

[43] IP 63, *Sulphur Content, Quartz Tube Method*, Institute of Petroleum, Great Britain, London.

[44] ASTM D 1552, *Standard Test Method for Sulphur in Petroleum Products (High Temperature Method)*, American Society for Testing Materials, Philadelphia.

[45] ASTM D 2764, *Standard Test Method for Sulphur in Liquefied Petroleum Gases (Oxy–Hydrogen Burner or Lamp)*, American Society for Testing Materials, Philadelphia.

[46] ASTM D 3227, *Standard Test Method for Mercaptan Sulphur in Gasoline, Kerosine, Aviation Turbine, and Distillate Fuels (Potentiometric Method)*, American Society for Testing Materials, Philadelphia.

[47] ASTM D 2385, *Standard Test Method for Hydrogen Sulphide and Mercaptan Sulphur in Natural Gas (Cadmium Sulphate Iodometric Titration Method*, American Society for Testing Materials, Philadelphia.

[48] ASTM D 96 (API Chapter 10.4), *Determination of Sediment and Water in Crude Oil by the Centrifuge Method (Field Procedure)*, American Society for Testing Materials, Philadelphia.

[49] ASTM D 4007 (API Chapter 10.3), *Determination of Water and Sediment in Crude Oil by the Centrifuge Method (Laboratory Procedure)*, American Society for Testing Materials, Philadelphia.

[50] Green, John E., Hall, A. H., and Luttrell, J. R., "Suspended Material Carried in Crude Oil Streams," *Proceedings of the American Petroleum Institute, Transportation Division*, November 1948.

[51] Thompson, D. D., and Nicksic, S. W., "Frequency Change Can Help Improve BS&W Monitor's Accuracy," *The Oil and Gas Journal*, 23 November 1970.

[52] Hanzevack, E. L., Martin, J. L., and Milliken, J. B., "Four Steps Are Required For Accurate BS&W Measurements," *The Oil and Gas Journal*, 21 July 1980.

[53] Stewart, T. L., "Improved BS&W Measurement System Compensates for Property Variations," *The Oil and Gas Journal*, 20 September 1982.

Computer Computation Template

CCT No. 9-1: Surface tension by the GPSA (Macleod and Sudgen) formula. Input data required is liquid density d_L and vapor density d_V in g/cm3 or kg/liter, and the molecular weight MW. If vapor density is not readily available, and pressure is not very high, it can be neglected.

References: Chapter 9. Equations 9-2 and 9-3.

Input

Input Data $MW := 115$ molecular weight $d_L := 0.707$ g/cm^3 $d_V := \frac{1.293}{1000} \cdot 3.943$ g/cm^3

Variables $P := 40 + 2.38 \cdot MW$ parachor

Equations

$$\sigma := \left[\left(\frac{P}{MW} \right) \cdot \left(d_L - d_V \right) \right]^4$$

Results

$\sigma = 13.4$ dynes / cm

Computer Computation Template

CCT No. 9-2: Dynamic surface tension by the Gomez formula. Requires input of the UOP K–Factor and the specific gravity at 60/60°F. The template as drawn plots a spread of points for surface tension over the temperature range 5°C> T > 95°C.

References: Chapter 9. Equation 9-4.

Input

Factors $i := 0, 1 .. 9$ $\quad T_{SeedC} := 5$ $\quad T_i := T_{SeedC} + i \cdot 10 + 273$

Input data $SpGr_{6060} := 0.799$ $\quad UOPK := 11.80$

Equations

$$\sigma_i := \frac{681.3}{UOPK} \cdot \left(1 - \frac{T_i}{13.448 \cdot UOPK^{1.7654} \cdot SpGr_{6060}^{\;2.1250}}\right)^{1.2056}$$

Results

$$T - 273 = \begin{bmatrix} 5 \\ 15 \\ 25 \\ 35 \\ 45 \\ 55 \\ 65 \\ 75 \\ 85 \\ 95 \end{bmatrix} \text{deg C} \quad \sigma = \begin{bmatrix} 29.5 \\ 28.6 \\ 27.6 \\ 26.7 \\ 25.8 \\ 24.8 \\ 23.9 \\ 23.0 \\ 22.1 \\ 21.2 \end{bmatrix} \text{surface tension dynes / cm}$$

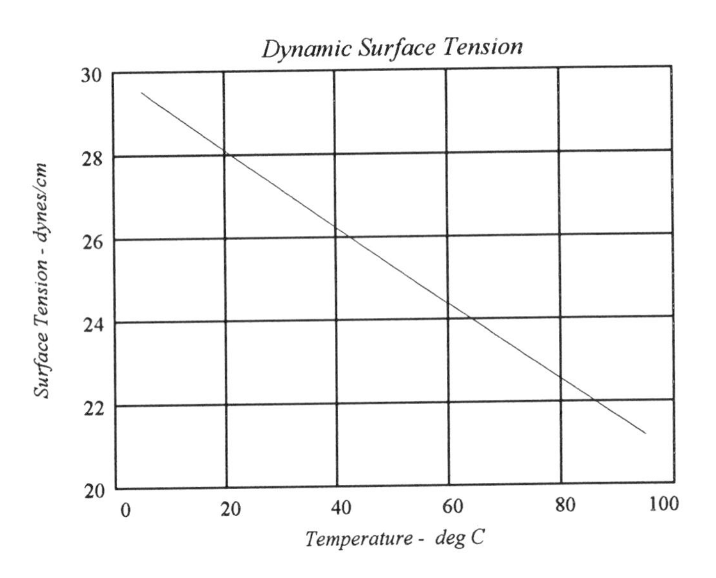

Computer Computation Template

CCT No. 9-3: Thermal conductivity of hydrocarbons using Cragoe's method. Calculation is for 60 °F. This can be changed in the t:= input line.

References: Chapter 9. Equation 9-11.

Input

Factors $i := 0, 1 .. 10$ $j := 0, 1 .. 2$

Input Data $SpGr_{6060Seed} := 0.50$ $t := 60$

Variables $\rho_i := SpGr_{6060Seed} + i \cdot 0.05$

Equations

$$k_i := \frac{0.813}{\rho_i} \cdot (1 - 0.0003 \cdot (t - 32))$$

Results

$$k = \begin{bmatrix} 1.612 \\ 1.466 \\ 1.344 \\ 1.240 \\ 1.152 \\ 1.075 \\ 1.008 \\ 0.948 \\ 0.896 \\ 0.849 \\ 0.806 \end{bmatrix}$$ Btu / [(hr x sq ft)(deg F / inch)]

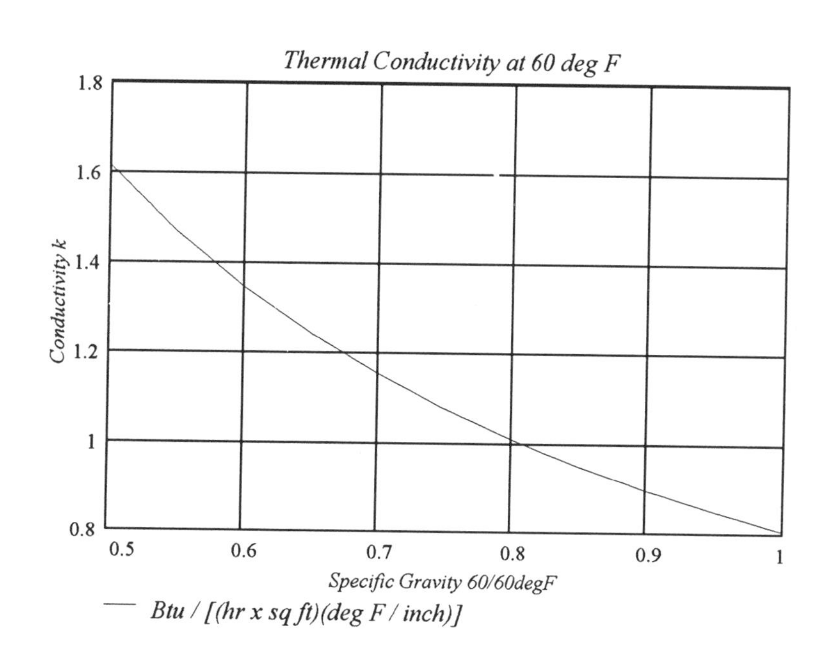

Computer Computation Template

CCT No. 9-4: Calculates the current in a pipeline (1) of infinite length, and (2) of finite length L, induced by a flowing hydrocarbon. The voltage generated in a tank receiving the stream is a function of of this current and the conductivity σ. The conductivity of black oils runs as high as 10^4 — 10^5 picomho / m; conductivity of clean oils and light products can run as low as 0.1 or 0.2 picomho / m and can generate very high kV / m values.

References: Chapter 9. Equations 9-14 and 9-15. Table 9-2.

Input

Factors	$i := 0, 1..3$	$n := 1, 2..10$	$\alpha := 15 \cdot 10^{-6}$	$e_o := 8.85 \cdot 10^{-12}$ pF / m
Input Data	$L_{Seed} := 100$ m	$V_{Seed} := 5$ m / sec	$\varepsilon_L := 2$	$\sigma_L := 1 \cdot 10^{-12}$ pmho / m
	$R := 0.50$ m	$V := 20$		
Variables	$\varepsilon_x := \varepsilon_L \cdot e_o$	$\tau := \dfrac{\varepsilon_x}{\sigma_L}$	$\tau = 17.700$	$\varepsilon_x = 1.770 \cdot 10^{-11}$
	$L := L_{Seed}$	$l_n := L_{Seed}\, n$	$v_i := V_{Seed} + i \cdot V_{Seed}$	$D := 2 \cdot R \cdot 1000$

Equations

(1) $$J_{Infinite} := \alpha \cdot R^2 \cdot V^2$$

(2) $$j_{i,n} := \alpha \cdot R^2 \cdot (v_i)^2 \cdot \left(1 - exp\left(-\frac{l_n}{v_i \cdot \tau}\right)\right)$$

Results

(1) $J_{Infinite} = 1.500 \cdot 10^{-3}$ current flow in an infinitely long 1 m diameter pipeline at a flow rate of 20 m / sec of dry oil

(2) $j_{i,n} =$ Current flow in finite pipelines

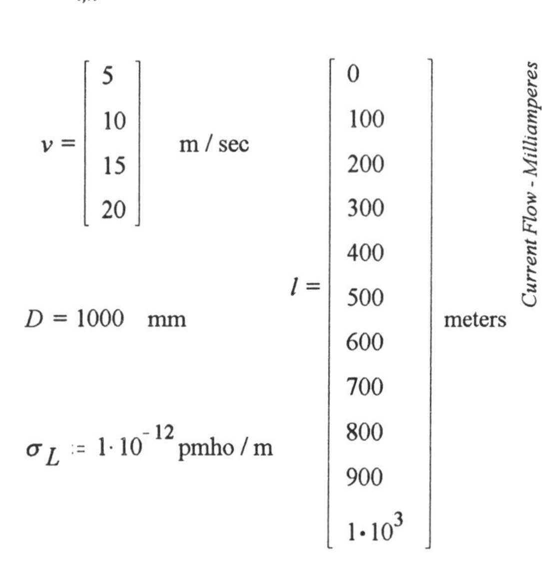

v = [5, 10, 15, 20] m / sec
l = [0, 100, 200, 300, 400, 500, 600, 700, 800, 900, 1·10^3] meters
D = 1000 mm
σL := 1·10^-12 pmho / m

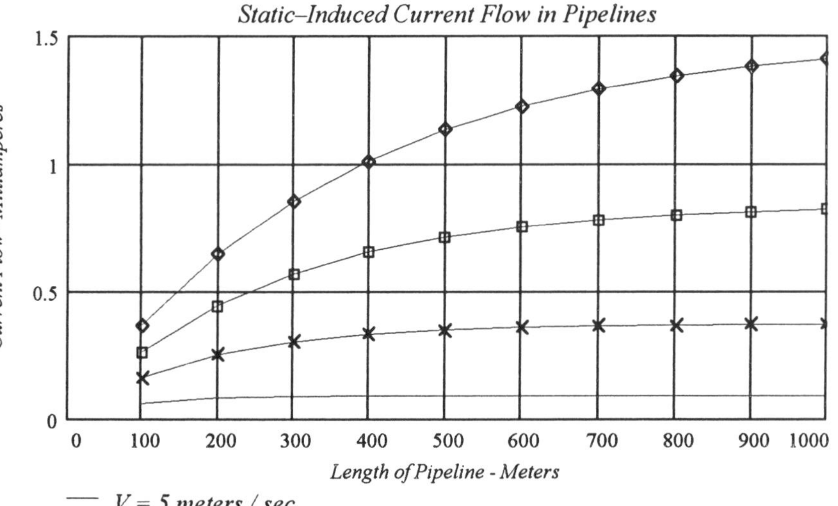

Computer Computation Template

CCT No. 9-5: Calculates the maximum electric field in the center of the surface of a liquid in a rectangular tank. To approximate a circular tank, assume a tank with equal length and breadth. The template provides for inputting the characteristics of the fluid, ε_V, ε_L, ε_0, and σ_L, the dielectric constants of the vapor, liquid, and vacuum phases, and the conductivity of the liquid phase. Tank dimensions are a, b, c and p, the length, breadth, height, and ullage, in meters. R is the radius of the filling pipe and V is the velocity of flow in the pipe, both in meters.

References: Chapter 9. Equation 9-18. Table 9-2.

Input

Factors $\alpha := 15 \cdot 10^{-6}$ $\quad i := 1, 2 .. 9$ $\quad \varepsilon_0 := 8.85 \cdot 10^{-12}$ $\quad \sigma_L := 0.1 \cdot 10^{-12}$

Input Data $\varepsilon_V := 1$ $\quad \varepsilon_L := 2$ $\quad V_{Seed} := 1$ $\quad D_{Meter} := 0.254$

$a := 6.0$ $\quad b := 6.0$ $\quad c := 10$ $\quad p := 0.3$ $\quad R := \dfrac{D_{Meter}}{2}$

Variables $V_i := V_{Seed} + i$ $\quad \beta := \pi \sqrt{\dfrac{1}{a^2} + \dfrac{1}{b^2}}$ $\quad d := c - p$ $\quad \tau := \dfrac{\varepsilon_L \cdot \varepsilon_0}{\sigma_L}$ $\quad D_{Inch} := \dfrac{D_{Meter}}{0.0254}$

Equations

$$E_{max_i} := \frac{16 \cdot \alpha \cdot R^2 \cdot \left(V_i\right)^2 \cdot \tau \left(\left(\cosh(\beta \cdot d) - 1\right) \cdot \cosh(\beta \cdot p)\right)}{\varepsilon_0 \cdot \left[\pi^2 \cdot \left[a \cdot b \cdot d \cdot \left[\left(\varepsilon_V \cdot \cosh(\beta \cdot p)\right) \cdot \sinh(\beta \cdot d) + \left(\left(\sigma_L \cdot \sinh(\beta \cdot p) \cdot \cosh(\beta \cdot d)\right)\right)\right]\right]\right]}$$

Results

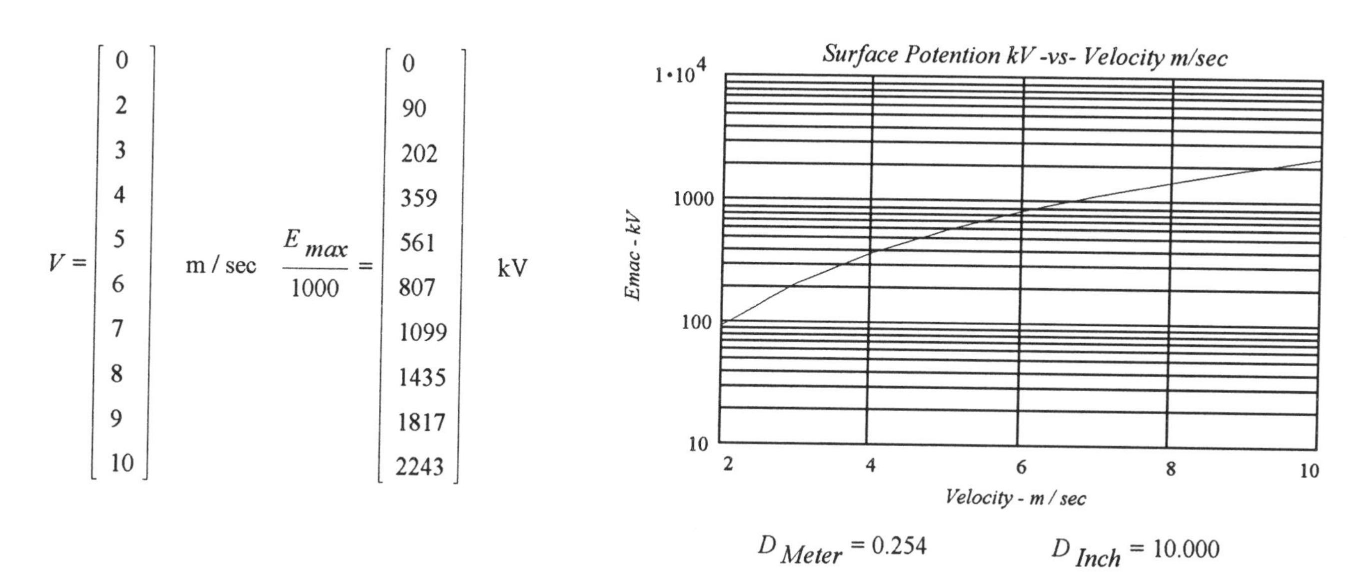

$D_{Meter} = 0.254$ $\qquad D_{Inch} = 10.000$

Computer Computation Template

CCT No. 9-6: Computes the Texas Railroad Commission's "Radius of exposure for 100 ppm H_2S concentration."

References: Chapter 9. Equation 9-20.

Input

Factors $i := 1, 2 .. 5$ $j := 1, 2 .. 10$

Input Data $MoleFractionH2S_{Seed} := 0.05$ $Q_{Seed} := 10000$

Variables $MoleFractionH2S_i := i \cdot MoleFractionH2S_{Seed}$ $Q_j := j \cdot Q_{Seed}$

Equations

$$R_{i,j} := \left(1.589 \cdot MoleFractionH2S_i \cdot Q_j\right)^{0.6258}$$

Results

$$MoleFractionH2S = \begin{bmatrix} 0.05 \\ 0.10 \\ 0.15 \\ 0.20 \\ 0.25 \end{bmatrix} \quad \frac{Q}{1000} = \begin{bmatrix} 10 \\ 20 \\ 30 \\ 40 \\ 50 \\ 60 \\ 70 \\ 80 \\ 90 \\ 100 \end{bmatrix} \text{ mcf / d}$$

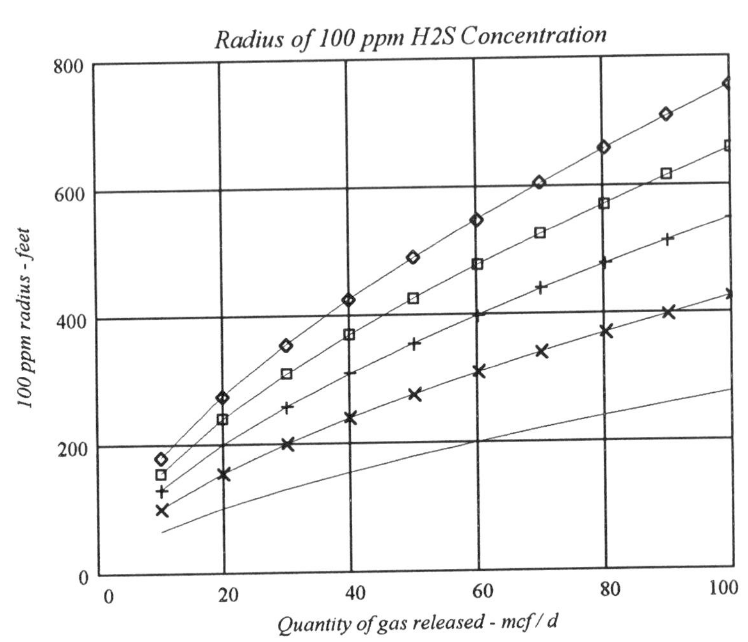

Part II
Review of Classical Hydrodynamics, Fluid Mechanics, and Hydraulics

Part II, which includes Chapters 10, 11, and 12, is a review of classical hydrodynamics, theoretical fluid mechanics, applied fluid mechanics, and hydraulics, as apply generally to the flow of fluids in round pipes. Hydrostatics is treated from the standpoint of standing liquids. Transient flows are not considered at this level of the book.

Emphasis is placed on the development of the energy, continuity, and momentum equations and, from these, the pipe friction equation.

Chapters 10 and 11 include tables summarizing the fundamental motion equations, and Chapter 12 contains an extensive tabulation of the fundamental equations of pipe flow.

10
Hydrodynamics: The Flow of Ideal Fluids

Introduction

This chapter, and the two chapters following, are review and reference chapters. There is nothing in them that can't be found in any good book on fluid mechanics. There are several of these included in the references to this chapter, and if you need reference material more complete than that included here, I recommend each of these books highly.

As to the reason for these chapters? It is simple. You can't remember everything, and sometimes you don't have your library with you.

Some History

The Forerunners

Archimedes is generally considered the founding father of *hydraulics*, defined as the practical application of *fluid statics* and *dynamics*. He found, *by experimentation only* (sitting in his bath tub, according to the story), the *theory of flotation buoyancy*, and went on to make a liar out of the kings jeweler who had mixed a little silver in with the kind's gold. This was sometime in the third century BC.

There wasn't a lot more technical writing about fluids until the early seventeenth century when Stevin published a treatise on *hydrostatics* which established that the force exerted on a submerged surface is a function of the height of the liquid above it and the area of the surface.

But in the meantime—written down or not—Roman engineers learned enough *hydraulics* to construct some very long *aqueducts*, any one of which could—and often did—incorporate bridges over rivers or inverted siphons under rivers, tunnels through intervening hills, and, at the end, a *castellum* to distribute the water to networks of pipelines carrying the water to the consumers. There are remnants of these ancient works all over the old Roman Empire, from the towering Pont du Gard, the almost 50 meters high triple–arched bridge on the aqueduct to Nîmes in France, to the small irrigation water aqueducts running along the coastal plain of Turkey. Hauck[1] and Hodge[2] provide enjoyable, readable analyses of these aqueducts.

The Romans piped this water around downtown Rome and some other cities in lead pipe distribution systems leading to continuously running water outlets in the home of the wealthy and frequent public outlets for the common citizen. No one knows how they sized these pipes, but they certainly knew big pipes carry more water than little pipes; that long pipes lost more pressure than short pipes; and that water could be flowed farther and farther as the inlet reservoir was raised higher and higher; and thus they must have known, even if they didn't put it into any surviving literature, that $Q = f(D, L, \Delta H)$.

They designed, built, and for centuries used perfectly acceptable water supply systems. For long–distance transport they used aqueducts; for short distances, and distribution, they used pipelines.

Then the Roman Empire slowly came to an end, and all the water works, aqueducts and pipelines alike, lacking skillful and knowledgeable maintenance, fell apart. They disintegrated or, in many cases, were savaged for their parts, and none exists today in a usable form.

It is hard to believe, but from the end of the days of Roman engineers until the mid–seventeenth century the *practice* of hydraulics and, for as much practical good as resulted from it, the *study* of hydraulics, stood still. The famous *Machine de Marly*, a set of positive displacement plunger pumps driven by undershot water wheels taking energy from a low head dam in the Seine, that delivered water from the Seine through a cast iron pipeline to the fountains and waterworks of Versailles, was not a great deal more advanced than the works of the Romans as at the time of Christ.

But the rapid advances of the industrial revolution brought a new impetus to the study of the flow of fluids in pipes. Pipelines carried water to and from all kinds of sources and sinks; networks of town–gas lines carried cooking, heating, and illuminating gas to residences and industries; compressed air lines started to appear; and, in some of the larger cities, energy was transported over considerable distances by pipelines carrying pressurized hot water.

Because of the number and extent of these systems, and because they were, for the most part, constructed in a profit–oriented, capitalistic, society, it became important from a business standpoint to be able to predict the transport capacity of a pipeline before it was built, and so the old *art* of hydraulics set out on the long road to converting itself into a *science*.

As it will be seen, in many ways it hasn't made it yet.

The Theorists

Nearly two millennia after Archimedes came Stevin; then came Leonardo da Vinci—the first to understand the principle of *continuity*, even though he conceived his explanation in terms of flow in rivers—and then came Pascal, Newton, Bernoulli, and Euler; and it was already the mid–eighteenth century. Daniel Bernoulli's equation was first published in 1738; Leonhard Euler's equation, based on the work of Bernoulli, dates from 1755. These are the basic equations of fluid flow, and are fundamental equations in hydrodynamics.

Navier and Stokes, who developed their equations for viscous flow in the first half of the 1800s (nobody can solve the equations analytically any better today than Navier and Stokes themselves could at the time of their invention), and Reynolds, who toward the end of the century, studied the transition from laminar to turbulent flow in viscous fluids and described the dimensionless number that bears his name, more or less wrapped up theoretical work for the nineteenth century.

Then came Prandtl in 1904 and van Karmen in 1924 to develop the boundary layer theory and that's about it for the theorists. While the world still continued to look into the theory of the flow of fluids—especially the flow of air at supersonic speeds and the dynamics of plasmas—these men were the last to become so important in the field of fluid flow that laws or equations were named for them.

After them came the world–wide expansion of the long–distance oil and gas pipeline industry, first with little–inch pipes and, later on, with pipes as large as 56 inches, and major emphasis was put on hydraulics, as differentiated from hydrodynamics or fluid mechanics, and the pipeline world has belonged to the pipeliners ever since.

The Tidewater Pipeline—a 6–inch crude line over 1000 miles long built in 1908 and 1909 from Stoy, Illinois to Bayonne, New Jersey, was the first long distance–oil pipe line. The hydraulics were rudimentary (an open system, with tankage floating on station suction); the pumps were formidable double–acting triplex plunger pumps running at 20 spm; the engines were 140 bhp horizontal hot–tube diesels with 25 000–pound flywheels; and, after a few years exposure to electrolysis corrosion from the stray currents from the tracks of dozens of electric interurban rail systems, the uncoated line leaked like a sieve.

And, you may ask, how do I know so much about this pioneer system? Because in 1946 it was part of the first pipeline system I ever managed, and you don't forget your first management assignment.

Hydrodynamics, Fluid Mechanics, and Hydraulics

One important thing regarding the definition of *hydrodynamics*, *fluid mechanics*, and *hydraulics*. There is no generally accepted definition of any of the three, especially at the boundary as between any two of them, and the following will have to do.

- *Hydrodynamics* is the study of *fluid statics*, and the dynamics of *inviscid fluids*. But except for super–cooled liquids there is no such thing as an inviscid fluid, and so hydrodynamics is generally considered as the study of *ideal fluids*.
- *Fluid mechanics* is the study of the flow of *real, viscous fluids*. It is based on hydrodynamics, and separated from hydraulics only by a fine line that separates primarily theoretical methods from primarily experimental methods.
- *Hydraulics* is the *engineering branch* of fluid mechanics; it is the artful kind of science aqueduct builders would have understood. It is

not always rigorous, as true fluid mechanics often tends to be; in hydraulics, if a method gets an *acceptable result* it is an *acceptable method.* And, if the method has a good theoretical basis, so much the better.

Comments on Systems of Units

Be forewarned: the most common mistake pipeliners make in dealing with the fundamental equations is to mix units. Table 10-1 lists the fundamental units for study of hydrodynamics in the gravitational FPS and MKS systems, as well as in SI. We will work for the most part in the first two.

Statics

Statics is the *hydrodynamics of still liquids*, in the sense that liquids are fluids that support an upper surface. A fluid may contain dissolved or entrained gases that would at atmospheric pressure—or less than atmospheric pressure in the case of a fluid under a vacuum—produce bubbles and foam; but if the pressure imposed on it is such that a still upper surface is maintained, all of the principles of statics apply without change except that submerged pressures and forces are increased to take into account the effect of the force impressed on the surface.

Fluid statics is simple, though to go back and bring the engineering equations up out of first assumptions is not a simple exercise. You can have a great time proving things with the vector calculus, defining streamlines, centers of pressure, and all that, but for our purposes we can take the point of view that users of this book already know what static head or pressure is and only need a review as to basics.

So, to commence. If a body is immersed in a fluid *that is at rest* and that has no outside pressure imposed on it, a force is applied to every surface of the body dependent on the depth of the overhead fluid and the specific weight of that fluid. If the force is F and the area is A then

$$p = \lim_{\Delta A \to 0} \frac{\Delta F}{\Delta A} = \frac{dF}{dA} \tag{10-1}$$

If we assume ΔA is a tiny circle lying h units below and parallel to the surface of a stilled liquid, the force bearing on it is the weight of a cylinder of fluid of area ΔA, height h, and specific weight γ, which equals $\Delta A h \gamma$. So

$$p = \lim_{\Delta A \to 0} \frac{\Delta F}{\Delta A} = \frac{\Delta A \gamma h}{\Delta A} = \gamma h \tag{10-2}$$

Table 10-1
Fundamental Units

Quantity	Symbol	Gravitational FPS	Gravitational MKS	Absolute SI
Base Quantities				
Length	L	foot	meter	meter
Time	T	second	second	second
Mass	M	slug=g·lb=32.174·lb	gkg=g·kg=9.806·kg	kg
Force	F	pound	kg	newton
Temperature	T or θ	°R or °F	°K or°C	K
Power	P	horsepower=550 ft·lb/s	cv=Ps=75 m·kg/s	watt=J/s=m²·kg/s³
Energy	E	ft·lb	kg·m	joule=N·m
Hydrodynamic Quantities				
Density	ρ	slug/ft³	gkg/m³	kg/m³
Pressure	p	lb/ft²	kg/m²	pascal=N/m²
Specific Weight	γ	lb/ft³	kg/m³	N/m³
Specific Volume	v	ft³/slug	m³/gkg	m³/kg
Specific Gravity	spgr or SG	density/62.43	(gkg/m³)/1000·g	(kg/m³)/1000

The unit gkg represents the metric slug, i.e., one kg multiplied by the acceleration due to gravity.
Numeric values are approximate, except for the multipliers for horsepower and cv=Ps. These are exact by definition.

This is the *pressure at a point*.

The *pressure* on a *horizontal surface* is the same value, and the force is directed at the centroid of the area. The *force on a horizontal area* immersed so that the height of liquid above the centroid is h is

$$F = pA = \gamma h A \tag{10-3}$$

The force on an *inclined surface* is again, the same, and such a plane can be rotated about its centroid without changing the magnitude of the force resulting from static pressure so long as the distance from the centroid to the surface remains constant and no part of the plane cuts the surface of the fluid. The vertical component of the force is equal to the weight of the fluid above it, i.e., the area as projected on the surface; the horizontal component is the same as that on a plane vertical surface which is the projected area on the surface. The force acts through the *center of pressure*, defined by

$$x_p = \frac{1}{F}\int_A xpdA = \frac{1}{F}\int_A xp\delta x\delta y \tag{10-4}$$

$$y_p = \frac{1}{F}\int_A ypdA = \frac{1}{F}\int_A yp\delta x\delta y \tag{10-5}$$

If knowing the line of action of a force on an inclined surface is important, there are ways of handling most shapes without going through the formal integration. See Streeter and Wylie[3] for ways to handle this problem. Or see Simon:[4] he points out that the center of pressure is removed from the centroid of the area by the distance

$$e = \frac{I_o}{\ell A} \tag{10-6}$$

where I_o = second moment of area A with respect to the centroid of A
ℓ = distance between the centroid and the line of intersection of the plane of A with the surface
e = distance between the centroid and the center of pressure measured along the surface of A

Then the mathematics become more complicated. The horizontal force on a *curved surface* is the pressure force exerted on the projection of the curved surface. Thus

$$F_x = \int_A p\cos\theta dA \tag{10-7}$$

where θ = the angle the normal of the element dA makes with the horizontal

The vertical force on a submerged curved surface is the weight of the liquid above the curved surface extending up to the surface of the liquid.

$$F_v = \int_A p\cos\theta dA \tag{10-8}$$

where θ = the angle the normal of the element dA makes with the vertical

For the total force on any submerged surface, there are always three force components, two horizontal and one vertical.

The total of forces exerted on a *submerged body* of any kind is the *buoyant force* of Archimedes.

$$F_B = V\gamma \tag{10-9}$$

where V = the volume of liquid displaced

For a body to float, F_B must be greater than the weight of the body itself; if the body is completely immersed, F_B is the total of forces on the body *acting vertically*.

There are other problems in fluid statics you may come up against in practice, but most of them have to do with *pressure*, and not *force*; and the *pressure at a point* is always as said in Equation 10-2: h, the *height of the column of liquid* above it, multiplied by γ, the *specific weight* of the liquid. There can be some funny shapes involved, and in some cases finding the total force, the direction in which it acts, and the center of pressure at which it acts, is more an exercise in geometry than in hydrostatics, but finding the pressure is simple.

Dynamics

Hydrodynamics is the study of *inviscid* fluids. These are fluids that have no viscosity, and no tendency to adhere to things but rather to slip by them. There are no such fluids in reality, but sometimes practical results can come out of the study of the flow of fluids that have no viscosity and, thereby, generate no friction losses from internal shear, and do not stick to things, so they flow with no shear at the pipe walls.

Flow of an inviscid fluid may be *steady* or *unsteady*, *uniform* or *non–uniform*, *rotational* or *irrotational*, and may be analyzed from a one–, two–, or three–dimensional standpoint. There are also *thermodynamic* considerations.

In *steady* flow there are no changes in conditions at a point with time; the converse is true for unsteady flow.

Uniform flow is flow in which the average flowing velocity over any cross section is invariable with time. Any other kind of flow is non–uniform.

Rotational flow, in which the velocity varies as some function of distance from a center of rotation, can exist in pumps, and at entrances to and exits from otherwise still reservoirs, but it does not naturally occur in pipe flow.

Though, on a microscopic basis, flow in a pipeline is three–dimensional—a pipe is, after all, an object of three dimensions—and on a not–so–microscopic basis pipe flow is two–dimensional; most analyses—even on curved pipes—are derived on a one–dimensional basis. These analyses are orders of magnitude simpler than three–dimensional analyses, and a great deal more simple than those for two dimension systems, and it has been learned, over time, that analyses prepared on such a basis yield acceptable results if the mass velocity over the cross section of the pipe is taken as the velocity of the individual packets of matter flowing down the stream tubes of a one–dimension system.

As to thermodynamics, flow may be either adiabatic or isothermal, or both. Inasmuch as we were dealing strictly with hydrodynamics, there are no losses, no heat input to or withdrawn from the system, and the temperature is everywhere constant. We can therefore either consider the flow as *frictionless adiabatic* (or *isentropic*, the same thing), *isothermal* flow, or forget about thermodynamics for a while. Either assumption is acceptable, and the mathematics is the same in either case. So, for purposes here, we forget about them.

The study of hydrodynamics is therefore the study of the uniform, steady, irrotational flow of inviscid fluids, or, as John von Neumann is reputed to have said, the study of *dry water*.[5]

Continuity

The *equation of continuity*—along with other defining equations considered here—can be in many mathematical forms. Benedict,[6] whose book is one of those I included in the references, is a distinctive teacher in that he starts out in his Chapter 1 with the conservation equations for ideal pipe flow and, for each of them, he gives the final, fully rigorous, equation first so that, as he says, "...we may know where we are headed...." Then he proceeds to develop the equations from fundamental concepts.

The continuity equation is almost always credited to Leonardo da Vinci. In its final form it is generally given as a partial differential equation such as

$$\frac{\partial \rho}{\partial t}+\frac{\partial}{\partial x}(\rho u_x)+\frac{\partial}{\partial y}(\rho u_y)+\frac{\partial}{\partial z}(\rho u_z)=0 \qquad (10\text{-}10)$$

Here u_x, u_y and u_z are the mutually perpendicular components of velocity, ρ is fluid density, and t is time.

In engineering terms the equation is usually written

$$\rho_1 A_1 V_1 = \rho_2 A_2 V_2 \qquad (10\text{-}11)$$

In this equation $\rho_1 A_1 V_1$, the mass of the fluid flowing into the system, is set equal to $\rho_2 A_2 V_2$, the mass flowing out of the system.

For a constant density system

$$A_1 V_1 = A_2 V_2 \qquad (10\text{-}12)$$

Since the product AV has the dimensions of a volume, we have *Harnett's First Law*: *Q = AV through–a–hole*.

In whatever form, the continuity equation expresses the principle that mass is conserved; it can neither be created nor destroyed: $dm/dt = 0$.

Energy

The *energy equation*, original with Daniel Bernoulli, in three–dimensional form is formidable; I won't put it here for that reason. If you need this equation, go to one of the references.

Bernoulli's equation, in *one*–dimensional form, can be derived from Euler's equation which, in one form, is

$$\frac{dp}{\rho} + g\,dz + v\,dv = 0 \qquad (10\text{-}13)$$

This can be integrated for the case of constant density to yield

$$gz + \frac{v^2}{2} + \frac{p}{\rho} = \text{constant} \tag{10-14}$$

Equation 10-14 is in terms of *energy per unit mass*. If this equation is divided by g it is in terms of *energy per unit weight*, thus

$$z + \frac{v^2}{2g} + \frac{p}{\gamma} = \text{constant} \tag{10–15}$$

Here each of the factors is in terms of energy per unit weight: ft·lb/lb, m·kg/kg, m·N/N, etc.

Equation 10-14 can be multiplied by ρ to yield

$$\gamma z + \frac{\rho v^2}{2} + p = \text{constant} \tag{10-16}$$

This puts the factors in terms of kg·m/m³, ft·lb/ft³, or m·N/m³—the energy equation in terms of unit volume.

In any of the above equations, the factor containing z is the *potential energy* of a unit of mass or weight raised to a height z above some datum; the factor containing v is the *kinetic energy* of a unit of mass or weight moving at a velocity v; and the factor containing p is the flow work or *flow energy* per unit of mass or weight.

Bernoulli's equation can be put in its most useful form as

$$z_1 + \frac{p_1}{\gamma} + \frac{v_1^2}{2g} = z_2 + \frac{p_2}{\gamma} + \frac{v_2^2}{2g} \tag{10-17}$$

This is usually seen in transposed form as

$$(z_1 - z_2) + \left(\frac{p_1 - p_2}{\gamma}\right) + \left(\frac{v_1^2 - v_2^2}{2g}\right) = 0 \tag{10-18}$$

Here, again, all factors are in terms of length (meters, feet), but all references to a datum have disappeared: $(z_1 - z_2)$ is the *acting static head*, $(p_1 - p_2)/\gamma$ is the *acting pressure head*, and the *head due to kinetic energy* is measured by the difference of the two squared quantities v_1^2 and v_2^2.

If work is done or by the system, or if heat is added to or drawn from the system, the zero on the right side of the equation will revert back to a real, but at any instant, constant value. For present purposes, Equation 10-18 is *Bernoulli's equation in terms of the conservation of energy*, and it is precise from the standpoint of classical hydrodynamics.

Momentum

The *momentum equation* is Newton's second law in algebraic notation, as compared to the original writings of Newton himself which were in terms of geometry. There is some evidence he used the calculus—which either he or Leibnitz invented, take your choice—to speed his work then converted the results to geometric proofs because his reading public didn't understand algebra, or analytical geometry, which Decartes had not yet invented. The conversion to algebraic notation is due to Euler.

We can start by looking at the energy equation, given above as

$$(z_1 - z_2) + \left(\frac{p_1 - p_2}{\gamma}\right) + \left(\frac{v_1^2 - v_2^2}{2g}\right) = 0 \tag{10-18}$$

Since momentum involves moving things, we can drop out the first term—the potential energy term—and rewrite the remaining equation in terms of Newton's Second Law $F = Ma$.

$$(p_1 A_1 - p_2 A_2) = (\rho A V)(V_2 - V_1) \tag{10-19}$$

Here the subscript 1 applies to the *p, A, V* components at the *entry to* a process, and subscript 2 at the *exit from* a process. The process can be almost anything that does not add energy to or take work from the flowing body; for purposes here we may think of a cone with the substance flowing into the larger diameter and out from the smaller diameter. By the law of continuity we know that $V_2 > V_1$.

Thus $p_1 A_1$ is the force at entry, $p_2 A_2$ is the force at exit, and the difference between the two is the change in total force ΔF. The factor $\rho A V$ is the total mass flowing; and this quantity, multiplied by the difference in V_2 and V_1, is the change in momentum ΔM. Thus

$$\Delta F = \Delta M \tag{10-20}$$

Table 10-2
Fundamental Equations of Hydrodynamics

Continuity Equation

$$\rho_1 A_1 V_1 = \rho_2 A_2 V_2 = \text{ constant (slugs/sec)}$$
$$\gamma_1 A_1 V_1 = \gamma_2 A_2 V_2 = \text{ constant (lbs/sec)}$$
$$Q = A_1 V_1 = A_2 V_2 = \text{ constant (ft}^3\text{/sec)(constant } \rho)$$

where $\rho =$ slugs/ft^3
$\gamma =$ lbs/ft^3
$A =$ ft^2
$V =$ ft/sec
$Q =$ ft^3/sec

Energy (Bernoulli's) Equation

$$\frac{p_1}{\gamma} + \frac{V_1^2}{2g} + z_1 = \frac{p_2}{\gamma} + \frac{V_2^2}{2g} + z_2$$

where $p =$ psf
$\gamma = lbs/ft^3$
$V =$ ft/sec
$z =$ ft
$g =$ 32.174 ft/sec^2
$\frac{p_1}{\gamma}, \frac{p_2}{\gamma} =$ pressure head, ft·lb/lb = ft
$\frac{V_1^2}{2g}, \frac{V_2^2}{2g} =$ velocity head, ft·lb/lb = ft
$z_1, z_2 =$ elevation head, ft

Momentum Equation

$$(\Sigma F)t = M(\Delta V)$$
$$(p_1 A_1 - p_2 A_2) = M(V_2 - V_1) = \rho Q(V_2 - V_1)$$

where $F =$ lb
$A =$ ft^2
$M =$ slugs
$p =$ psf
$\rho =$ slugs/ft^3
$Q =$ ft
$V =$ ft/sec

In this equation F is terms of a force, i.e., lb, kg, or N in the FPS, KGS, and SI systems, and M is in terms of mass, i.e., slugs, gkg, or kg. This is the *conservation of momentum equation* which states, in simplest form, that *in a flowing fluid system momentum cannot be lost*, though it can be converted to *impulse forces*. The garden hose with its nozzle closed off does not move, but if the nozzle is open it will writhe like a snake—impulse.

Working Equations

Table 10-2 shows the continuity, energy, and momentum equations in their usual engineering forms. The units shown are for the gravimetric FPS system; for the gravimetric MKS system they are in kg, m, sec and gkg.

References

1. Hauck, George F. W., "The Roman Aqueduct of Nîmes," *Scientific American*, March 1989. Includes references to four other of Hauck's writings.
2. Hodge, Trevor A., "Siphons in Roman Aqueducts." (Author's note: I believe this is from *Scientific American*, but my tear sheets contain no reference to the publication in which they appeared. The page size, and art work, *look* like *Scientific American* quality.)
3. Streeter, Victor L., and Wylie, Benjamin E., *Fluid Mechanics*, 8th. ed., p. 42. New York: McGraw–Hill Inc., 1993.
4. Simon, Andrew L., *Practical Hydraulics*, 2nd. ed. New York: John Wiley & Sons, 1981.
5. Author's Note: This comment is credited to von Neumann by Richard Feynman who, besides winning a Nobel prize, is famous himself for making the statement that "We can do all kinds of things mathematically, but what we can't do is compute flow through a pipe." He was speaking of the mathematics of turbulence, but he was right. See *Genius, The Life and Science of Richard Feynman*, and *Chaos, Making a New Science*, both by James Gleick.
6. Benedict, Robert P., *Fundamentals of Pipe Flow*, New York: John Wiley & Sons, 1980.

Suggested Reading

Fluid Mechanics, Streeter and Wylie, McGraw–Hill

Fundamentals of Pipe Flow, Benedict, John Wiley & Sons

Physics, Halliday and Resnick, John Wiley & Sons

Elementary Fluid Mechanics, Vennard and Street, John Wiley & Sons

Fluid Dynamics, Hughes and Brighton *Schaum's Outline Series*, McGraw–Hill

11
Fluid Mechanics: The Flow of Real Fluids

Introduction

The preceding chapter was concerned with the flow of *ideal* fluids. These fluids were assumed to be inviscid and incompressible, flowing in steady, uniform, irrotational, one–dimensional, frictionless adiabatic, isothermal conditions. Ideal fluids do not exist in nature, though at slow rates of flow of low pressure gases or at high rates of flow for low viscosity gases and liquids, real fluids may approximate an ideal fluid. But it is no wonder that with all the limitations placed on the characteristics of the fluid flow that Nobel laureate John von Neumann called hydrodynamics "...the study of *dry water*."

The study of the flow of *real* fluids has none of the simplifying characteristics of the ideal fluid or of the ideal flow regime; and while that is good, in that real fluids are just that—they are *real*—another comment by another Nobel winner, Richard Feynman, points out the problem we face:

> "...(one of the unsolved problems is) the analysis of circulating or turbulent fluids. If we watch the evolution of a star, there comes a point where we can deduce that it is going to start convection, and thereafter we can no longer deduce what should happen...We cannot analyze the weather. We do not know the patterns of motion that there should be inside the earth.
>
> "What we really cannot do is deal with actual, wet water running through a pipe. That is the central problem which we ought to solve some day.[1]

In the review of *fluid mechanics*—the study of the flow of real fluids—that follows, it would be well to remember Feynman's comments. Also, refer to Professor Fazarinc's analysis in **Introduction–Pipeline Hydraulics in a Computer Environment.**

The Concept of the Steady Flow Machine

In most texts on fluid mechanics the basics of flow of real fluids is taught in terms of a *process*. I once studied a thermodynamics text which taught in terms of a *machine*. I prefer this concept, because a process can be almost anything involving the pressure–volume–temperature relations for a substance, while a machine is set up to *do* something, and pipelines do something. Figure 11-1 was drawn with this concept in mind.

The machine has an input, and every parameter concerned with the characteristics of the fluid or the conditions of flow at the input is clearly identified by a subscripted 1. Similarly, the parameters and conditions at the output of the machine are subscripted with 2. Everything that happens between input and output—in the machine itself—is grouped into

- Net energy *inputted* to the machine
- Net energy *outputted* by the machine
- Energy *lost* by the machine in doing its work

Energy *input* can be in the form of heat or mechanical energy, as can energy *output*, but energy *lost* is always in the form of friction generated as the machine does its work and appears as heat that is transferred from the machine by conduction, convection or radiation.

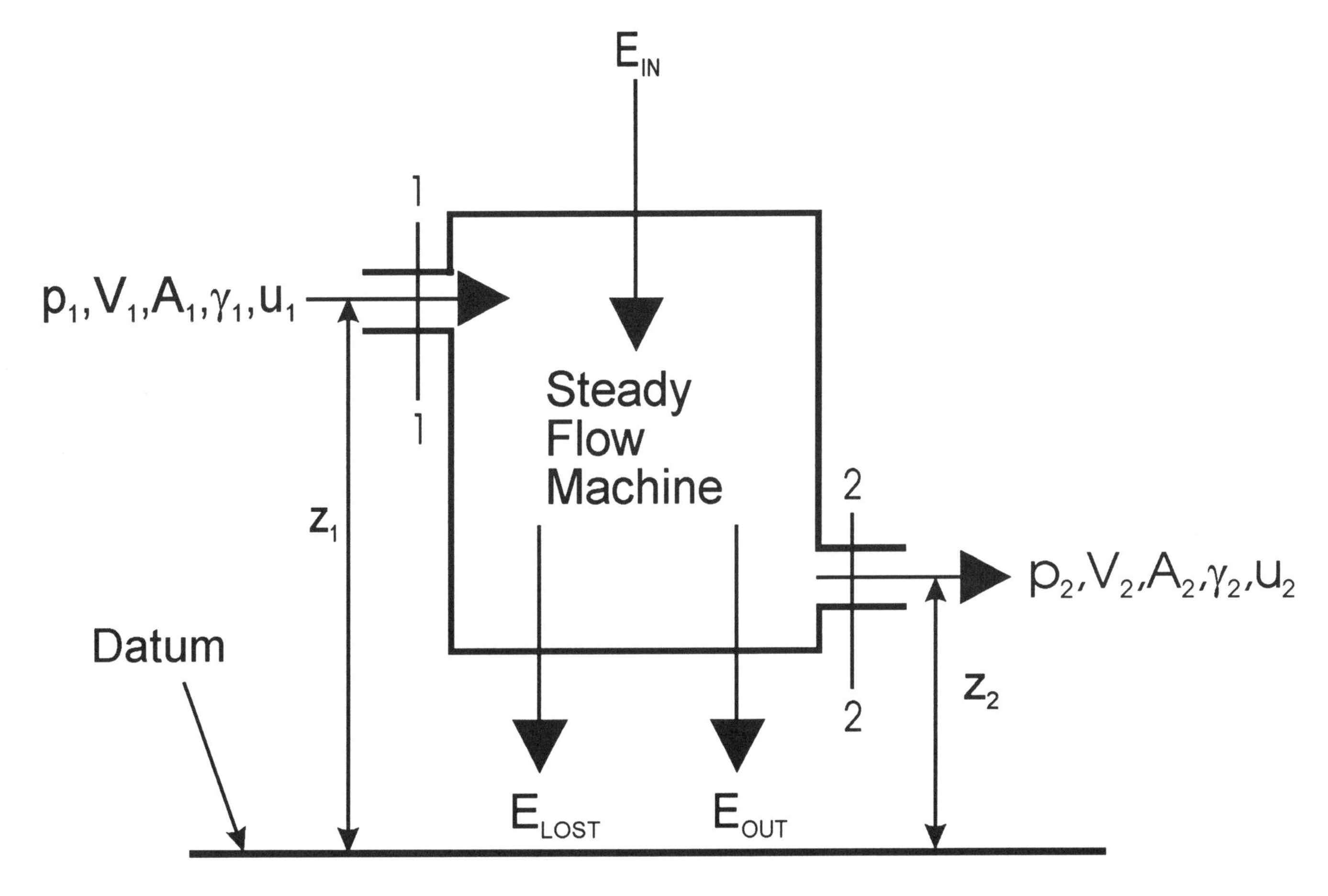

Figure 11-1. Energy diagram for a steady flow machine.

The flowing substance also has an *internal energy* due to molecular action within the fluid. Internal energy is usually symbolized as U or, for a unit quantity, u. These also have values at the input and output of the machine.

The General Energy Equation

An equation can be written to describe the energy relations between the input and output of the machine in the same way that the energy equation was developed for ideal fluids, but now there are three more variables to be taken into account: *energy input*, *energy output*, and *energy lost*, and the *internal energy* of the flowing substance has to be considered.

Equation 11-1 is written in the same terms as Equation 10-17. This equation, in whatever terms or dimensions, is usually called the *general energy equation*. It is not *completely* general, of course, but a completely general equation would be major overkill in pipelining.

$$\frac{p_1}{\gamma_1}+\frac{V_1^2}{2g}+z_1+u_1+E_{IN}= \frac{p_2}{\gamma_2}+\frac{V_2^2}{2g}+z_2+u_2+E_{OUT}+E_{LOST} \qquad (11\text{-}1)$$

Each term has the dimensions of ft·lb/ft or m·kg/kg in gravimetric systems and N·m/m in SI, so in any case the result is each term represents a *head* in feet or meters. The terms can be divided and considered separately as, indeed, they have to be in solving any practical problem.

- z is the *potential energy* of *position*, always referenced to a datum. It is the change in z, $z_2 - z_1$, that is important.
- p/γ, sometimes $p\omega$, is the *flow work energy* due to *pressure*. The term *flow work* is used to distinguish this parameter from *shaft work*, the mechanical work done *by* or *on* the machine. I

use γ, the specific weight, instead of ω, the specific volume, purely because of personal preference. However, I have a thermodynamics reference written entirely in terms of ω; specific weight is not mentioned anywhere in the index.

- $V^2/2g$ is the *kinetic energy* due to *velocity*.
- u is the *internal energy* of the substance.
- E_{IN}, E_{OUT}, and E_{LOST} are as described.

Note this equation can handle compressible as well as incompressible flow. If $\gamma_1 \neq \gamma_2$, it is obvious that energy has been absorbed or given up by the fluid to account for the change in specific weight. In fact, Equation 11-1 is in enough detail that if each of the factors involved can be evaluated with some precision the general equation can be satisfactorily solved. That doesn't help much with evaluation of flow of real fluids in pipes, however, without a further look into the equation and its constituent factors.

The Machine as a Pipeline

In most fields of endeavor the most important variables in Equation 11-1 are E_{IN} and E_{OUT}, and E_{LOST} is only of interest as relates to the calculation of the efficiency of the machine.

Also, especially in processes involving major changes in *state*, u_1 and u_2 are important; if a substance enters as a vapor and exits as a liquid, there has been a major change in unit internal energy.

In the flow of fluids in pipes, however, and especially in the study of flow in long pipelines, E_{IN} and E_{OUT} are not important and may be disregarded or, which is the same thing, assigned the value of zero. All the energy to be added to the machine from outside sources has already been added before the fluid reaches the input, and there is no mechanical work drawn from the fluid as it flows down the pipeline to the output. There could be some amount of E_{OUT} to be considered if the pipeline loses a large amount of heat in flowing between the input and output, but for purposes here this kind of loss can be withheld until later on in the book when hot oil lines are under consideration. The value of E_{LOST}, however, is compelling; essentially all of the energy inputted to a long oil pipeline exits as E_{LOST} due to friction of the flowing fluid in the pipe and very little of it appears in the output.

Most of the time, especially for high-pressure trunk pipelines, the kinetic energy $V^2/2g$ can be neglected as compared to the flow work energy p/γ, but in suction lines from tanks, operating at low pressures or even partial vacuums, the value of the kinetic energy can be crucial so it must be left in. Also, as will be found later on, the $V^2/2g$ factor is very important in the flow equation for pipes.

In further simplification, if temperature changes are not of substantial magnitude there will be no substantial change in internal energy and so u_1 and u_2 can also be disregarded.

Figure 11-2 shows the machine with the above simplifications. Equation 11-2 is Equation 11-1 with the described simplifications.

$$\frac{p_1}{\gamma_1}+\frac{V_1^2}{2g}+z_1=\frac{p_2}{\gamma_2}+\frac{V_2^2}{2g}+z_2+H_{LOST} \tag{11-2}$$

Note E_{LOST} has been changed to H_{LOST} to be sure the friction loss is understood in terms of head. In further simplification H_{LOST} will hereafter be written as h_L to assure friction loss is understood in terms of unit quantities. Thus

$$\frac{p_1}{\gamma_1}+\frac{V_1^2}{2g}+z_1=\frac{p_2}{\gamma_2}+\frac{V_2^2}{2g}+z_2+h_L \tag{11-3}$$

Or, letting $\gamma_1=\gamma_2$ and solving in terms of h_L

$$\frac{(p_1-p_2}{\gamma}+\frac{\left(V_1^2-V_2^2\right)}{2g}+(z_1-z_2)=h_L \tag{11-4}$$

This gets closer to the solution. If the value of h_L can be determined from the energy equation, can the value be predicted? The answer, of course, is no; if all of the values at 1 and 2 are known, obviously h_L can be obtained by subtraction; one or more of the values at 1 or 2 must be missing or there is no problem.

We can again start by simplifying our assumptions; we can assume there is no static head, i.e., $z_1=z_2$. This brings the equation down to

$$\frac{(p_1-p_2}{\gamma}+\frac{(V_1^2+V_2^2)}{2g}=h_L \tag{11-5}$$

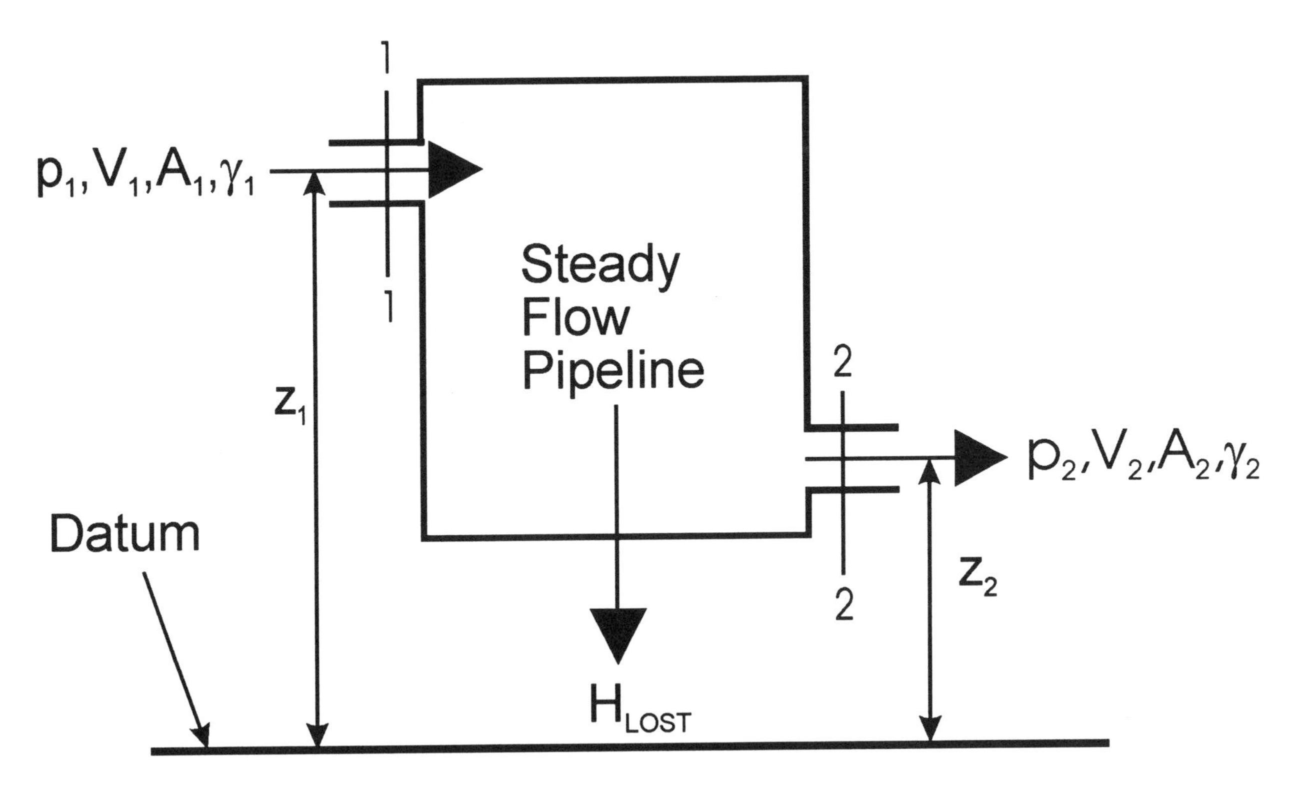

Figure 11-2. Energy diagram for steady flow pipeline.

If the pipeline is of constant diameter, $V_1 = V_2$, and the equation is finally reduced to

$$\frac{(p_1 - p_2)}{\gamma} = h_L \qquad (11\text{-}6)$$

And this is as simple as the general energy equation can be written. In a constant diameter pipeline flowing an incompressible fluid in isothermal adiabatic conditions, if the output is at the same elevation as the input, the total head loss $(p_1 - p_2)/\gamma$ is the friction head loss h_L.

This is not particularly strange, however, because there is no other place for the energy to go. And this is very important: *a pipeline will use up all of the energy inputted to it by increasing the rate of flow until the energy inputted equals the energy lost in friction.*

The Pipe Friction Equation

The philosophy behind both diagrams in Figure 11-3 is taken from Daugherty,[2] my favorite reference on fluid mechanics from an engineering standpoint. Figure 11-3-A is a representation of the conditions of flow for the machine of Figure 11-2 with constant γ, and Equation 11-4 is the applicable description of the flow regime.

This diagram introduces for the first time the concept of the *hydraulic grade line*, usually called the *hydraulic gradient*, or just the *gradient*, by pipeliners, which is everywhere $z + p/\gamma$ above the datum. (Note: The gradient can *attempt* to go below the datum, but just as soon as the value of the negative gradient exceeds the vapor pressure of the fluid, fluid column separation will take place and any calculation having to do with continuous flow becomes invalid.)

It is Figure 11-3-B that is of interest for the present. This is a free–body diagram of a fluid flowing in a closed conduit. The conduit is drawn as a cylinder, but this is only for convenience; for purposes here the cross section of the conduit can be anything reasonable and even need not be symmetrical. If the element really were a cylinder the *radius* would be *r*, the *periphery* would be $2\pi r$, and the *area* would be πr^2 and but since the cross section is not defined the area is symbolized as *A* and the perimeter is *P*.

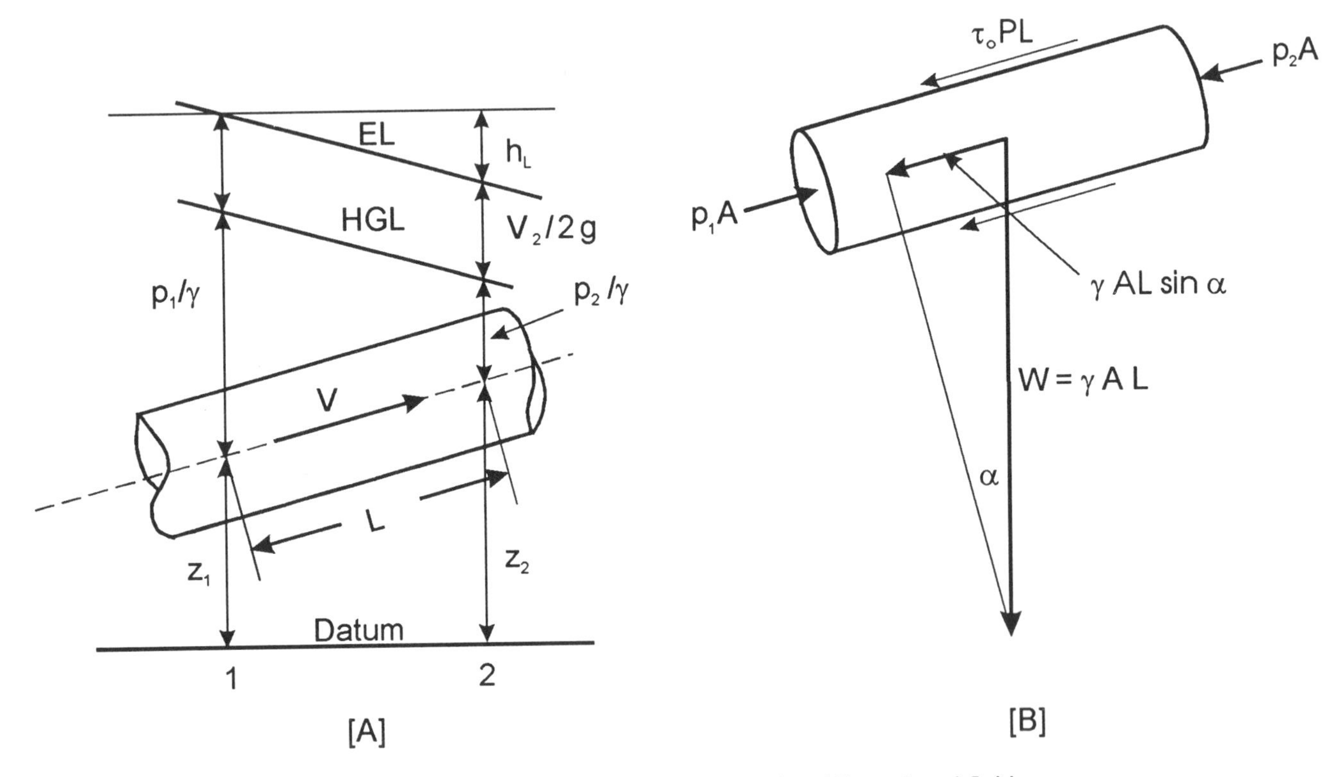

Figure 11-3. Basic considerations for analysis of flow of real fluids

This leads to one of the most famous fictitious factors in fluid mechanics, **R**, the *hydraulic radius*, which is defined for any cross section as **R** = A/P. This makes **R** for a round pipe $\pi r^2/2\pi r = r/2$, which is illogical, inconvenient and inconsistent but buried so thoroughly in the language of fluid mechanics it has to be put up with.

While the shape of the cross section is not important, it is very important that the sides of the conduit be smooth and straight—which demands a constant cross section—or the analysis will become very messy, not to say difficult or impossible.

Figure 11-3-B also introduces a new factor into analyses of flow, *fluid shear*, almost always symbolized as γ. Shear anywhere in the fluid is γ. Shear along the exterior boundary of the conduit, or, in other analyses, along the wall of the pipe, is γ_0.

A balance of forces yields

$$p_1 A - p_2 A - \gamma LA \sin\alpha - \tau_0 (PL) = 0 \tag{11-7}$$

Substituting $\sin\alpha = (z_2 - z_1)/L$ and dividing by γA yields the simpler

$$\frac{p_1}{\gamma} - \frac{p_2}{\gamma} - z_2 - z_1 = \tau_o \left(\frac{PL}{\gamma A} \right) \tag{11-8}$$

Note from Figure 11-3-A that

$$h_L = (z_1 + p_1/\gamma) - (z_2 + p_2/\gamma) \tag{11-9}$$

This is precisely the left side of Equation 11-8, and by substituting **R** for A/P

$$h_L = \tau_0 \frac{L}{\mathbf{R}\gamma} \tag{11-10}$$

L, **R**, and γ are known; if τ_0 is known, h_L can be calculated.

There are many ways of arriving at τ_0. I prefer to use the Reynolds method of analysis followed by Daugherty. The derivation below follows Daugherty. A more modern classical approach using the Buckingham Π Theorem can be found in Giles.[3]

For a *smooth* conduit it can be assumed that τ_0 is some function of ρ, μ, V and **R**. Thus

$$\tau_0 = K\mathbf{R}^a \rho^b \mu^c V^d$$

where K is dimensionless though not necessarily a constant.

Substituting for the real values of the variables their F, L, T dimensions

$$FL^{-2} = K(L)^a\left(FL^{-4}T^2\right)^b\left(FL^{-2}T\right)^c\left(LT^{-1}\right)^d$$

Since the dimensions of the two sides must be the same, the exponents of each dimension must be the same. Therefore,

For F: $1 = b + c$
For L: $-2 = a - 4b - 2c + d$
For T: $0 = 2b + c - d$

In terms of an undetermined variable n the solution of these three simultaneous equations is

$a = n - 2$
$b = n - 1$
$c = 2 - n$

Rewriting, in terms of n

$$\tau_0 = K\mathbf{R}^{n-2}\rho^{n-1}\mu^{2-n}V^n$$

Rearranging

$$\tau_0 = K\left(\frac{\mathbf{R}V\rho}{\mu}\right)^{n-2}\rho V^2 = 2KN_R^{n-2}\rho\frac{V^2}{2}$$

where the simplification arises in realizing that the factors inside the parentheses are the *Reynolds Number*, N_R, and the right hand factor has been both multiplied and divided by 2. When all of the dimensionless numbers are collected into a single term C_f

$$C_f = 2KN_R^{n-2}$$

and

$$\tau_0 = C_f\rho\frac{V^2}{2}$$

If this value for τ_0 is inserted in Equation 11-11 and $\gamma = \rho g$

$$h_L = C_f\left(\frac{L}{\mathbf{R}}\right)\left(\frac{V^2}{2g}\right)$$

Note this equation is applicable for any flow regime, i.e., laminar, transitional or turbulent, and for any cross section: the value of C_f has been derived in a perfectly general way.

The only problem with C_f is that it isn't a constant for any flow regime other than for laminar flow.

The result of all the above, in familiar form, is derived by setting $\mathbf{R} = r/2 = D/4$ for circular conduits and the pipe friction equation appears.

$$h_L = f\frac{L}{D}\frac{V^2}{2g} \tag{11-11}$$

Note that since both f and L/D are dimensionless —they are pure numerics—h_L can be seen as a numeric multiplier times the velocity head $V^2/2g$.

Note also, however, that while f is dimensionless it is not necessarily constant; f varies with flow rate and with the flow regime as well as with the physical parameters of the problem, and in its present stage of development here is only valid for smooth conduits. However, for smooth conduits it is completely defined as a function of N_R.

Darcy, Weisbach, and Fanning

I personally refuse to be drawn into the argument of whether it's Darcy, Darcy–Weisbach, Fanning or some other name or set of names that is properly attached to Equation 11-11. There is no doubt all three of those named did, at one time or another, develop a formula with

Table 11-1
Fundamental Equations of Fluid Mechanics

Continuity Equation

Same as for hydrodynamics; see Figure 10-2.

General Energy Equation

$$\frac{p_1}{\gamma_1}+\frac{V_1^2}{2g}+z_1+u_1+E_{IN}=\frac{p_2}{\gamma_2}+\frac{V_2^2}{2g}+z_2+u_2+E_{OUT}+E_{LOST}$$

where $p=$ psf
$\gamma = lbs/ft^3$
$V=$ ft / sec
$z=$ ft
$g=$ 32.174 ft / sec^2

$\frac{p_1}{\gamma},\frac{p_2}{\gamma}=$ pressure head, ft · lb/lb = ft
$\frac{V_1^2}{2g},\frac{V_2^2}{2g}=$ velocity head, ft · lb / lb = ft
$z_1, z_2=$ elevation head, ft
$u_1, u_2=$ internal energy, ft · lb / lb = ft
$E_{IN}=$ net energy input, ft · lb / lb = ft
$E_{OUT}=$ net energy output, ft · lb / lb = ft
$E_{LOST}=h_L=$ energy lost to friction, ft · lb / lb = ft

Pipe Friction Equation

$$h_L=C_f\left(\frac{L}{\mathbf{R}}\right)\left(\frac{V^2}{2g}\right)$$ conduit of hydraulic radius **R**, ft

$$h_L=f\frac{L}{D}\frac{V^2}{2g}$$ any round pipe of diameter D, ft

where f and $C_f=f(N_R)$
other parameters are as defined for energy equation

Momentum Equation

Same as for hydrodynamics; see Table 10-2.

the same philosophical outlook on friction as that of Equation 11-11, and their names have been justly preserved in the literature.

However, just for the record:

$$f_{D'Arcy-Weisbach}=4\times f_{Fanning}$$

Tip: If you are working with an unfamiliar friction–loss problem and wonder if you're working with the Fanning or the Darcy–Weisbach version, compute f for a N_R of 4000. If the result is an f of about 0.04, you've got a Darcy–Weisbach kind of friction factor; if it's about 0.01, it's a Fanning factor.

Most oil pipeliners use the Darcy–Weisbach factors, and gas pipeliners—who do their calculations upside down with a *transmission factor* that is the reciprocal of the friction factor—generally use Fanning factors. There is no reason for this; it is just so.

Working Equations

Table 11-1 shows the general energy equation and the two versions of the pipe friction equation in their usual engineering forms. The units shown are for the gravimetric FPS system; for the gravimetric MKS system they would be in kg, m, sec and gkg. The continuity and momentum equations are as shown in Table 10-2.

References

1 Gleick, James, *Genius*, p. 360. New York: Pantheon Books, 1992.
2 Daugherty, R. L., and Franzini, J. B., *Fluid Mechanics With Engineering Applications*, 6th ed. New York: McGraw–Hill Book Company, 1965. The 6th edition is a revision of the 5th edition by Daugherty and Ingersoll. The first edition, then titled *Hydraulics*, was published in 1916 with Professor Daugherty as sole author.
3 Giles, R. V., *Fluid Mechanics and Hydraulics*, 2nd. ed., p. 56, Schaum's Outline Series. New York: McGraw–Hill Book Company, 1962.

12
Hydraulics: Flow of Liquids in Pipes

Introduction

Chapter 10 reviewed *hydrodynamics*, the study of ideal fluids. Hydrostatic relations of still fluids were discussed from the standpoint of ideal liquids, and then the continuity, energy, and momentum equations of ideal fluids in motion were developed from the standpoint of internal flow in pipes. Table 10-2 summarizes the fundamental equations of hydrodynamics.

Then, in Chapter 11, emphasis was changed from the purely theoretical approach of hydrodynamics to that of *fluid mechanics*, which recognizes that a fluid may have a viscosity, and that the flow of such a fluid can be in a form other than the steady, uniform, irrotational, one–dimensional, frictionless adiabatic, isothermal form that comprise the restrictive limitations of hydrodynamics, and the general energy and pipe friction equations were introduced. The fundamental equations of fluid mechanics are included in Table 11-1.

In this chapter, which is the last of the three chapters devoted to a review of the classical approach to the study of pipe flow, the concepts of fluid mechanics as carried over to the study of the flow of real liquids in pipes in the workplace—*hydraulics*—will be summarized. Hydraulics can and does recognize experimental data and empirical formulas, and is thus quite far removed from the present trend to concentrate on the mathematics of the problem and disregard data from the real world, especially if the data conflict with the mathematics. Hydraulics considers flow in open channels as well as in closed conduits of any shape. Emphasis here will be on flow in round pipes. There is no other configuration of interest to a pipeliner except, perhaps, the annular flows of some heated piping systems, but that development is not for this book.

A Short History of Pipe Flow

It was pointed out in Chapter 11 that the pipe friction equation was derived without any consideration of the flow regime. In fact, when Darcy, Weisbach, and Fanning were doing their work, Osborne Reynolds had not yet performed his work on the modes of fluid flow and Lord Rayleigh had not yet developed the laws of dynamic similarity. But the investigators who followed had, with each passing year, a continuously increasing body of theoretical and experimental work to draw from, and over the years a consensus developed along the following lines:

- There is a particular regime of flows for which the *friction factor is a function of* N_R *only*, and the head lost due to friction is *proportional to the velocity of flow to the first power.*
- There is a second particular regime of flows, greater than the first but less than the third, for which the *friction factor is a function of* N_R *only*, and the head lost due to friction is *proportional to the velocity of flow to some power greater than one but less than two.*
- There is a third particular regime of flows, greater than the second but less than the fourth, for which the *friction factor is a function of both* N_R *and the hydraulic roughness of the pipe*, and the head lost due to friction is *proportional to the velocity of flow to some power greater than one but less than two.*
- There is a fourth regime of flows, greater than all those described above, for which the *friction factor is a function only of the hydraulic roughness of the pipe*, and the head lost due to

friction is *proportional to the velocity of flow to the second power*.

In today's terminology we call these flow regimes, in order,

- Laminar flow
- Turbulent flow, smooth pipe
- Transitional turbulent flow, rough pipe
- Fully developed turbulent flow, rough pipe

These conceptual bases of the mechanisms of fluid flow developed slowly over about fifty years. Consider the following chronology:

- 1892—Reynolds[1] publishes his work on the flow regime criterion that bears his name and also introduces the use of dimensional analysis in fluid mechanics.
- 1904—Prandtl's[2] work on the boundary layer theory appears. The theory, which presumes that all friction losses are confined to a very thin boundary layer lying adjacent to the wall of the pipe, offers for the first time a solid basis for linking the flow of real and ideal fluids.
- 1911—Blasius,[3] while a student of Prandtl, proves that the friction factor for smooth pipe flow is a function of Reynolds number only, and derives the exponential formula that bears his name.
- 1914—Stanton and Pannell[4] publish their experimental data on the flow of water and air in smooth drawn brass tubing. Their precise, consistent results gave investigators their first reliable data on the f-vs-N_R relationship for smooth pipe.
- 1915—Buckingham[5] develops and publishes the Π Theorem, which formalizes dimensional analysis and offers a powerful new tool for the theoretical analysis of flow patterns.
- 1921—von Kármán[6] publishes the results of his work on laminar and turbulent flow in the boundary layer, which lays out theoretically sound methods for such analyses.
- 1925—Prandtl[7] publishes his mixing length theory, which is an effort to do for the theory of energy transport in liquids what the mean free path of *molecules* provides for the kinetic theory of gases—offer a method of calculating the exchange of energy due to colliding *particles* of liquid.
- 1930—Heltzel[8] produces the first major oil industry study of fluid flow in pipelines and presents a recommended curve of f-vs-N_R that was picked up by several of the U.S. pipemakers for inclusion in their "pipe manuals," and the curve went into wide-spread use.
- 1932—Nikuradse,[9] a student of Prandtl and associate of von Kármán, publishes his results of experiments on flow in sand-roughened pipes, and thus provides reliable data on the f-vs-N_R relationship for rough pipe. A peculiarity of these data is that the friction factor falls, as would be expected, with increasing N_R in the smooth pipe flow regime and then, separating from the smooth pipe curve, *increases*—which is *not* expected nor predicted by theory—before stabilizing at a constant value in rough pipe flow. This behavior, later attributed to (a) the fact that the sand grains were essentially identical and distributed uniformly over the surface of the pipe, or (b) the fact that the *smoothest* pipe used, with an ε/D of 0.001, is rougher than any commercially available pipe, is not seen in later work, including the analyses of Colebrook or in the U.S. Bureau of Mines Monograph No. 9[10] and IGT TR-10[11] tests with natural gas.
- 1933—Pigott[12] summarizes the work of Kemler[13] who studied the results of thousands of tests including, for the first time, many tests based on the flow of petroleum oils. Pigott's family of curves for oil pipelines included six separate plots of f-vs-N_R for pipes 0.5 to 42 inches in diameter (two curves covered the range 6–42 inches, however). The Pigott curves seem to have been used mostly by pipeliners in the California oil fields; they did not make a large impression on the mid-west U.S. oil fraternity.
- 1934—von Kármán,[14] working with the laws of simularity, provides a working equation for the mixing length. It becomes possible to set up and integrate expressions that could reasonably be expected to describe flowing conditions in the boundary layer.
- 1939—Colebrook,[15] with the acknowledged help of White, introduces the relationship now called the Colebrook-White function for flow in the

transition region between the smooth and rough pipe laws. Colebrook's paper includes revised (corrected) data on the 216 inch diameter Ontario tunnel which extends experimental proof of the smooth pipe law out to $\mathbf{N_R}$ = 28 000 000. It also includes, almost as an aside, an approximation of the smooth pipe law that allows calculation of f explicitly; all other versions of the smooth pipe law till that time were such that f could only be obtained implicitly or by empirical exponential formulas explicit in f over narrow ranges of $\mathbf{N_R}$.

- 1943—Rouse[16] develops his equation for the Rouse Line—a curve in the f–$\mathbf{N_R}$ plane that defines the end of the transition region and the start of the fully turbulent region.
- 1944—Moody[17] publishes his famous curves, yielding f as a function of $\mathbf{N_R}$ with ε/D as a parameter, on log–log axes, and thereby provides a way to use the Colebrook–White function which stood untouched until around 1975 when the HP-65 programmable calculator appeared and pipeline engineers immediately programmed an iterative solution for the Colebrook–White function simply because they had never been able to do it before, short of a mainframe computer.

By 1935 there existed three theoretically sound, and experimentally checked, equations for flow:

- The Hagen–Poiseuille law, and its pipe friction equation equivalent, for laminar flow.
- The Nikuradse equation for turbulent flow in hydraulically smooth pipes.
- The von Kármán equation for turbulent flow in hydraulically rough pipes.

There did not exist—and there does not exist today as this book is written—a theoretically sound equation for flow in rough pipes in the transitional region. Nikuradse's experiments suggest one form of this transition, but other data—especially data on gas flow—suggest other forms, and it is an interesting commentary on the state of the art that the present day, generally accepted, form of transitional flow equation is the White function—usually called the Colebrook–White function even though Colebrook went out of his way in his 1939 publication to give credit to White—which combines the smooth and rough pipe laws by simply adding them together. There is absolutely no theoretical basis for doing this, but it does provide a transition equation which is conservative and so all who use it are protected from the sin of *underdesign*. The obverse, the sin of *overdesign*, is built in and cannot be avoided.

Since all liquid pipelines that do not operate in laminar flow operate in turbulent smooth pipe flow or transitional turbulent rough pipe flow, it is really important that pipeliners understand smooth and transitional rough pipe flow and how to calculate flow rates and friction losses in these flow regimes. There is an extended look at flow of real liquids in long pipes commencing in Chapter 13; the object of the present chapter is to get from Osborne in 1892 to Moody in 1944 in as few pages as possible.

The Basic Friction Factor Equations

Laminar Flow

In Chapter 3 the Hagen–Poiseuille law for laminar flow was described in its original form as

$$V = \left[\left(\pi p r^4\right) / \left(8 \ell \mu\right)\right] t \qquad (3\text{-}44)$$

where V = volume of liquid flowing in time t
p = pressure difference between inlet and outlet
r = radius of the bore of the tube
ℓ = length of the tube
μ = absolute viscosity of the liquid

This can be rewritten in terms of h_L as

$$h_L = \frac{32\,\nu L V}{g D^2} \qquad (12\text{-}1)$$

If Equation 12-1 is multiplied and divided by $2V$ and terms collected

$$h_L = \frac{64\,\nu L V^2}{2 g V D^2} = \frac{64}{\frac{VD}{\nu}} \frac{L}{D} \frac{V^2}{2g} = \frac{64}{\mathbf{N_R}} \frac{L}{D} \frac{V^2}{2g} \qquad (12\text{-}2)$$

If Equation 12-2 is set equal to the pipe friction equation

$$h_L = f \frac{L}{D}\frac{V^2}{2g} \qquad (11\text{-}11)$$

the usual form of the pipe friction loss equation for laminar flow follows

$$h_L = \frac{64}{\mathbf{N_R}}\frac{L}{D}\frac{V^2}{2g} \qquad (12\text{-}3)$$

Thus, to use the pipe friction equation for laminar flow,

$$f_{LAMINAR} = f(\mathbf{N_R}) = \frac{64}{\mathbf{N_R}} \qquad (12\text{-}4)$$

This is the usual form of the pipe friction equation used by pipeliners. It is not necessary to compute the value of $\mathbf{N_R}$, however, and Equation 12-1 can be used to solve for h_L directly. An advantage to this equation is that it is a reminder that the unit *friction loss is directly proportional to the flowing velocity* in the laminar flow regime.

It is also important to note here that the solution to the pipe friction equation for laminar flow is *the only one of the pipe friction equations that can be derived entirely by theoretical methods from fundamental bases*; all of the others have some element of empiricism (or experimental data) in them.

Flow in Smooth Pipe

The laws of dynamic similarity offered investigators a solid way of grouping variables to arrive at a pattern for a working equation. The first of these was the exponential form

$$f = k\mathbf{N_R}^{n} \qquad (12\text{-}5)$$

Many later equations were written with an additive constant in the form

$$f = k\mathbf{N_R}^{n} + m \qquad (12\text{-}6)$$

Blasius wrote this equation in the form

$$f = K\frac{V^x}{D^y} \qquad (12\text{-}7)$$

and showed that the sum of exponents x and y must equal three. The Blasius equation for head loss in flow in smooth pipes, valid for $4000 > \mathbf{N_R} > 100\,000$, is still in use, especially in Europe.

$$f_{BLASIUS} = \frac{0.3164}{\mathbf{N_R}^{0.25}} \qquad (12\text{-}8)$$

Over the years after Blasius, however, there developed a body of evidence, experimental as well as theoretical, that *if there was such a thing as smooth pipe flow*, which some investigators doubted, the friction factor would be of the form

$$\frac{1}{\sqrt{f}} = A_{SMOOTH} + B_{SMOOTH}\log\left(\mathbf{N_R}\sqrt{f}\right) \qquad (12\text{-}9)$$

Stanton and Pannell, in addition to their pioneer work on flow in smooth pipes, also proved that the $f = \varphi(\mathbf{N_R})$ relationship did not depend on the kind of fluid, and that it was equally valid for gases or liquids. Then Nikuradse experimentally determined A_{SMOOTH} and B_{SMOOTH} as –0.913 and 2.035, respectively, so

$$\frac{1}{\sqrt{f_{SMOOTH}}} = 2.035\log\left(\mathbf{N_R}\sqrt{f_{SMOOTH}}\right) - 0.913 \qquad (12\text{-}10)$$

This equation, in rounded form, is generally known as Prandtl's, though sometimes Nikuradse's, equation. It is shown below in two forms, the first in the $A + \log B$ form, and the second in $\log(C/D)$ form. The two are equal. It is interesting, if you have time, to derive one from the other. It will test your knowledge of the theory of logarithms.

$$\frac{1}{\sqrt{f_S}} = 2\log\left(\mathbf{N_R}\sqrt{f}\right) - 0.8 = -2\log\left(\frac{2.51}{\mathbf{N_R}\sqrt{f}}\right) \qquad (12\text{-}11)$$

This equation is as valid today as when it was first written. Nikuradse believed it applied to $\mathbf{N_R} < 3\,000\,000$, which was about the limit of his data. Later work—most importantly that of the U.S. Bureau of Mines—proved that the smooth pipe law holds in mirror—finished real pipes at least to $\mathbf{N_R} = 10\,800\,000$. It is certainly valid for any $\mathbf{N_R}$ an oil pipeliner is apt to encounter.

Flow in Rough Pipe

Rough pipe flow or, really, *fully developed rough pipe flow*, describes a kind of flow regime in which N_R has no function—the friction factor is strictly a function of the hydraulic roughness—usually symbolized as ε—and the diameter of the pipe, thus

$$f = \varphi\left(\frac{\varepsilon}{D}\right) \tag{12-12}$$

The experimental work of Nikuradse, who worked with artificially roughened pipes using as a roughening agent a graded sand from local sources in Gottingen, Germany, is as important in the development of the rough pipe flow equation as the work of Stanton and Pannell was for the smooth pipe flow equation; it was a body of experimental data that could be used as a sure basis of comparison for theoretical or empirically developed formulations for the friction factor.

von Kármán, working with the boundary layer theory and mixing length theory of Prandtl, and the various theories of the distribution of the point velocity over the flowing face of the fluid, provided a working equation for the mixing length and, subsequently, the equation which is named after him:

$$\frac{1}{\sqrt{f}} = 2\log_{10}\left(\frac{R}{\varepsilon}\right) + 1.74 = -2\log\left(\frac{\varepsilon/D}{3.76}\right) \tag{12-13}$$

Here, as in Equation 12-11, the first is the straight–line form and the second is the single–function equivalent.

This equation, with some minor variations in the y–axis intercept value, has been shown experimentally to be valid for all reasonable ε/D and N_R to over 20 000 000, and is the generally accepted form for the friction factor for rough pipe flow.

Transitional Flow

With the development of the laminar flow, smooth pipe turbulent flow, and rough pipe turbulent flow equations, it would seem that determining the relations defining the remaining flow regime—that covering the transition from turbulent flow in smooth pipes, where the friction factor is a function of N_R only, to turbulent flow in rough pipes, where the friction factor is a function of ε/D only—would be simple: Connect the curves of the bounding regimes with another smooth curve and then define the new curve.

The problem with this concept is that there is a large body of experimental data available, and a number of theories to account for these data, and none is really solidly connected with the other.

Also, until Rouse developed the *Rouse Line*—a curve in the f–N_R plane that establishes the upper limit of transitional flow and, thereby, the lower limit of fully turbulent flow—there was no general agreement even as to the *extent* of the transitional region.

Figure 12-1, drawn on the log–log Stanton (or Moody) axes, shows the situation prior to Colebrook, but with the Rouse Line, which came after Colebrook's paper, added for comparison. Several facts can be derived from the information contained on this plot.

- The laminar flow equation is generally considered valid up to about $N_R = 2000$.
- The region $2000 > N_R > 4000$ (sometimes the upper limit is taken as 3000) is unstable, and is to be avoided; there is no good way to compute the friction factor in this region.
- The smooth pipe law commences at about $N_R = 4000$ with $f =$ (very nearly) 0.04. Figure 12-1 carries the curve out to only about $N_R =$ 2 000 000 but it is known to carry to at least 28 000 000 and probably can be extended indefinitely.
- The *Rouse Line*, defined as

$$N_R = \frac{200}{\sqrt{f}(\varepsilon/D)} \quad \text{or} \quad \frac{1}{\sqrt{f}} = \frac{N_R}{200}\frac{\varepsilon}{D}$$

defines the line of demarcation between the region where $f = \varphi(N_R, \varepsilon/D)$ and where $f = \varphi(\varepsilon/D)$ only.

The Moody coordinates are deceptive:

- The smooth pipe law and the Rouse Line appear to be curved. They are, but very slightly. The physical distance along the axis of N_R to double the flow from any given flow to another (remember these are logarithmic coordinates) is only about 0.3 inches—about 8 mm. Few pipelines are purposefully operated over a

$Q_{MAX}/Q_{MIN} \geq 2$, so the range of operation of a commercially viable pipeline will probably cover considerably less than 0.5 inches or about 12 mm horizontally on the curves. The smooth pipe and Rouse lines are essentially straight over such a short distance.

- Similarly, the transition zone—between the smooth pipe law and the Rouse Line—appears to be about the same *length*, and it is, more or less. But along the $f = 0.04$ line its span in $\mathbf{N_R}$ units is in the order of 100 000, along $f = 0.02$ it is about 1 000 000, and along $f = 0.01$ it is in the order of 100 000 000.

Colebrook made an intensive study of flow in the transition region using the best available experimental data. His data ran from the Stanton and Pannell data for 0.5–inch diameter brass pipe to the 216–inch diameter concrete–lined Ontario tunnel. He concluded, along with most of those who preceded him, that *many kinds of pipes are functionally smooth at commercial rates of flow with liquids*. He also included, almost as an afterthought, an explicit relation for f in smooth pipe flow that is not, but ought to be, also named for him.

$$\frac{1}{\sqrt{f}} = 1.8 \log\left(\frac{\mathbf{N_R}}{7}\right) \qquad (12\text{-}14)$$

This equation follows the implicit, theoretically correct, Equation 12-11, over $5000 > \mathbf{N_R} > 100\,000\,000$, to within ±0.5%, far better than the spread any set of experimental data on smooth pipe can claim.

Colebrook also accepted the rough pipe law as correct, though data available to him did not enable him to make a solid experimental verification of the law. This was a common failing of most of the experimental data gathered in the 50 years 1890–1940; there were very few measured values that would logically fall in the fully developed turbulent rough pipe regime of flow.

The data available to Colebrook, without regard to its source, seemed to trend toward the smooth pipe law at one extreme and the rough pipe law at the other, without any particular evidence of the sagging profile of the f–$\mathbf{N_R}$ *uniform sand gain* curve of the Nikuradse data which Colebrook clearly believed to be a special kind of pattern different from the irregular, random, roughness patterns of commercial pipes and conduits.

Colebrook then reasoned that, in the transitional zone, the friction factor would be higher than it would be if only the smooth or the rough pipe laws applied because both viscous and mechanical mixing were occurring at the wall in each instance and so each would contribute to the total.

The Colebrook–White function is exactly as described; it is the sum of the smooth pipe and the rough pipe laws. Refer to Equation 12-11 and Equation 12-13.

$$\frac{1}{\sqrt{f}} = -2\log\left(\frac{\varepsilon/D}{3.7} + \frac{2.51}{\mathbf{N_R}\sqrt{f}}\right) \qquad (12\text{-}15)$$

Figure 12-2 shows the basic background curves of Figure 12-1 with two computations for the C–W function

- For $\varepsilon/D = 0.0004$, corresponding to 1890 4½–inch wrought iron pipe with $\varepsilon = 0.0018$ inch
- For $\varepsilon/D = 0.00001$, corresponding to 1980 48–inch steel pipe with $\varepsilon = 0.0005$ inch

Note the curve for $\varepsilon/D = 0.0004$—the 4½-inch pipe with 0.0018 in roughness—the curve cuts the Rouse Line at $\mathbf{N_R} \approx 3\,000\,000$ just as it is supposed to do, but is already, to the eye, above the smooth pipe law at $\mathbf{N_R} = 4000$.

The curve for $\varepsilon/D = 0.00001$ is still with the smooth pipe curve to the limit of the eye at $\mathbf{N_R} = 700\,000$, and differs less than 2% at $\mathbf{N_R} = 1\,000\,000$.

Comments on 0.0018 inch

Colebrook's data for steel pipes was non–existent. Only some data on wrought iron pipe, taken from Freeman,[18] was available. It was used and highly praised by Colebrook.

But Freeman's data, in turn, were based on tests made in 1892—*over 50 years before Colebrook, and over 100 years before the writing of this book*—and the famous roughness value $\varepsilon = 0.0018$ inch (Colebrook actually used 0.0017; Rouse[19] bumped this up to 0.0018 a few years later)—which was to later on achieve the status of a standard—was the result of taking the mean value (not even the average value) of a set of only 18 points.

To repeat: the value $\varepsilon = 0.0018$ inch, which has been cited thousands of times as being representative of steel pipe, is based on the *mean*

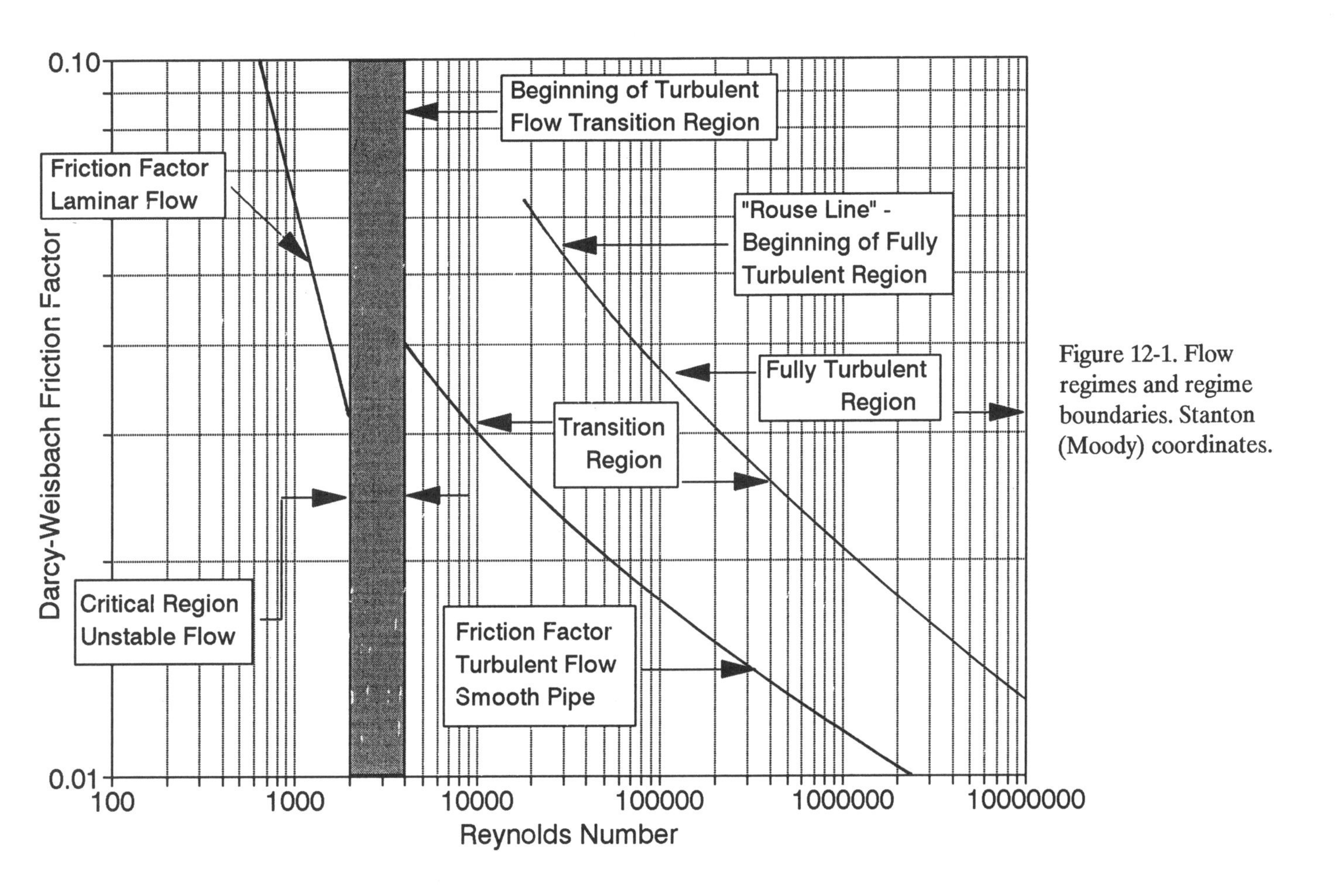

Figure 12-1. Flow regimes and regime boundaries. Stanton (Moody) coordinates.

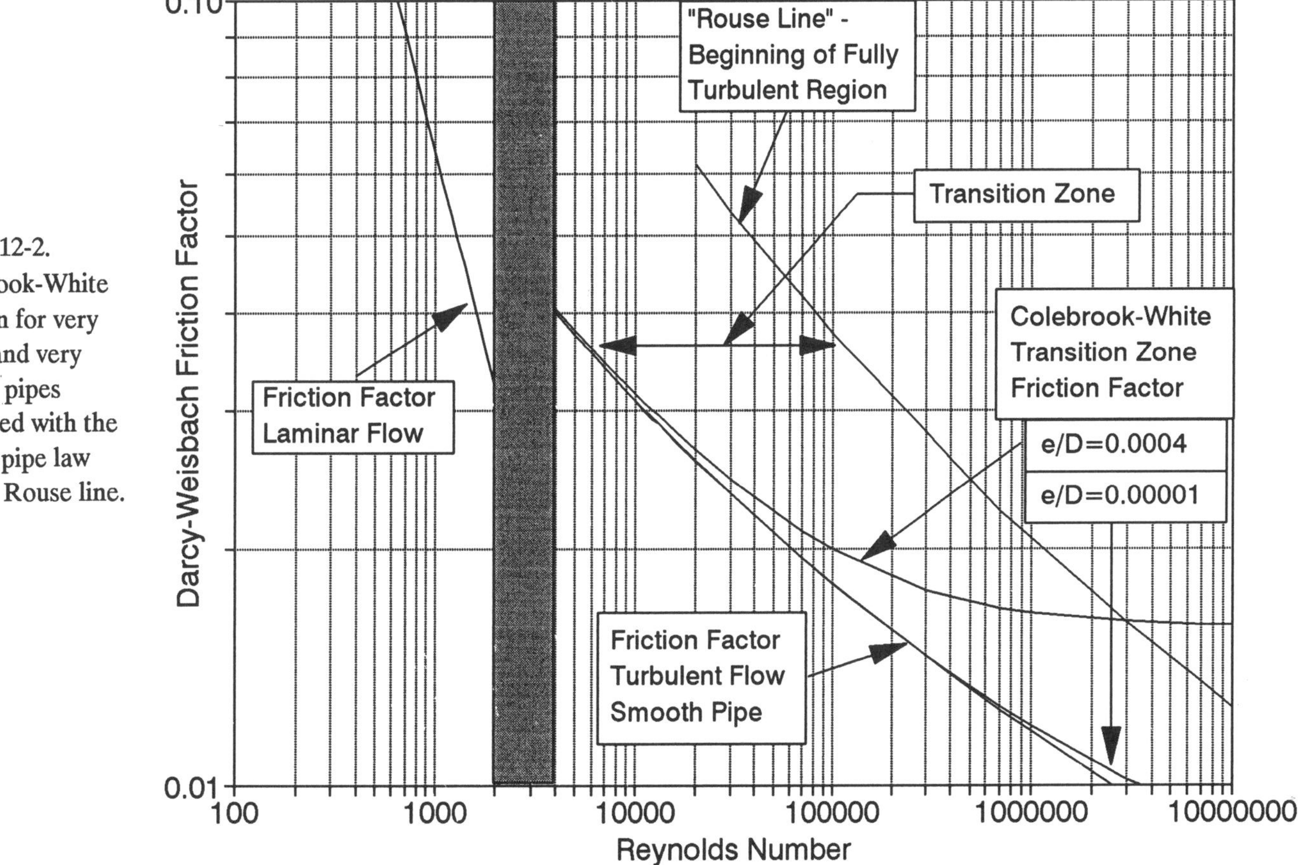

Figure 12-2. Colebrook-White function for very rough and very smooth pipes compared with the smooth pipe law and the Rouse line.

value of 18 test points taken on wrought–iron pipe in 1892; there is absolutely no indication of when the *pipe* itself was made.

Thus do standards become standards. After a time, no one looks into their bases for being, much less their bases for becoming a standard.

The world oil pipeline industry has suffered from *this* standard for 50 years.

Commercial Application of Friction Factors

The Colebrook–White function was little used at first because it contained f implicitly and the equation could not be solved for f except by iterative methods. There are many methods available for solving such equations, of course, but the function was unwieldy, perhaps unnecessarily precise for water and sewer line design (because the rate of flow and roughness are never accurately known for these lines), and the petroleum industry—which did need a reasonable accuracy—had come up with some alternates.

The most famous of the petroleum industry friction factor curves is the *Recommended Curve for the Coefficient of Friction*, from Heltzel,[8] shown in Figure 12-3. This curve, which was the basis for the friction factor curves of *Hydraulics for Pipeliners*, First Edition, was derived by Heltzel from the Stanton and Pannell data and was recommended by Heltzel for oil pipelines larger than 6 inches in diameter; Heltzel believed a *family* of curves existed, with higher friction factors, for smaller pipes. Though he didn't say as much in the reference, he also believed in lower friction factors for larger pipes. It is important to note that Heltzel's curve extends only to $\mathbf{N_R}$ = 1 000 000. For the size and flow rates of pipelines in the U.S.—which was really all Heltzel was interested in—even this figure was too high; the first *high pressure* 12–inch trunk oil pipeline—the largest at the time—was built in the U.S. in 1938, and it had a $\mathbf{N_R} \approx$ 70 000 at the maximum flow of 75 000 bpd.

Figure 12-4 shows the Moody curves of the Colebrook–White function as they appeared in his presentation before the Hydraulic Division of the ASME in 1944. The reproduction here is from yet another reproduction nearly 50 years old, and is not up to modern reproduction standards, but it does, however, have the advantage that it presents Moody's curves in their original form.

Note the very wide spread of ε / D values, from 0.05, which would correspond to 0.5 inch obstructions in a

FRICTION CURVE FOR STREAMLINE AND TURBULENT FLOW

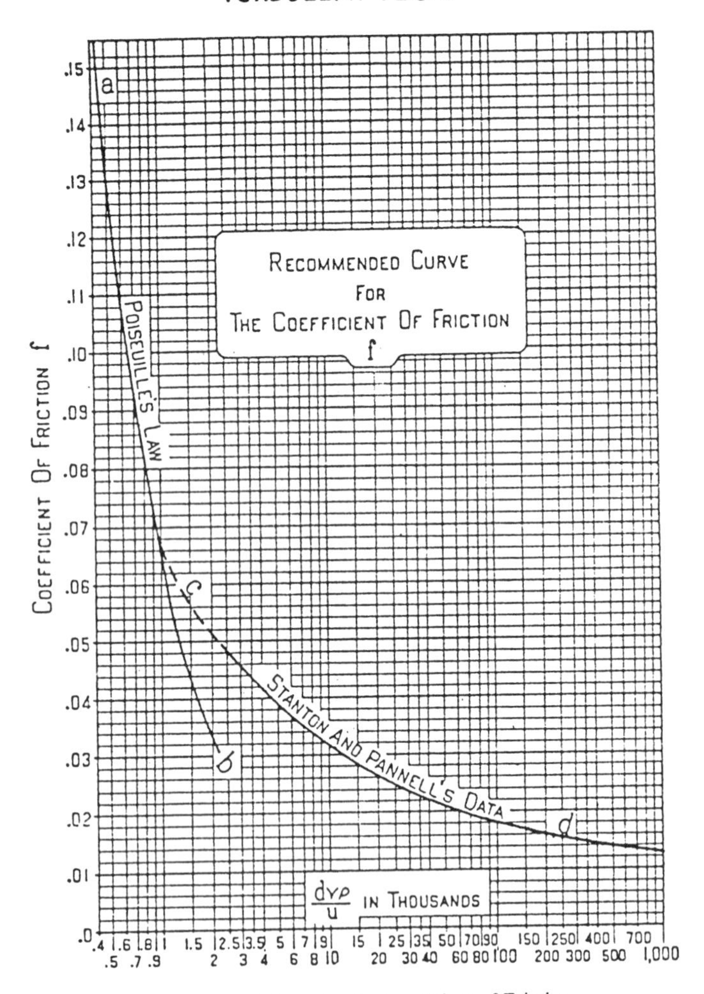

Recommended Curve for the Coefficient of Friction

Figure 12-3. Photocopy of the original Heltzel diagram.

10-inch pipe, down to the mirror–finish smooth pipe where $\varepsilon / D \rightarrow 0$.

Also, note the spread of $\mathbf{N_R}$ is very wide, commencing at 600 (so as to be able to plot $f = 0.1$ at $\mathbf{N_R}$ = 640) and extending out to $\mathbf{N_R}$ = 100 000 000 where the ε / D = 0.00001 line approaches $f = 0.008$, the lower limit of the plot. (The smooth pipe curve had already gone below $f = 0.008$ at a little more than $\mathbf{N_R}$ = 10 000 000.)

Summary of Pipe Flow 1945

This chapter concludes the three–chapter review of the classical approaches to pipe flow from the *dry water* of

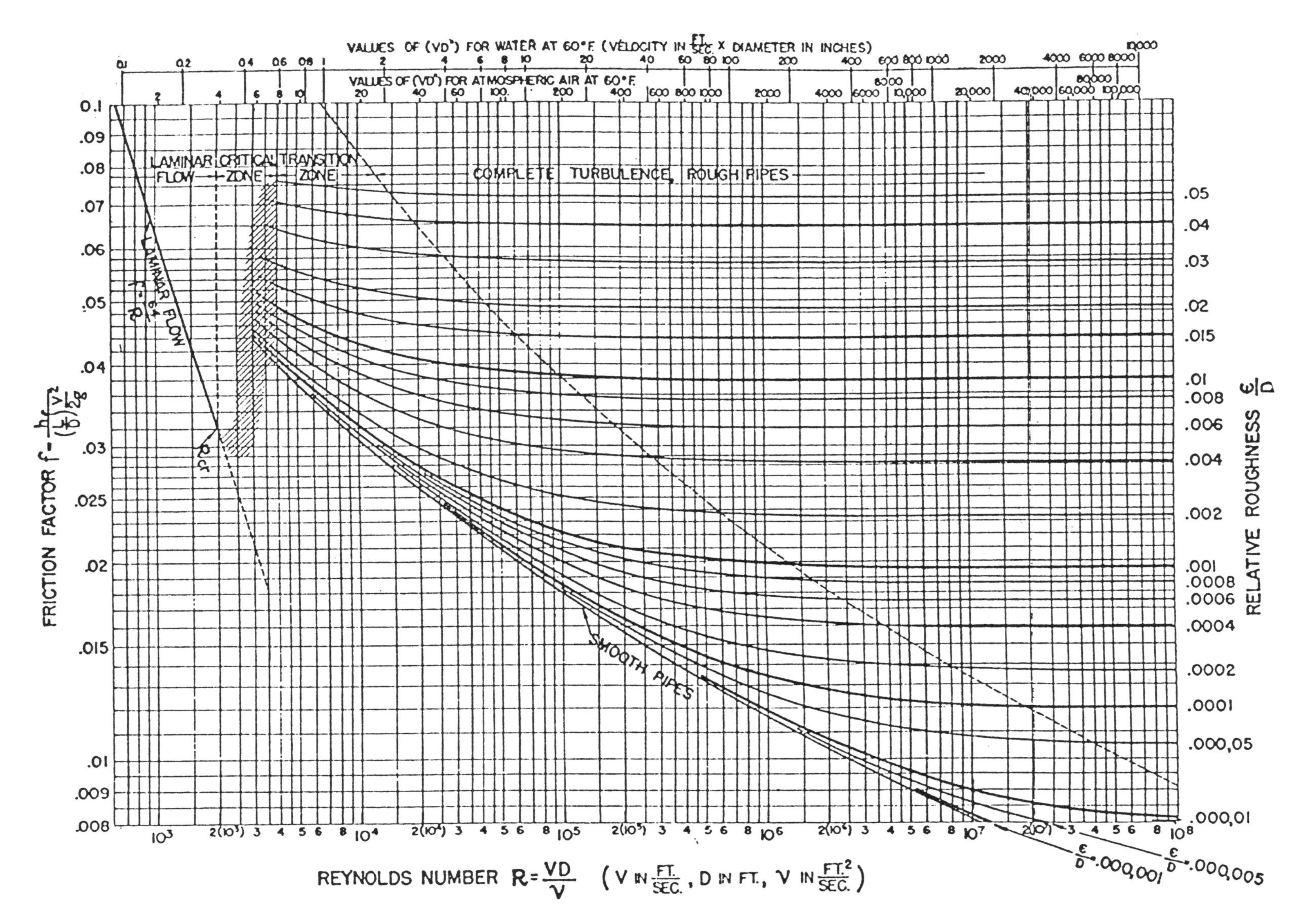

Figure 12-4. Photocopy of original Moody diagram.

Table 12-1

Fundamental Equations for Pipe Flow

Item	Quantity or Identity	Description	Comments
1	$\tau_0 = f\rho\frac{V^2}{8}$	viscous shear at pipe wall; valid for any flow regime	see development of pipe friction equation in Chapter 10
2	$\sqrt{\frac{\tau_0}{\rho}} = \sqrt{\frac{f}{8}}V$	τ_0 -vs- f and V; valid for any flow regime	relates shear at pipe wall to friction factor and velocity; an important identity for correlating experimental results
3	$\tau = \mu\frac{du}{dr}$	internal viscous shear; valid for laminar flow	τ is internal shear, μ is absolute viscosity, u is velocity in direction of flow, r is distance measured along radius
4	$\tau = \mu\frac{du}{dr} + \eta\frac{du}{dr}$	internal viscous shear; valid for turbulent flow	η is a fictitious value, sometimes called eddy viscosity
5	$\tau = \mu\frac{du}{dr} + \rho l^2\left(\frac{du}{dr}\right)^2$	Prandtl's mixing length equation for internal shear; valid for turbulent flow	l is Prandtl's mixing length, the distance a particle of liquid travels before encountering and transferring a part of its momentum to another particle
6	$l = k\frac{du/dr}{d^2r/dr^2}$	von Karman's equation for mixing length; valid for turbulent flow	k is von Karman's (almost) constant, usually taken as 0.40
7	$\mathbf{N_R} = \varphi(V, D, \mu)$	Reynolds Number	valid for any flow regime
8	$f_L = K/\mathbf{N_R}$	form of friction factor	valid for laminar flow in any pipe
9	$f_S = \varphi(\mathbf{N_R})$	form of friction factor	valid for flow in smooth pipe
10	$f_T = \varphi(\mathbf{N_R}, \varepsilon/D)$	form of friction factor	valid for transitional flow in rough pipe
11	$f_R = \varphi\left(\frac{\varepsilon}{D}\right)$	form of friction factor	valid for turbulent flow in rough pipe; ε is a measure of equivalent sand–grain roughness
12	$f_L = 64/\mathbf{N_R}$	friction factor-laminar flow	valid for N_R<2000; from Poisuille's law
13	$f_T = 0.316/\mathbf{N_R}^{0.25}$	Blasius' friction factor	valid for smooth pipe flow for N_R<100 000
14	$\frac{1}{\sqrt{f_S}} = A_S + B_S\log\left(\mathbf{N_R}\sqrt{f_S}\right)$	form for smooth pipe friction factor equations	A and B are constants to be determined
15	$\frac{1}{\sqrt{f_R}} = A_R\left(m, \frac{e'}{D}\right) + B_R\log\left(\frac{e}{D}\right)$	form for rough pipe friction factor equations	m, e and e' account for placement, form, and configuration of internal roughness; usually combined in ε (see [11] above.
16	$\frac{1}{\sqrt{f_S}} = 2\log_{10}\left(\mathbf{N_R}\sqrt{f_S}\right) - 0.8 = -2\log_{10}\left(\frac{2.51}{\mathbf{N_R}\sqrt{f_S}}\right)$		Prandtl's (sometimes called Nikuradse's) friction factor equation for smooth pipe flow; based on Nikuradse's data.
17	$\frac{1}{\sqrt{f_R}} = 2\log_{10}\left(\frac{r}{\varepsilon}\right) + 1.74 = -2\log_{10}\left(\frac{\varepsilon/D}{3.76}\right)$		von Karman's friction factor equation for rough pipe flow
18	$\frac{1}{\sqrt{f_T}} = -2\log_{10}\left(\frac{\varepsilon/D}{3.7} + \frac{2.51}{\mathbf{N_R}\sqrt{f_T}}\right)$	Colebrook–White function	widely considered to be valid for all flow in the transitional range between smooth and rough pipe turbulent flow; it may not be.

hydrodynamics through the mathematically–oriented *fluid dynamics* to the primarily experimental methods of *hydraulics*. At this point a short philosophical commentary is in order.

The Moody curves, which were included, in some form, in every book on fluid mechanics or hydraulics for nearly 50 years, received widespread distribution; and when the Hydraulics Institute[20] made them the foundation of the *Pipe Friction Manual*, they achieved the status of a standard. They were not widely used in the oil industry, but were considered to be, theoretically, a *better body of data* than the Heltzel data. It wasn't; it just looked that way. The following is quoted from Moody.[15]

> "It must be recognized that any high degree of accuracy in determining f is not to be expected. With smooth tubing, it is true, good degrees of accuracy are obtainable; a probable variation in f of about ±5%,[21] and for commercial steel and wrought–iron tubing, a variation of about ±10%. But, in the transition and rough–pipe regions, we lack the primary and obvious essential, a technique for measuring the roughness of a pipe mechanically. Until such a technique is developed, we have to get along with descriptive terms to specify the roughness; and naturally this leaves much latitude…
>
> "Even with this handicap, however, fairly reasonable estimates of friction can be made, *and, fortunately, engineering problems rarely require more than this*."

Unfortunately, engineering problems involving long oil and gas pipelines *do require a better accuracy*, and the world's oil and—especially gas—transmission industries did the research and made the major advances required in the years immediately after the end of World War II.

It is unfortunate that the Colebrook–White function (and, notwithstanding the programmable calculators, the corresponding Moody curves) is still considered as a standard in almost every field of endeavor involving the flow of fluids except in the gas trunk line transmission industry, which almost always operates in the rough pipe regime, and for some—but not many—trunk oil pipelines.

Chapter 13 looks at this phenomenon and makes some comments about it.

Working Equations

The fundamental equations developed in the study of pipe flow in the field of hydraulics are included in Table 12-1.

References

1 Reynolds, O., "An experimental investigation of the circumstances which determine whether the motion of water will be direct or sinuous, and the laws of resistance in parallel channels," *Philosophical Transactions of the Royal Society of London*, 1883.

2 Prandtl, L., "Über Flussigkeitsbewegnung bei sehr kleiner Reibung," International Mathematical Congress, Heidelberg, 1904. (translated) "On fluid motion with very small friction," *NACA TM* 435, 1927.

3 Blasius, P. R. H., "Das Achnlichkeitsgesetz bei Reibungsvorgangen in Flussigkeiten," Physikalischer Zeitung, Vol. 12, 1911 (translated) "The law of similarity applied to friction phenomena."

4 Stanton, P. E., and Pannell, J. R., "Similarity of motion in relation to the surface friction of fluids," *Philosophical Transactions of the Royal Society,* Vol. 214, London, 1914.

5 Buckingham, E., "Model experiments and forms of empirical equations," *Transactions of the ASME,* Vol. 35, 1915.

6 von Kármán, T., "Über laminare und turbulente Reibung," 1921, (translated) "On laminar and turbulent friction," NACA TM 1092, 1946.

7 Prandtl, T., "Über die Ausgebildete Turbulenz," *Proceedings of the International Congress on Applied Mechanics,* Zurich, 1926. (translated) "On the development of turbulence," *NACA TM* 425, 1927.

8 Heltzel, W. G., "Fluid Flow and Friction in Pipelines," *Oil and Gas Journal,* 05 June 1930 (a revision of an earlier article on the same subject published in *Oil and Gas Journal*, 07 October 1926.) Another article, "Derivation of Equivalent Length Formulae for Multiple Parallel Oil Pipe Line Systems" appeared in the *Oil and Gas Journal*, 10 May 1934.

9 Nikuradse, J., "Laws of flow in rough pipes," 1933 (translation) *NACA TM* 1291, 1950.

10 Smith, R. V., Miller, J. S., and Ferguson, J. W., "Flow of Natural Gas Through Experimental Pipe Lines and Transmission Lines," Monograph No. 9., *U. S. Bureau of Mines,* (published by) *American Gas Association,* New York, 1956.

[11] Uhl, A. E., *et al*, "Steady Flow In Gas Pipelines," *Institute of Gas Technology Report No. 10*, (published by) *American Gas Association*, New York, 1965.

[12] Pigott, R. J. S., "The Flow of Fluids in Closed Conduits," *Mechanical Engineering*, August, 1933.

[13] Kemler, E., "A Study of the Flow of Fluids in Pipes," *Transactions of ASME*, 1933, *Hydraulics*, 55.2.

[14] von Kármán, T., "Aspects of Turbulence Problems," *Proceedings of the Fourth International Congress on Applied Mechanics*, Cambridge, England, 1934.

[15] Colebrook, C. F., "Turbulent Flow in Pipes, with Particular Reference to the Transition Region Between the Smooth Pipe and Rough Pipe Laws," *Journal of the Institute of Civil Engineers*, Vol. 11, London, 1938–1939.

[16] Rouse, H., "Evaluation of Boundary Roughness," Proceedings of the Second Hydraulics Conference, *University of Iowa Bulletin 27*, 1943.

[17] Moody, L. F., "Friction Factors for Pipe Flow," *Transactions of ASME*, November 1944.

[18] Freeman, J. R., *Experiments Upon the Flow of Water in Pipe and Pipe Fittings*. New York: American Society of Mechanical Engineers, 1941.

[19] Rouse, H., Bulletin 27, University of Iowa Studies in Engineering, 1943.

[20] "Pipe Friction Manual," 1st. ed., *Hydraulics Institute*, Cleveland, Ohio, 1961.

[21] Drew, T. B., Koo, E. C., and McAdams, W. H., "The Friction Factors for Clean Round Pipes," *Transactions of A.I.Ch.E.*, Vol. 28, 1932.

Part III
The Working Equations of Pipeline Hydraulics

Part I of this book was dedicated to the characteristics of fluids, usually from the standpoint of those characteristics that affect fluid flow though, for completeness, certain other characteristics important when the fluid is being stored or becomes involved in custody transfer were included.

Part II was a review of the classical developments of hydrodynamics, fluid mechanics, and the hydraulics of pipe flow.

The four chapters of this Part III are devoted to the study and development of working equations for modern pipelining.

Chapter 13 describes the extensive data behind the development of:

- A very accurate, non–iterative friction factor function, and
- An extremely accurate non–iterative transmission factor function.

and includes summaries of the work done by the oil and gas industries that support these functions.

Chapter 14 develops the unit friction equation, Chapter 15 the flow rate equation, and Chapter 16 provides working equations for the solution of the six major (common) configurations of oil pipelines.

Each chapter contains a summary tabulation of working equations and such Computer Computation Templates as are necessary to illustrate recommended procedures.

13

Friction Factors: f_{oil}, T_{oil}, $T_{FSmooth}$

Introduction

The world's pipeline industry has long been saddled with a friction factor formulated over 50 years ago from sound fluid mechanics concepts modified by some very suspicious experimental data. Some of these data are now over 100 years old.

The objective of this chapter is to, again, start with the proved concepts of fluid mechanics, but to come forward with a much later, and much better documented, body of data to arrive at a friction factor suitable for the present day field of pipeline hydraulics.

Background

The Status Quo

The friction factor recommended in the first edition of *Hydraulics for Pipeliners*, which was based on a kind of transformation of Heltzel's curve (which was itself drawn from the work of Stanton and Pannell), was adequate for its time—which was the time of the ten–inch log–log slide rule—but it isn't adequate for use by pipeliners carrying around calculators that can do amazing calculations in a few seconds and desk–top computers that can make the identical calculations seemingly instantaneously.

Note that no statement has been made that the Heltzel or *Hydraulics for Pipeliners* curves are not *accurate*; they are just, in today's engineering world, *archaic*.

Also, between the time the first edition of *Hydraulics for Pipeliners* was written and the present time, a major move has been made away from the Heltzel curve and, as a consequence, the Stanton and Pannell data. Later work than Heltzel's, generally that of Colebrook, Moody, and the followers of their philosophy, was based on the theory of a long and deep *transition curve* which connected the *smooth pipe law*, typically the Prandtl–von Kármán–Nikuradse functions, and the *rough pipe law*, typified by the functions of van Kármán, and, (probably) primarily because these works were available to academia whereas the work done by the U.S. oil and gas industry was not, the Colebrook–White function, as calculated by iterating calculator and/or computer algorithms, or as picked off a Moody curve in a textbook, became the friction factor taught and, thereby, the friction factor used.

This should not be thought in any way to be an attempt to disparage the Colebrook–White function, because with the data on pipe roughness Colebrook had to work with, dating back to tests made over 100 years ago, there was no way some kind of transition function would not be promulgated to explain the erratic spread of the data, and the Colebrook–White function just happened to be the one that took hold in the collective mind of the fluid mechanics practitioners and became the accepted method of calculation.

A check of 16 English–language fluid mechanics and hydraulics texts published in the period 1950–1985 showed not a single instance of the Heltzel's Stanton and Pannell curve, but every book, *without exception*, carried a copy—often a reproduction—of the Moody diagram for the Colebrook–White function. Thus new U.S. engineers, and most all non–U.S. engineers, not having the Heltzel curve available, and with no prior experience to back them up, were naturally attracted to the Moody diagram and the Colebrook–White function. And with the arrival of the programmable hand–held calculator, and thus the ability to program the Colebrook–White function and thereby eliminate the necessity of having to carry a hard copy of either the Heltzel curve or the Moody plot along, the Colebrook–White function became routinely used by new U.S. and most non–U.S. pipeline engineers.

It should be noted that the cited texts also carried the curves of relative roughness for steel (read: wrought iron) pipe based on the old Colebrook data in the Moody graph form, and the fiction of 0.0018 inch roughness for line pipe was thereby perpetuated.

So students learned the method in class, and applied it in their later professional practice, with the result that by 1970 reference to the Heltzel curve became professionally unacceptable, and the Colebrook–White function—with no competition—was enthroned in all English language fluid mechanics and hydraulics textbooks of the world and most of those in other languages.

Limiting the Problem

The pipe friction equations and their associated friction factor functions presented in previous chapters were those developed by mathematicians and engineers interested in finding a *general* solution to the problem of pipe flow. To date they have not been successful, because *there is no rigorous mathematical description of turbulent pipe flow*. Lacking this, there is no mathematically correct solution to *any* turbulent flow problem; there are generalizations, there are empirical equations, and there are the records of carefully controlled experiments, but the working equations for turbulent flow are at best approximations of some as yet unknown law (perhaps) embedded in chaos theory.

Since there are no mathematically *correct* solutions, workers in the field have turned to mathematically *sound* solutions; and while these are not mathematically correct in the sense that they are *rigorously* correct, they are based on lines of reasoning from mathematical principles *believed* to be correct. This has led to a kind of "be all things to all people approach" in the field, with no emphasis placed on any particular aspect of the pipe flow problem.

Take, for instance, the Moody diagram. It covers

- N_R from 600 to 100 000 000, a range of over 160 000:1. For any given pipe diameter this is about the same as a variation in flow rate of 160 000:1; the variation of N_R as among oil pipelines *may be* as much as 100:1.
- ε/D from 0.05 to 0.000001, a range of 50 000:1. For a given pipe wall roughness, this is the same as a variation in diameter of 50 000:1. Since any pipemaking process can be expected to produce very nearly the same *absolute* roughness over its entire range of production, the ε/D range of pipes from a modern mill may be 8:1, no more.

It is obvious that the problem facing pipeliners, whose work is in the small ranges described, is much less challenging than that faced by those workers in the field trying to generate functions and equations equally valid over the very wide ranges of (for instance) the Moody diagram.

But, on the other hand, pipeliners need a degree of *accuracy* not found in the work of the generalists: to badly miss a design for a 1000 ft water line is one thing; to miss it for a 1000 mile oil line is clearly another. In the development here a *precision* of ±0.5%, and an *accuracy* of ±2.0%, will be the aiming points.

So the emphasis in this chapter will be on that part of the $f = \varphi(N_R)$ function applicable to oil pipelines and, in particular, trunk oil pipelines, which are defined for purposes here as cross–country pipelines eight inches and larger in diameter. It is not intended that this approach will slight smaller, shorter, lines; it is just that such lines need not be designed to the same standards as the larger, longer lines and thus need not bear the same intense scrutiny.

To further restrict—refine, perhaps, is a better word—the definition, it is in order to look at the range of N_R that can be expected within the above broader definition.

Table 13-1, compiled from published information, is representative of the kind of flow patterns that have been publicly reported for successful commercial pipelines. Some of the examples are somewhat out of the ordinary; the 8–inch products line carrying 60 MBD with a N_R of some 3 500 000 is a propane line, and propane—which has a viscosity in the order of 0.2 cSt—produces the very high N_R. If the line flowed gasoline at the same rate the N_R would be about 750 000.

If the development of a friction factor is confined to crude lines the search can be limited to $N_R < 750\,000$. For products and LPG lines, where N_R is determined more by the viscosity *dividing* the fraction than by the diameter *multiplying* it, it may be necessary to go as high as N_R = 3–4 000 000. For either, a suitable lower limit is about N_R = 30 000.

Table 13-1

Typical Maximum N_R for Trunk Pipelines

Diameter	Category	Flow Rate MBD	Reynolds Number
48	Crude	1 850	600 000
42	Crude	1 030	400 000
40	Crude	1 050	740 000
34	Crude	800	640 000
30	Crude	800	360 000
24	Crude	350	630 000
20	Crude	275	290 000
16	Crude	160	100 000
12	Crude	90	70 000
8	Crude	30	30 000
40	Products	1 300	3 500 000
36	Products	1 000	2 500 000
24	Products	500	2 000 000
16	Products	200	1 100 000
12	Products	115	900 000
8	Products	60	3 500 000

So, on an arbitrary but reasonably founded basis, the following friction factor development will be dedicated to

- A precision of ±0.5% and an accuracy of ±2.0%
- For cross–country oil pipelines
- Eight inches and larger in diameter
- Flowing with 30 000 > N_R > 3 000 000

This will take care of any crude line, all but the largest, highest–capacity, products lines, and most of the LPG lines.

The intended precision and accuracy are better than anything else available at the present time.

And, before the development is complete, there will be a $T_{FSmooth}$ function that can calculate the smooth pipe law function to any reasonable accuracy without iteration valid over the entire range 4000 > N_R >10 000 000.

Historical Background

It is difficult to remember, looking back from the standpoint of the present–day world's pipeline industry, that most of the early oil pipelines were built in the U.S and most of them were small. If a field needed a higher capacity transportation outlet, more pipelines were built; there were systems with as many as six parallel lines. This was partly because the fields were small and didn't need large lines to carry their production, and partly because the world's capacity to build high–pressure large–diameter pipe was minimal; it did not exist anywhere but in the U.S. and even there the capacity of the pipe mills was not large and the quality was not that desired. There were some large diameter gas pipelines, to be sure, but these were medium pressure lines joined with a patented mechanical coupling of some kind—not the kind of construction one would expect for a high pressure oil pipeline. Even the East Texas field, the only field in the contiguous U.S. to rank with the fields of the Middle East, was serviced by 6–, 8– and 10–inch pipelines —more than three dozen of them—operated by many different companies. Given this kind of economic environment it is not unusual that the study of the flow of oil in pipelines really never got out of the laboratory; with the small lines involved, it wasn't worth it. Until the *Big Inch* and *Little Inch* pipelines of World War II.

The Big Inch was a 1254–mile 770–psig 350 000–b/d 24–inch welded crude oil pipeline from Longview, Texas to a terminal near Philadelphia; the Little Inch line was a 1475–mile long 780–psig 250 000–b/d 20–inch welded products that ran from Beaumont, Texas to a distribution point in New Jersey. These were truly pioneering efforts, and the entire U.S. industry put its best engineering and construction personnel into their design and construction. The engineers responsible for designing these lines used what they had—which sometimes wasn't very good—and they designed for a sure result because, in wartime, errors could cost more than money. But the Big Inch and, not withstanding its unexpected severe problems with the electric–resistance–welded pipes themselves, the Little Inch, were successful, and they demonstrated the kind of economical transportation costs that could be achieved with a large, long, oil pipeline. Not as low per ton–mile, perhaps, as transport by ocean going tankers, or even as low as tow boats and barge trains operating on the inland waterways, but the pipelines went directly to their end customers—they weren't constrained by coastlines or the meanders of rivers—and their shorter transport distances often made the difference.

Thus in the first years after the end of World War II the U.S. saw a boom in long pipelines in the 20–inch to 30–inch range, such as the Basin–Ozark 20/22/24–inch, the Mid–Valley 20/22–inch, and the West Texas Gulf 24/26–inch; and some of the Middle East pipelines, most notably the 30/32–inch IPC Kirkuk–Banias line and the

30/31–inch Abqaiq–Sidon TAPLine, were built during this period.

What all of these lines had in common was U.S. or European major oil company ownership, and the activity in the world's pipeline industry finally reached a point in the early 1950's where these owners decided they *had* to do something about the specifications to which they were building these major, *costly*, projects.

One immediately identified task was the determination and codification of the allowable working stresses in pipes. (This subject took up two full chapters in the original *Hydraulics for Pipeliners*.)

Another—which is the subject of this chapter—was the development of methods for accurately calculating the carrying capacity of large oil lines.

And, at almost the same time, there began a twenty year period of intense building for the U.S. gas industry. The success of the welded designs of the Big Inch and Little Inch lines suddenly made feasible the construction of large, high pressure, gas lines covering great distances; and the surge of long big–inch transmission lines from the U.S. southwest to the U.S. midwest and east coast states occupied the best technical skills of the industry.

But the gas industry, too, realized that it did not have in place the codes and the specifications that could be used as structural design standards, and the equations of flow used in low pressure lines did not properly describe the flow of long, large diameter, high pressure trunk lines.

So the U.S. oil and gas pipeline industries, separately, simultaneously, and united only by their use of the same basic building block—pipe—started taking the steps to establish codes and specifications for line pipe and the construction of pipelines. In this endeavor the results, for all practical purposes, were identical. There is no fundamental difference in the way stresses are calculated in ANSI B31.4 for oil lines or ANSI B31.8 for gas lines. Nor should there be.

The studies of gas and oil flow were also carried out apart—perhaps imprudently, as it results—because of a then perceived difference of the flow mechanisms of oil and gas pipelines. But both industries suffered from the same problem: it didn't have a suitable, accurate method for calculating pressure drops and flows for long pipelines, and they set out to develop it.

Connecting Oil and Gas Research

In Chapter 12 a note was made that the U.S. gas pipeline industry uses the *Fanning equation*, while the U.S.—and most of the world's—oil pipelines use the *Darcy–Weisbach equation*. While these are simply two ways of saying the same thing, there is a numerical difference between the two that makes it difficult to compare the results of oil and gas pipeline industry flow research if it is not understood.

The Darcy–Weisbach equation for pipe friction is

$$h_L = f_{Darcy}\left(\frac{L}{D}\right)\left(\frac{V^2}{2g}\right) \tag{13-1}$$

The Fanning equation is

$$h_L = 2 \cdot f_{Fanning}\left(\frac{L}{D}\right)\left(\frac{V^2}{g}\right) \tag{13-2}$$

If these two equations in h_L are equated there results

$$f_{Darcy} = 4 f_{Fanning} \tag{13-3}$$

Also note that in any of the functions for calculating the friction factor (see Chapter 12) the variable is $1/\sqrt{f}$, not just f alone. This variable has been used by the gas industry to define a parameter called the *transmission factor* T (sometimes, F) that is, for either function,

$$T_{Fanning} = \frac{1}{\sqrt{f_{Fanning}}} \text{ and } T_{Darcy} = \frac{1}{\sqrt{f_{Darcy}}} \tag{13-4}$$

Setting the Fanning and Darcy–Weisbach versions of this equation equal to each other, and understanding that in the following text, comparisons will continuously be made between oil and gas industry investigations, the following connecting equations appear.

$$f_{Darcy} = \frac{4}{T^2_{Fanning}} \text{ and } T_{Fanning} = 2\,T_{Darcy} \tag{13-5}$$

The relations expressed by Equations 13-3 and 13-5 are so important in the following text they are displayed, separately, as follows:

$$f_{Darcy} = 4\, f_{Fanning}$$

$$f_{Darcy} = \frac{4}{T^2_{Fanning}}$$

$$T_{Darcy} = \frac{T_{Fanning}}{2}$$

These relations are simple, and once learned are not forgotten.

Gas Pipeline Investigations

The gas pipeline industry pipe flow investigations were, at the outset, limited to the precise laboratory work done by the U.S. Bureau of Mines at the Bureau of Mines' Petroleum Experiment Station at Bartlesville, Oklahoma. Later, as the Bureau of Mines work proved that gas could flow in transmission lines in the smooth pipe or transitional flow regimes, additional work was carried out, including field work performed under the auspices of the Institute of Gas Technology on more than 40 operating gas transmission lines.

The following summarizes these efforts.

1950–1965
Tests and Analyses

The U.S. gas industry had a period of intense work on the mechanism of flow in long gas transmission pipelines that was carried out under the sponsorship of the *Pipeline Research Committee* of the *American Gas Association.*

The first of two large, continuing, studies was carried out by the U.S. Bureau of Mines and resulted in the publication in 1956 of the now justly famous Monograph No. 9,[1] the *Flow of Natural Gas Through Experimental Pipe Lines and Transmission Lines*, a follow–up of a previous Bureau of Mines study, Monograph No. No. 6.[2]

The second study was carried out and reported by the Institute of Gas Technology in Technical Report No. 10,[3] *Steady Flow in Gas Pipelines*, with active assistance of the operators of gas transmission pipeline companies.

The Monograph No. 9 work commenced with the idea that gas flow in commercial pipelines is fully developed turbulent flow in accord with the rough pipe law (it turned out that this was not true; for low flow rates in very smooth pipes, gas flows in the smooth pipe regime), but the program also provided for a smooth pipe that was far beyond anything ever seen before; a 122–ft long test bed of 3.5–inch o.d., superfinished stainless steel pipe that the U.S. Naval Gun Factory had machined and then honed to a mirror smooth finish with wood laps until a 3.008–inch i.d. pipe, uniformly circular to 0.001 inch, was produced. The greatest surface roughness found was 15 μ inches rms; some joints had a maximum roughness of only 3 μ inches.

Laboratory tests were also performed on 2–, 4–, 6– and 8–inch inch commercial seamless pipes, on an 8–inch commercial welded pipe, and on a composite test bed of 3–inch welded and 3–inch seamless commercial pipes.

The tests proved that the rough pipe law was valid, assuming the appropriate flowing conditions, for all pipes tested except the superfinished 3–inch; the testers were never able to get a N_R high enough in the 3–inch test bed to take this line into the transition rough pipe regime, much less into fully developed turbulent flow, though an N_R of 10 800 000 was achieved.

The testers also could take the 2–inch seamless test line down into the smooth pipe law range, and they found the results followed the theoretical predictions with excellent accuracy.

Figure 13-1 is a plot of points tabulated in Monograph No. 9 for the 2–inch commercial seamless pipe, while in the smooth pipe regime, and for the 3–inch superfinished line, which was always in the smooth pipe regime, for all $N_R < 7\,000\,000$. The testers saw no reason this set of points could not be extended almost indefinitely; the superfinished pipe really *was* hydraulically smooth. The spread of points for both the 2–inch commercial seamless and the 3–inch superfinished stainless steel rifle barrel is remarkably small, and the *Monograph No. 9 Smooth Pipe Function* holds with the test data out to the maximum N_R.

Note the comparison is with the *Monograph 9 Smooth Pipe Function*. This was developed separately by the testers after all the experimental work had been done, and resulted in the smooth pipe function being taken back to the original van Kármán values of 4 and –0.6 from the Nikuradse values of 4 and –0.4, thus making the smooth pipe law in Fanning terms

$$\frac{1}{\sqrt{f_F}} = 4\log\left(N_R\sqrt{f_F}\right) - 0.6 \qquad (13\text{-}6)$$

This can be written in terms of the transmission factor T_F as

$$T_F = 4\log\left(\frac{\mathbf{N_R}}{T_F}\right) - 0.6 \qquad (13\text{-}7)$$

There is not a large difference between the values calculated by Equation 13-7 and Equation 12-11 (which is in Darcy notation; in Fanning notation the function would read $T_F = 4\log(\mathbf{N_R}/T_F) - 0.4$), but Figure 13-2 shows that the departure at low $\mathbf{N_R}$ can reach 4%, and at best is almost 2% at $\mathbf{N_R} > 1\,000\,000$. (*An important point here is that the Nikuradse smooth pipe law used by Colebrook and Moody produces an f_D that is smaller than it should be; i.e., smooth pipe is not as smooth as Nikuradse, Colebrook, and Moody made it out to be.*)

Figure 13-3 is a replotting of the same data in the usual $f_D - vs - \mathbf{N_R}$ form. The spread of points appears to be somewhat less than on Figure 13-1, but this is merely the way the plot appears; exactly the same points were input to the computer to make the two diagrams.

It is interesting to look at the curve defining these data with the curves seen on the Moody diagram. Figure 13-4 shows a $f_D - vs - \mathbf{N_R}$ plot for the Nikuradse smooth pipe function and for the Colebrook–White function for five other pipes:

- roughest pipe on Moody's plot, $\varepsilon/D = 0.05$
- smoothest Nikuradse pipe, $\varepsilon/D = 0.001$
- 1890 4–inch, $\varepsilon/D = 0.0004$
- 1950 20–inch, $\varepsilon/D = 0.000035$
- 1980 48–inch, $\varepsilon/D = 0.00001$

Note that the Colebrook–White curve for the Nikuradse pipes is beyond anything apt to be found commercially, and the 1890 4–inch Colebrook–White curve *starts* above the smooth pipe curve, i.e., it has already reached the Colebrook–White transition zone at $\mathbf{N_R} = 4000$. The 1950 20–inch Colebrook–White curve stays with Nikuradse's

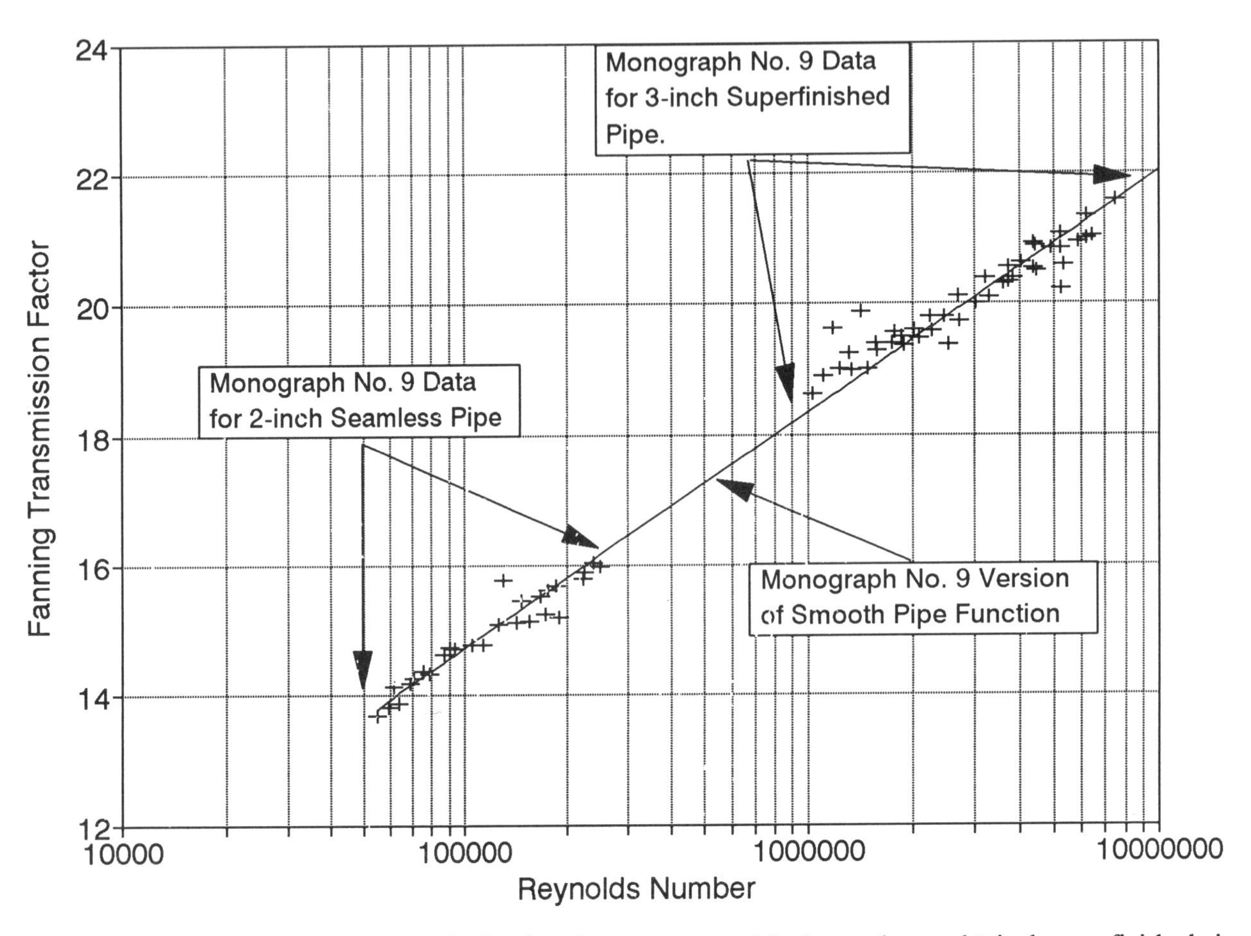

Figure 13-1. Monograph No. 9 smooth pipe function -vs- tests on 2-inch seamless and 3-inch superfinished pipe.

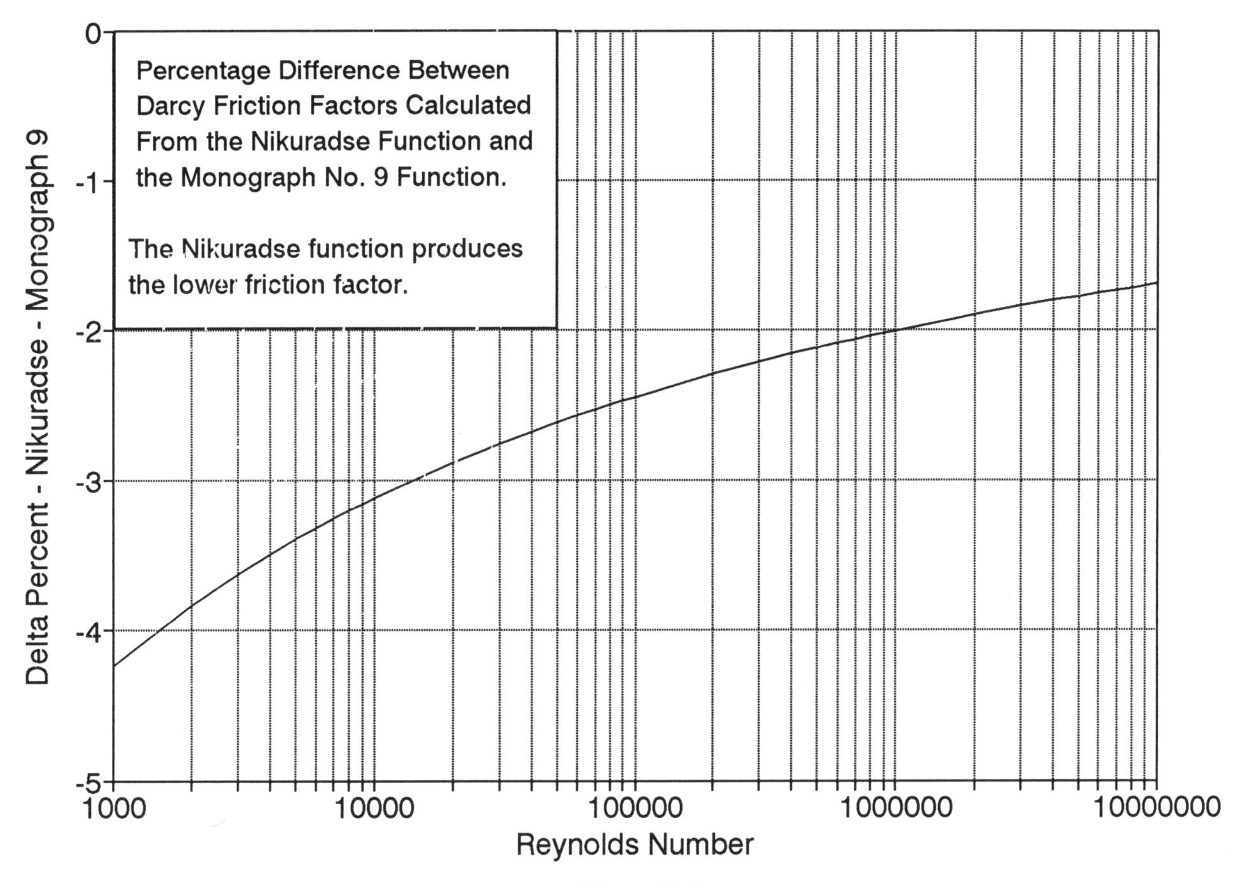

Figure 13-2.

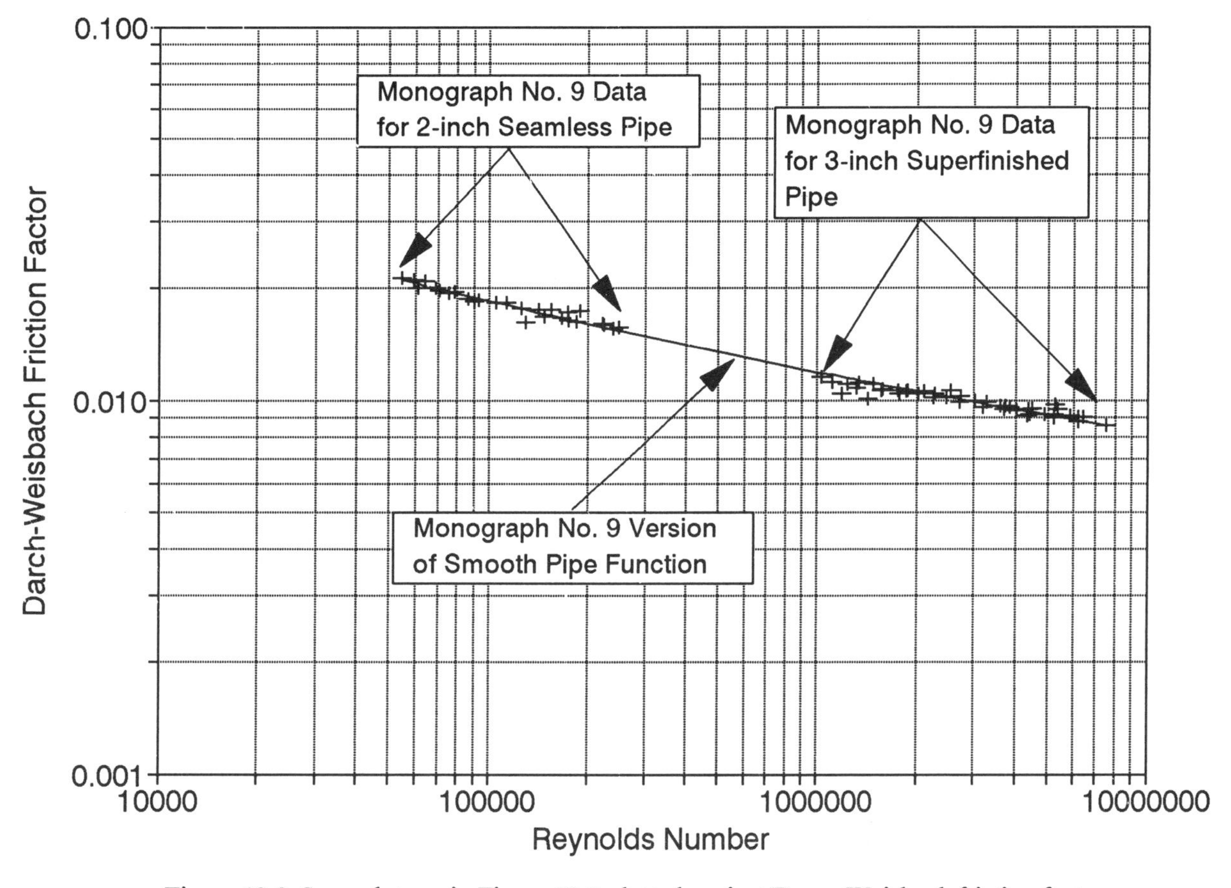

Figure 13-3. Same data as in Figure 13-1 plotted against Darcy-Weisbach friction factor.

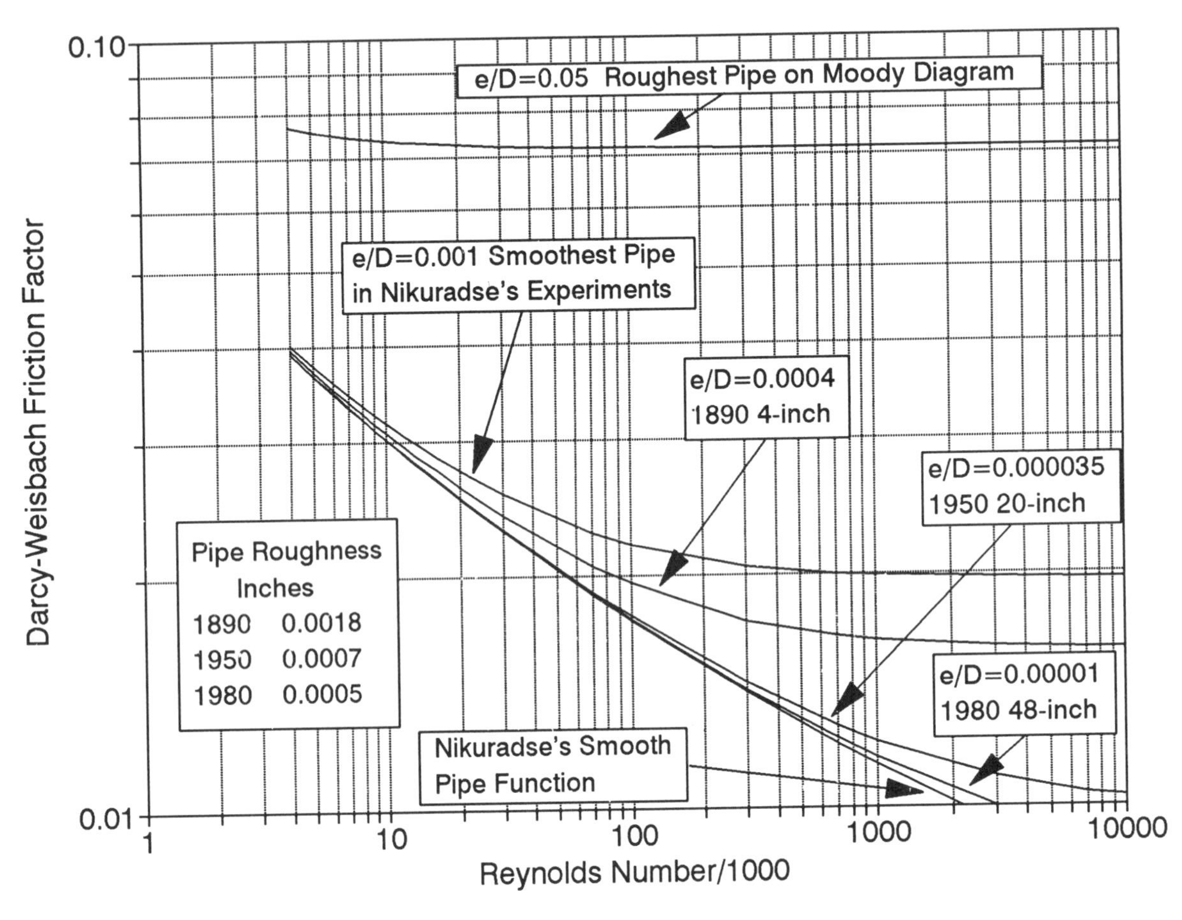

Figure 13-4. Defining curves, Colebrook-White function on Moody coordinates.

smooth pipe function out to about N_R = 100 000; and the 1980 48–inch pipe curve doesn't leave the smooth pipe curve until $N_R > 300\ 000$.

Figure 13-5 is the same data as a $T_F - vs - N_R$ plot. It is easier here to see how the Colebrook–White function curves depart from the smooth pipe law function because in the $T_F - vs - N_R$ form the smooth pipe law is very nearly a straight line.

The Monograph No. 9 testers also took pains to make accurate measurements of ε and ε / D. This was an easy test for gas experimenters, however; take the flow regime into the fully turbulent rough pipe region, measure flow rate and pressure drop, calculate T_F, and back–out ε / D from the rough pipe law. In their summary, the testers recommended a design roughness of 0.0007–inches for gas transmission pipelines.

Later tests, some under the auspices of TR 10 and some simply made by interested companies, found that ε could vary down to as low as 80 μ inches for sand–blasted pipe, and as high as 1850 μ inches for some small, older, pipe (actually, one of the pipes used in the Monograph No. 9 tests), but most of the findings for pipe of the kind apt to be in an oil pipeline—modern, well–pigged, welded line pipe—were in the range 400–800 μ inches. In fact, a well–pigged oil line carrying a crude that tends to deposit paraffin on the pipewall can actually exhibit a pipe wall roughness very nearly as smooth as the superfinished 3-inch.

Table 13-2 is a tabulation of typical measurements and summaries of measurements. The references are shown as footnotes to the table and are not necessarily in the references to this chapter.

Figure 13-6 shows how the fully developed rough pipe and the smooth pipe regimes compare for a wide range of commercial pipe sizes. Based on an ε of 0.0007 inch, even a 6–inch pipe would have to reach an N_R of at least 800 000 to go into completely turbulent rough pipe flow, and there is no possibility a commercial line would reach this unless it was a very highly pumped propane or LPG pipeline. A 48–inch line would have to reach an N_R ·of

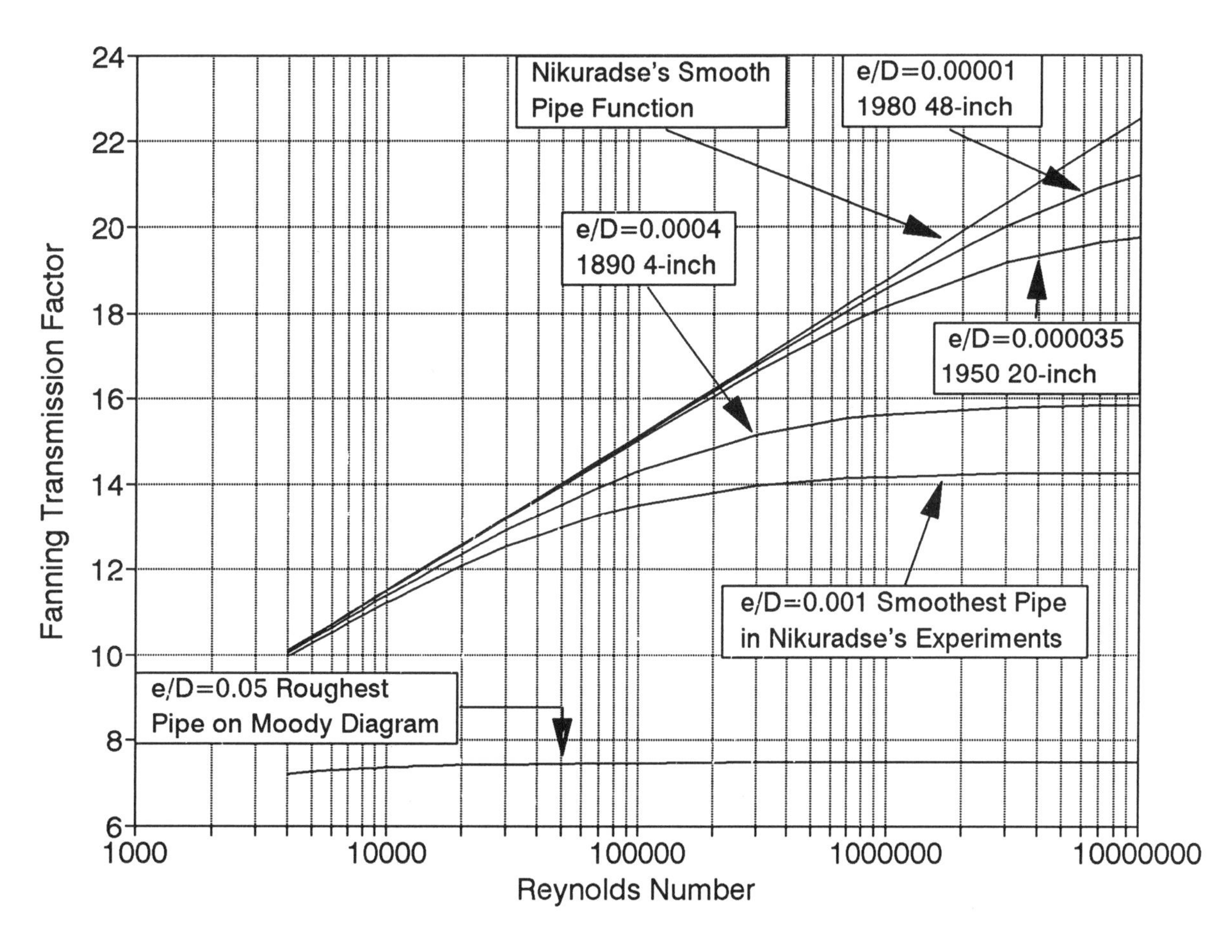

Figure 13-5. Same data as in Figure 13-4 plotted against Fanning transmission factor.

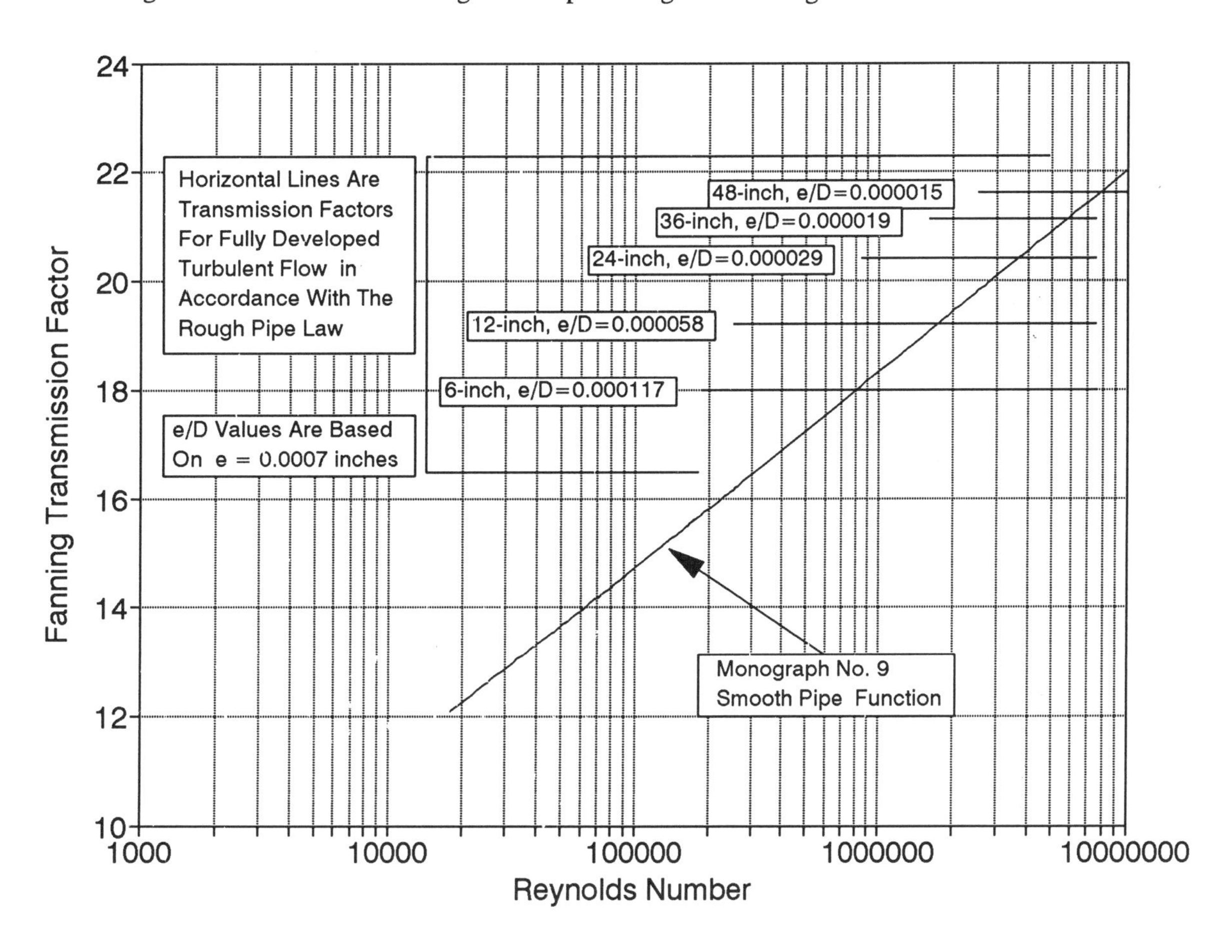

Figure 13-6.

Table 13-2

Effective Roughness Values for Modern Pipes Natural Gas Industry Statistics

Year Reported	Pipe Class	Kind of Pipeline	Roughness Microinches	Conditions	Ref. No.	Reference
1956	Recommended	gas	700	weighted average	1	Monograph 9
	2-8 inch	gas	553-1850	laboratory tests		
	1¼-7 inch tubulars	gas	600-650	field tests		
	12-inch	gas	450-680	field test		
	20-inch	gas	760-860	field tests		
	12-inch	gas	920	operating test		
1959	24-inch No. 1	gas	912	field tests, 10 year old pipe	2	Klohn
			520	same pipe, pigged twice		
			180	same pipe, epoxy coated		
	24-inch No. 2	gas	952	field tests, 10 year old pipe		
			573	same pipe, pigged twice		
			413	same pipe, epoxy coated		
1960	36-inch	gas	280	average, coated pipe	3	Crowe
			450	average, new smooth pipe		
			700	average, new average pipe		
			1300	average, weathered pipe		
1960	big inch	gas	450	new pipe at mill	4	Ellington and Staats
			760	pipe in use		
			1100-1300	weathered 6 mos		
			1400	weathered 12 mos		
			1600	weathered 24 mos		
			250-300	epoxy internally coated		
1961	30-inch	gas	200	field tests, 1 year old pipe	5	McAnneny and Sullivan
			80	same pipe, sand blasted		
1964	big inch	gas	500-750	new bare pipe	6	Uhl
			1000-1250	new pipe weathered 6 mos		
			1500	new pipe weathered 12 mos		
			1750	new pipe weathered 24 mos		
			200-300	plastic lined pipe		
			200-300	sand blasted pipe		
			300-500	pig burnished pipes		
1965	big inch	gas	700	clean new pipe	7	Shriver
			912	uncleaned service line		
			520	pigged pipe		
			180	coated pipe		
1979	24-inch	gas	936	field test, 10 year old pipe	8	Popan
			543	same pipe, pigged twice		
			297	same pipe, epoxy coated		

[1] Smith et al, Flow of Natural Gas Through Experimental Pipelines and Transmission Lines, U.S. Bureau of Mines Monograph No. 9, 1956.

[2] Klohn, C. H., "Flow Tests on Internally In-Place Coated Pipe," *Pipe Line Industry*, July 1959.

[3] Crowe, R. H., "Internal Coating of Gas Pipelines," Flow Calculations in Pipelining, *Oil and Gas Journal*, 1960.

[4] Ellington, R. T., and Staats, W. R., "Flow Calculation Manual for Natural Gas Transmission Lines in Under Way," *Pipe Line Industry,* July 1960.

[5] McAnneny, A. W., and Sullivan, L. C., "Trunkline Reports on Low–Cost Sandblasting That Increases Flow," *Pipe Line Industry*, May and June 1961.

[6] Uhl, A. E., Pipeline Efficiency Testing," *Oil and Gas Journal*, 03 May 1965.

[7] Shriver, W. B., "Gas Flow in Resin–Coated Pipelines," *Pipe Line News*, October 1965.

[8] Popan, V. A., "Pipeline Efficiency Testing: Measurements and Calculations," *Pipeline and Gas Journal*, May 1979.

8 000 000 to go into fully turbulent rough pipe flow.

And in the end, the TR 10 investigators verified with large diameter transmission lines what the Monograph No. 9 testers had found in the laboratory: *there is no deep, broad, Colebrook–White transition zone*. The tests show that a pipe stays in smooth flow according to the smooth pipe law until some point when—in a way even not today determined—over a fairly broad range of $\mathbf{N_R}$ but a *quite narrow range of* T_F—it commences flowing in turbulent rough pipe flow. Monograph No. 9 reached a conclusion that the T_F at the lower limit would be about where $T_F = 4\log D + 14$, and the upper limit of the range about where $\mathbf{N_R} = 180\,000\,D\log D + 670000D$. These equations are valid for diameters in inches and ε = 0.0007 inch. Table 13-3 gives typical limiting values for a range of pipe diameters. Figure 13-7 shows the schematic representation of the limits on the smooth and rough pipe laws for a 24–inch with an absolute roughness ε of 0.0007 inches.

Table 13-3

Upper and Lower Limits of Transition Zone

Based on: Monograph No. 9, page 69
Absolute roughness = 700 μ in.

Equations: Upper limit = $180\,000D\log D = 14.9$
Lower limit = $1/\sqrt{f} = 4\log D + 14$

Pipe Diameter Inches	Lower Limit = Upper End of Smooth Pipe Flow	Upper Limit = Lower End of Rough Pipe Flow
	T(Fanning)	Reynolds Nr
6	17.1	4 900 000
12	18.3	10 400 000
24	19.5	22 000 000
36	20.2	34 200 000
48	20.7	46 700 000

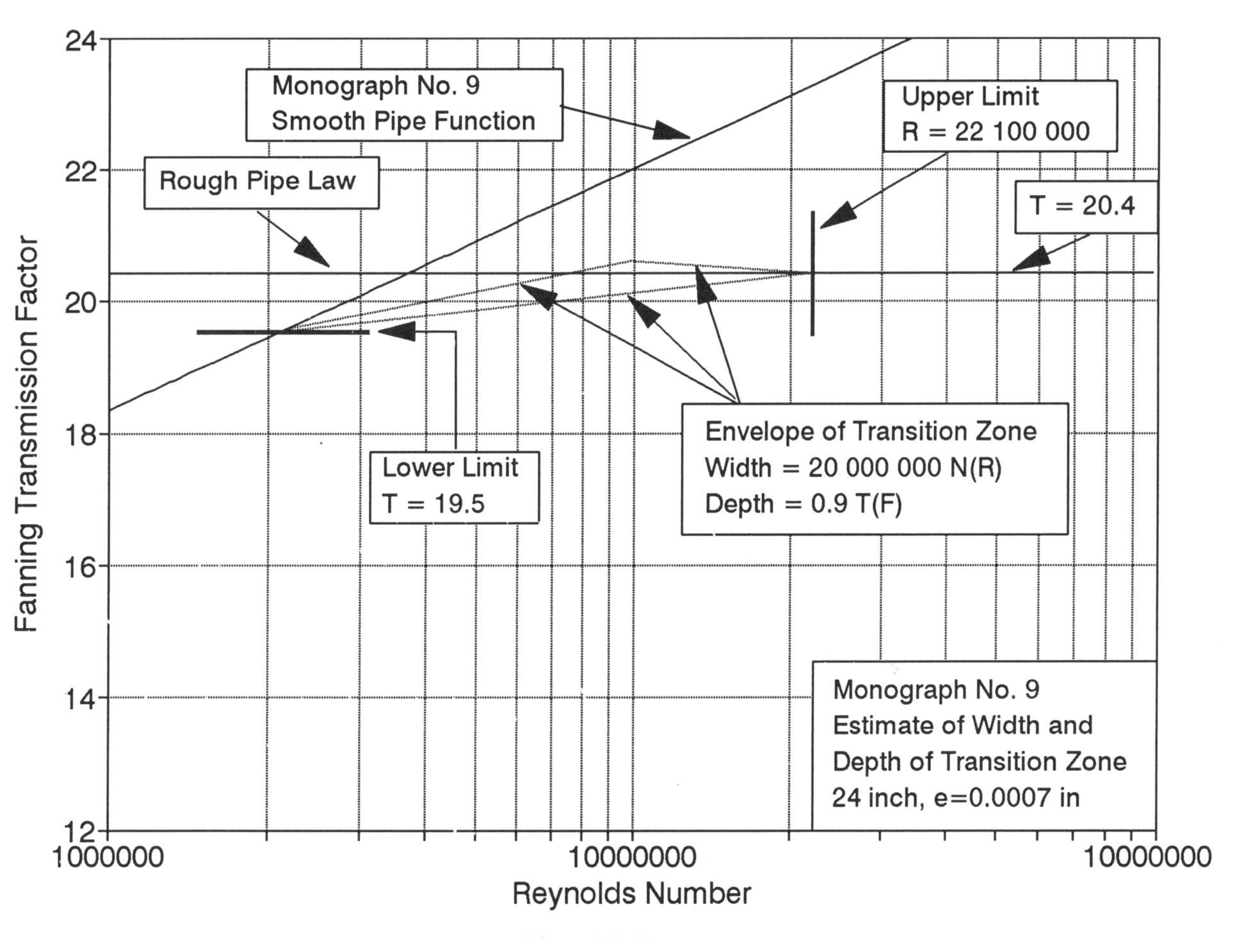

Figure 13-7.

The graphical representation of the *theory* of this phenomenon is shown in Figure 13-8 which is taken from the AGA *Steady State Flow Computation Manual*.[4] This document doesn't support even the Monograph No. 9 kind of transition, and trends toward a very narrow–$\mathbf{N_R}$ narrow–T_F kind of transition.

Summary of Gas Industry Findings

Those findings of the research reported in Monograph No. 9, and TR 10 work that followed, that are important to the development of friction factors for oil pipelines are:

- The *smooth pipe law* is $T_F = 4\log(R/T_F) - 0.6$
- The *roughness* of line pipe ≈ 0.0005 inch
- The *transition* from *smooth pipe flow* to *rough pipe flow*—in whatever form appears to be—may be fairly wide as measured over the spread of $\mathbf{N_R}$ but is quite narrow as measured over T_F and is not of major importance in pipeline flow.

Oil Pipeline Investigations

1950 and Earlier

Not all of the work done prior to 1950 was done by academics working in university laboratories; there were some who worked in the oil pipeline industry, and among these Heltzel was without doubt the leader. Heltzel's papers, published in 1930[5] and 1934,[6] laid a basis for the U.S. oil pipeline industry that served the industry—at that time just beginning to use low–pressure 12–inch pipe—up to the time of the Big Inch pipeline. Heltzel's basic premise was that the data taken by Stanton and Pannell[7], which he analyzed as having a spread of no more than ±2% from the mean value at any $\mathbf{N_R}$, was not only applicable to the drawn brass tubes of the Stanton and Pannell experiments but to the 6–inch and larger pipes being used by the oil industry at the time. His curve,[1] herein shown in Figure 12-3, was widely published, and widely used. Heltzel's work, completed before the papers of Colebrook and Moody et al., for all practical purposes

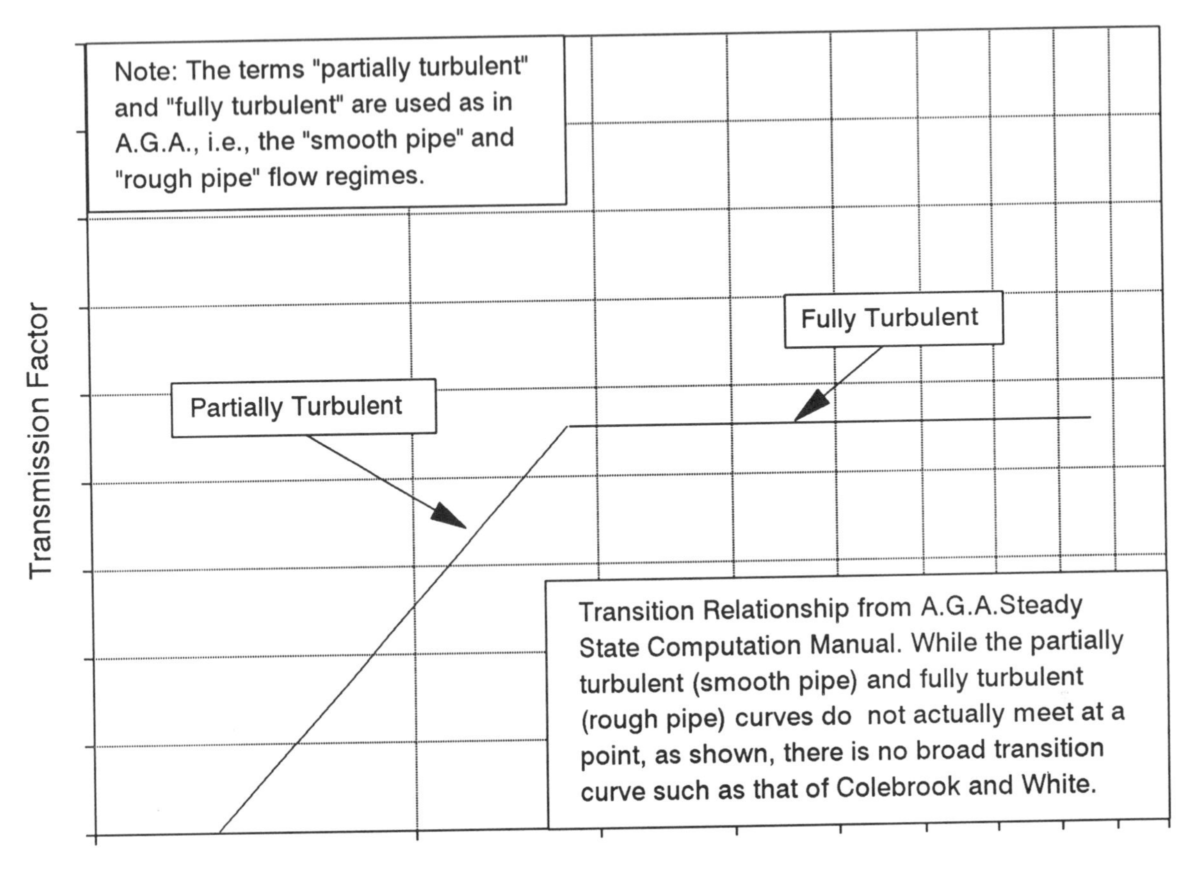

Figure 13-8.

proved the reliability of the Stanton and Pannell smooth pipe data for oil industry pipelines of the size being built and operated when Heltzel was doing his investigations.

Another somewhat later body of work that received a great deal of attention, and rightfully so, was that of Miller, which was summarized and published in 1947.[8]

Miller believed in the smooth pipe law for oil pipelines, and he

- *Did not believe* in the Colebrook–White transition function
- *Did not believe* oil industry pipes had a roughness of 0.0018–inch

Miller not only dwelled at considerable length on the very poor data backing the Colebrook–White function, but also went to great lengths to disparage—to in fact *deride*—the standard of .0018 inch for pipe roughness. He quotes Freeman, who made the tests referenced by Colebrook when arriving at the 0.0018–inch number, who described his (i.e., Freeman's) pipes (in 1892) as follows:

> "In the whole lot of 6–inch pipe almost every long piece, in addition to the scabs and scales, contains a few bunchy spots. Some of these appear to be patches of fused scale or slag, others are small bunches of iron. Hardly a bunch or protuberance in the whole 200 feet projects over 1/8–inch from the general surface, but quite a number of rounded bunches with a base of 1–inch to 2–inch diameter, project 1/8–inch perhaps.
>
> "There are a dozen of these in the whole 200 feet of pipe, located very irregularly, and generally the sides are of such easy slopes that they would not cause any contraction beyond their height.
>
> "There are a considerable number of slag spots and iron bunches with base diameter about 1/2–inch to 1–inch and which project 1/16–inch. These two types of bunches are irregularly located and scarcely any section 5–inch long is free from one or more such.
>
> "The general surface (approximately the whole area has been *ironed out* and leveled off by the contact of the plug and the welding rolls) is by no means smooth.
>
> "One–half the surface is covered by ridges almost all of which run along the pipe. The other half is covered with pits and patches of scale, giving a rough surface with hills and valleys 1/50–inch in depth, and spaced from 1/4–inch to 1–inch apart. The scabs of scale generally have abrupt edges."

Miller follows this quotation from Freeman with a personal comment:

> "...Surely no one would seriously contend that the foregoing description could apply to present–day wrought–iron or steel pipe such as would be used in oil or gas pipe lines."

And, Miller's comment was made in 1947—nearly 50 years ago, and *his* present–day pipes were the pipes of the U.S. pipemaking industry immediately after World War II.

Miller, as did Heltzel, believed oil flow in commercial pipes was smooth pipe flow, and in 1937 he developed a function,[9] still in use by come companies and called the *Miller Formula*, explicit in his (Miller's) C_f function:

$$C_f = 1.80 \log \mathbf{N_R} - 1.53 \tag{13-8}$$

This is an excellent approximation of the Nikuradse function and, when written in $\log(C/D)$ form, is almost exactly the same as the function published by Colebrook in 1939—some two years later. (In 1950 Konakov,[10] reported by Round,[11] produced yet another variation of this formula, as shown below.)

$$\begin{aligned}
&\text{Miller}_{1937} \quad \frac{1}{\sqrt{f_D}} = 1.8 \log\left(\frac{\mathbf{N_R}}{7.079}\right) \\
&\text{Colebrook}_{1939} \quad \frac{1}{\sqrt{f_D}} = 1.8 \log\left(\frac{\mathbf{N_R}}{7.000}\right) \\
&\text{Konakov}_{1950} \quad \frac{1}{\sqrt{f_D}} = 1.8 \log\left(\frac{\mathbf{N_R}}{6.813}\right)
\end{aligned} \tag{13-9}$$

After the publication of the Miller papers[7] cited, there is very little in the pipeline industry technical literature about the smooth pipe law except as it is the lower limit of f_D in the Colebrook–White function; all references to smooth pipe have reference to the Heltzel curve of the

Stanton–Pannell data, and that is the major reference in the oil industry studies reported in the following.

1951–1979 Test and Analyses

One of the first accurate sets of field tests on big–inch lines reported were those run by Creole Petroleum.

- The 1951 Creole Petroleum tests on 24– and 26–inch lines were reported by Redmond.[12] These data, except for *two rogue* points reported but not plotted by Redmond, tracked the Heltzel curve within ±3%.

Redmond concluded:

> "If the profile of the wall remained the same ...we would expect lower friction factors for larger pipe sizes. The results do not support this idea. Since the steel is fairly smooth, it may be that the change in relative roughness is not great enough to affect the friction factor. It may also be possible the experimental error involved is large enough to obscure such evidence.
>
> *"At any rate, if any such correlation between relative roughness and pipe diameter exists, it is not of practical importance if it cannot be detected in normal operation."*

Another set of tests were those run by Mid Valley Pipe Line Company which had been puzzled as to why its 20/22–inch system seemingly had more capacity than it should have had and set up to run some very exacting tests to find how the lines really were performing.

- The 1953 Mid–Valley[13] tests were run on a highly pumped, high pressure 20–inch crude line. These very carefully instrumented tests tracked the Heltzel curve within ±2%, and the test sections were obviously operating as smooth pipelines.

There was in the industry, however, a feeling that there had to be something better than the Heltzel representation of the Stanton and Pannell smooth pipe data, and so in 1953 the *Subcommittee on Pipe Line Hydraulics* of the *American Petroleum Institute* set up a series of flow tests that eventually involved 15 oil companies operating in 12 U.S. states, Canada and Saudi Arabia, that tested 36 separate sections of pipeline from 12– to 31–inches in nominal diameter. These data were correlated, analyses prepared, and presented before the API Division of Transportation in 1954 by Kennedy.[14] There were two *rogue points* that didn't fit with the other data but, in general, deviations from the Heltzel curve were no more than ±5%, about equally divided between tests that gave a friction factor higher or lower than the Heltzel curve. Considering that these tests were run before any kind of crude oil meters were available—flow rates had to be calculated from tank gages—it is remarkable the spread is not larger. As a result of these data Kennedy, who had a wide background in fluid flow analysis, found

> "There is not a direct relationship between the friction factor and the Reynolds number for the turbulent–flow region.
>
> "...design of crude–oil pipelines, in particular 6–inch and larger, based on data of Stanton and Pannell should be sound and reliable.
>
> "In addition, it is believed that products lines, 6–inch and larger, operating at much higher Reynolds numbers, further will substantiate Stanton and Pannell's data, provided their internal smoothness has been maintained. Consequently, this same basis of design (Stanton and Pannell's data) should be reliable for new products lines."

It is interesting to note that neither the Nikuradse smooth pipe function, the Colebrook–White function, nor the Moody diagram is mentioned anywhere in Kennedy's text: *the U.S. oil pipeline industry didn't use them.* Heltzel had considered boundary roughness and discarded it as not applicable to oil pipelines, and the industry agreed.

There were other tests run, but not many of them were made public. The following tests are from private files:

- The 1964 SOPEG crude line tests reported by Dreyfuss.[15] These tests, run on SOPEG's 24–inch line at $N_R > 500\,000$, compared test results with Colebrook–White and estimated a wall roughness between 0.0007 inch and zero inches (smooth pipe).

- The 1979 Trans–Alpine tests reported by Uhde.[16] These tests, run on two sections of the 40–inch crude line with a wide variety of crudes, generally tracked the Nikuradse smooth pipe law in the Moody diagram out to about N_R = 200 000 and then trend upward to reach an absolute roughness of some 0.0008 inches for N_R > 800 000.

Note that all of the above–described tests, except the SOPEG and TAL tests, were compared with the Heltzel curve. In general, it was found that the data tracked the Heltzel curve with fair accuracy.

It is interesting to analyze these data with today's tools and data.

Figure 13-9 is a plot prepared by Landers et al.[17] to illustrate the relation between the Stanton & Pannell data and the Monograph No. 9 smooth pipe function. It is very apparent that the same kind of law lies behind these two streams of data, and it may be assumed for convenience further herein that the Stanton & Pannell data do follow the Monograph No. 9 smooth pipe law.

Figure 13-10 is a plot of all the data from all the above described oil industry field tests. There is a shortage of data in the range of N_R > 600 000 (the SOPEG tests fill in the range 500 000 > N_R > 600 000) but there are well over 100 points in the range N_R <300 000.

Summary of Oil Industry Findings

While the scatter in the field tests of the oil industry is greater than in the laboratory tests of the gas industry investigations this is to be expected because it is nearly impossible to bring a long oil line to quiescent, steady flow for any extended period and test measurements are made, therefore, on a moving target. However, the scatter is small enough, and the trend of the data so certain, that the two requisite conclusions important for developments further herein can be reached without ambiguity.

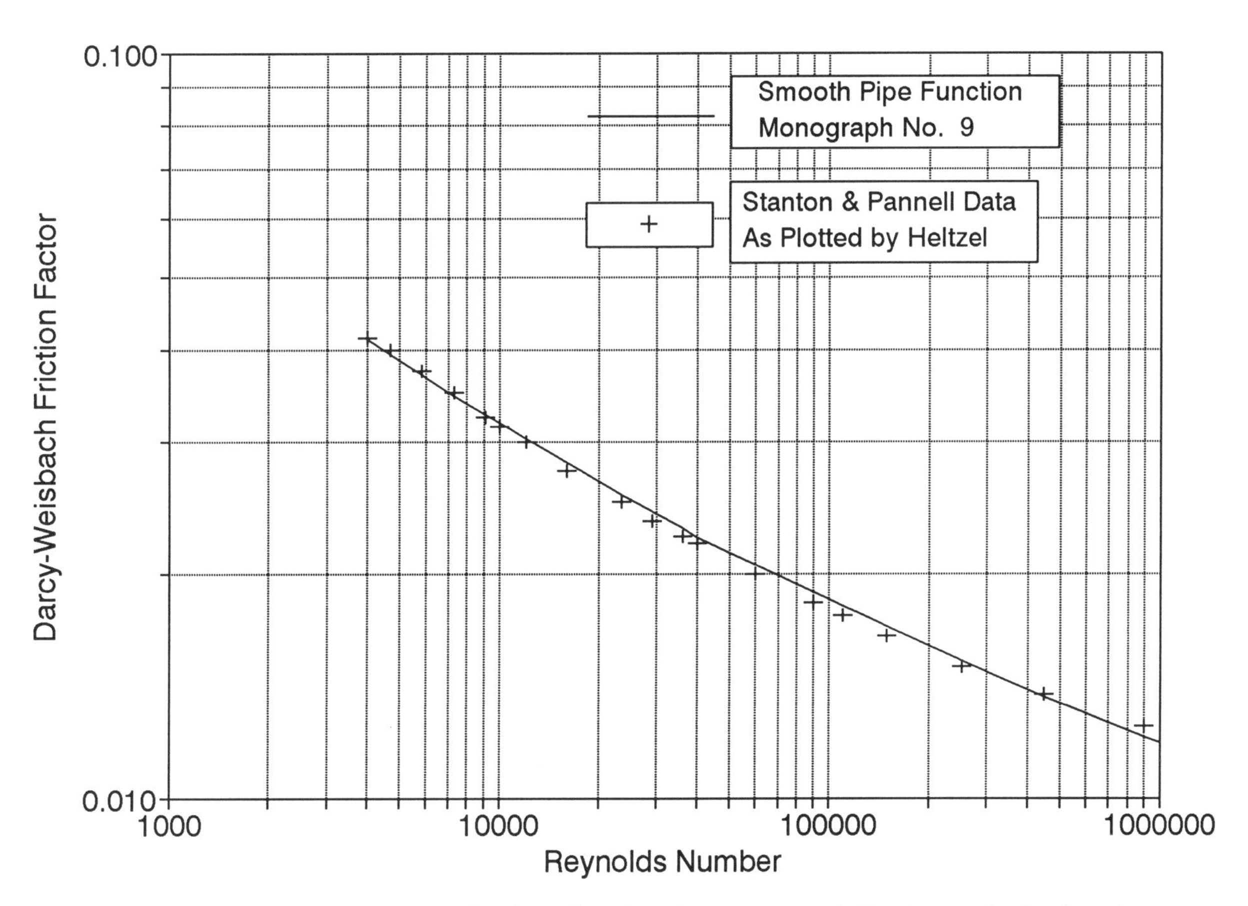

Figure 13-9. Stanton & Pannell points plotted against Monograph No. 9 smooth pipe function.

- Oil pipelines flow in *smooth pipe flow*
- The smooth pipe flow follows Monograph No. 9

Derivation of the Friction Factor

A convincing demonstration of the validity of the experimental data, starting with the low values of $\mathbf{N_R}$ of some of the Stanton and Pannell data up to the nearly 10 000 000 found in the Monograph No. 9 superfinished pipeline tests, is provided in Figure 13-11. This mass of data, complied from the $\mathbf{N_R}$–vs–f_D or $\mathbf{N_R}$–vs–T_F points of the original references, is probably the largest accurate plotting of friction factor and/or transmission factor data yet made available for the purpose of analyzing oil line flowing friction.

The data provide these bases for deriving the friction factor for oil flowing in modern linepipe:

- Flow is according to the *smooth pipe law*
- $T_F = 4\log(\mathbf{N_R} / T_F) - 0.6$.

It is obvious that since $f_D = 4/(T_F)^2$ the value of f_D can be derived by computing T_F, squaring it, and dividing this value into 4.

This method is technically correct, but it is intrinsic in T_F and has all the problems of the Colebrook–White function except its inaccuracy; it has to be programmed for iterative solution or drawn up in a curve of some form to be useful.

The T_{OIL} and f_{OIL} Functions

A numerical analysis of the Monograph No. 9 smooth pipe law showed that a function of the Miller–Colebrook–Kanakov form in Equation 13-9 used to approximate the Nikuradse smooth pipe law could approximate the Monograph No. 9 smooth pipe law to an unusual degree of accuracy in the 30 000 > $\mathbf{N_R}$ > 3 000 000 range chosen and, in the end, the function was far simpler than had ever been thought possible. This function is called $\mathbf{T_{OIL}}$ and is written

$$\mathbf{T_{OIL}} = 3.6\log(\mathbf{N_R} / 8) \qquad (13\text{-}10)$$

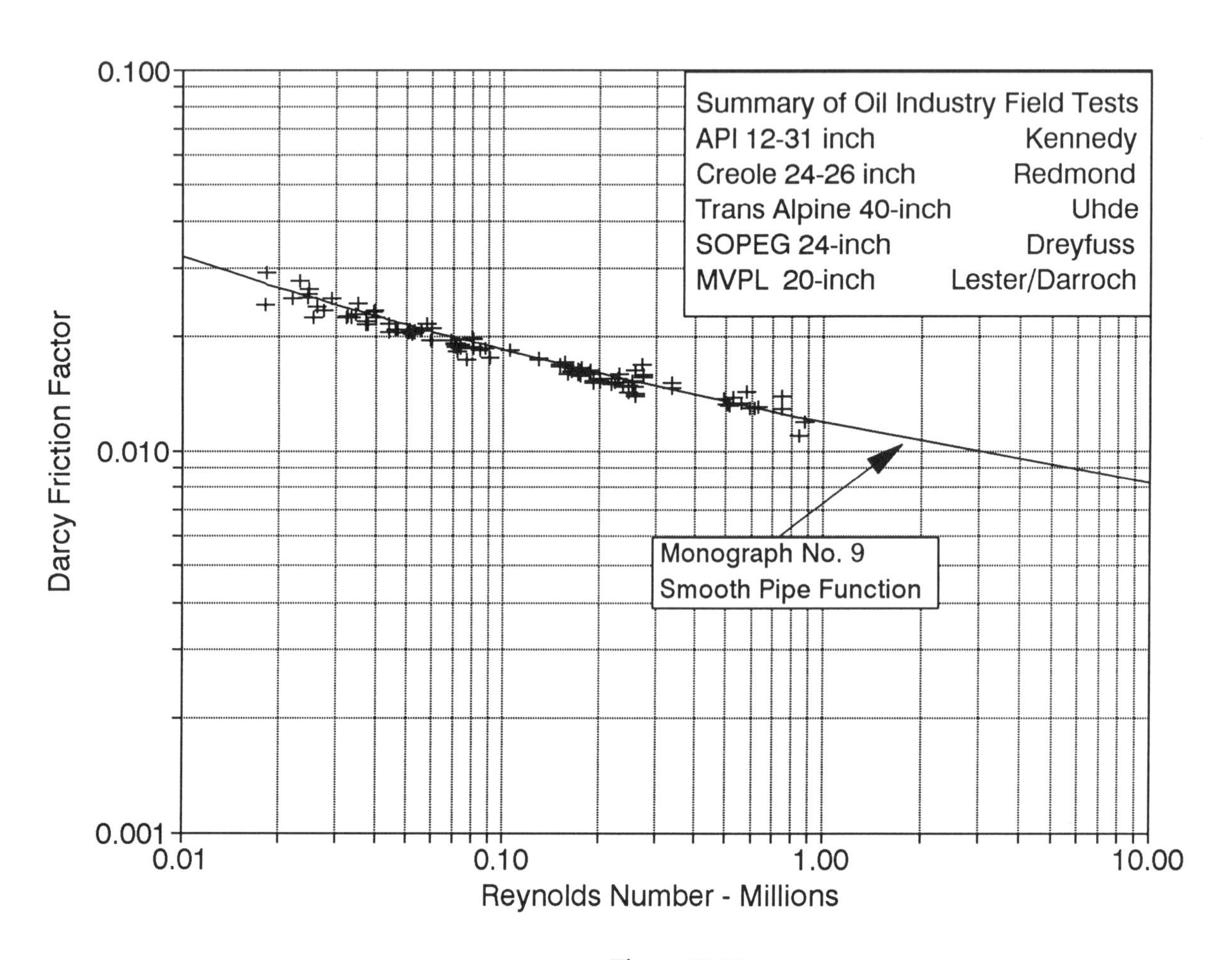

Figure 13-10.

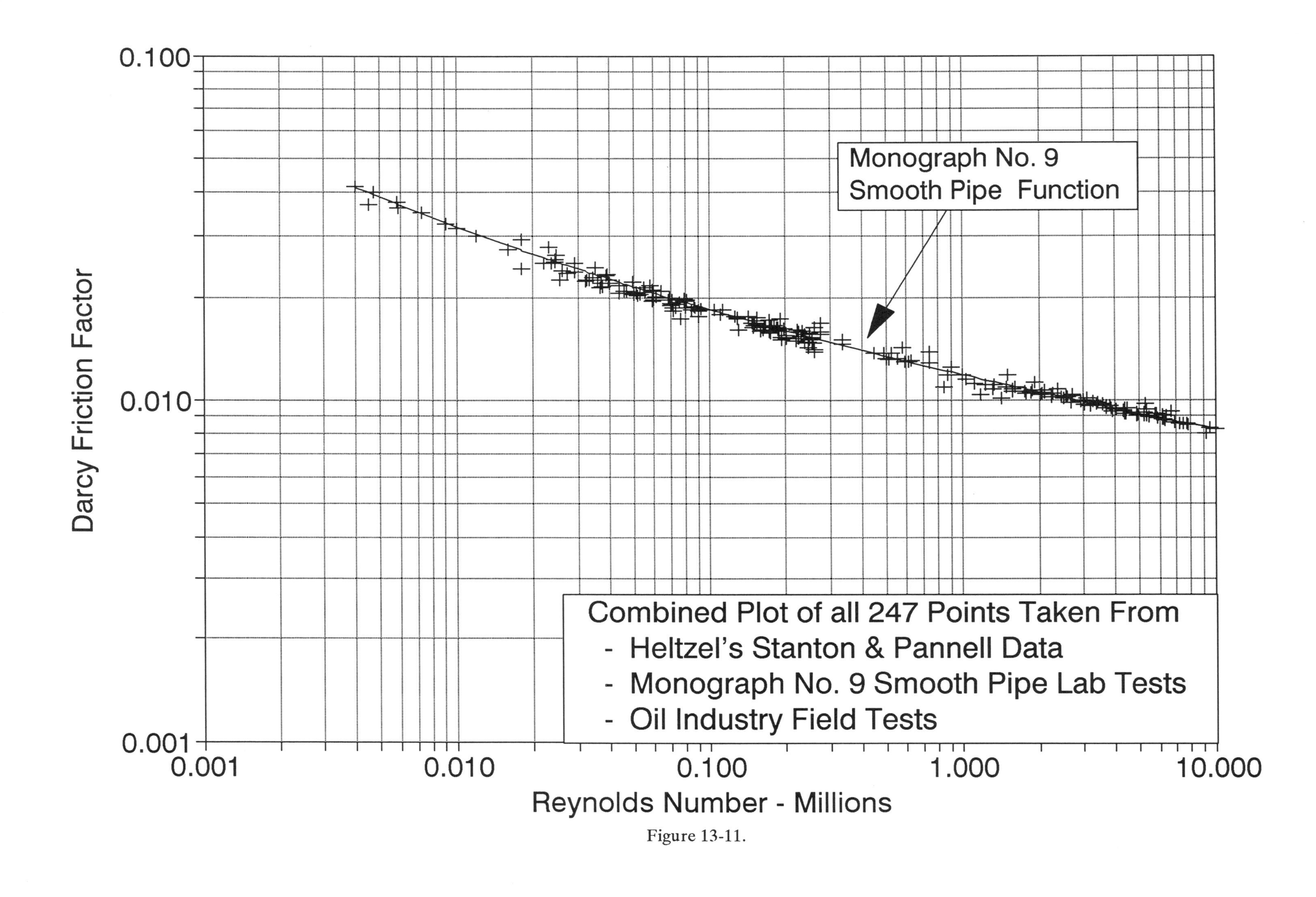
Monograph No. 9
Smooth Pipe Function
Combined Plot of all 247 Points Taken From
- Heltzel's Stanton & Pannell Data
- Monograph No. 9 Smooth Pipe Lab Tests
- Oil Industry Field Tests
Darcy Friction Factor
0.100
0.010
0.001
0.001
0.010
0.100
1.000
10.000
Reynolds Number - Millions

Figure 13-11.

$\mathbf{T_{OIL}}$ is in terms of T_F, the Fanning friction factor. To compute f_D to use in the Darcy–Weisbach formula, simply compute $\mathbf{f_{OIL}}$ as

$$\mathbf{f_{OIL}} = 4/(\mathbf{T_{OIL}})^2 \tag{13-11}$$

The accuracy of these two functions is amazing; and they certainly meet the criterion of being very accurate tools over the range 30 000 > $\mathbf{N_R}$ 3 000 000. They are so close to the Monograph No. 9 smooth pipe function over this range that no variation in the numeric constants to improve the results could justify making the constants some kind of a more complex, harder–to–remember decimal fraction than the cited 3.6 and 8. In fact, one of the beauties of $\mathbf{T_{OIL}}$ is that it is simple and easy to memorize.

Figure 13-12 shows that $\mathbf{T_{OIL}}$ is within 0.25% of the Monograph No. 9 smooth pipe law from $\mathbf{N_R}$ = 30 000 to $\mathbf{N_R}$ = 3 000 000; this is certainly accurate enough for all commercial calculations where the input data for $\mathbf{N_R}$ are not known to anything near this kind of precision.

Figure 13-13 shows the same kind of calculation for the $\mathbf{f_{OIL}}$ function. Note here the delta is 0.50% instead of the 0.25% for $\mathbf{T_{OIL}}$, but the accuracy is strictly relative, since $\mathbf{f_{OIL}} = \varphi(\mathbf{T_{OIL}})^{-2}$.

The Friction Factor

Thus ends the search for a friction factor for trunk oil pipelines. It is not perfect—nor did it need to be—but it is technically sound and numerically accurate, and nothing more is required within its range of application. By selecting one of the following versions, and using the connecting equations of Equations 13-3, 13-4, and 13-5, a friction or transmission factor can be written for any head loss equation.

$$T_{Fanning} = \mathbf{T_{Oil}} = 3.6\log\left(\frac{\mathbf{N_R}}{8}\right) \tag{13-12}$$

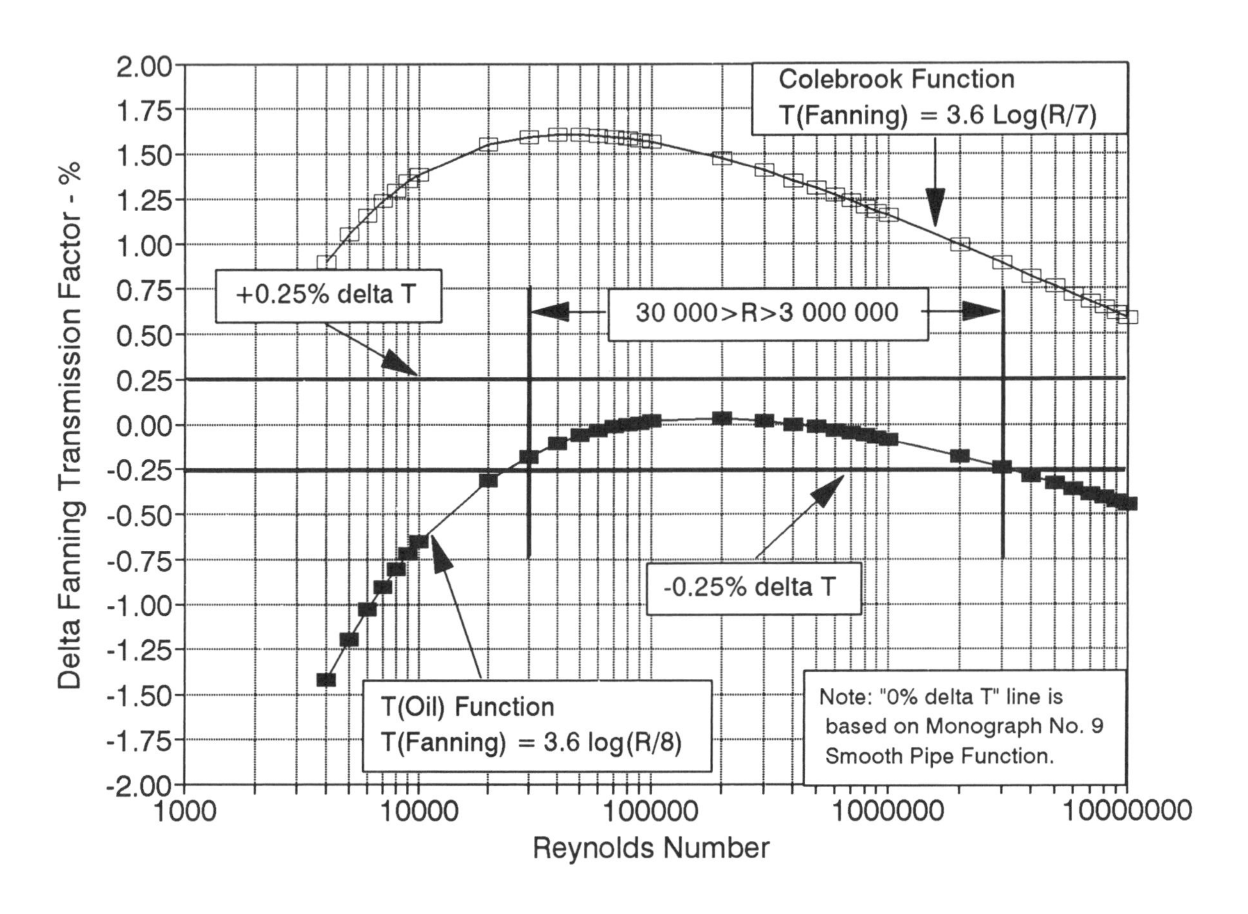

Figure 13-12.

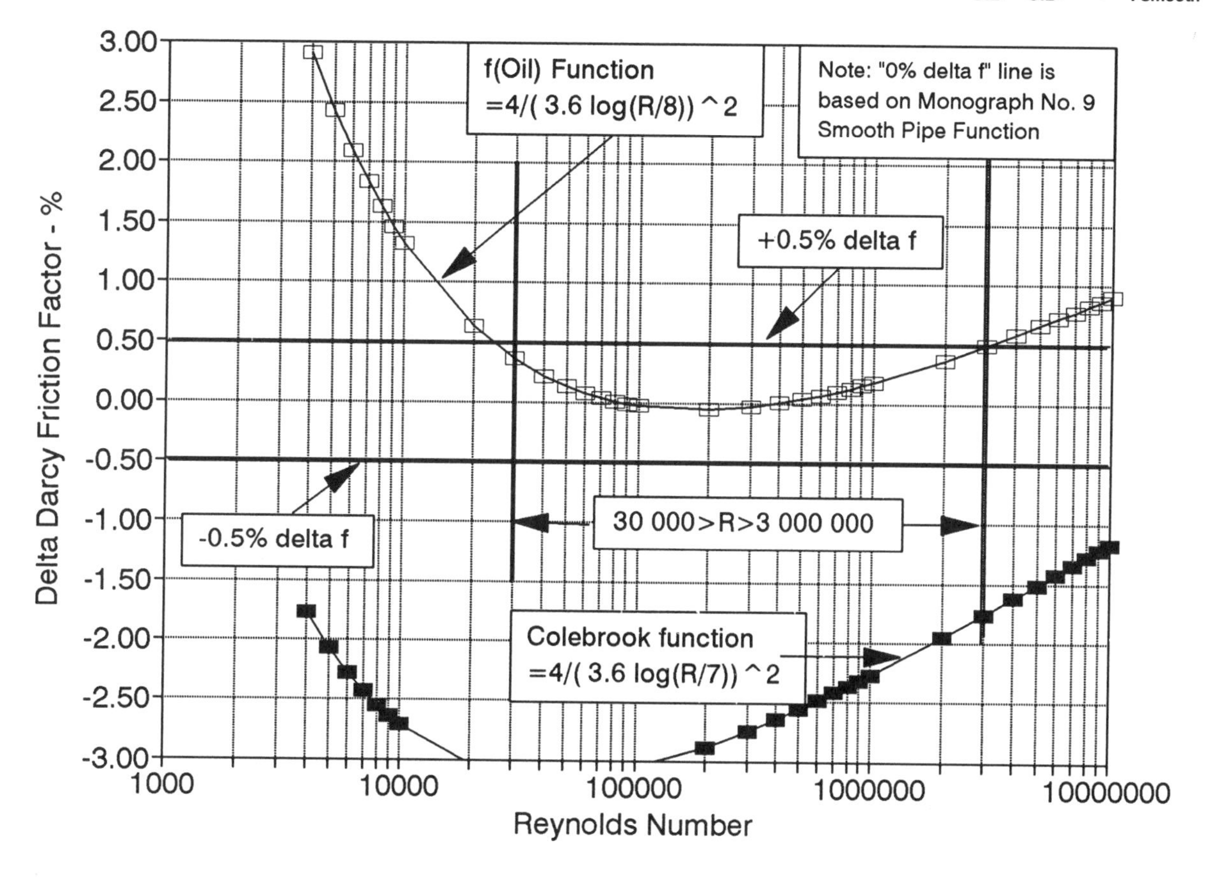

Figure 13-13.

$$f_{Darcy} = \mathbf{f}_{OIL} = \frac{4}{(\mathbf{T}_{OIL})^2} \tag{13-13}$$

A New Smooth Pipe Function: $T_{FSmooth}$

In doing numerical analyses leading to $\mathbf{T}_{OIL}$ it became evident that $\mathbf{T}_{OIL}$, itself, is an excellent approximation of the Monograph No. 9 smooth pipe law over a very wide range of $\mathbf{N}_R$.

And other analyses, not used and so not presented here, had shown that the Newton–Raphson function, if given an accurate initial seed, can close very rapidly on the value of an intrinsically described variable such as T_F and will, in fact, quite often obtain a sufficiently accurate answer *without iteration*; i.e., the initial calculation based on the seed is sufficiently close to the desired solution.

The Newton–Raphson function is described in any text on numerical analysis, but two references which are exceptionally useful to pipeliners are Jeppson[18] and Streeter and Wylie.[19] Both contain examples of computer programs using the function, Jeppson in FORTRAN and Streeter in BASICA. Another good reference, with some excellent, very simple, FORTRAN illustrations is Stanton and Lawson,[20] one of the *AIChE Today Series*.

The fundamental methods of approximation taught in the earlier courses of mathematics are either that of linear interpolation or, since computers are nearly everywhere available, the bi–section method: split the difference, try again, and split the remaining difference. This sometimes yields a *linear closure*, in which the answer is improved by about one digit with each trial, and sometimes closes more slowly, but with the computer making thousands of trials per second the rate of closure for solving a single problem doesn't matter too much.

There are more complex methods that yield *quadrature closure*, in which each trial produces approximately *twice as many correct digits* as the previous one; if the first trial was correct to three digits, the next trial will produce about six accurate digits, etc. One of the more popular of these is the Newton–Raphson method based on Newton's *methods of differences*. In simplest form,

$$x^{(m+1)} = x^{(m)} - \frac{\varphi\left(x^{(m)}\right)}{d\left[\varphi\left(x^{(m)}\right)\right]/dx} \qquad (13\text{-}14)$$

Note that the values m are not exponents, but denote the number of iterations. If the function has been iterated five times then $m = 5$, x^5 is in hand, and the value sought is x^6. For the method to be successful, the equation containing the unknown must be expressed as a function which equals zero when the correct solution is substituted for the unknown, i.e., $\varphi(x) = 0$.

The method works only if the function $\varphi(x)$ can be differentiated symbolically; if otherwise, another method, such as the *regula falsi* method, that relies on a numerical derivative of the function rather than a symbolic one, is required.

In the present instance, any of the friction factor or transmission factor functions that may be of interest can be differentiated at least once and, for purposes here, this is sufficient. The functions of Table 13-4 were determined using the symbolic derivative operator of *Mathcad 4.0*,[21] a mathematical analysis program published by MathSoft. Note in addition to the derivative of the Monograph No. 9 smooth pipe function, which is needed in the following, there are several other T and f functions evaluated; these may come in handy someday.

Also note the derivatives have not been simplified numerically. It is considerably easier to remember $\ln(10)$ than 2.302585093, for instance, and the symbolic form is also much easier to input to a calculator or computer.

The following illustrates the non-iterative $\mathbf{T_{FSmooth}}$ function applying the Newton-Raphson algorithm to solving the Monograph No. 9 smooth pipe function for $\mathbf{N_R}$ = 100 000.

Compute $\mathbf{T_{OIL}}$ = 3.6 log (100 000/8) = 14.7488

Compute $\varphi(SPL) = 0$ using $\mathbf{T_{OIL}}$ as T
= [4 log (100 000/14.7488)–0.6]–14.7488
= –0.0239

Compute $\varphi'(SPL) = 0$ using $\mathbf{T_{OIL}}$ as T
= [–4/(14.7488·ln(10))]–1
= –1.1118

Compute T_F = 14.7488–(–0.0239/–1.1118) = 14.7274.

The accepted value of T_F from AGA[4] is 14.727.

For extreme accuracy where, for instance, certain new calculations are to be compared with numerically precise calculations taken from another source, $\mathbf{T_{FSmooth}}$ can be programmed in spreadsheet form to any desired degree of accuracy. Table 13-5 shows the output of such a small block of calculations, and Table 13-6 is a *cell–formulas* printout of the block. Note that $\mathbf{T_{OIL}}$ is very close to the four–place value of T_F, and that $\mathbf{T_{FSmooth}}$ is exact for all but the $\mathbf{N_R}$ = 4000 value.

The formula for $\mathbf{T_{FSmooth}}$ in terms of $\mathbf{T_{OIL}}$ and the Monograph No. 9 smooth pipe function is:

$$\mathbf{T_{FSmooth}} = \mathbf{T_{OIL}} - \left[\frac{4\log\left(\frac{\mathbf{N_R}}{\mathbf{T_{OIL}}}\right) - 0.6 - \mathbf{T_{OIL}}}{\left(\frac{-4}{\mathbf{T_{OIL}}\cdot\ln(10)}\right) - 1}\right] \qquad (13\text{-}15)$$

Working Equations

Calculators

Table 13-7 is a listing of the working equations for friction factor. The algorithm for the Monograph No. 9 smooth pipe law equation can be solved by any of the single–variable solvers, or can be simply programmed to be solved iteratively using a bi–section kind of closure.

Computers

There are two **CCT**s supplied with this chapter. **CCT** No. 13-1 contains the algorithms for the Monograph No. 9 smooth pipe law Fanning transmission factor, $\mathbf{T_{Oil}}$, and $\mathbf{T_{FSmooth}}$, and plots the variation between the first with the latter two equations. **CCT** No. 13-2 provides the algorithm for $\mathbf{f_{Oil}}$, and plots the $\mathbf{f_{Oil}}$ Darcy friction factor against the Darcy factor obtained from the Monograph No. 9 smooth pipe law.

Table 13-4
T_F and f_D Functions and Their Derivatives

Name of Function	Function	Derivative With Respect to T or f, as is applicable.
Monograph No. 9 Smooth Pipe Law	$4\cdot\log\left(\frac{R}{T}\right) - 0.6 - T$	$\frac{-4}{(T\cdot\ln(10))} - 1$
Nikuradse Smooth Pipe Law	$4\cdot\log\left(\frac{R}{T}\right) - 0.4 - T$	$\frac{-4}{(T\cdot\ln(10))} - 1$
Monograph No. 9 Smooth Pipe Law	$4\cdot\log\left(R\cdot\sqrt{f}\right) - 0.6 - \frac{1}{\sqrt{f}}$	$\frac{2}{(f\cdot\ln(10))} + \frac{1}{\left[2\cdot f^{\left(\frac{3}{2}\right)}\right]}$
Nikuradse Smooth Pipe Law	$4\cdot\log\left(R\cdot\sqrt{f}\right) - 0.4 - \frac{1}{\left(\sqrt{f}\right)}$	$\frac{2}{(f\cdot\ln(10))} + \frac{1}{\left[2\cdot f^{\left(\frac{3}{2}\right)}\right]}$
T_{OIL}	$T - 3.6\cdot\log\left(\frac{R}{8}\right)$	1
$T_{Colebrook}$	$T - 3.6\cdot\log\left(\frac{R}{7}\right)$	1
Colebrook-White Transition Function Format No. 1	$\frac{1}{\sqrt{f}} - 1.14 + 2\cdot\log\left[\left(\frac{e}{D}\right) + \left(\frac{9.35}{R\cdot\sqrt{f}}\right)\right]$	$\frac{-1}{\left[2\cdot f^{\left(\frac{3}{2}\right)}\right]} - \frac{9.35}{\left[R\cdot\left[f^{\left(\frac{3}{2}\right)}\cdot\left[\left[\frac{e}{D} + \frac{9.35}{\left(R\cdot\sqrt{f}\right)}\right]\cdot\ln(10)\right]\right]\right]}$
Colebrook-White Transition Function Format No. 2	$\frac{1}{\sqrt{f}} + 2\cdot\log\left[\left(\frac{\frac{10\cdot e}{D}}{37}\right) + \frac{2.51}{R\cdot\sqrt{f}}\right]$	$\frac{-1}{\left[2\cdot f^{\left(\frac{3}{2}\right)}\right]} - \frac{2.51}{\left[R\cdot\left[f^{\left(\frac{3}{2}\right)}\cdot\left[\left[\frac{10}{37}\cdot\frac{e}{D} + \frac{2.51}{\left(R\cdot\sqrt{f}\right)}\right]\cdot\ln(10)\right]\right]\right]}$

Table 13-5
Spreadsheet Demonstration of Precision of T(OIL) and T(FSmooth) Functions

Reynolds Number 1	T(OIL) Seed 2	T(FSmooth Function 3	Delta % T(OIL) minus T(FSmooth 4	Iteration 1 T(Fanning) 5	Delta % T(FSmooth) minus Iteration 1 6	Iteration 2 T(Fanning) 7	Delta % Iteration 1 minus Iteration 2 9
4000	9.7163	9.8367	-1.2242	9.8368	-0.0011	9.8368	-0.0000
10000	11.1489	11.2027	-0.4804	11.2027	-0.0002	11.2027	-0.0000
100000	14.7489	14.7275	0.1453	14.7275	-0.0000	14.7275	-0.0000
1000000	18.3489	18.3458	0.0165	18.3458	-0.0000	18.3458	0.0000
10000000	21.9489	22.0281	-0.3596	22.0281	-0.0000	22.0281	-0.0000

Table 13-6
Cell-Formula Listing
Spreadsheet Form for Calculating
T_{OIL} and $T_{FSmooth}$

```
A20: [W10] 'Column A contains the Reynolds Number.
B20: [W10] 'Column B contains the T(OIL) function which acts as the seed.
C20: [W10] 'Column C contains the T(FSmooth) function, seeded by T(OIL).
D20: [W10] 'Column D shows delta % between T(OIL) and T(FSmooth).
E20: [W12] 'Column E contains the 1st Newton-Raphson iteration of T(FSmooth).
F20: [W12] 'Column F shows delta % betwen T(FSmooth) and the 1st N-R iteration.
G20: [W12] 'Column G contains the 2nd Newton-Raphson iteration of T(FSmooth).
H20: [W12] 'Column H shows the delta % between the 1st and 2nd N-R iterations.
A21: [W10] 4000
B21: (F4) [W10] 3.6*@LOG($A21/8)
C21: (F4) [W10] +B21-(4*@LOG($A21/B21)-0.6-B21)/(-4/(B21*@LN(10))-1)
D21: (F4) [W10] 100*(B21-C21)/C21
E21: (F4) [W12] +C21-(4*@LOG($A21/C21)-0.6-C21)/(-4/(C21*@LN(10))-1)
F21: (F4) [W12] 100*(C21-E21)/E21
G21: (F4) [W12] +E21-(4*@LOG($A21/E21)-0.6-E21)/(-4/(E21*@LN(10))-1)
H21: (F4) [W12] 100*(E21-G21)/G21
A22: [W10] 10000
B22: (F4) [W10] 3.6*@LOG($A22/8)
C22: (F4) [W10] +B22-(4*@LOG($A22/B22)-0.6-B22)/(-4/(B22*@LN(10))-1)
D22: (F4) [W10] 100*(B22-C22)/C22
E22: (F4) [W12] +C22-(4*@LOG($A22/C22)-0.6-C22)/(-4/(C22*@LN(10))-1)
F22: (F4) [W12] 100*(C22-E22)/E22
G22: (F4) [W12] +E22-(4*@LOG($A22/E22)-0.6-E22)/(-4/(E22*@LN(10))-1)
H22: (F4) [W12] 100*(E22-G22)/G22
A23: [W10] 100000
B23: (F4) [W10] 3.6*@LOG($A23/8)
C23: (F4) [W10] +B23-(4*@LOG($A23/B23)-0.6-B23)/(-4/(B23*@LN(10))-1)
D23: (F4) [W10] 100*(B23-C23)/C23
E23: (F4) [W12] +C23-(4*@LOG($A23/C23)-0.6-C23)/(-4/(C23*@LN(10))-1)
F23: (F4) [W12] 100*(C23-E23)/E23
G23: (F4) [W12] +E23-(4*@LOG($A23/E23)-0.6-E23)/(-4/(E23*@LN(10))-1)
H23: (F4) [W12] 100*(E23-G23)/G23
A24: [W10] 1000000
B24: (F4) [W10] 3.6*@LOG($A24/8)
C24: (F4) [W10] +B24-(4*@LOG($A24/B24)-0.6-B24)/(-4/(B24*@LN(10))-1)
D24: (F4) [W10] 100*(B24-C24)/C24
E24: (F4) [W12] +C24-(4*@LOG($A24/C24)-0.6-C24)/(-4/(C24*@LN(10))-1)
F24: (F4) [W12] 100*(C24-E24)/E24
G24: (F4) [W12] +E24-(4*@LOG($A24/E24)-0.6-E24)/(-4/(E24*@LN(10))-1)
H24: (F4) [W12] 100*(E24-G24)/G24
A25: [W10] 10000000
B25: (F4) [W10] 3.6*@LOG($A25/8)
C25: (F4) [W10] +B25-(4*@LOG($A25/B25)-0.6-B25)/(-4/(B25*@LN(10))-1)
D25: (F4) [W10] 100*(B25-C25)/C25
E25: (F4) [W12] +C25-(4*@LOG($A25/C25)-0.6-C25)/(-4/(C25*@LN(10))-1)
F25: (F4) [W12] 100*(C25-E25)/E25
G25: (F4) [W12] +E25-(4*@LOG($A25/E25)-0.6-E25)/(-4/(E25*@LN(10))-1)
H25: (F4) [W12] 100*(E25-G25)/G25
```

Table 13-7

Working Equations for Friction Factor

The Pipe Friction Equations

$$h_L = 2 \cdot f_{Fanning}\left(\frac{L}{D}\right)\left(\frac{V^2}{g}\right) = f_{Darcy}\left(\frac{L}{D}\right)\left(\frac{V^2}{2g}\right)$$

The Connecting Relationships

$$f_{Darcy} = 4 f_{Fanning} \qquad f_{Darcy} = \frac{4}{T^2_{Fanning}}$$

$$T_{Darcy} = \frac{T_{Fanning}}{2}$$

The Monograph No. 9 Smooth Pipe Law

$$T_F = 4\log\left(\frac{\mathbf{N_R}}{T_F}\right) - 0.6$$

The $\mathbf{T_{OIL}}$ and $\mathbf{f_{OIL}}$ Functions

$$\mathbf{T_{OIL}} = 3.6\log(\mathbf{N_R}/8)$$

$$\mathbf{f_{OIL}} = 4/(\mathbf{T_{OIL}})^2$$

Friction Factors for Oil Trunk Pipelines
$30\,000 < \mathbf{N_R} < 3\,000\,000$
$D \geq 8$ inches

$$T_F = \mathbf{T_{OIL}} = 3.6\log\left(\frac{\mathbf{N_R}}{8}\right) \qquad \delta = \pm 0.25\%$$

$$f_D = \mathbf{f_{OIL}} = \frac{4}{(\mathbf{T_{OIL}})^2} \qquad \delta = \pm 0.50\%$$

The $\mathbf{T_{FSmooth}}$ Function
$4000 < \mathbf{N_R} < 100\,000\,000$

$\delta = \pm 0.01\%$ (max)

$$\mathbf{T_{FSmooth}} = \mathbf{T_{OIL}} - \left[\frac{4\log\left(\frac{\mathbf{N_R}}{\mathbf{T_{OIL}}}\right) - 0.6 - \mathbf{T_{OIL}}}{\left(\frac{-4}{\mathbf{T_{OIL}} \cdot \ln(10)}\right) - 1}\right]$$

References

1 Smith, R. V., Miller, J. S., and Ferguson, J. W., "Flow of Natural Gas Through Experimental Pipe Lines and Transmission Lines," Monograph No. 9, U.S. Department of the Interior, Bureau of Mines, Division of Petroleum. New York: *American Gas Association*, 1956.

2 Johnson, T. W., and Berwald, W. B., "Flow of Natural Gas Through High–Pressure Transmission Lines," Monograph No. 6, U.S. Department of the Interior, Bureau of Mines. New York: *American Gas Association*, 1935.

3 Uhl, A. E., "*Steady Flow in Gas Pipelines*," Institute of Gas Technology Technical Report No. 10. New York: American Gas Association, 1960.

4 *Steady State Computation Manual for Natural Gas Transmission Lines*. New York: American Gas Association, 1964.

5 Heltzel, W. G., "Fluid Flow and Friction in Pipelines," *Oil and Gas Journal*, 05 June 1930.

6 Heltzel, W. G., "Development of Equivalent Length Formulae for Multiple Parallel Oil Pipe Line Systems," *Oil and Gas Journal*, 10 May 1934.

7 Stanton, W. E., and Pannell, T. R., "Similarity of Motion in Relation to Surface Friction of Fluids," *National Physical Laboratory, London, collected.*

8 Miller, Benjamin, "Pipe–Line Flow Formulas," (in three parts), *Oil and Gas Journal*, 20 September 1947, 27 September 1947, and 04 October 1947. These papers were reprinted in several collections of pipeline engineering papers published over the years by the Petroleum Publishing Company, among which was *Fluid Flow Formulas for the Gas Pipeline Engineer* in 1956.

9 Miller, Benjamin, *Chemical and Metallurgical Engineering*, Vol. 44, No. 10, October 1937.

10 Konakov, V. K., *Dokl. Akad. Nauk SSSR*, 24, 14, 1950.

11 Round, G. E., "Accurate, Fast Friction Factor Computation, Turbulent Flow in Pipes," *Pipeline & Gas Journal*, December 1985.

12 Redmond, W. P., "Friction Losses in Large–Diameter Crude–Oil Pipe Lines," *Oil and Gas Journal*, 04 October 1951.

13 Lester, C. B., and Darroch, R. H., MVPL internal report on tests run for API testing program, 1953. (See Kennedy, below)

14 Kennedy, W. L., Jr., "Pressure Loss in Oil Pipelines," *Proceedings of the API, Division of Transportation*, Vol. 34, (V), 1954.

15 Dreyfuss, G., "Exploitation du rapport sur la campagne de Mesures de pertes de charge SOPEG," O.T.P. Report 64–487–EG, 15 Avril 1964.

16 Uhde, A., private communication, 09 February 1979.

[17] Landers, C. R., and Compton, C. A., private investigation of smooth pipe flow in large pipes, 1977.

[18] Jeppson, R. W., *Analysis of Flow in Pipe Networks*, 1st ed., 5th printing. Ann Arbor: Ann Arbor Science Publishers, Inc., 1982.

[19] Streeter, V. L., and Wylie, E. B., *Fluid Mechanics*, 8th ed. New York: McGraw–Hill Book Company, 1985.

[20] Stanton, R. G., and Lawson, J. D., *Numerical Analysis*, AIChE Today Series. New York: American Institute of Chemical Engineers, 1968.

[21] *Mathcad 4.0*. Cambridge, Massachusetts: MathSoft Inc.

Computer Computation Template

CCT No. 13-1: Reference algorithms for $\mathbf{T_{OIL}}$, $\mathbf{T_{FSmooth}}$, and the Monograph No. 9 smooth pipe law function. The plot shows the very small variations of $\mathbf{T_{Oil}}$ and $\mathbf{T_{FSmooth}}$ from the Monograph No. 9 smooth pipe law. Equation A uses the *MathCad* root function, but any other solver can be used or an iterative solution can be programmed.

References: Chapter 13. Equations 13-6 and 13-7 (Monograph No. 9 smooth pipe law), Equation 13-10 ($\mathbf{T_{OIL}}$), and Equation 3-15 ($\mathbf{T_{FSmooth}}$).

Input

Seeds and Factors $T := 20$ $N_0 := 20000$ $i := 1, 2..12$ $N_i := N_{i-1} + 0.5 \cdot N_{i-1}$

Equations

A T_F according to Monograph No. 9 $f(N, T_{Mon}) := root\left(4 \cdot log\left(\frac{N}{T_{Mon}}\right) - 0.6 - T_{Mon}, T_{Mon}\right)$

B T_F according to $\mathbf{T_{OIL}}$ $T_{Oil_i} := 3.6 \cdot log\left(\frac{N_i}{8}\right)$

C T_F according to $\mathbf{T_{FSmooth}}$ $T_{FS_i} := \left(T_{Oil_i}\right) - \dfrac{4 \cdot log\left(\frac{N_i}{T_{Oil_i}}\right) - 0.6 - T_{Oil_i}}{\left(\frac{-4}{T_{Oil_i} \cdot ln(10)}\right) - 1}$

Results

$\frac{N_i}{1000}$	$f(N_i, T)$	T_{FS_i}	T_{Oil_i}
30	12.87	12.87	12.87
45	13.49	13.49	13.50
68	14.12	14.12	14.13
101	14.75	14.75	14.77
152	15.38	15.38	15.40
228	16.01	16.01	16.04
342	16.65	16.65	16.67
513	17.29	17.29	17.30
769	17.93	17.93	17.94
1153	18.57	18.57	18.57
1730	19.22	19.22	19.21
2595	19.86	19.86	19.84

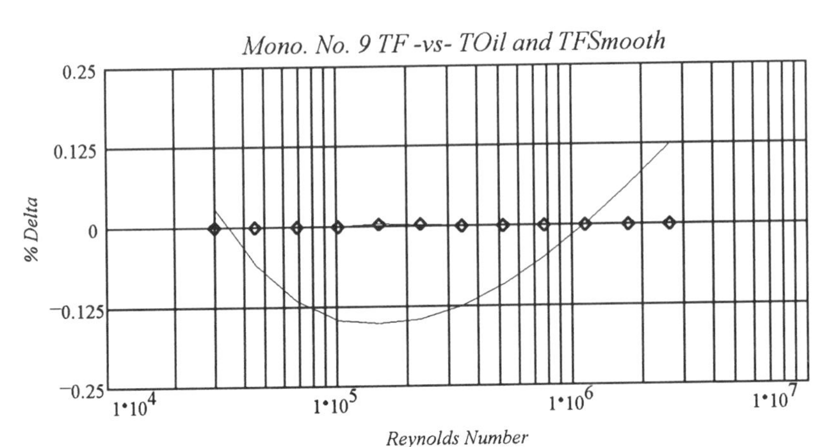

Computer Computation Template

CCT No. 13-2: Reference algorithm for $\mathbf{f_{OIL}}$ and comparison with f_{Darcy} derived from the Monograph No. 9 smooth pipe law. See **CCT** No. 13-1 for comparison of the smooth pipe law with $\mathbf{T_{Oil}}$ and $\mathbf{T_{FSmooth}}$.

References: Chapter 13. Equations 13-12 and 13-13.

Input

Factors $T := 20$ $N_0 := 20000$ $i := 1, 2..13$ $N_i := N_{i-1} + 0.5 \cdot N_{i-1}$

Equations

$$T_{Oil_i} := 3.6 \cdot \log\left(\frac{N_i}{8}\right) \qquad f(N, T_{Mon}) := root\left(4 \cdot \log\left(\frac{N}{T_{Mon}}\right) - 0.6 - T_{Mon}, T_{Mon}\right)$$

<== See **CCT** 13-1

$$F_{Oil_i} := \frac{4}{\left(T_{Oil_i}\right)^2} \qquad F_{Mon_i} := \frac{4}{f(N_i, T)^2}$$

Results

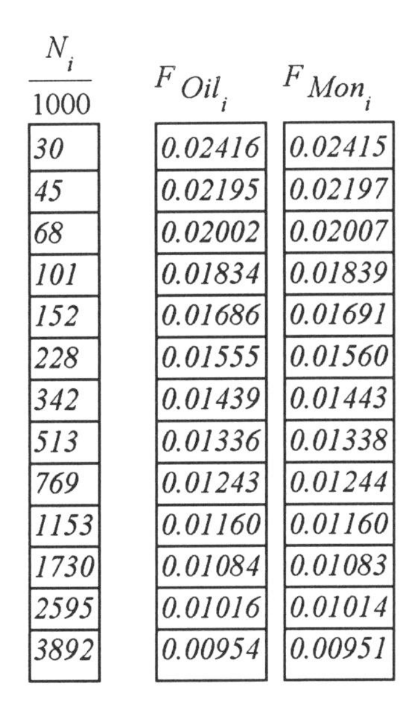

$\frac{N_i}{1000}$	F_{Oil_i}	F_{Mon_i}
30	0.02416	0.02415
45	0.02195	0.02197
68	0.02002	0.02007
101	0.01834	0.01839
152	0.01686	0.01691
228	0.01555	0.01560
342	0.01439	0.01443
513	0.01336	0.01338
769	0.01243	0.01244
1153	0.01160	0.01160
1730	0.01084	0.01083
2595	0.01016	0.01014
3892	0.00954	0.00951

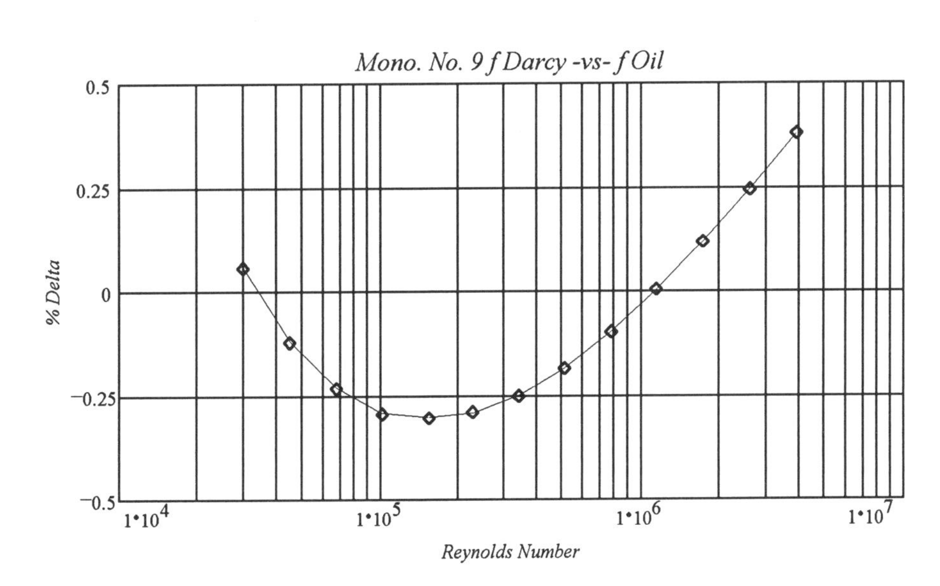

$\mathbf{f_{Oil}}$ is within 0.5% of Monograph No. 9 for 30 000 > $\mathbf{N_R}$ > 3 000 000

14

Unit Friction Losses: N_R, S_f, and Δp_f

Introduction

The calculation of unit friction loss is the fundamental calculation of pipeline hydraulics. In this and subsequent chapters these losses will be calculated in terms of the Darcy–Weisbach equation, repeated here for reference

$$h_f = f\left(\frac{L}{D}\right)\left(\frac{V^2}{2g}\right) \tag{14-1}$$

Note that f_D has been converted into f because use of the Darcy–Weisbach friction factor is understood. Thus f equates to the $\mathbf{f_{OIL}}$ function of Chapter 13.

Also, h_L has been changed to h_f to clearly denote that it is the head lost due to friction that is in question.

To compute $\mathbf{f_{OIL}}$ requires first computing $\mathbf{N_R}$ which, in general form, is

$$\mathbf{N_R} = \frac{VD}{\nu} \tag{14-2}$$

These two equations, while fundamental to the practice of pipeline hydraulics, have two common characteristics that prevent their routine use:

- They are stated in terms of coherent FPS or MKS units, which are not the working units of the industry
- The flow rates are in terms of velocity, while in practice the unit is quantity/time

Taking $(q/a)/t = V$, and setting $D = d$, the bases for two working equations can be derived as follows:

$$\mathbf{N_R} = \mathrm{C_1}\left(\frac{\left(\frac{q}{a \cdot t}\right)d}{\nu}\right) \tag{14-3}$$

$$\frac{h_f}{L} = \mathrm{C_2}\, f\left(\frac{\left(\frac{q}{a \cdot t}\right)^2}{2g \cdot d}\right) \tag{14-4}$$

The obvious problem is to compute $\mathrm{C_1}$ and $\mathrm{C_2}$.

Reynolds Number

The numeric value of the Reynolds Number is the same without regard to the units used to compute it as long as these units are coherent. In the FPS system the units would be ft/s, ft, and ft²/s; in the MKS system the units would be m/s, m, and m²/s.

In actual practice, however, pipeliners speak in terms of volume, diameter and viscosity in the terms of the trade; i.e., for the FPS system flow is (usually) in barrels per day, diameter in inches, and viscosity in cSt. Since these units are not coherent—they are not even consistent—any equation for computing the Reynolds Number from such input variables will of necessity have to contain a factor of proportionality, viz. $\mathrm{C_1}$.

N_R

C_1, itself, is comprised of several conversion factors and can be visualized as follows

$$N_R = \left(\frac{\dfrac{K_1\ q}{K_2\ t}}{K_3 d \cdot K_4 \nu} \right) \frac{4}{\pi} \qquad (14\text{-}5)$$

Table 14-1 contains a selection of K factors suitable for computing the value of C_1 for making conversions into the FPS system of units.

Table 14–1

Conversion Factors for C_1, C_2, and C_3.

To Convert	Into	Multiply By
Length		
mm	ft	3.280 840E–3
cm	ft	3.280 840E–2
m	ft	3.280 840E00
in	ft	8.333 333E–2
ft	ft	1.000 000E00
mile	ft	5.280 000E+3
Volume		
liter	ft³	3.531 468E–2
US gal	ft³	1.336 806E–1
UK gal	ft³	1.605 440E–1
US oil bbl	ft³	5.614 583E00
m³	ft³	3.531 468E+1
Viscosity		
ft²/s	mm²/s	1.076 391E–5
cSt	mm²/s	1.000 000E00
Weight		
kg	pound	4.535 924E–1
Time		
year	sec	3.153 600E+7
day	sec	8.640 000E+4
hour	sec	3.600 000E+3
min	sec	6.000 000E+1

To convert the FPS equation to Q in bpd, d in inches, and ν in cSt

K_1= 5.614 583E+0
K_2= 8.640 000E+4
K_3= 1.076 391E–5
K_4= 8.333 333E–2

Making the indicated substitutions in Equation 14-5 yields the working equations for N_R

$$N_R = (92.241\,232)\left(\frac{Q}{d\nu}\right) = \text{(use) } 92.2\left(\frac{Q}{d\nu}\right) \qquad (14\text{-}6)$$

where Q = flow rate in oil barrels per day
d = inside diameter of pipe, inches
ν = kinematic viscosity, cSt

Similarly, using the conversion factors of Table 14-2, C_1 for computing N_R for the usual MKS units is 3.536 777E+5, thus working equations

$$N_R = (3.536\,777 \times 10^5)\left(\frac{Q}{d\nu}\right) = \text{(use) } 354\,000\left(\frac{Q}{d\nu}\right) \qquad (14\text{-}7)$$

where Q = flow rate, m^3/hr
d = inside diameter of pipe, mm
ν = kinematic viscosity, cSt

Equation 14-7 is more easily remembered if diameter is expressed in meters instead of millimeters because the numeric is easier to remember: 354 instead of 354 000.

Note that the equations marked (use) contain rounded values for C_1. This truncation is acceptable for all except computer work, where many intermediate products may be formed in an algorithm, because since N_R enters into the pipe flow equation only as a variable behind the logarithm, using C_1 to only three significant figures is logical.

There are many other combinations of units that can be plugged into Equation 14-5 to yield the value of N_R. In the FPS system flow is often quoted for gathering lines in bph instead of bpd, and some products line still run on the basis of gpm. In the MKS sector some still calculate on the basis of m³/s; and there are countries where the legal flow rate must be expressed in terms of metric tons, whether per hour, day, or (as a rate) per year; metric tons per annum = mta. The numerics for any of these are easily calculated having in hand a calculator and a table of equivalents. In SI the *correct* values are m³/s, m, and m²/sec, but practically the values given above are used. Note that 1 mm²/s = 1 cSt.

Unit Friction Loss

Slope

Unit friction losses can be stated in terms of pressure or head loss per unit length. In pipelining the unit length is almost always the mile or kilofoot (FPS) or kilometer (MKS). For reasons that will appear a few paragraphs later, unit friction loss equations herein will be based on the *km* and the *kft*, where kft stands for 1000 ft.

Head loss will be expressed in terms of unit length, so in FPS head loss will be in terms of feet of liquid column, abbreviated *flc*, and in MKS in meters of liquid column, abbreviated *mlc*. The reason for using flc and mlc instead of just *ft* or *m* is that in setting up equations involving both active and static heads it is important to be able to differentiate between the two. Thus *flc* and *mlc* are *active* heads, either positive (generated by a pump, perhaps) or negative (friction or conversion loss), and *m* and *ft* are always *static* heads, either positive or negative. Both represent energy in the form of m·kg/kg, ft·lb/lb, or similar.

In working between the systems of units (setting up a computer program to plot the hydraulic gradient against a ground profile is an excellent example) it is important to have as many of the inputs to the algorithms as possible be numerically identical. In the case of the unit friction loss this identity is perfect because a slope of 1 mlc/km = 1 flc/kft, and it is therefore logical to establish a standard method of calculating friction loss based on a 1:1000 ratio, i.e., 1 flc/1000 ft, 1 mlc/1000 m, etc.

This relationship depends on the concept of the slope S of the *energy grade line*, a concept used in the Chezy formula, $V = C\sqrt{RS}$, and the similar Manning formula $V = (C/n)\,R^{2/3}S^{1/2}$, for flow in *open channels*. The energy grade line is a plot of the terms

$$\frac{V^2}{2g} + \frac{p}{\gamma} + z \tag{14-8}$$

The first term is the *kinetic energy*, the second is the *pressure energy* or *pressure head*, and the third is the *static head*, all expressed as head in terms of energy per unit of mass.

The *hydraulic grade line* is a plot of the terms

$$\frac{p}{\gamma} + z \tag{14-9}$$

and thus differs from the energy grade line by the amount of the kinetic energy.

Since pipeline problems usually involve at least several hundred flc or mlc of combined pressure and static heads, neglecting a kinetic energy which may be only a few flc or mlc is acceptable, and so slope as used herein will be the slope of the hydraulic grade line or, as it is more commonly named, the *hydraulic gradient*. In addition, if a pipeline in steady flow has a constant diameter, the hydraulic grade line will have the same slope as the energy grade line; the kinetic energy is everywhere the same. The *magnitude* of the two gradients won't be the same, but their *slopes* will be identical.

Since the value of flc/kft and mlc/km are the same, it is in order to name this particular slope, and hereafter it will be called $\mathbf{S_f}$. It signifies the slope of the flc/kft and/or the mlc/km function, and its value is the difference in the hydraulic gradient at the inlet and outlet over the length of measurement, in each case 1000 linear units.

$\mathbf{S_f}$

Equation 14-4 can be rewritten for L = 1 as

$$\begin{aligned} h_f &= \left(\frac{1}{2g}\right)\left(\frac{16}{\pi^2}\right) f \left(\frac{q^2}{t^2 d^5}\right) \\ &= (2.519\,327 \times 10^{-2})\, f \left(\frac{K_1^{\,2}(q)}{K_2^{\,2}(t) K_3^{\,5}(d)}\right) \end{aligned} \tag{14-10}$$

Using the same kind of $K_1, K_2 \ldots$ calculation described earlier in developing working equations for $\mathbf{N_R}$, if Q is in bpd and d is in inches

$$h_f = C_2\, f\left(\frac{Q^2}{d^5}\right) = (2.647\,271 \times 10^{-5})\, f \left(\frac{Q^2}{d^5}\right) \tag{14-11}$$

If L = 1000 ft

$$\begin{aligned} \mathbf{S_f} &= (2.647\,271 \times 10^{-2})\, f\left(\frac{Q^2}{d^5}\right) \\ &= (\text{use})\ 2.647 \times 10^{-2}\, f\left(\frac{Q^2}{d^5}\right) \end{aligned} \tag{14-12}$$

If Q is in m³/h, d is in mm, and L = 1000 m

$$\mathbf{S_f} = (6.377\,707\times 10^9)\, f\left(\frac{Q^2}{d^5}\right) = \text{(use) } 6.378\times 10^9 f\left(\frac{Q^2}{d^5}\right) \qquad (14\text{-}13)$$

To repeat: each equation gives a value of the slope of the hydraulic gradient on a 1:1000 basis, flc/kft or mlc/km; the values are numerically the same.

Note that the values of C_2 (use) for $\mathbf{S_f}$ are carried to four significant figures instead of the three figures recommended for $\mathbf{N_R}$. This is not to indicate the expected *accuracy* of the calculation is in the order of ±0.001—a properly carried out set of measurements and calculations can yield a computed result within ±2% of the measured value, and anything better is happenstance—but the *precision* of the calculation is worth four figures; the $\mathbf{S_f}$ is the most used parameter in pipelining and so even with hand calculators it should be carried to four figures for purposes of minimizing accidental or coincidental errors, and, for the same reason, the full seven–figure values should be used in computer algorithms.

Δp_f

There are times when it is important to have the friction loss in terms of lost pressure instead of lost head. If the working equations were in terms of the FPS or MKS base units this would be simple:

$$p = \gamma h \qquad (14\text{-}14)$$

It is not so simple, however, because in the FPS system p is in psf and γ is in slugs/ft³, and in the MKS system in kg/m² and kg/m³, not one of which is a working unit.

The more usual procedure is to convert $\mathbf{S_f}$ from mlc/km or flc/kft to ksc/km or psi/kft by assuming that the head loss $\mathbf{S_f}$ is a head loss of *water*, and then multiplying this value by the *specific gravity* (or density, referred to water as 1.000) of the *oil*. Let $\mathbf{S_f}$ =5.00 and ρ = 0.85:

5 mlc/km = 5 × 1.000E–1 ksc/mlc × 0.85 = 0.425 ksc/km

Or

5 flc/kft = 5 × 4.335E–1 psi/flc × 0.85 = 1.843 psi/kft

Table 14-2 is a compilation of conversion factors that are useful in working with pressures, heads, and friction loss calculations in general. These factors can be used to generate C_3 for the following general equation to yield friction loss in terms of pressure instead of head.

$$\Delta \mathbf{p_f} = C_3 \rho f\left(\frac{Q^2}{d^5}\right) \qquad (14\text{-}15)$$

Two specific formulas are of interest, inasmuch as they use the most common FPS and MKS pipeline units:

$$\Delta \mathbf{p_f} = 2.647\,271\times 10^{-2}\times 0.433\,514\times\rho\times f\left(\frac{Q^2}{d^5}\right) = 1.147\,629\times 10^{-2}\times\rho\times f\left(\frac{Q^2}{d^5}\right) = \text{(use) } 1.148\times 10^{-2}\times\rho\times f\left(\frac{Q^2}{d^5}\right) \qquad (14\text{-}16)$$

where $\Delta \mathbf{p_f}$ = pressure loss, psi/kft
Q = bbd
d = inches
ρ = density (or spgr) referred to water as 1.000

$$\Delta \mathbf{p_f} = 6.377\,707\times 10^{9}\times 1.000\times 10^{-1}\times\rho\times f\left(\frac{Q^2}{d^5}\right) = 6.377\,707\times 10^{8}\times\rho\times f\left(\frac{Q^2}{d^5}\right) = \text{(use) } 6.378\times 10^{8}\times\rho\times f\left(\frac{Q^2}{d^5}\right) \qquad (14\text{-}17)$$

where $\Delta \mathbf{p_f}$ = pressure loss , ksc / km
Q = m³/ hr
d = mm
ρ = density or spgr referred to water as 1.000

Working Equations

Calculators

Table 14-3 is a compilation of working equations for unit friction loss suitable for use with hand calculators:

Table 14-2
Standard Conversion Factors For Pressure, Head, and Friction Loss Calculations

Value	=	Value	Unit	Comments
Length				
1 ft	=	0.3048	m	Exact
	=	12	in	Exact
1 in	=	2.54	cm	Exact
1 m	=	3.280 84	ft	Use 3.281E00
	=	39.370 0	in	Use 3.937E+1
Volume				
1 UK gal	=	277.42	in^3	Use 2.774E+2
1 US gal	=	231.	in^3	Exact
1 oil bbl	=	42.	US gal	Exact
	=	0.158 987	m^3	Use 1.590E–1
	=	5.614 58	ft^3	Use 5.615E00
1 m^3	=	6.289 82	b	Use 6.290E00
	=	35.314 7	ft^3	Use 3.531E+1
Weight				
1 lb	=	0.453 592	kg	Use 4.536E–1
1 kg	=	2.204 62	lb	Use 2.205E00
Flowrate				
1 m^3/h	=	150.956	b/d	Use 1.510E+2
1 b/d	=	0.006 624	m^3/h	Use 6.624E–3
Pressure				
1 psi	=	0.070 307 0	ksc	Use 7.031E–2
	=	0.068 947 7	bar	Use 6.895E–2
1 ksc	=	14.223 3	psi	Use 1.422E+1
	=	0.980 665	bar	Use 9.807E–1
1 bar	=	14.503 8	psi	Use 1.450E+1
	=	1.019 72	ksc	Use 1.020E00
	=	100	kPa	Exact
Gravity*				
9.806 65 m/s/s	=	32.174 0	ft/s/s	Use 3.217E+1
Head				
1 mlc H_2O	=	0.099 997 3	ksc	Use 1.000E–1
	=	0.098 0638	bar	Use 9.806E–2
	=	1.422 29	psi	Use 1.422E00
1 flc H_2O	=	0.030 479	ksc	Use 3.048E–2
	=	0.029 890	bar	Use 2.989E–2
		0.433 514	psi	Use 4.335E–1
1 ksc	=	10.000 27	mlc H_2O	Use 1.000E+1
1 bar	=	10.197 4	mlc H_2O	Use 1.020E+1
1 psi	=	2.306 73	flc H_2O	Use 2.307E00
1 mlc(x)	=	$1\times\rho(x)$	mlc H_2O	(Use $\rho(x)$ for
1 flc(x)	=	$1\times\rho(x)$	flc H_2O	$\rho(H_2O)$=1.00)

* *Standard* gravity is accepted as 9.806 65 m/s/s exactly

compute $\mathbf{N_R}$, then $\mathbf{f_{OIL}}$, and then $\mathbf{S_f}$ or $\Delta \mathbf{p_f}$ as applies. The equations for S_f and p_f, unit head and unit pressure loss, 1:1000 basis, using the Colebrook–White function are not included because they require an iterative solution for the friction factor and the way this algorithm is best coded depends on the programming language used. Problems using the Colebrook–White function are included in the **CCT**s, however.

Computers

The Computer Computation Templates provided in this chapter use the C1, C2, C3 concept to produce unit friction loss equations for the pipeline FPS and pipeline MKS systems of units that differ only in the values of the C-factors.

Since the calculation of unit friction loss is the seminal calculation of pipeline hydraulics, there are twelve **CCT**s provided to calculate not only head and pressure losses using the $\mathbf{f_{OIL}}$ function but using the Colebrook–White function as well. While the Colebrook–White function has outlived its usefulness in oil pipeline hydraulics there is always the chance that the pressure or head loss in an encrusted water line will have to be calculated and the Colebrook–White function is as good as any for such a problem. The examples in the C–W templates are the same oil pipeline examples as those in the $\mathbf{f_{OIL}}$ templates, however.

Templates **CCT** Nos. 14-1, 14-1M, 14-2, and 14-2M are primarily reference templates, in that they include not only the $\mathbf{S_f}$ and $\Delta \mathbf{p_f}$ functions but the derivatives of these functions as well. The derivatives can be very useful in quickly determining the effect of a change of an input variable on an operating pipeline. Comparable templates for the Colebrook–White alternative were not prepared. The templates with the M indicate a MKS template.

The other eight templates are working templates. By selecting the proper template, data can be input in either the pipeline FPS or pipeline MKS system, the friction loss output can be in either FPS or MKS head or pressure loss terms, and the friction factor can be either that of the $\mathbf{f_{OIL}}$ function or the Colebrook–White function.

Table 14-3

Working Equations For Unit Friction Loss

Reynolds Number

$$N_R = C_1\left(\frac{Q}{dv}\right)$$

$$N_R = (92.241232)\left(\frac{Q}{dv}\right) = \text{(use) } 92.2\left(\frac{Q}{dv}\right)$$

where N_R = Reynolds Number
Q = flow rate in oil barrels per day
d = inside diameter of pipe, inches
v = kinematic viscosity, cSt

$$N_R = (3.536\,777\times10^5)\left(\frac{Q}{dv}\right) = \text{(use) } 354\,000\left(\frac{Q}{dv}\right)$$

where N_R = Reynolds Number
Q = flow rate, m^3 / hr
d = inside diameter of pipe, mm
v = kinematic viscosity, cSt

Unit Friction Loss in Terms of Head

$$S_f = C_2 f\left(\frac{Q^2}{d^5}\right)$$

$$S_f = 2.647\,271\times10^{-2} f\left(\frac{Q^2}{d^5}\right) = \text{(use) } 2.647\times10^{-2} f\left(\frac{Q^2}{d^5}\right)$$

where S_f = head loss, ft / kft
Q = flow rate in oil barrels per day
d = inside diameter of pipe, inches
v = kinematic viscosity, cSt

$$S_f = 6.377\,707\times10^{9} f\left(\frac{Q^2}{d^5}\right) = \text{(use) } 6.378\times10^{9} f\left(\frac{Q^2}{d^5}\right)$$

where S_f = head loss, mlc / km
Q = flow rate, m^3 / hr
d = inside diameter of pipe, mm
v = kinematic viscosity, cSt

Unit Friction Loss in Terms of Pressure

$$\Delta p_f = C_3 \rho f\left(\frac{Q^2}{d^5}\right)$$

$$\Delta p_f = 2.647\,271\times10^{-2}\times0.433\,514\times\rho\times f\left(\frac{Q^2}{d^5}\right)$$
$$= 1.147\,629\times10^{-2}\times\rho\times f\left(\frac{Q^2}{d^5}\right)$$
$$= \text{(use) } 1.148\times10^{-2}\times\rho\times f\left(\frac{Q^2}{d^5}\right)$$

where Δp_f = pressure loss, psi / kft
Q = bbd
d = inches
ρ = density or spgr, $H_2O = 1.000$

$$\Delta p_f = 6.377\,707\times10^{9}\times1.000\times10^{-1}\times\rho\times f\left(\frac{Q^2}{d^5}\right)$$
$$= 6.377\,707\times10^{8}\times\rho\times f\left(\frac{Q^2}{d^5}\right)$$
$$= \text{(use) } 6.378\times10^{8}\times\rho\times f\left(\frac{Q^2}{d^5}\right)$$

where Δp_f = pressure loss, ksc / km
Q = m^3/hr
d = mm
ρ = density or spgr, $H_2O = 1.000$

These templates are numbered as follows:

CCT No.	$\mathbf{S_f}$ / Sf	$\Delta \mathbf{p_f}$ / p_f	$\mathbf{f_{OIL}}$	C-W	FPS	MKS
14-3	♦		♦		♦	
14-3M	♦		♦			♦
14-3C	♦			♦	♦	
14-3MC	♦			♦		♦
14-4		♦	♦		♦	
14-4M		♦	♦			♦
14-4C		♦		♦	♦	
14-4MC		♦		♦		♦

Columns 2 and 3 indicate whether the computation is for unit head or unit pressure loss, columns 4 and 5 note whether the $\mathbf{f_{OIL}}$ or the Colebrook–White friction factor is used, and columns 6 and 7 mark the use of the pipeline FPS or pipeline MKS system of units in input and output values.

Note that each of these templates accepts a single input value for flow rate and then generates a range of flow rates used to plot the friction loss–flow rate curve. If only a single point is desired, input it as the Q_0 value, and the top values on the numerical readouts will provide the answer required.

Also note that templates using the Colebrook–White function are written for *MathCad's root* function, a *solver* that solves single–variable equations by iteration, to produce the friction factor. If using another language, the bi–section technique, or a *regula falsi* algorithm, can be used to solve for friction factor. Substitute the output of this computation for the a_i variable in the Colebrook–White templates.

Computer Computation Template

CCT No. 14-1: Examples of applications and functions of $\mathbf{S_f}$. Equation (1) is the equation for $\mathbf{S_f}$. Equations (2), (3), and (4) are the derivatives of $\mathbf{S_f}$ with respect to flow rate (Q), diameter (d=D-2*t), and viscosity (ν), respectively. $\mathbf{S_f}$ calculated with the C–factors shown will yield pipeline FPS output for pipeline FPS input data.

References: Chapter 14. Equation 14-12.

Input

Factors $C_1 := 92.241232$ $C_2 := 0.02647271$ $C_3 := 0.01147629$ $K := \dfrac{3.6}{2.302585}$ $k := \dfrac{2.0}{2.302585}$

Input Data $Q := 301910$ bpd $D := 24.00$ inches $t := 0.375$ inches $\nu := 7.0$ cSt

Variables $d := D - 2 \cdot t$

Equations

1. $\mathbf{S_f}(Q, d, \nu)$

$$C_2 \cdot \left[\frac{4}{\left(K \cdot \ln\left(\dfrac{C_1 \cdot Q}{8 \cdot d \cdot \nu} \right) \right)^2} \right] \cdot \frac{Q^2}{d^5} = 5.846 \quad \text{flc / kft}$$

2. $d\,\mathbf{S_f} / d\,Q$

$$-8 \cdot \frac{C_2}{\left(K^2 \cdot \ln\left(\dfrac{1}{8} \cdot C_1 \cdot \dfrac{Q}{d \cdot \nu} \right)^3 \right)} \cdot \frac{Q}{d^5} + 8 \cdot \frac{C_2}{\left(K^2 \cdot \ln\left(\dfrac{1}{8} \cdot C_1 \cdot \dfrac{Q}{d \cdot \nu} \right)^2 \right)} \cdot \frac{Q}{d^5} = 3.484 \cdot 10^{-5} \quad \text{(flc / kft) / bpd}$$

3. $d\,\mathbf{S_f} / d\,d$

$$8 \cdot \frac{C_2}{\left(K^2 \cdot \ln\left(\dfrac{1}{8} \cdot C_1 \cdot \dfrac{Q}{d \cdot \nu} \right)^3 \right)} \cdot \frac{Q^2}{d^6} - 20 \cdot \frac{C_2}{\left(K^2 \cdot \ln\left(\dfrac{1}{8} \cdot C_1 \cdot \dfrac{Q}{d \cdot \nu} \right)^2 \right)} \cdot \frac{Q^2}{d^6} = -1.207 \quad \text{(flc / kft) / inch}$$

4. $d\,\mathbf{S_f} / d\,\nu$

$$8 \cdot \frac{C_2}{\left(K^2 \cdot \ln\left(\dfrac{1}{8} \cdot C_1 \cdot \dfrac{Q}{d \cdot \nu} \right)^3 \right)} \cdot \frac{Q^2}{d^5 \cdot \nu} = 0.168 \quad \text{(flc / kft) / cSt}$$

Computer Computation Template

CCT No. 14-1M: Examples of applications and functions of $\mathbf{S_f}$. Equation (1) is the equation for $\mathbf{S_f}$. Equations (2), (3), and (4) are the derivatives of $\mathbf{S_f}$ with respect to flow rate (Q), diameter (d=D-2*t), and viscosity (ν), respectively. $\mathbf{S_f}$ calculated with the C–factors shown will yield pipeline MKS output for pipeline MKS input data.

References: Chapter 14. Equation 14-12.

Input

Factors $C_1 := 3.536777 \cdot 10^5$ $C_2 := 6.377707 \cdot 10^9$ $C_3 := 6.377707 \cdot 10^8$ $K := \dfrac{3.6}{2.302585}$ $k := \dfrac{2.0}{2.302585}$

Input Data $Q := 2000$ m³/h $D := 609.6$ mm $t := 9.525$ mm $\nu := 7.0$ cSt

Variables $d := D - 2 \cdot t$

Equations

1. $\mathbf{S_f}(Q, d, \nu)$

$$C_2 \cdot \left[\frac{4}{\left(K \cdot ln\left(\dfrac{C_1 \cdot Q}{8 \cdot d \cdot \nu}\right)\right)^2}\right] \cdot \frac{Q^2}{d^5} = 5.846 \quad \text{mlc / km}$$

2. $d\,\mathbf{S_f} / dQ$

$$-8 \cdot \frac{C_2}{\left(K^2 \cdot ln\left(\dfrac{1}{8} \cdot C_1 \cdot \dfrac{Q}{d \cdot \nu}\right)^3\right)} \cdot \frac{Q}{d^5} + 8 \cdot \frac{C_2}{\left(K^2 \cdot ln\left(\dfrac{1}{8} \cdot C_1 \cdot \dfrac{Q}{d \cdot \nu}\right)^2\right)} \cdot \frac{Q}{d^5} = 0.005 \quad \text{(mlc / km) / m}^3\text{h}$$

3. $d\,\mathbf{S_f} / d\,d$

$$8 \cdot \frac{C_2}{\left(K^2 \cdot ln\left(\dfrac{1}{8} \cdot C_1 \cdot \dfrac{Q}{d \cdot \nu}\right)^3\right)} \cdot \frac{Q^2}{d^6} - 20 \cdot \frac{C_2}{\left(K^2 \cdot ln\left(\dfrac{1}{8} \cdot C_1 \cdot \dfrac{Q}{d \cdot \nu}\right)^2\right)} \cdot \frac{Q^2}{d^6} = -0.048 \quad \text{(mlc / km) / mm}$$

4. $d\,\mathbf{S_f} / d\,\nu$

$$8 \cdot \frac{C_2}{\left(K^2 \cdot ln\left(\dfrac{1}{8} \cdot C_1 \cdot \dfrac{Q}{d \cdot \nu}\right)^3\right)} \cdot \frac{Q^2}{d^5 \cdot \nu} = 0.168 \quad \text{(mlc / km) / cSt}$$

Computer Computation Template

CCT No. 14-2: Examples of application and functions of $\Delta\mathbf{p_f}$. Equation (1) is the equation for $\Delta\mathbf{p_f}$. Equations (2), (3), (4), and (5) are the derivatives of $\Delta\mathbf{p_f}$ with respect to flow rate (Q), diameter (d=D-2*t), viscosity (ν), and density (ρ, reference H_2O=1.000). $\Delta\mathbf{p_f}$ calculated with the C-factors shown will yield pipeline FPS output for pipeline FPS input data.

References: Chapter 14. Equation 14-16.

Input

Factors $C_1 := 92.241232$ $C_2 := 0.02647271$ $C_3 := 0.01147629$ $K := \dfrac{3.6}{2.302585}$ $k := \dfrac{2.0}{2.302585}$

Input Data $Q := 1250000$ bpd $D := 36.00$ inches $t := 0.375$ inches $\rho := 0.845$ $\nu := 7.0$ cSt

Variables $d := D - 2\cdot t$

Equations

1. $\Delta\mathbf{p_f}$ (Q, ν, d, ρ)

$$C_3\cdot\rho\cdot\left[\frac{4}{\left(K\cdot ln\left(\frac{C_1\cdot Q}{8\cdot d\cdot \nu}\right)\right)^2}\right]\cdot\frac{Q^2}{d^5} = 3.782 \quad \text{psi / kft}$$

2. $d\,\Delta\mathbf{p_f}\,/\,d\,Q$

$$-8\cdot C_3\cdot\frac{\rho}{\left(K^2\cdot ln\left(\frac{1}{8}\cdot C_1\cdot\frac{Q}{d\cdot\nu}\right)^3\right)}\cdot\frac{Q}{d^5} + 8\cdot C_3\cdot\frac{\rho}{\left(K^2\cdot ln\left(\frac{1}{8}\cdot C_1\cdot\frac{Q}{d\cdot\nu}\right)^2\right)}\cdot\frac{Q}{d^5} = 5.500\cdot 10^{-6} \quad \text{(psi / kft) / bpd}$$

3. $d\,\Delta\mathbf{p_f}\,/\,d\,d$

$$8\cdot C_3\cdot\frac{\rho}{\left(K^2\cdot ln\left(\frac{1}{8}\cdot C_1\cdot\frac{Q}{d\cdot\nu}\right)^3\right)}\cdot\frac{Q^2}{d^6} - 20\cdot C_3\cdot\frac{\rho}{\left(K^2\cdot ln\left(\frac{1}{8}\cdot C_1\cdot\frac{Q}{d\cdot\nu}\right)^2\right)}\cdot\frac{Q^2}{d^6} = -0.517 \quad \text{(psi / kft) / inch}$$

4. $d\,\Delta\mathbf{p_f}\,/\,d\,\nu$

$$8\cdot C_3\cdot\frac{\rho}{\left(K^2\cdot ln\left(\frac{1}{8}\cdot C_1\cdot\frac{Q}{d\cdot\nu}\right)^3\right)}\cdot\frac{Q^2}{d^5\cdot\nu} = 0.098 \quad \text{(psi / kft) / cSt}$$

5. $d\,\Delta\mathbf{p_f}\,/\,d\,\rho$

$$4\cdot\frac{C_3}{\left(K^2\cdot ln\left(\frac{1}{8}\cdot C_1\cdot\frac{Q}{d\cdot\nu}\right)^2\right)}\cdot\frac{Q^2}{d^5} = 4.476 \quad \text{(psi / kft) / }\rho$$

Computer Computation Template

CCT No. 14-2M: Examples of application and functions of Δp_f. Equation (1) is the equation for Δp_f. Equations (2), (3), (4), and (5) are the derivatives of Δp_f with respect to flow rate (Q), diameter (d=D-2*t), viscosity (ν), and density (ρ, reference H_2O=1.000). Δp_f calculated with the C-factors shown will yield pipeline MKS output for pipeline MKS input data.

References: Chapter 14. Equation 14-16.

Input

Factors $C_1 := 3.536777 \cdot 10^5$ $C_2 := 6.377707 \cdot 10^9$ $C_3 := 6.377707 \cdot 10^8$ $K := \frac{3.6}{2.302585}$ $k := \frac{2.0}{2.302585}$

Input Data $Q := 8281$ m^3/h $D := 914.4$ mm $t := 9.525$ mm $\rho := 0.845$ $\nu := 7.0$ cSt

Variables $d := D - 2 \cdot t$

Equations

1. Δp_f (Q, ν, d, ρ) $$C_3 \cdot \rho \cdot \left[\frac{4}{\left(K \cdot ln\left(\frac{C_1 \cdot Q}{8 \cdot d \cdot \nu} \right) \right)^2} \right] \cdot \frac{Q^2}{d^5} = 0.873$$ ksc / km

2. $d\Delta p_f / dQ$ $$-8 \cdot C_3 \cdot \frac{\rho}{\left(K^2 \cdot ln\left(\frac{1}{8} \cdot C_1 \cdot \frac{Q}{d \cdot \nu} \right) \right)^3} \cdot \frac{Q}{d^5} + 8 \cdot C_3 \cdot \frac{\rho}{\left(K^2 \cdot ln\left(\frac{1}{8} \cdot C_1 \cdot \frac{Q}{d \cdot \nu} \right) \right)^2} \cdot \frac{Q}{d^5} = 1.915 \cdot 10^{-4}$$ (ksc / km) / m^3 / h

3. $d\Delta p_f / dd$ $$8 \cdot C_3 \cdot \frac{\rho}{\left(K^2 \cdot ln\left(\frac{1}{8} \cdot C_1 \cdot \frac{Q}{d \cdot \nu} \right) \right)^3} \cdot \frac{Q^2}{d^6} - 20 \cdot C_3 \cdot \frac{\rho}{\left(K^2 \cdot ln\left(\frac{1}{8} \cdot C_1 \cdot \frac{Q}{d \cdot \nu} \right) \right)^2} \cdot \frac{Q^2}{d^6} = -0.005$$ (ksc / km) / mm

4. $d\Delta p_f / d\nu$ $$8 \cdot C_3 \cdot \frac{\rho}{\left(K^2 \cdot ln\left(\frac{1}{8} \cdot C_1 \cdot \frac{Q}{d \cdot \nu} \right) \right)^3} \cdot \frac{Q^2}{d^5 \cdot \nu} = 0.023$$ (ksc / km) / cSt

5. $d\Delta p_f / d\rho$ $$4 \cdot \frac{C_3}{\left(K^2 \cdot ln\left(\frac{1}{8} \cdot C_1 \cdot \frac{Q}{d \cdot \nu} \right) \right)^2} \cdot \frac{Q^2}{d^5} = 1.033$$ (ksc / km) / ρ

Computer Computation Template

CCT No. 14-3: This template is a working template for $\mathbf{S_f}$ with input in the pipeline FPS system. The plot shows the value of $\mathbf{S_f}$ for a range of Q from Q_0 to 2 x Q_0.

References: Chapter 14. Equation 14-12.

Input

Factors $i := 0, 1 .. 10$

$C_1 := 92.241232$ $C_2 := 0.02647271$ $C_3 := 0.01147629$ $K := \frac{3.6}{2.302585}$ $k := \frac{2.0}{2.302585}$

Input Data $Q_0 := 200000$ bpd $D := 24.00$ inches

$t := 0.375$ inches $\nu := 7.0$ cSt

Variables $d := D - 2 \cdot t$ $Q_i := Q_0 + 0.1 \cdot i \cdot Q_0$

Equations

$$S_{f_i} := C_2 \cdot \left[\frac{4}{\left(K \cdot ln\left(\frac{C_1 \cdot Q_i}{8 \cdot d \cdot \nu} \right) \right)^2} \right] \cdot \frac{(Q_i)^2}{d^5}$$

Results

Q_i	S_{f_i}
200000	2.791
220000	3.311
240000	3.871
260000	4.469
280000	5.105
300000	5.780
320000	6.492
340000	7.241
360000	8.027
380000	8.849
400000	9.707

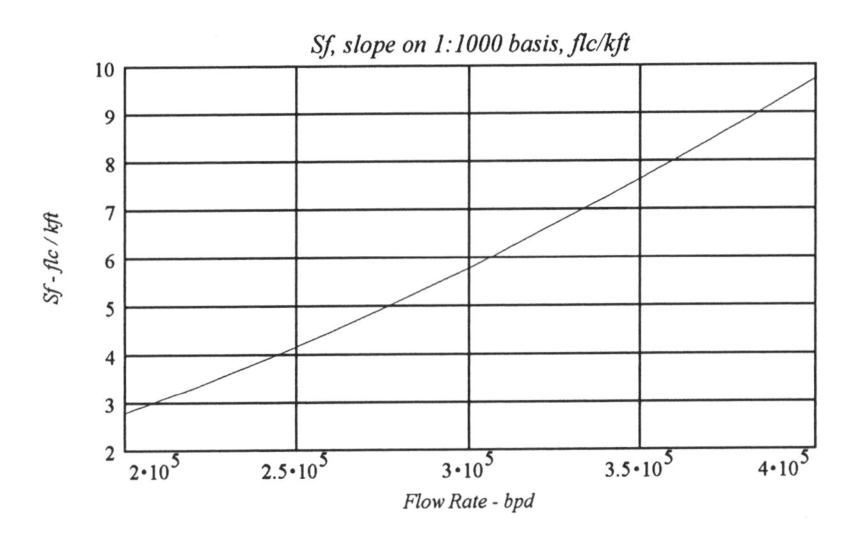

Computer Computation Template

CCT No. 14-3M: This template is a working template for $\mathbf{S_f}$ with input in the pipeline MKS system. The plot shows the value of $\mathbf{S_f}$ for a range of Q from Q_0 to 2 x Q_0.

References: Chapter 14. Equation 14-12.

Input

Factors $i := 0, 1 .. 10$

$C_1 := 3.536777 \cdot 10^5$ $C_2 := 6.377707 \cdot 10^9$ $C_3 := 6.377707 \cdot 10^8$ $K := \dfrac{3.6}{2.302585}$ $k := \dfrac{2.0}{2.302585}$

Input Data $Q_0 := 1325$ m3 / h $D := 609.6$ mm

$t := 9.525$ mm $\nu := 7.0$ cSt

Variables $d := D - 2 \cdot t$ $Q_i := Q_0 + 0.1 \cdot i \cdot Q_0$

Equations

$$S_{f_i} := C_2 \cdot \left[\frac{4}{\left(K \cdot \ln\left(\frac{C_1 \cdot Q_i}{8 \cdot d \cdot \nu} \right) \right)^2} \right] \cdot \frac{(Q_i)^2}{d^5}$$

Results

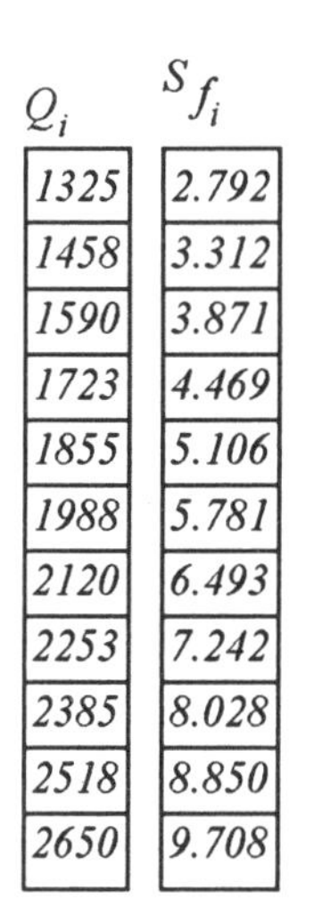

Q_i	S_{f_i}
1325	2.792
1458	3.312
1590	3.871
1723	4.469
1855	5.106
1988	5.781
2120	6.493
2253	7.242
2385	8.028
2518	8.850
2650	9.708

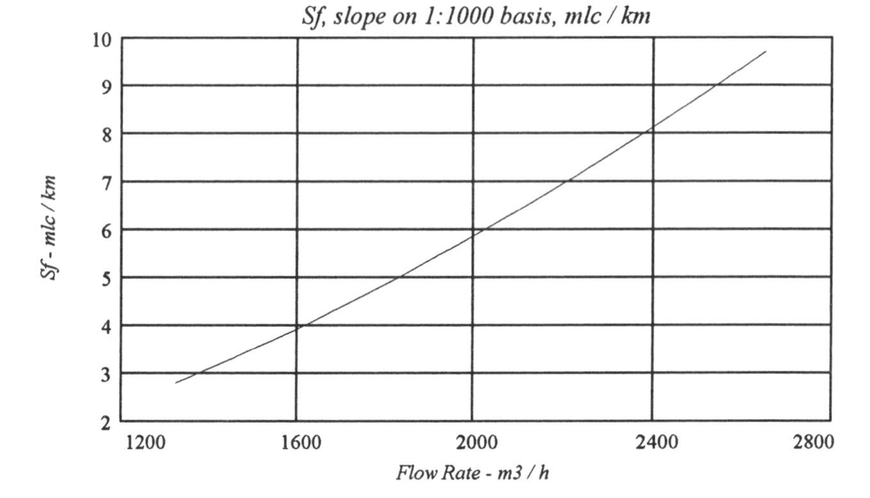

Computer Computation Template

CCT No. 14-3C: This template is a working template for S on a 1:1000 basis (flc / kft) calculated using the Colebrook-White function with input in the pipeline FPS system. The plot shows the value of S for a range of Q from Q_0 to 2 x Q_0

References: Chapter 14. Equations 14-11, 14-12, and 12-15.

Input

Factors $i := 0, 1 .. 10$

$C_1 := 92.241232$ $C_2 := 0.02647271$ $C_3 := 0.01147629$ $K := \frac{3.6}{2.302585}$ $k := \frac{2.0}{2.302585}$

Input Data $Q_0 := 200000$ bpd $D := 24.00$ inches

$t := 0.375$ inches $\nu := 7.0$ cSt

$\varepsilon := 0.0005$ inches

Variables $d := D - 2 \cdot t$ $Q_i := Q_0 + 0.1 \cdot Q_0 \cdot i$ $R_i := \frac{C_1 \cdot Q_i}{d \cdot \nu}$

Equations

$f_{CW} := 0.03$ $f(R, f_{CW}) := root\left[\left(-2 \cdot \log\left(\frac{\varepsilon d^{-1}}{3.7} + \frac{2.51}{R \cdot \sqrt{f_{CW}}}\right) - \frac{1}{\sqrt{f_{CW}}}\right), f_{CW}\right]$ $a_i := f(R_i, f_{CW})$ $S_i := C_2 \cdot a_i \cdot \frac{(Q_i)^2}{d^5}$

Results

i	Q_i	R_i	$f(R_i, f_{CW})$	a_i	S_i
0	200000	113353	0.01765	0.01765	2.752
1	220000	124689	0.01732	0.01732	3.267
2	240000	136024	0.01703	0.01703	3.821
3	260000	147359	0.01676	0.01676	4.415
4	280000	158695	0.01652	0.01652	5.047
5	300000	170030	0.01631	0.01631	5.718
6	320000	181365	0.01611	0.01611	6.427
7	340000	192701	0.01592	0.01592	7.173
8	360000	204036	0.01575	0.01575	7.956
9	380000	215371	0.01560	0.01560	8.776
10	400000	226707	0.01545	0.01545	9.633

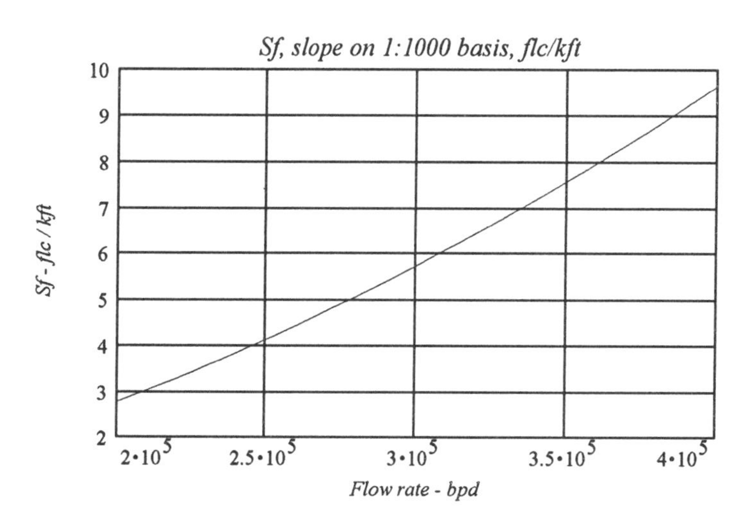

Computer Computation Template

CCT No. 14-3MC: This template is a working template for S on a 1:1000 basis (mlc / km) calculated using the Colebrook-White function with input in the pipeline MKS system. The plot shows the value of S for a range of Q from Q_0 to 2 x Q_0.

References: Chapter 14. Equations 14-11, 14-12, and 12-15.

Input

Factors $i := 0, 1 .. 10$

$C_1 := 3.536777 \cdot 10^5$ $C_2 := 6.377707 \cdot 10^9$ $C_3 := 6.377707 \cdot 10^8$ $K := \frac{3.6}{2.302585}$ $k := \frac{2.0}{2.302585}$

Input Data $Q_0 := 1325$ m3 / h $D := 609.6$ mm

$t := 9.525$ mm $\nu := 7.0$ cSt

$\varepsilon := 0.0127$ mm

Variables $d := D - 2 \cdot t$ $Q_i := Q_0 + 0.1 \cdot Q_0 \cdot i$ $R_i := \frac{C_1 \cdot Q_i}{d \cdot \nu}$

Equations

$f_{CW} := 0.03$ $f(R, f_{CW}) := root\left[\left(-2 \cdot log\left(\frac{\varepsilon \cdot d^{-1}}{3.7} + \frac{2.51}{R \cdot \sqrt{f_{CW}}}\right) - \frac{1}{\sqrt{f_{CW}}}\right), f_{CW}\right]$ $a_i := f(R_i, f_{CW})$ $S_i := C_2 \cdot a_i \cdot \frac{(Q_i)^2}{d^5}$

Results

i	Q_i	R_i	$f(R_i, f_{CW})$	a_i	S_i
0	1325	113362	0.01765	0.01765	2.752
1	1458	124699	0.01732	0.01732	3.268
2	1590	136035	0.01702	0.01702	3.822
3	1723	147371	0.01676	0.01676	4.416
4	1855	158707	0.01652	0.01652	5.048
5	1988	170044	0.01630	0.01630	5.719
6	2120	181380	0.01611	0.01611	6.428
7	2253	192716	0.01592	0.01592	7.174
8	2385	204052	0.01575	0.01575	7.957
9	2518	215388	0.01560	0.01560	8.777
10	2650	226725	0.01545	0.01545	9.635

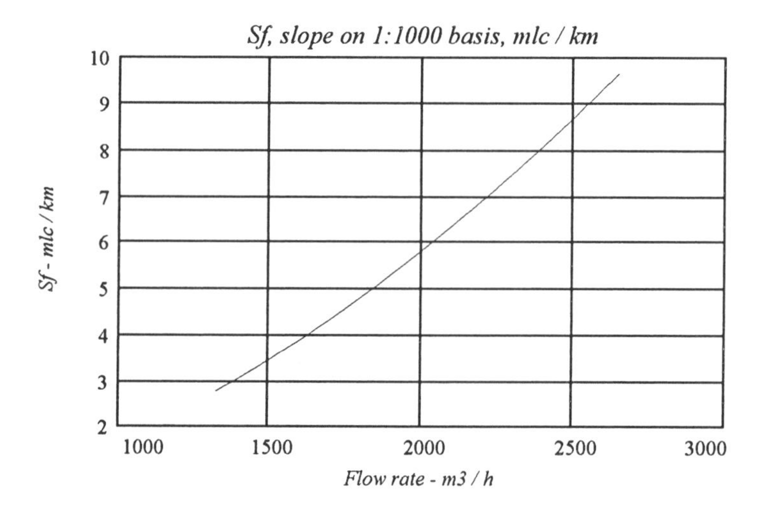

Computer Computation Template

CCT No. 14-4: This template is a working template for $\Delta \mathbf{p_f}$ with input in the pipeline FPS system. The plot shows the value of $\Delta \mathbf{p_f}$ for a range of Q from Q_0 to 2 x Q_0.

References: Chapter 14. Equation 14-12.

Input

Factors $i := 0, 1 .. 10$

$C_1 := 92.241232$ $C_2 := 0.02647271$ $C_3 := 0.01147629$ $K := \frac{3.6}{2.302585}$ $k := \frac{2.0}{2.302585}$

Input Data $Q_0 := 200000$ bpd $D := 24.00$ inches

$t := 0.375$ inches $\nu := 7.0$ cSt $\rho := 0.845$

Variables $d := D - 2 \cdot t$ $Q_i := Q_0 + 0.1 \cdot i \cdot Q_0$

Equations

$$pf_i := C_3 \cdot \rho \cdot \left[\frac{4}{\left(K \cdot \ln\left(\frac{C_1 \cdot Q_i}{8 \cdot d \cdot \nu} \right) \right)^2} \right] \cdot \frac{(Q_i)^2}{d^5}$$

Results

Q_i	pf_i
200000	1.023
220000	1.213
240000	1.418
260000	1.637
280000	1.870
300000	2.117
320000	2.378
340000	2.652
360000	2.940
380000	3.241
400000	3.556

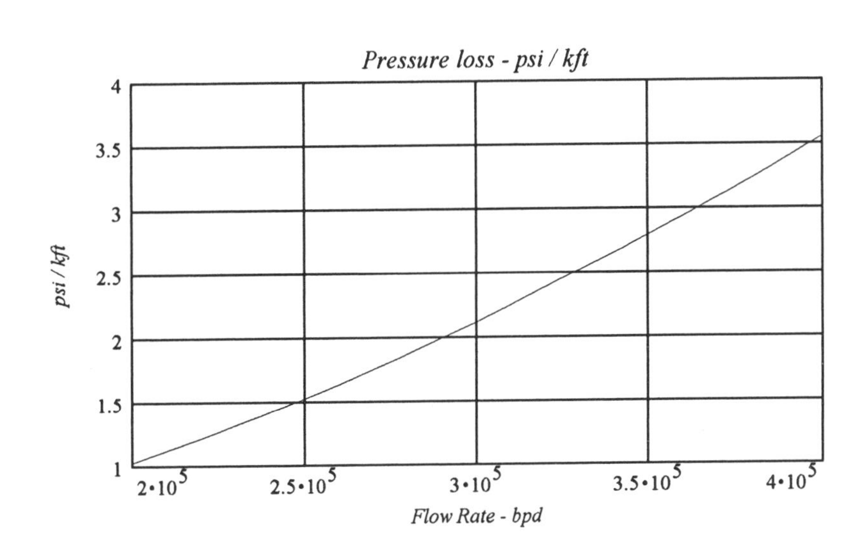

Computer Computation Template

CCT No. 14-4M: This template is a working template for Δp_f with input in the pipeline MKS system. The plot shows the value of Δp_f for a range of Q from Q_0 to 2 x Q_0.

References: Chapter 14. Equation 14-12.

Input

Factors $i := 0, 1 .. 10$

$C_1 := 3.536777 \cdot 10^5$ $C_2 := 6.377707 \cdot 10^9$ $C_3 := 6.377707 \cdot 10^8$ $K := \frac{3.6}{2.302585}$ $k := \frac{2.0}{2.302585}$

Input Data $Q_0 := 1325$ m3 /h $D := 609.6$ mm

$t := 9.525$ mm $\nu := 7.0$ cSt $\rho := 0.845$

Variables $d := D - 2 \cdot t$ $Q_i := Q_0 + 0.1 \cdot i \cdot Q_0$

Equations

$$p_{f_i} := C_3 \cdot \rho \cdot \left[\frac{4}{\left(K \cdot \ln\left(\frac{C_1 \cdot Q_i}{8 \cdot d \cdot \nu} \right) \right)^2} \right] \cdot \frac{(Q_i)^2}{d^5}$$

Results

Q_i	p_{f_i}
1325	0.236
1458	0.280
1590	0.327
1723	0.378
1855	0.431
1988	0.488
2120	0.549
2253	0.612
2385	0.678
2518	0.748
2650	0.820

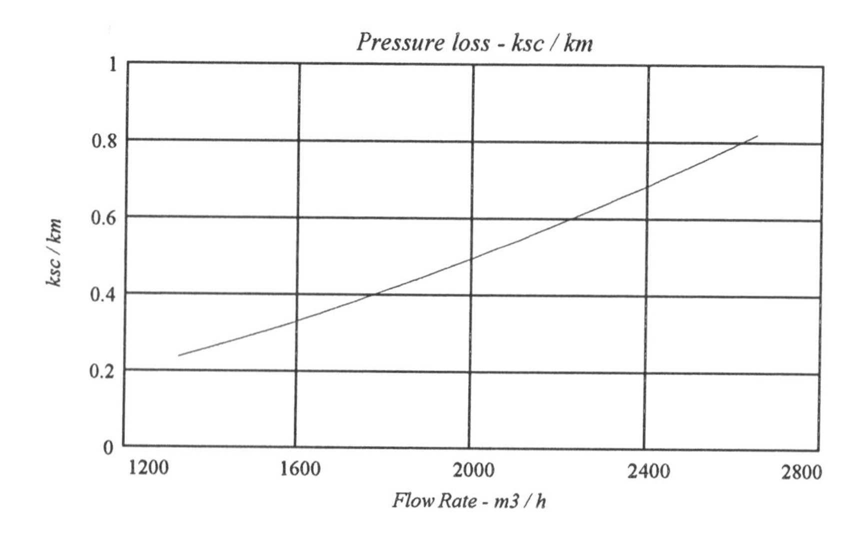

Computer Computation Template

CCT No. 14-4C: This template is a working template for p_f calculated using the Colebrook-White function with input in the pipeline FPS system. The plot shows the value of p_f for a range of Q from Q_0 to 2 x Q_0

References: Chapter 14. Equations 14-11, 14-12, and 12-15.

Input

Factors $i := 0, 1 .. 10$

$C_1 := 92.241232$ $C_2 := 0.02647271$ $C_3 := 0.01147629$ $K := \dfrac{3.6}{2.302585}$ $k := \dfrac{2.0}{2.302585}$

Input Data $Q_0 := 200000$ bpd $D := 24.00$ inches

$t := 0.375$ inches $\nu := 7.0$ cSt $\rho := 0.845$

$\varepsilon := 0.0005$ inches

Variables $d := D - 2 \cdot t$ $Q_i := Q_0 + 0.1 \cdot Q_0 \cdot i$ $R_i := \dfrac{C_1 \cdot Q_i}{d \cdot \nu}$

Equations

$f_{CW} := 0.03$ $f(R, f_{CW}) := root\left[\left(-2 \cdot log\left(\dfrac{\varepsilon d^{-1}}{3.7} + \dfrac{2.51}{R \cdot \sqrt{f_{CW}}}\right) - \dfrac{1}{\sqrt{f_{CW}}}\right), f_{CW}\right]$ $a_i := f(R_i, f_{CW})$ $p_{f_i} := C_3 \cdot \rho \cdot a_i \cdot \dfrac{(Q_i)^2}{d^5}$

Results

i	Q_i	R_i	$f(R_i, f_{CW})$	a_i	p_{f_i}
0	200000	113353	0.01765	0.01765	1.008
1	220000	124689	0.01732	0.01732	1.197
2	240000	136024	0.01703	0.01703	1.400
3	260000	147359	0.01676	0.01676	1.617
4	280000	158695	0.01652	0.01652	1.849
5	300000	170030	0.01631	0.01631	2.095
6	320000	181365	0.01611	0.01611	2.354
7	340000	192701	0.01592	0.01592	2.627
8	360000	204036	0.01575	0.01575	2.914
9	380000	215371	0.01560	0.01560	3.215
10	400000	226707	0.01545	0.01545	3.529

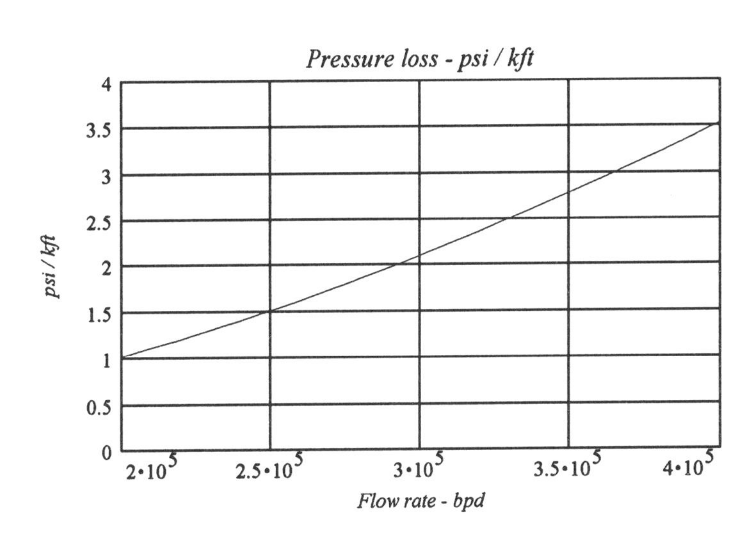

Computer Computation Template

CCT No. 14-4MC: This template is a working template for p_f calculated using the Colebrook-White function with input in the pipeline MKS system. The plot shows the value of p_f for a range of Q from Q_0 to 2 x Q_0

References: Chapter 14. Equations 14-11, 14-12, and 12-15.

Input

Factors $i := 0, 1 .. 10$

$C_1 := 3.536777 \cdot 10^5$ $C_2 := 6.377707 \cdot 10^9$ $C_3 := 6.377707 \cdot 10^8$ $K := \dfrac{3.6}{2.302585}$ $k := \dfrac{2.0}{2.302585}$

Input Data $Q_0 := 1325$ m3 / h $D := 609.6$ mm

$t := 9.525$ mm $\nu := 7.0$ cSt $\rho := 0.845$

$\varepsilon := 0.0127$ mm

Variables $d := D - 2 \cdot t$ $Q_i := Q_0 + 0.1 \cdot Q_0 \cdot i$ $R_i := \dfrac{C_1 \cdot Q_i}{d \cdot \nu}$

Equations

$$f_{CW} := 0.03 \qquad f(R, f_{CW}) := root\left[\left(-2 \cdot \log\left(\frac{\varepsilon d^{-1}}{3.7} + \frac{2.51}{R \cdot \sqrt{f_{CW}}}\right) - \frac{1}{\sqrt{f_{CW}}}\right), f_{CW}\right] \qquad a_i := f(R_i, f_{CW}) \qquad p_{f_i} := C_3 \cdot \rho \cdot a_i \cdot \frac{(Q_i)^2}{d^5}$$

Results

i	Q_i	R_i	$f(R_i, f_{CW})$	a_i	p_{f_i}
0	1325	113362	0.01765	0.01765	0.233
1	1458	124699	0.01732	0.01732	0.276
2	1590	136035	0.01702	0.01702	0.323
3	1723	147371	0.01676	0.01676	0.373
4	1855	158707	0.01652	0.01652	0.427
5	1988	170044	0.01630	0.01630	0.483
6	2120	181380	0.01611	0.01611	0.543
7	2253	192716	0.01592	0.01592	0.606
8	2385	204052	0.01575	0.01575	0.672
9	2518	215388	0.01560	0.01560	0.742
10	2650	226725	0.01545	0.01545	0.814

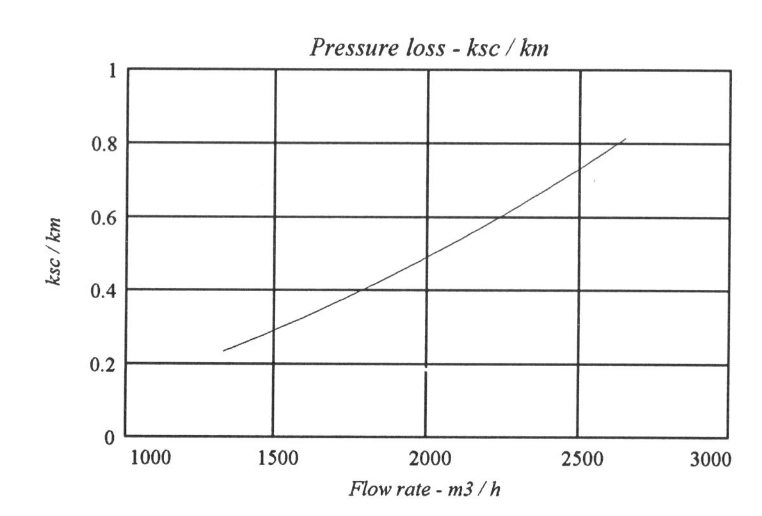

15 The Flow Rate Equation: Q_f and Q_p

Introduction

The oil pipeline industry has long used the *unit friction loss* equation as its fundamental equation of hydraulics:

- Input the characteristics of the oil
- Input the characteristics of the pipe
- Input the flow rate
- Calculate

The result is the *unit friction loss* in flc/kft, mlc/km, psi/mile,...etc. To obtain *total* friction loss, multiply by length.

The gas pipeline industry, however, has always used another fundamental equation—the *flow rate* equation—as its basic formula:

- Input the characteristics of the gas
- Input the characteristics and *length* of the pipe
- Input pressure loss over the said *length*
- Calculate

The result is *flow rate* in mmcfd, m^3/hr,...etc. For some special cases the result may be expressed in lb/hr or similar.

There is a reason for this difference in philosophies between the two similar industries. The unit friction loss in a liquid pipeline in steady flow is a valuable parameter, whereas the term *unit friction loss* in a gas line is meaningless; the rate of friction loss per unit of length in gas pipe flow varies from point to point along the length of the pipe as the density and flowing velocity of the gas vary. The meaningful pressure parameter in gas pipeline flow is the end–to–end loss of pressure over the length of the section.

In contrast, however, in steady state flow, the *flow rate* is a constant by definition, so it is this value that is calculated.

There is another good reason why oil pipeliners usually do not use the flow rate equation.

Gas pipeliners most often use some form of the friction factor that does not depend on flow rate, and the formula for flow rate is therefore a simple, *explicit* equation.

In oil pipeline terms, however, the friction factor—in *whatever* form—is dependent in some way on the flow rate, and so the flow rate formula is *implicit* in flow rate and a straightforward solution would ordinarily appear to be impossible without some kind of iterative algorithm.

There are certain non–iterative solutions, but they are reasonably complex—especially when the friction factor is stated in implicit form—and it is easy to make numeric mistakes when using them. However, today's pipeliner doesn't have this worry, because any flow rate equation, once it is programmed into a calculator or computer, is always available; and a correct answer only awaits a correct set of inputs.

Therefore, this chapter is devoted to the flow rate equation for oil pipelines.

The Flow Rate Formula

Derivation of Q_f

Inasmuch as the friction factor is expressed in terms of the Reynolds number N_R which, in turn, is a function of Q, the flow rate, it would seem that an explicit equation for flow rate would be impossible. However, solving the $T = A + B\log(N_R T)$ form of the pipe law in terms of $N_R T$ instead of T leads to a solution for $N_R T$ which, if divided by T, yields N_R which in turn yields Q. The derivation, in terms of T_F (Equation 13-7), $\mathbf{N_R}$ (Equation 14-3), and $\mathbf{S_f}$ (Equation 14-11), follows:

Take

$$\mathbf{N_R} = \frac{C_1 Q}{d\nu}$$

and

$$\mathbf{S_F} = \frac{C_2 Q^2}{T_D{}^2 d^5} \text{ or } Q = T_D\sqrt{\frac{\mathbf{S_f} d^5}{C_2}}$$

Substitute for Q in $\mathbf{N_R}$

$$\mathbf{N_R} = \frac{C_1}{d\nu}\left(T_D\sqrt{\frac{\mathbf{S_f} d^5}{C_2}}\right)$$

and

$$\frac{T_D}{\mathbf{N_R}} = \frac{d\nu}{C_1}\sqrt{\frac{C_2}{\mathbf{S_f} d^5}} = \frac{\nu}{C_1}\sqrt{\frac{C_2}{\mathbf{S_f} d^3}}$$

But

$$T_F = 4\log\left(\frac{\mathbf{N_R}}{T_F}\right) - 0.6 = -4\log\left(\frac{1.4125 \cdot T_F}{\mathbf{N_R}}\right)$$

and

$$T_D = -2\log\left(2.8250\frac{T_D}{\mathbf{N_R}}\right) = -2\log\left(2.8250\frac{\nu}{C_1}\sqrt{\frac{C_2}{\mathbf{S_f} d^3}}\right)$$

Thus

$$\mathbf{Q_f} = \sqrt{\frac{\mathbf{S_f} d^5}{C_2}} \times T_D$$

$$= \sqrt{\frac{\mathbf{S_f} d^5}{C_2}} \times \left(-2\log\left(2.8250\frac{\nu}{C_1}\sqrt{\frac{C_2}{\mathbf{S_f} d^3}}\right)\right) \tag{15-1}$$

Note this equation has been given a function name, $\mathbf{Q_f}$. As used herein, $\mathbf{Q_f}$ should be understood to mean the flow rate calculated from $\mathbf{S_f}$ in flc/kft or mlc/km.

For the Colebrook–White function, the formula is

$$Q_f = \sqrt{\frac{S_f d^5}{C_2}} \times \left(-2\log\left(\frac{e}{3.7d} + \frac{2.51\nu}{C_1}\sqrt{\frac{C_2}{S_f d^3}}\right)\right) \tag{15-2}$$

Note that the constants C_1 and C_2 are the same as for Equation 15-1 if S_f is given in terms of flc/kft or mlc/km.

Derivation of Q_p

A similar equation can be written if the unit friction loss is given in terms of pressure instead of head, i.e., as $\mathbf{\Delta p_f}$ instead of $\mathbf{S_f}$.

Replacing $\mathbf{S_f}$ with $\mathbf{\Delta p_f}$ and C_2 with $C_3\rho$ in Equation 15-1 yields

$$\mathbf{Q_p} = \sqrt{\frac{\mathbf{\Delta p_f} d^5}{C_3\rho}} \times \left(-2\log\left(2.8250\frac{\nu}{C_1}\sqrt{\frac{C_3\rho}{\mathbf{\Delta p_f} d^3}}\right)\right) \tag{15-3}$$

For the Colebrook–White function, the equation is

$$Q_p = \sqrt{\frac{\Delta p_f d^5}{C_3\rho}} \times \left(-2\log\left(\frac{e}{3.7d} + \frac{2.51\nu}{C_1}\sqrt{\frac{C_3\rho}{\Delta p_f d^3}}\right)\right) \text{(15-4}$$

Note

It should be pointed out that the derivation of the flow rate formula depends on having a transmission factor function involving both N_R and T, so the $\mathbf{f_{OIL}}$ function is not used in the derivation. The Monograph No. 9 smooth pipe function, $T_F = 4\log(N_R / T_F) - 0.6$, however, does satisfy this requirement and so it is used in the derivation. As pointed out in Chapter 13, the two equations yield nearly identical answers.

Table 15-1
Working Equations for Flow Rate Calculations

Equations Explicit in Unit Head Loss S_f

For unit friction loss expressed in terms of $\mathbf{S_f}$

$$\mathbf{Q_f} = \sqrt{\frac{\mathbf{S_f} d^5}{C_2}} \times \left(-2\log\left(2.8250\frac{\nu}{C_1}\sqrt{\frac{C_2}{\mathbf{S_f} d^3}}\right)\right)$$

For unit friction loss expressed in terms of the Colebrook–White function.

$$Q_f = \sqrt{\frac{S_f d^5}{C_2}} \times \left(-2\log\left(\frac{e}{3.7d} + \frac{2.51\nu}{C_1}\sqrt{\frac{C_2}{S_f d^3}}\right)\right)$$

Equations Explicit in Unit Pressure Loss Δp_f

For unit friction loss expressed in terms of $\mathbf{\Delta p_f}$

$$\mathbf{Q_p} = \sqrt{\frac{\mathbf{\Delta p_f} d^5}{C_3\rho}} \times \left(-2\log\left(2.8250\frac{\nu}{C_1}\sqrt{\frac{C_3\rho}{\mathbf{\Delta p_f} d^3}}\right)\right)$$

For unit friction loss expressed in terms of the Colebrook–White function.

$$Q_p = \sqrt{\frac{\Delta p_f d^5}{C_3\rho}} \times \left(-2\log\left(\frac{e}{3.7d} + \frac{2.51\nu}{C_1}\sqrt{\frac{C_3\rho}{\Delta p_f d^3}}\right)\right)$$

Working Equations

Calculators

Table 15-1 is a compilation of working equations for flow rate functions $\mathbf{Q_f}$ and $\mathbf{Q_p}$ based on the $\mathbf{f_{OIL}}$ function. Using the proper C–constants, the result will be in bpd or m³/hr.

Similar equations are provided for Q_f and Q_p based on the Colebrook–White function.

All of these equations accept input on a 1:1000 basis, i.e., flc/kft, mlc/km, psi/kft, or ksc/km.

Computers

The Computer Computation Templates included in this chapter use the C_1, C_2, C_3 concept to produce flow rate equations for the pipeline FPS and pipeline MKS systems of units that are the same except for the values of the C–factors.

There are twelve **CCT**s provided, along the same lines as the twelve templates included in the previous chapter. **CCT**s Nos. 15-1, 15-1M, 15-2, and 15-2M are the reference templates. These include not only the basic flow rate functions $\mathbf{Q_f}$ and $\mathbf{Q_p}$ in terms of unit head loss and unit pressure loss, respectively, but also the derivatives of these functions.

The remaining eight templates are working templates. Each template accepts either FPS or MKS units, in terms of head loss or pressure loss on a 1:1000 basis. Four of the templates are based on the $\mathbf{Q_f}$ and $\mathbf{Q_p}$ functions which use the $\mathbf{f_{OIL}}$ function, and the other four templates provide the Q_f and Q_p functions which are based on the Colebrook–White function. Note the results of $\mathbf{Q_f}$ and $\mathbf{Q_p}$ compare very closely with the results of Q_f and Q_p if the pipe roughness is taken as 0.0005 inch = 0.0127 mm as recommended in Chapter 13. The C–W function can be used to solve flow problems in very rough or corroded, encrusted, pipelines; it is not needed for modern trunk line pipe.

The eight **CCT**s from No. 15-3 through 15-4MC are organized as follows:

CCT No.	$\mathbf{Q_f}$ / Q_f	$\mathbf{Q_p}$/ Q_f	$\mathbf{f_{OIL}}$	C-W	FPS	MKS
15-3	♦		♦		♦	
15-3M	♦		♦			♦
15-3C	♦			♦	♦	
15-3MC	♦			♦		♦
15-4		♦	♦		♦	
15-4M		♦	♦			♦
15-4C		♦		♦	♦	
15-4MC		♦		♦		♦

Columns 2 and 3 indicate whether the input is in terms of unit head or pressure loss, columns 4 and 5 whether the flow rate equation is based on the $\mathbf{f}_{\mathbf{OIL}}$ function or the Colebrook–White function, and columns 6 and 7 indicate whether input and output are in the pipeline FPS or MKS system of units.

In Chapter 14 a note was made that the functions using the Colebrook–White function had to be coded for an iterative solution or entered into a *solver* program. This is not necessary for the flow rate equations; each of them is an ordinary function that accepts several variables but, in most cases, a solution is desired while varying only one of these, the others remaining fixed. Thus all the plots are in terms of varying friction loss, the characteristics of the oil and the pipe being taken as constant.

Computer Computation Template

CCT No. 15-1: Example of application and functions of $\mathbf{Q_f}$. Equation (1) is the equation for $\mathbf{Q_f}$. Equations (2), (3), and (4) are the derivatives of $\mathbf{Q_f}$ with respect to the slope ($\mathbf{S_f}$), diameter (d=D-2*t), and viscosity (ν). Input and output are in the pipeline FPS system of units.

References: Chapter 15. Equation 15-1.

Input

Factors $C_1 := 92.241232$ $C_2 := 0.02647271$ $C_3 := 0.01147629$ $K := \dfrac{3.6}{2.302585}$ $k := \dfrac{2.0}{2.302585}$

Input Data $S_f := 5.846$ flc / kft $D := 24.00$ inches $t := 0.375$ inches $\nu := 7.0$ cSt

Variables $d := D - 2 \cdot t$ inches

Equations

1. $\mathbf{Q_f}\ (\mathbf{S_f}, \nu, d)$

$$\sqrt{\frac{S_f d^5}{C_2}} \cdot \left(-k \cdot ln\left(2.8250 \cdot \frac{\nu}{C_1} \cdot \sqrt{\frac{C_2}{S_f d^3}}\right)\right) = 3.014 \cdot 10^5$$ bpd

2. $d\,\mathbf{Q_f} / d\,\mathbf{S_f}$

$$\frac{-1}{2 \cdot \sqrt{S_f}} \cdot \frac{d^{\left(\frac{5}{2}\right)}}{\sqrt{C_2}} \cdot k \cdot ln\left[2.825 \cdot \frac{\nu}{C_1} \cdot \frac{\sqrt{C_2}}{\left[\sqrt{S_f} d^{\left(\frac{3}{2}\right)}\right]}\right] + \frac{.5}{\sqrt{S_f}} \cdot \frac{d^{\left(\frac{5}{2}\right)}}{\sqrt{C_2}} \cdot k = 2.865 \cdot 10^4$$ bpd / flc / kft

3. $d\,\mathbf{Q_f} / d\,d$

$$\frac{-5}{2} \cdot \sqrt{S_f} \cdot \frac{d^{\left(\frac{3}{2}\right)}}{\sqrt{C_2}} \cdot k \cdot ln\left[2.825 \cdot \frac{\nu}{C_1} \cdot \frac{\sqrt{C_2}}{\left[\sqrt{S_f} d^{\left(\frac{3}{2}\right)}\right]}\right] + 1.5 \cdot \sqrt{S_f} \cdot \frac{d^{\left(\frac{3}{2}\right)}}{\sqrt{C_2}} \cdot k = 3.458 \cdot 10^4$$ bpd / inch

4. $d\,\mathbf{Q_f} / d\,\nu$

$$-1.0 \cdot \sqrt{S_f} \cdot \frac{d^{\left(\frac{5}{2}\right)}}{\sqrt{C_2}} \cdot \frac{k}{\nu} = -4.806 \cdot 10^3$$ bpd / cSt

Computer Computation Template

CCT No. 15-1M: Example of application and functions of $\mathbf{Q_f}$. Equation (1) is the equation for $\mathbf{Q_f}$. Equations (2), (3), and (4) are the derivatives of $\mathbf{Q_f}$ with respect to the slope ($\mathbf{S_f}$), diameter (d=D-2*t), and viscosity (ν). Input and output are in the pipeline MKS system of units.

References: Chapter 15. Equation 15-1.

Input

Factors $C_1 := 3.536777 \cdot 10^5$ $\quad C_2 := 6.377707 \cdot 10^9$ $\quad C_3 := 6.377707 \cdot 10^8$ $\quad K := \dfrac{3.6}{2.302585}$ $\quad k := \dfrac{2.0}{2.302585}$

Input Data $S_f := 5.846$ mlc / km $\quad D := 609.6$ mm $\quad t := 9.525$ mm $\quad \nu :=$ [illegible]

Variables $d := D - 2 \cdot t$ mm

Equations

1. $\mathbf{Q_f}\,(\mathbf{S_f}, \nu, d)$

$$\sqrt{\frac{S_f d^5}{C_2}} \cdot \left(-k \cdot \ln\left(2.8250 \cdot \frac{\nu}{C_1} \cdot \sqrt{\frac{C_2}{S_f d^3}}\right)\right) = 1.997 \cdot 10^3 \quad \text{m3 / h}$$

2. $d\,\mathbf{Q_f} / d\,\mathbf{S_f}$

$$\frac{-1}{2 \cdot \sqrt{S_f} \sqrt{C_2}} \cdot d^{\left(\frac{5}{2}\right)} \cdot k \cdot \ln\left[2.825 \cdot \frac{\nu}{C_1} \cdot \frac{\sqrt{C_2}}{\left[\sqrt{S_f}\, d^{\left(\frac{3}{2}\right)}\right]}\right] + \frac{.5}{\sqrt{S_f}} \cdot \frac{d^{\left(\frac{5}{2}\right)}}{\sqrt{C_2}} \cdot k = 189.82 \quad \text{m3 / h / mlc / km}$$

3. $d\,\mathbf{Q_f} / d\,d$

$$\frac{-5}{2} \cdot \sqrt{S_f} \frac{d^{\left(\frac{3}{2}\right)}}{\sqrt{C_2}} \cdot k \cdot \ln\left[2.825 \cdot \frac{\nu}{C_1} \cdot \frac{\sqrt{C_2}}{\left[\sqrt{S_f}\, d^{\left(\frac{3}{2}\right)}\right]}\right] + 1.5 \cdot \sqrt{S_f} \frac{d^{\left(\frac{3}{2}\right)}}{\sqrt{C_2}} \cdot k = 9.018 \quad \text{m3 / h / mm}$$

4. $d\,\mathbf{Q_f} / d\,\nu$

$$-1.0 \cdot \sqrt{S_f} \frac{d^{\left(\frac{5}{2}\right)}}{\sqrt{C_2}} \cdot \frac{k}{\nu} = -31.839 \quad \text{m3 / h / cSt}$$

Computer Computation Template

CCT No. 15-2: Examples of application and functions of $\mathbf{Q_p}$. Equation (1) is the equation for $\mathbf{Q_p}$. Equations (2), (3), (4) and (5) are the derivatives of $\mathbf{Q_p}$ with respect to the unit pressure loss ($\Delta \mathbf{p_f}$), diameter (d=D-2*t), viscosity (ν), and density (ρ, reference H_2O = 1.000). Input and output are in the pipeline FPS system of units.

References: Chapter 15. Equation 15-3.

Input

Factors $C_1 := 92.241232$ $C_2 := 0.02647271$ $C_3 := 0.01147629$ $K := \frac{3.6}{2.302585}$ $k := \frac{2.0}{2.302585}$

Input Data $\Delta p_f := 3.782$ psi / kft $D := 36.00$ inches $t := 0.375$ inches $\rho := 0.845$ $\nu := 7.0$ cSt

Variables $d := D - 2 \cdot t$ inches

Equations

1. $\mathbf{Q_p}(\Delta \mathbf{p_f}, \nu, d, \rho)$ $\sqrt{\frac{\Delta p_f d^5}{C_3 \cdot \rho}} \cdot \left(-k \cdot ln\left(2.8250 \cdot \frac{\nu}{C_1} \cdot \sqrt{\frac{C_3 \cdot \rho}{\Delta p_f d^3}} \right) \right) = 1.249 \cdot 10^6$ bpd

2. $d\,\mathbf{Q_p} / d\,\Delta \mathbf{p_f}$ $\frac{-1}{2 \cdot \sqrt{\Delta p_f}} \cdot \frac{d^{\left(\frac{5}{2}\right)}}{\left(\sqrt{C_3} \cdot \sqrt{\rho}\right)} \cdot k \cdot ln\left[2.825 \cdot \frac{\nu}{C_1} \cdot \sqrt{C_3} \cdot \frac{\sqrt{\rho}}{\left[\sqrt{\Delta p_f} d^{\left(\frac{3}{2}\right)} \right]} \right] + \frac{.5}{\sqrt{\Delta p_f}} \cdot \frac{d^{\left(\frac{5}{2}\right)}}{\left(\sqrt{C_3} \cdot \sqrt{\rho}\right)} \cdot k = 1.818 \cdot 10^5$ bpd / psi / kft

3. $d\,\mathbf{Q_p} / d\,d$ $\frac{-5}{2} \cdot \sqrt{\Delta p_f} \cdot \frac{d^{\left(\frac{3}{2}\right)}}{\left(\sqrt{C_3} \cdot \sqrt{\rho}\right)} \cdot k \cdot ln\left[2.825 \cdot \frac{\nu}{C_1} \cdot \sqrt{C_3} \cdot \frac{\sqrt{\rho}}{\left[\sqrt{\Delta p_f} d^{\left(\frac{3}{2}\right)} \right]} \right] + 1.5 \cdot \sqrt{\Delta p_f} \cdot \frac{d^{\left(\frac{3}{2}\right)}}{\left(\sqrt{C_3} \cdot \sqrt{\rho}\right)} \cdot k = 9.394 \cdot 10^4$ bpd / inch

4. $d\,\mathbf{Q_p} / d\,\nu$ $-1.0 \cdot \sqrt{\Delta p_f} \cdot \frac{d^{\left(\frac{5}{2}\right)}}{\left(\sqrt{C_3} \cdot \sqrt{\rho}\right)} \cdot \frac{k}{\nu} = -1.808 \cdot 10^4$ bpd / cSt

5. $d\,\mathbf{Q_p} / d\,\rho$ $\frac{1}{2} \cdot \sqrt{\Delta p_f} \cdot \frac{d^{\left(\frac{5}{2}\right)}}{\left[\sqrt{C_3} \cdot \rho^{\left(\frac{3}{2}\right)} \right]} \cdot k \cdot ln\left[2.825 \cdot \frac{\nu}{C_1} \cdot \sqrt{C_3} \cdot \frac{\sqrt{\rho}}{\left[\sqrt{\Delta p_f} d^{\left(\frac{3}{2}\right)} \right]} \right] - .5 \cdot \sqrt{\Delta p_f} \cdot \frac{d^{\left(\frac{5}{2}\right)}}{\left[\sqrt{C_3} \cdot \rho^{\left(\frac{3}{2}\right)} \right]} \cdot k = -8.137 \cdot 10^5$ bpd / ρ

Computer Computation Template

CCT No. 15-2M: Examples of application and functions of $\mathbf{Q_p}$. Equation (1) is the equation for $\mathbf{Q_p}$. Equations (2), (3), (4) and (5) are the derivatives of $\mathbf{Q_p}$ with respect to the unit pressure loss ($\Delta \mathbf{p_f}$), diameter (d=D-2*t), viscosity (ν), and density (ρ, reference H_2O = 1.000). Input and output are in the pipeline MKS system of units.

References: Chapter 15. Equation 15-3.

Input

Factors $C_1 := 3.536777 \cdot 10^5$ $C_2 := 6.377707 \cdot 10^9$ $C_3 := 6.377707 \cdot 10^8$ $K := \frac{3.6}{2.302585}$ $k := \frac{2.0}{2.302585}$

Input Data $\Delta p_f := 0.8725$ ksc / km $D := 914.4$ mm $t := 9.525$ mm $\nu := 7.0$ cSt $\rho := 0.845$

Variables $d := D - 2 \cdot t$ mm

Equations

1. $\mathbf{Q_p}$ ($\Delta \mathbf{p_f}$, ν, d, ρ)

$$\sqrt{\frac{\Delta p_f d^5}{C_3 \cdot \rho}} \cdot \left(-k \cdot ln \left(2.8250 \cdot \frac{\nu}{C_1} \cdot \sqrt{\frac{C_3 \cdot \rho}{\Delta p_f d^3}} \right) \right) = 8.272 \cdot 10^3 \quad \text{m3 / h}$$

2. d $\mathbf{Q_p}$ / d $\Delta \mathbf{p_f}$

$$\frac{-1}{2 \cdot \sqrt{\Delta p_f}} \cdot \frac{d^{\left(\frac{5}{2}\right)}}{\left(\sqrt{C_3} \cdot \sqrt{\rho}\right)} \cdot k \cdot ln \left[2.825 \cdot \frac{\nu}{C_1} \cdot \sqrt{C_3} \cdot \frac{\sqrt{\rho}}{\left[\sqrt{\Delta p_f} d^{\left(\frac{3}{2}\right)} \right]} \right] + \frac{.5}{\sqrt{\Delta p_f}} \cdot \frac{d^{\left(\frac{5}{2}\right)}}{\left(\sqrt{C_3} \cdot \sqrt{\rho}\right)} \cdot k = 5.221 \cdot 10^3 \quad \text{m3 / h / ksc / km}$$

3. d $\mathbf{Q_p}$ / d d

$$\left[\frac{-5}{2} \cdot \sqrt{\Delta p_f} \cdot \frac{d^{\left(\frac{3}{2}\right)}}{\left(\sqrt{C_3} \cdot \sqrt{\rho}\right)} \cdot k \right] \cdot ln \left[2.825 \cdot \frac{\nu}{C_1} \cdot \sqrt{C_3} \cdot \frac{\sqrt{\rho}}{\left[\sqrt{\Delta p_f} d^{\left(\frac{3}{2}\right)} \right]} \right] + 1.5 \cdot \sqrt{\Delta p_f} \cdot \frac{d^{\left(\frac{3}{2}\right)}}{\left(\sqrt{C_3} \cdot \sqrt{\rho}\right)} \cdot k = 24.501 \quad \text{m3 / h / mm}$$

4. d $\mathbf{Q_p}$ / d ν

$$-1.0 \cdot \sqrt{\Delta p_f} \cdot \frac{d^{\left(\frac{5}{2}\right)}}{\left(\sqrt{C_3} \cdot \sqrt{\rho}\right)} \cdot \frac{k}{\nu} = -119.762 \quad \text{m3 / h / cSt}$$

5. d $\mathbf{Q_p}$ / d ρ

$$\frac{1}{2} \cdot \sqrt{\Delta p_f} \cdot \frac{d^{\left(\frac{5}{2}\right)}}{\left[\sqrt{C_3} \cdot \rho^{\left(\frac{3}{2}\right)} \right]} \cdot k \cdot ln \left[2.825 \cdot \frac{\nu}{C_1} \cdot \sqrt{C_3} \cdot \frac{\sqrt{\rho}}{\left[\sqrt{\Delta p_f} d^{\left(\frac{3}{2}\right)} \right]} \right] - .5 \cdot \sqrt{\Delta p_f} \cdot \frac{d^{\left(\frac{5}{2}\right)}}{\left[\sqrt{C_3} \cdot \rho^{\left(\frac{3}{2}\right)} \right]} \cdot k = -5.391 \cdot 10^3 \quad \text{m3 / h / } \rho$$

Computer Computation Template

CCT No. 15-3: This template is a working template for $\mathbf{Q_f}$ with input in the pipeline FPS system. The plot shows the value of $\mathbf{Q_f}$ for a range of $\mathbf{S_f}$ from S_0 to 10 x S_0.

References: Chapter 15. Equation 15-1.

Input

Factors $i := 0, 1 .. 9$

$C_1 := 92.241232$ $C_2 := 0.02647271$ $C_3 := 0.01147629$ $K := \dfrac{3.6}{2.302585}$ $k := \dfrac{2.0}{2.302585}$

Input Data $S_0 := 1$ flc / kft $D := 24.00$ inches

$t := 0.375$ inches $\nu := 7.0$ cSt

Variables $d := D - 2 \cdot t$ $S_{f_i} := S_0 + S_0 \cdot i$

Equations

$$Q_i := \sqrt{\frac{S_{f_i} \cdot d^5}{C_2}} \cdot \left(-k \cdot \ln\left(2.8250 \cdot \frac{\nu}{C_1} \cdot \sqrt{\frac{C_2}{S_{f_i} \cdot d^3}} \right) \right)$$

Results

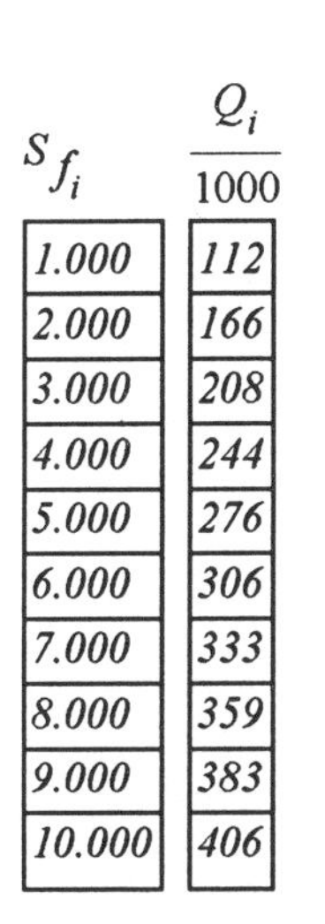

S_{f_i}	$\frac{Q_i}{1000}$
1.000	112
2.000	166
3.000	208
4.000	244
5.000	276
6.000	306
7.000	333
8.000	359
9.000	383
10.000	406

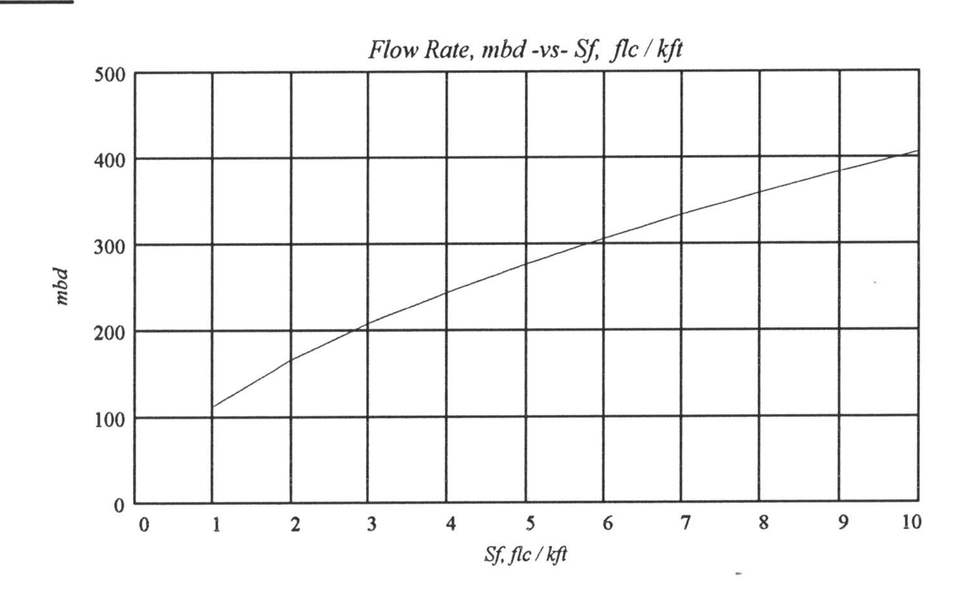

Computer Computation Template

CCT No. 15-3M: This template is a working template for $\mathbf{Q_f}$ with input in the pipeline MKS system. The plot shows the value of $\mathbf{Q_f}$ for a range of $\mathbf{S_f}$ from S_0 to 10 x S_0.

References: Chapter 15. Equation 15-1.

Input

Factors $i := 0, 1 .. 9$

$C_1 := 3.536777 \cdot 10^5$ $C_2 := 6.377707 \cdot 10^9$ $C_3 := 6.377707 \cdot 10^8$ $K := \frac{3.6}{2.302585}$ $k := \frac{2.0}{2.302585}$

Input Data $S_0 := 1$ mlc / km $D := 609.6$ mm

$t := 9.525$ mm $\nu := 7.0$ cSt

Variables $d := D - 2 \cdot t$ $S_{f_i} := S_0 + S_0 \cdot i$

Equations

$$Q_i := \sqrt{\frac{S_{f_i} \cdot d^5}{C_2}} \cdot \left(-k \cdot \ln \left(2.8250 \cdot \frac{\nu}{C_1} \cdot \sqrt{\frac{C_2}{S_{f_i} \cdot d^3}} \right) \right)$$

Results

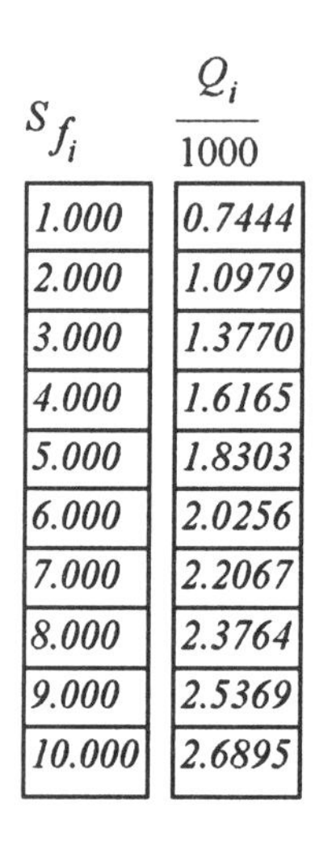

S_{f_i}	$\frac{Q_i}{1000}$
1.000	0.7444
2.000	1.0979
3.000	1.3770
4.000	1.6165
5.000	1.8303
6.000	2.0256
7.000	2.2067
8.000	2.3764
9.000	2.5369
10.000	2.6895

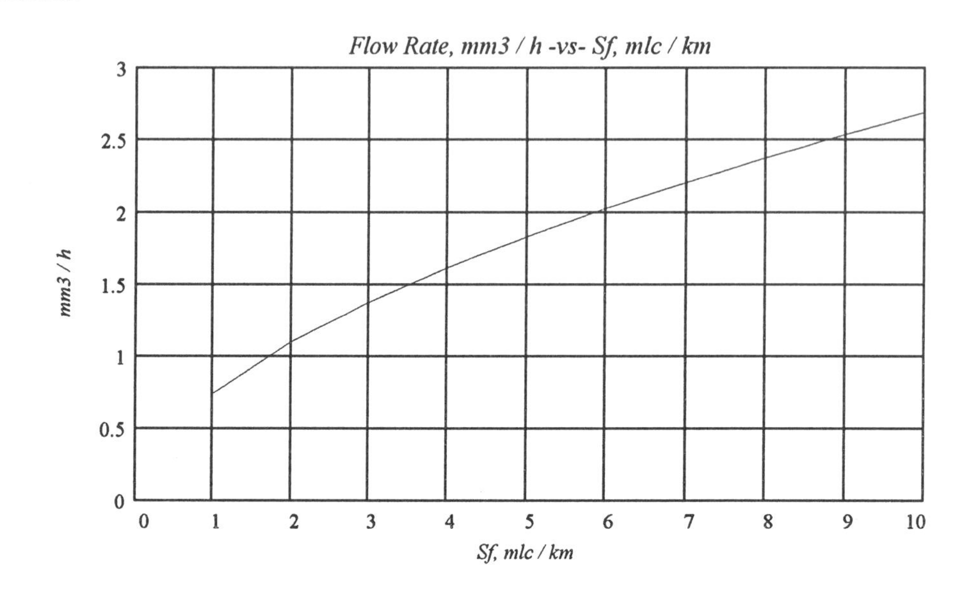

Computer Computation Template

CCT No. 15-3C: This template is a working template for Q_f using the Colebrook–White function with input in the pipeline FPS system. The plot shows the value of Q_f for a range of S_f from S_0 to 10 x S_0. Note S_f is on a 1:1000 basis, i.e., flc / kft.

References: Chapter 15. Equation 15-3.

Input

Factors $i := 0, 1 .. 9$

$C_1 := 92.241232$ $C_2 := 0.02647271$ $C_3 := 0.01147629$ $K := \frac{3.6}{2.302585}$ $k := \frac{2.0}{2.302585}$

Input Data $S_0 := 1$ flc / kft $D := 24.00$ inches

$t := 0.375$ inches $\nu := 7.0$ cSt

$\varepsilon := 0.0005$ inches

Variables $d := D - 2 \cdot t$ $S_{f_i} := S_0 + S_0 \cdot i$

Equations

$$Q_i := \sqrt{\frac{S_{f_i} \cdot d^5}{C_2}} \cdot \left(-k \cdot \ln \left(\frac{\varepsilon}{3.7 \cdot d} + \frac{2.51 \cdot \nu}{C_1} \cdot \sqrt{\frac{C_2}{S_{f_i} \cdot d^3}} \right) \right)$$

Results

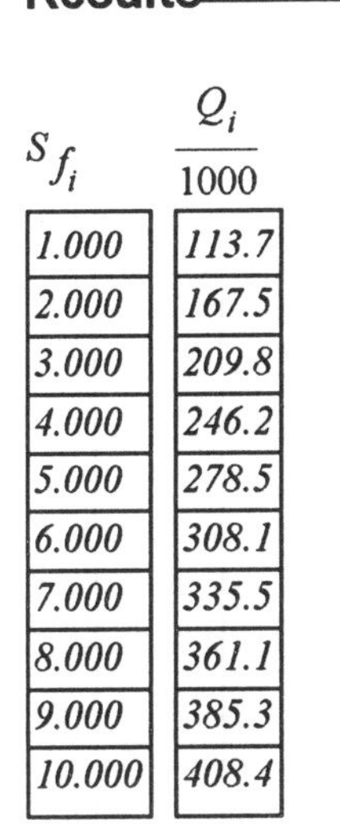

S_{f_i}	$\frac{Q_i}{1000}$
1.000	113.7
2.000	167.5
3.000	209.8
4.000	246.2
5.000	278.5
6.000	308.1
7.000	335.5
8.000	361.1
9.000	385.3
10.000	408.4

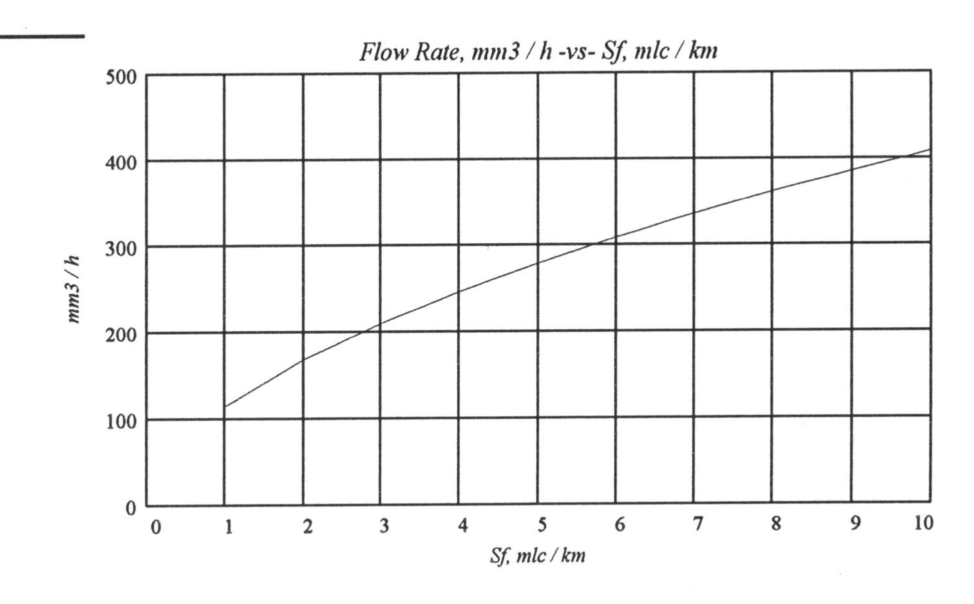

Computer Computation Template

CCT No. 15-3MC: This template is a working template for Q_f using the Colebrook–White function with input in the pipeline MKS system. The plot shows the value of Q_f for a range of S_f from S_0 to 10 x S_0. Note S_f is on a 1:1000 basis, i.e., mlc / km.

References: Chapter 15. Equation 15-2.

Input

Factors $i := 0, 1 .. 9$

$C_1 := 3.536777 \cdot 10^5$ $C_2 := 6.377707 \cdot 10^9$ $C_3 := 6.377707 \cdot 10^8$ $K := \frac{3.6}{2.302585}$ $k := \frac{2.0}{2.302585}$

Input Data $S_0 := 1$ mlc / km $D := 609.6$ mm

$t := 9.525$ mm $\nu := 7.0$ cSt

$\varepsilon := 0.0127$ mm

Variables $d := D - 2 \cdot t$ $S_{f_i} := S_0 + S_0 \cdot i$

Equations

$$Q_i := \sqrt{\frac{S_{f_i} \cdot d^5}{C_2}} \cdot \left(-k \cdot \ln \left(\frac{\varepsilon}{3.7 \cdot d} + \frac{2.51 \cdot \nu}{C_1} \cdot \sqrt{\frac{C_2}{S_{f_i} \cdot d^3}} \right) \right)$$

Results

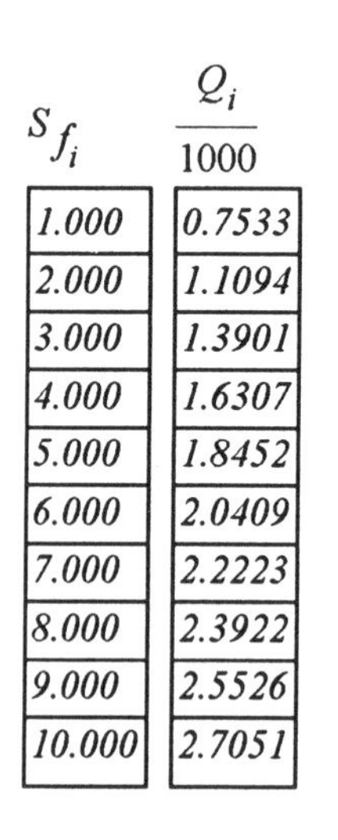

S_{f_i}	$\frac{Q_i}{1000}$
1.000	0.7533
2.000	1.1094
3.000	1.3901
4.000	1.6307
5.000	1.8452
6.000	2.0409
7.000	2.2223
8.000	2.3922
9.000	2.5526
10.000	2.7051

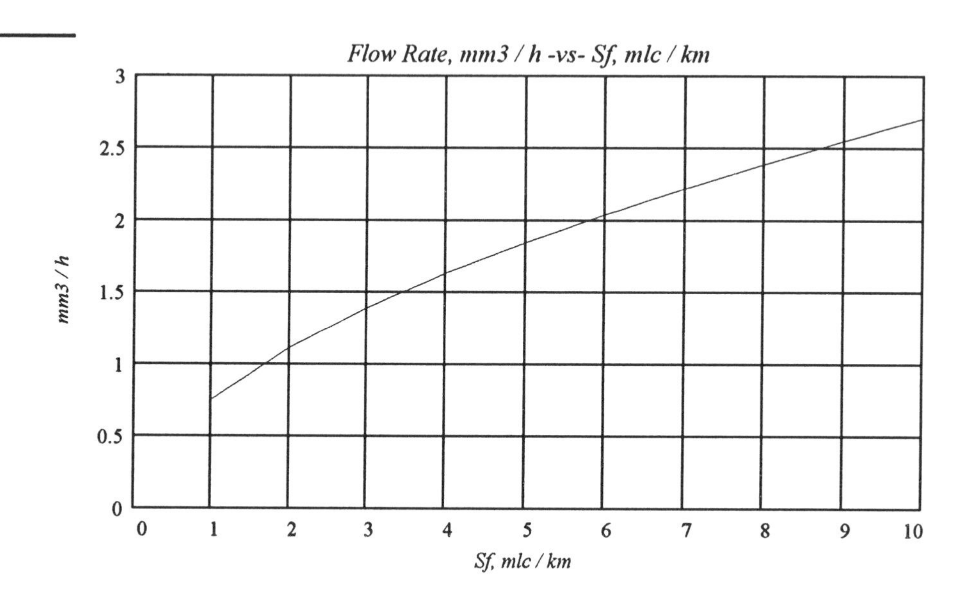

Computer Computation Template

CCT No. 15-4: This template is a working template for $\mathbf{Q_p}$ with input in the pipeline FPS system. The plot shows the value of $\mathbf{Q_p}$ for a range of $\mathbf{\Delta p_f}$ from Δp_0 to 10 x Δp_0.

References: Chapter 15. Equation 15-3.

Input

Factors $\quad i := 0, 1 .. 9$

$C_1 := 92.241232 \quad C_2 := 0.02647271 \quad C_3 := 0.01147629 \quad K := \frac{3.6}{2.302585} \quad k := \frac{2.0}{2.302585}$

Input Data $\quad \Delta p_0 := 1.000$ psi / kft $\quad D := 36.00$ inches

$t := 0.375$ inches $\quad \nu := 7.0$ cSt $\quad \rho := 0.845$

Variables $\quad d := D - 2 \cdot t \quad \Delta p_{f_i} := \Delta p_0 + \Delta p_0 \cdot i$

Equations

$$Q_i := \sqrt{\frac{\Delta p_{f_i} \cdot d^5}{C_3 \cdot \rho}} \cdot \left[-k \cdot ln \left[2.8250 \cdot \frac{\nu}{C_1} \cdot \sqrt{\frac{C_3 \cdot (\rho)}{\Delta p_{f_i} \cdot d^3}} \right] \right]$$

Results

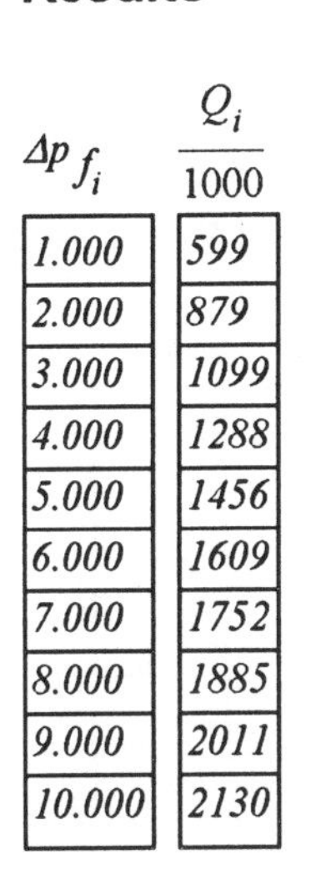

Δp_{f_i}	$\frac{Q_i}{1000}$
1.000	599
2.000	879
3.000	1099
4.000	1288
5.000	1456
6.000	1609
7.000	1752
8.000	1885
9.000	2011
10.000	2130

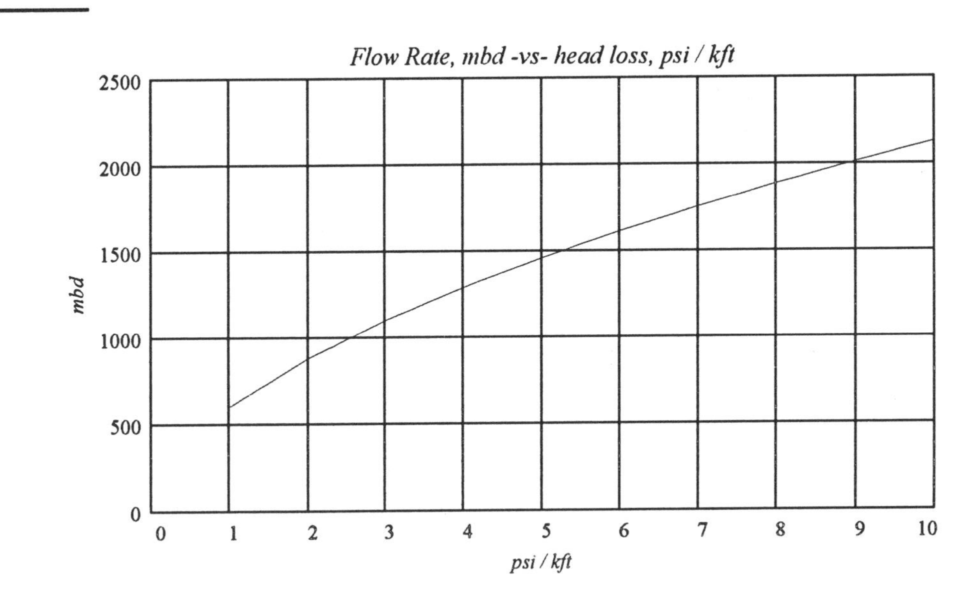

Computer Computation Template

CCT No. 15-4M: This template is a working template for $\mathbf{Q_p}$ with input in the pipeline MKS system. The plot shows the value of $\mathbf{Q_p}$ for a range of $\Delta \mathbf{p_f}$ from Δp_0 to 10 x Δp_0.

References: Chapter 15. Equation 15-3.

Input

Factors $i := 0, 1 .. 9$

$C_1 := 3.536777 \cdot 10^5$ $C_2 := 6.377707 \cdot 10$ $C_3 := 6.377707 \cdot 10^8$ $K := \frac{3.6}{2.302585}$ $k := \frac{2.0}{2.302585}$

Input Data $p_0 := 0.20$ ksc / km $D := 914.4$ mm

$t := 9.525$ mm $\nu := 7.0$ cSt $\rho := 0.845$

Variables $d := D - 2 \cdot t$ $p_{f_i} := p_0 + p_0 \cdot i$

Equations

$$Q_i := \sqrt{\frac{p_{f_i} \cdot d^5}{C_3 \cdot \rho}} \cdot \left(-k \cdot \ln \left(2.8250 \cdot \frac{\nu}{C_1} \cdot \sqrt{\frac{C_3 \cdot \rho}{p_{f_i} \cdot d^3}} \right) \right)$$

Results

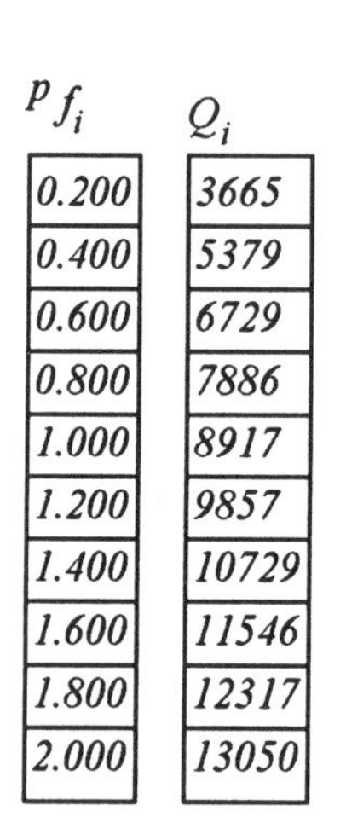

p_{f_i}	Q_i
0.200	3665
0.400	5379
0.600	6729
0.800	7886
1.000	8917
1.200	9857
1.400	10729
1.600	11546
1.800	12317
2.000	13050

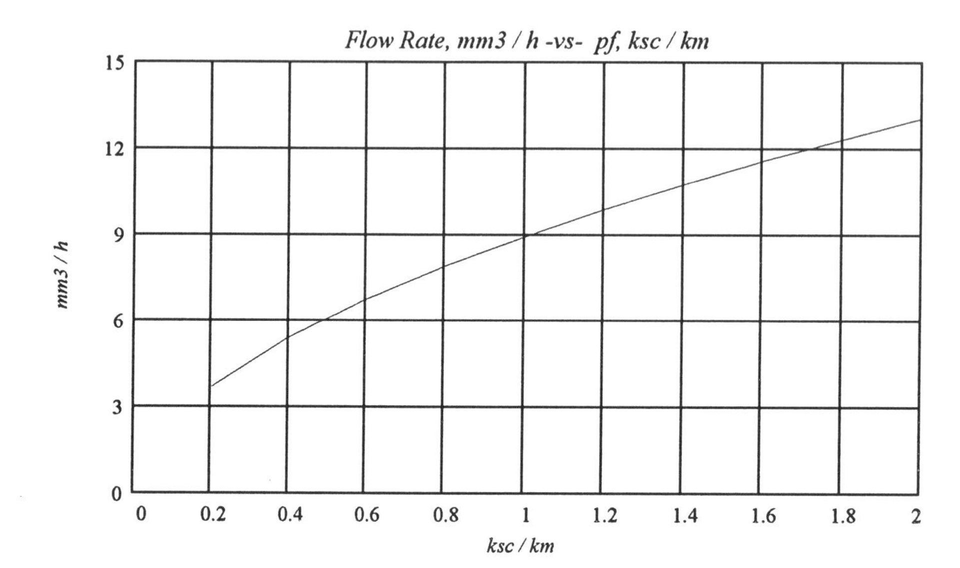

Computer Computation Template

CCT No. 15-4C: This template is a working template for Q_p using the Colebrook–While function with input in the pipeline FPS system. The plot shows the value of Q_p for a range of p_f from p_0 to 10 x p_0. Note p_f is on a 1:1000 basis, i.e., psi / kft.

References: Chapter 15. Equation 15-4.

Input

Factors $i := 0, 1 .. 9$

$C_1 := 92.241232$ $C_2 := 0.02647271$ $C_3 := 0.01147629$ $K := \frac{3.6}{2.302585}$ $k := \frac{2.0}{2.302585}$

Input Data $p_0 := 1.000$ psi / kft $D := 36.00$ inches

$t := 0.375$ inches $\nu := 7.0$ cSt $\rho := 0.845$

$\varepsilon := 0.0005$ inches

Variables $d := D - 2 \cdot t$ $p_{f_i} := p_0 + p_0 \cdot i$

Equations

$$Q_i := \sqrt{\frac{p_{f_i} \cdot d^5}{C_3 \cdot \rho}} \cdot \left[-k \cdot ln \left[\left(\frac{\varepsilon}{3.7 \cdot d} + \frac{2.51 \cdot \nu}{C_1} \right) \cdot \sqrt{\frac{C_3 \cdot (\rho)}{p_{f_i} \cdot d^3}} \right] \right]$$

Results

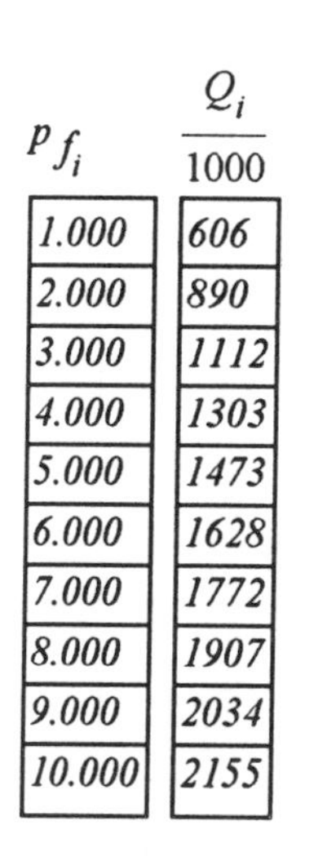

p_{f_i}	$\frac{Q_i}{1000}$
1.000	606
2.000	890
3.000	1112
4.000	1303
5.000	1473
6.000	1628
7.000	1772
8.000	1907
9.000	2034
10.000	2155

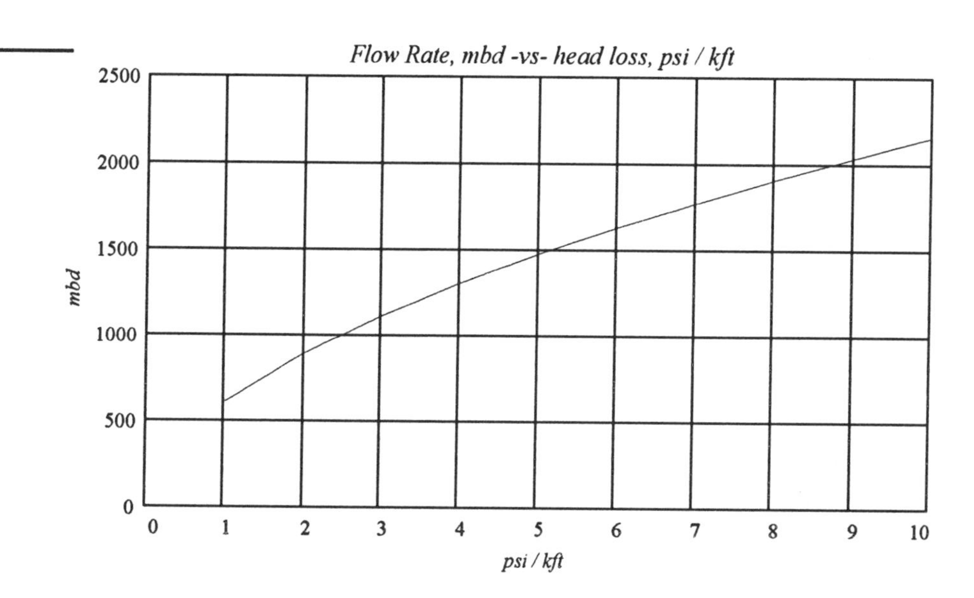

Computer Computation Template

CCT No. 15-4MC: This template is a working template for Q_p using the Colebrook–White function with input in the pipeline MKS system. The plot shows the value of Q_p for a range of p_f from p_0 to 10 x p_0. Note p_f is on a 1:1000 basis, i.e., ksc / km.

References: Chapter 15. Equation 15-4.

Input

Factors $i := 0, 1 .. 9$

$C_1 := 3.536777 \cdot 10^5$ $C_2 := 6.377707 \cdot 10^9$ $C_3 := 6.377707 \cdot 10^8$ $K := \dfrac{3.6}{2.302585}$ $k := \dfrac{2.0}{2.302585}$

Input Data $p_0 := 0.2$ ksc / km $D := 914.4$ mm

$t := 9.525$ mm $\nu := 7.0$ cSt $\rho := 0.845$

$\varepsilon := 0.0127$ mm

Variables $d := D - 2 \cdot t$ $\Delta p_{f_i} := p_0 + p_0 \cdot i$

Equations

$$Q_i := \sqrt{\frac{\Delta p_{f_i} \cdot d^5}{C_3 \cdot \rho}} \cdot \left[-k \cdot \ln \left[\left(\frac{\varepsilon}{3.7 \cdot d} + \frac{2.51 \cdot \nu}{C_1} \right) \cdot \sqrt{\frac{C_3 \cdot \rho}{\Delta p_{f_i} \cdot d^3}} \right] \right]$$

Results

Δp_{f_i}	Q_i
0.200	3682
0.400	5404
0.600	6760
0.800	7921
1.000	8956
1.200	9901
1.400	10776
1.600	11595
1.800	12370
2.000	13106

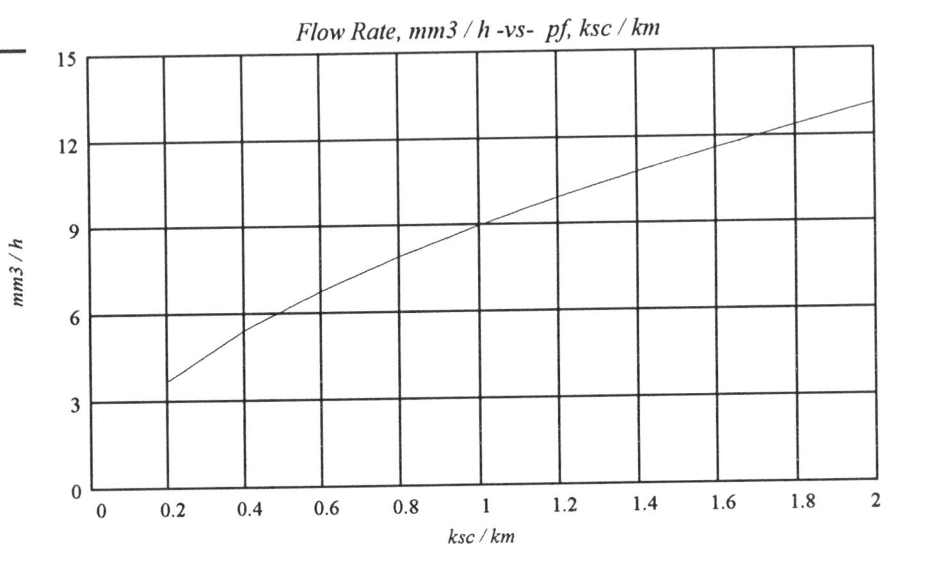

16

The Fundamental Calculations of Pipeline Hydraulics

Introduction

This is the last chapter of Volume I of *Hydraulics for Pipeliners*, Second Edition, and completes the chapters devoted to *fundamentals* of pipeline hydraulics. Volume II is devoted to *special topics*.

Chapters 1–9 covered the characteristics of fluids, with special emphasis on petroleum liquids; Chapters 10, 11 and 12 provided a review of classical hydrodynamics, theoretical fluid mechanics, and applied fluid mechanics (read *hydraulics*); Chapter 13 summarized the research leading to a new, more accurate, friction factor for trunk oil pipelines constructed with the kind of high quality pipe now routinely produced by the world's pipe mills; and, in Chapters 14 and 15, the two most basic pipeline working equations—the *unit friction loss* and the *flow rate* equations—were derived, and each was provided with constants of proportionality that allow using the equation in terms of the working units of both FPS and MKS pipeliners.

The concept of *slope:1000*, introduced in Chapter 14, provides for computation of head loss on a 1:1000 basis that yields the same numeric answer in both measurement systems, i.e., flc/kft or mlc/km.

With all that in hand, it is now possible to look at the different kinds of hydraulic circuits—there are only six of these—that are apt to be encountered by the working pipeliner and to develop methods to solve the problems presented by each of them—how to compute the unknown heads (or pressures) and flows in a given configuration of pipes.

Pipeline Hydraulics Units FPS and MKS Systems

The recommended units for use in pipeline hydraulic calculations are given for both the FPS and MKS systems in Table 16-1.

The MKS units can be converted into SI units if necessary; coherent SI MKS units do not fit the pipeline industry.

Table 16-1

Pipeline Computation Parameters FPS and MKS Systems

Unit	Symbol	Pipeline Unit FPS	Pipeline Unit MKS
Physical			
length	L	kft	km
diameter	d	in	mm
pipe wall	t	in	mm
Fluid			
density	ρ	H_2O=1.000	H_2O=1.000
viscosity	ν	cSt	cSt
Hydraulic			
volume	V	bbl	m^3
flow rate	Q	bpd	m^3/hr
head	h	flc	mlc
slope 1:1000	h/L	flc/kft	mlc/km
pressure	p	psi	ksc
Δp 1:1000	p/L	psi/kft	ksc/km

There are many sets of equivalencies included in the chapters on fluid characteristics that can be used if the original input data for a problem is in other than the recommended pipeline units.

The Tools

Table 16-2 is a concise summary of the functions and procedures developed in Chapters 13, 14, and 15. These are:

- The Reynolds number $\mathbf{N_R}$
- The Fanning transmission factor $\mathbf{T_{OIL}}$ and the Darcy friction factor $\mathbf{f_{OIL}}$ in terms of $\mathbf{N_R}$
- The unit head loss $\mathbf{S_f}$ and unit pressure loss $\mathbf{\Delta p_f}$ in terms of $\mathbf{f_{OIL}}$
- The flow rate $\mathbf{Q_f}$ and $\mathbf{Q_p}$ in terms of $\mathbf{S_f}$ and $\mathbf{\Delta p_f}$

The functions in boldface are functions defined earlier herein, and all are designed to work on the basis of the slope, S_f, on a 1:1000 basis, and, by incorporating the proper values of C_1, C_2, and C_3, can be used in the pipeline FPS system or the pipeline MKS system.

Table 16-3 is identical to Table 16-2 except that the transcendental equations have been written in terms of natural logarithms, which computers prefer, instead of common logarithms. The two constants K and k, being numerically 3.6 and 2 divided by ln 10, make the required conversions.

Essentially all of the tools required to solve almost any *steady–state* pipeline flow problem are included on Tables 16-2 or 16-3.

Basic Configurations of Oil Pipelines

Pipelines are non–linear devices. The flow of fluids in pipes resembles, to a certain degree, the flow of electric current in non–linear resistances. Kirchoff's Laws and Ohm's Law apply directly; and in the study of transient flow, especially pressure surges in long pipelines, there are close equivalents to the transmission line telegraph and telephone equations.

While there can be almost infinitely complex networks of pipes—the usual example is the domestic water distribution system of a large city—just as there can be infinitely complex networks of electrical devices—you may think about the logic circuits in the CPU chip of a computer—oil pipelines are usually built in rather simple configurations. Figure 16-1 shows the most common of these. There are six, keyed to examples of Figure 16-1:

Example 1: A *single constant diameter pipeline* running from one source to one sink

Example 2: A *single pipeline that is extended by one or more pipelines of different diameters in series* running from one source to one sink.

Example 3: *Two or more pipelines connected in parallel* running from one source to one sink.

Example 4: A *single pipeline that is extended by two or more pipelines in parallel* (loops) running from one source to one sink.

Example 5: A *single pipeline that divides flow into two or more pipelines* (branches) running from one source to two or more sinks.

Example 6: A *network of three or more pipelines*, which may run from one or more sources to one or more sinks, and which may have one or more cross-connections.

In these descriptions the terms *source* and *sink* are introduced without definition. The meanings should be clear.

The remainder of this chapter is dedicated to methods of solving each of the above configurations.

The parameters *Ha*, *Hb*,... and *Qa*, *Qb*,...etc., used in the following are taken from Figure 16-1.

Kirchoff's Laws for Pipelines

Kirchoff is credited with two laws that bear his name. Figure 16-2 is a comparison of the liquid flow and dc electric circuit forms of the two laws, together with a short writing of the laws in words. Law No. 1 is the continuity equation: *what goes in comes out unless consumed in the process*. Law No. 2 is the conservation of energy equations: *the sum of the energies injected into and withdrawn from a closed process is zero*. Note the common term for a closed *loop* in hydraulics is *mesh* in electrical engineering.

Kirchoff's laws govern the solution of flow problems in pipelines. In transient flow problems there are times the laws in their simple form shown do not apply, but in steady state flow the laws must be satisfied.

Solving for Heads and Flows in Complex Pipelines

Problems in pipeline hydraulics can always be resolved into a question of determining the unknown values of

(text continued on page 309)

Table 16-2

Review of Friction Loss and Flow Rate Functions From Chapters 13, 14, and 15.

Reynolds Number (dimensionless)

$$N_R = C_1\left(\frac{Q}{dv}\right)$$

Fanning Transmission Factor (dimensionless)

$$T_{OIL} = 3.6\log(N_R/8)$$

$$T_{OIL} = 3.6\log(C_1 Q/8dv)$$

Darcy Friction Factor (dimensionless)

$$f_{OIL} = 4/(T_{OIL})^2$$

$$f_{OIL} = 4/\left(3.6\log(C_1 Q/8dv)\right)^2$$

$$f = f_{OIL} = 4/\left(3.6\log(C_1 Q/8dv)\right)^2$$

Unit friction loss, flc/kft or mlc/km

$$S_f = C_2 f\left(\frac{Q^2}{d^5}\right)$$

$$S_f = C_2\left(\frac{4}{(3.6\log(N_R/8))^2}\right)\left(\frac{Q^2}{d^5}\right)$$

$$S_f = C_2\left(\frac{4}{\left(3.6\log(C_1 Q/8dv)\right)^2}\right)\left(\frac{Q^2}{d^5}\right)$$

Unit pressure loss, psi/kft or ksc/km

$$\Delta p_f = C_3\rho f\left(\frac{Q^2}{d^5}\right)$$

$$\Delta p_f = C_3\rho\left(\frac{4}{(3.6\log(N_R/8))^2}\right)\left(\frac{Q^2}{d^5}\right)$$

$$\Delta p_f = C_3\rho\left(\frac{4}{\left(3.6\log(C_1 Q/8dv)\right)^2}\right)\left(\frac{Q^2}{d^5}\right)$$

Flow Rate in Terms of S_f, bpd or m³/h

$$Q_f = \sqrt{\frac{S_f d^5}{C_2}} \times T_D$$

$$Q_f = \sqrt{\frac{S_f d^5}{C_2}} \times \left(-2\log\left(2.8250\frac{v}{C_1}\sqrt{\frac{C_2}{S_f d^3}}\right)\right)$$

Flow Rate in Terms of Colebrook-White Slope, bpd or m³/h

$$Q_f = \sqrt{\frac{S_f d^5}{C_2}} \times \left(-2\log\left(\frac{e}{3.7d} + \frac{2.51v}{C_1}\sqrt{\frac{C_2}{S_f d^3}}\right)\right)$$

Flow Rate in Terms of Δp_f, bpd or m³/h

$$Q_p = \sqrt{\frac{\Delta p_f d^5}{C_3\rho}} \times \left(-2\log\left(2.8250\frac{v}{C_1}\sqrt{\frac{C_3\rho}{\Delta p_f d^3}}\right)\right)$$

Flow Rate in Terms of Colebrook-White Δp_f, bpd or m³/h

$$Q_p = \sqrt{\frac{\Delta p_f d^5}{C_3\rho}} \times \left(-2\log\left(\frac{e}{3.7d} + \frac{2.51v}{C_1}\sqrt{\frac{C_3\rho}{\Delta p_f d^3}}\right)\right)$$

Numerical Values for C_1, C_2, and C_3

FPS	$C_1 = 9.224123 \times 10$	Q in bpd, d in inches
	$C_2 = 2.647271 \times 10^{-2}$	v in centistokes
	$C_3 = 1.147629 \times 10^{-2}$	ρ ref $H_2O = 1.000$
MKS	$C_1 = 3.536777 \times 10^5$	Q in m³/hr, d in mm
	$C_2 = 6.377707 \times 10^9$	v in centistokes
	$C_3 = 6.377707 \times 10^8$	ρ ref $H_2O = 1.000$

Table 16-3
Friction Loss and Flow Rate Functions in Terms of Natural Logarithms
Values for K and k Included With Values for C_1, C_2, and C_3.

Reynolds Number (dimensionless)

$$N_R = C_1\left(\frac{Q}{d\nu}\right)$$

Fanning Transmission Factor (dimensionless)

$$T_{OIL} = K\ln(N_R/8)$$

$$T_{OIL} = K\ln(C_1Q/8d\nu)$$

Darcy Friction Factor (dimensionless)

$$f_{OIL} = 4/(T_{OIL})^2$$

$$f_{OIL} = 4/(K\ln(C_1Q/8d\nu))^2$$

$$f = f_{OIL} = 4/(K\ln(C_1Q/8d\nu))^2$$

Unit friction loss, flc/kft or mlc/km

$$S_f = C_2 f\left(\frac{Q^2}{d^5}\right)$$

$$S_f = C_2\left(\frac{4}{(K\ln(N_R/8))^2}\right)\left(\frac{Q^2}{d^5}\right)$$

$$S_f = C_2\left(\frac{4}{(K\ln(C_1Q/8d\nu))^2}\right)\left(\frac{Q^2}{d^5}\right)$$

Unit pressure loss, psi/kft or ksc/km

$$\Delta p_f = C_3\rho f\left(\frac{Q^2}{d^5}\right)$$

$$\Delta p_f = C_3\rho\left(\frac{4}{(K\ln(N_R/8))^2}\right)\left(\frac{Q^2}{d^5}\right)$$

$$\Delta p_f = C_3\rho\left(\frac{4}{(K\ln(C_1Q/8d\nu))^2}\right)\left(\frac{Q^2}{d^5}\right)$$

Flow Rate in Terms of S_f, bpd or m³/h

$$Q_f = \sqrt{\frac{S_f d^5}{C_2}} \times T_D$$

$$Q_f = \sqrt{\frac{S_f d^5}{C_2}} \times \left(-k\ln\left(2.8250\frac{\nu}{C_1}\sqrt{\frac{C_2}{S_f d^3}}\right)\right)$$

Flow Rate in Terms of Colebrook-White Slope, bpd or m³/h

$$Q_f = \sqrt{\frac{S_f d^5}{C_2}} \times \left(-k\ln\left(\frac{e}{3.7d} + \frac{2.51\nu}{C_1}\sqrt{\frac{C_2}{S_f d^3}}\right)\right)$$

Flow Rate in Terms of Δp_f, bpd or m³/h

$$Q_p = \sqrt{\frac{\Delta p_f d^5}{C_3\rho}} \times \left(-k\ln\left(2.8250\frac{\nu}{C_1}\sqrt{\frac{C_3\rho}{\Delta p_f d^3}}\right)\right)$$

Flow Rate in Terms of Colebrook-White Δp_f, bpd or m³/h

$$Q_p = \sqrt{\frac{\Delta p_f d^5}{C_3\rho}} \times \left(-k\ln\left(\frac{e}{3.7d} + \frac{2.51\nu}{C_1}\sqrt{\frac{C_3\rho}{\Delta p_f d^3}}\right)\right)$$

Numerical values for constants K and k, and for factors of proportionality C_1, C_2, and C_3. K and k are the same value in either the FPS or MKS system of units.

FPS	$C_1 = 9.224123 \times 10$	Q in bpd, d in inches
	$C_2 = 2.647271 \times 10^{-2}$	ν in centistokes
	$C_3 = 1.147629 \times 10^{-2}$	ρ ref H_2O = 1.000
MKS	$C_1 = 3.536777 \times 10^5$	Q in m³/hr, d in mm
	$C_2 = 6.377707 \times 10^9$	ν in centistokes
	$C_3 = 6.377707 \times 10^8$	ρ ref H_2O = 1.000

Factors to allow use of natural logarithms in computations

$K = 3.6/2.302585 = 1.563460$

$k = 2.0/2.302585 = 0.868589$

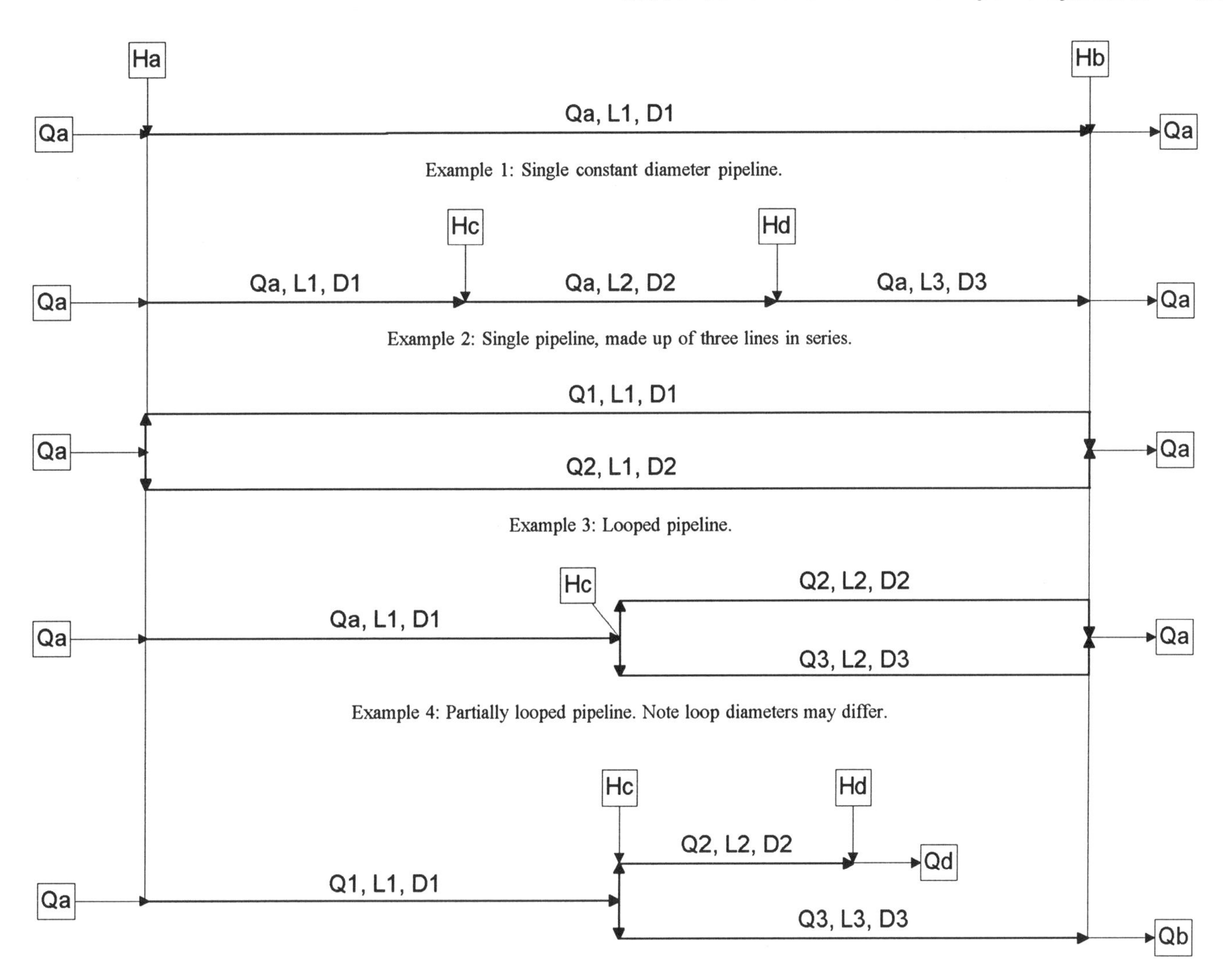

Example 1: Single constant diameter pipeline.

Example 2: Single pipeline, made up of three lines in series.

Example 3: Looped pipeline.

Example 4: Partially looped pipeline. Note loop diameters may differ.

Example 5: Branching pipeline. Branches may be different diameters and lengths.

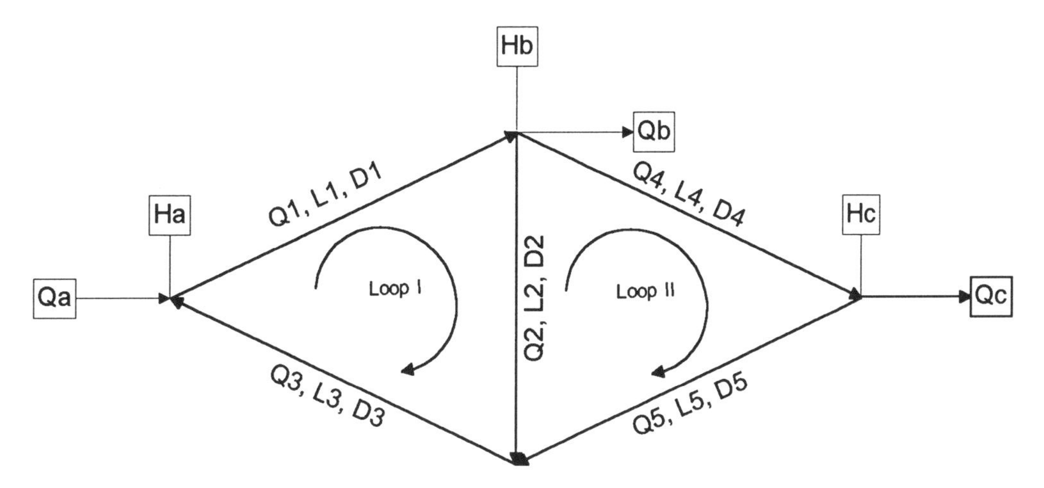

Example 6: Simple two-loop network with one source, two sinks, four junctions, and five pipes.

Figure 16-1. Some typical piping configurations

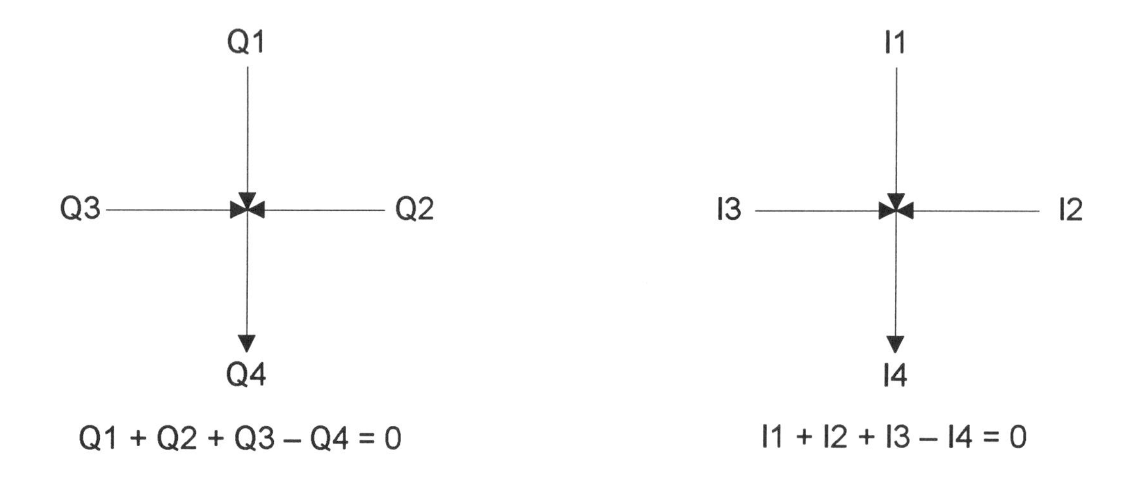

First Law: The algebraic sum of flow into and out of a a junction is zero.

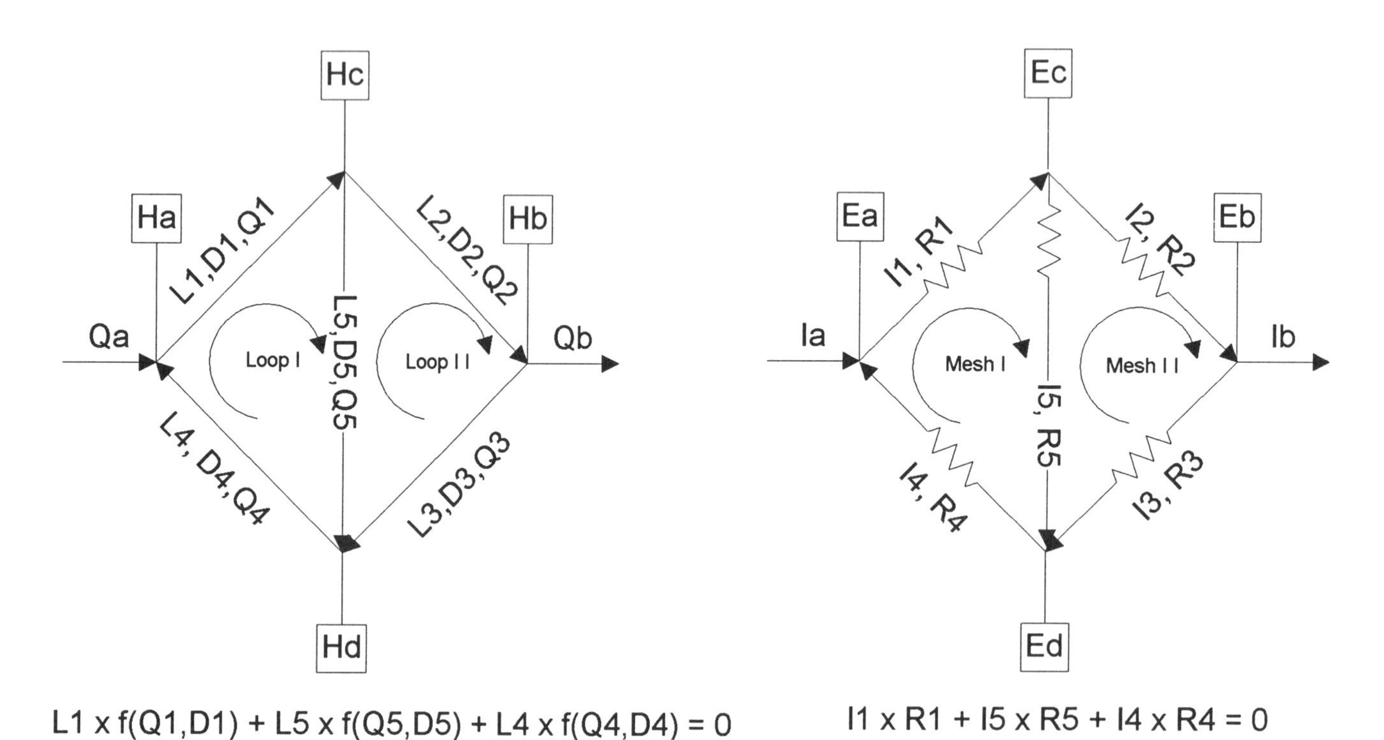

Second Law: The algebraic sum of the potential drops around a closed loop is zero.

Figure 16-2. Kirchoff's laws as applied to liquid flow (left) and direct current electrict circuits (right).

(text continued from page 304)

flow and head from the geometry of the piping system, the characteristics of the flowing fluid, and at least two given parameters. Given $\mathbf{S_f}$ or $\Delta \mathbf{p_f}$ is actually given H_{in} (or P_{in}) and H_{out} (or P_{out}) of a pipe segment of length L, so in reality there are two parameters given in $\mathbf{Q_f}$ and $\mathbf{Q_p}$ problems.

The six examples of Figure 16-1 are solved in the following.

Example 1: The Single Pipeline

The unit friction loss in a single, constant–diameter, pipeline in steady flow is $\mathbf{S_f}$.

The friction loss over the length L of the line is h_f, defined as

$$h_f = L \times \mathbf{S_f} \tag{16-1}$$

This equation is highly non–linear in all its variables except L itself. This can be seen if $\mathbf{S_f}$ is expanded.

$$h_f = L \times C_2 \left(\frac{4}{\left(3.6 \log\left(C_1 Q / 8 d \nu\right)\right)^2} \right) \left(\frac{Q^2}{d^5} \right) \tag{16-2}$$

Figure 16-3 shows the usual way $h_f - Q$ diagrams are plotted. Plots of $p_f - Q$ are of exactly the same form.

Figure 16-4 shows how such curves can be plotted on log–log coordinates. The data here are the two decades of flow rate from 10 000 bpd to 1 000 000 bpd. The plot is, for practical purposes, a straight line. (It really isn't; the plot of $h_f = L \times f(\mathbf{N_R}/T)$ *is* a straight line, but the plot of $h_f = L \times f(\mathbf{N_R})$ is *not*.)

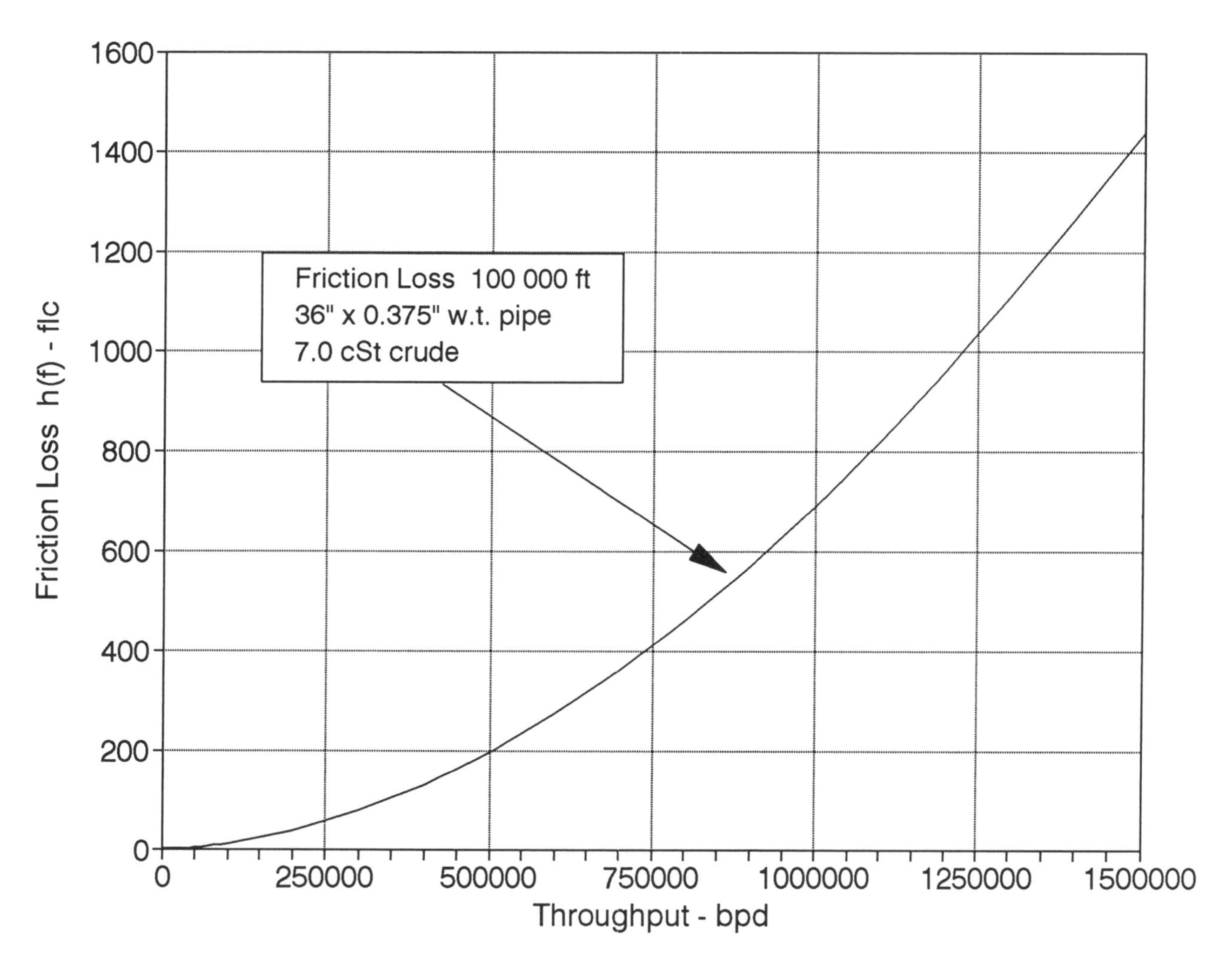

Figure 16-3.

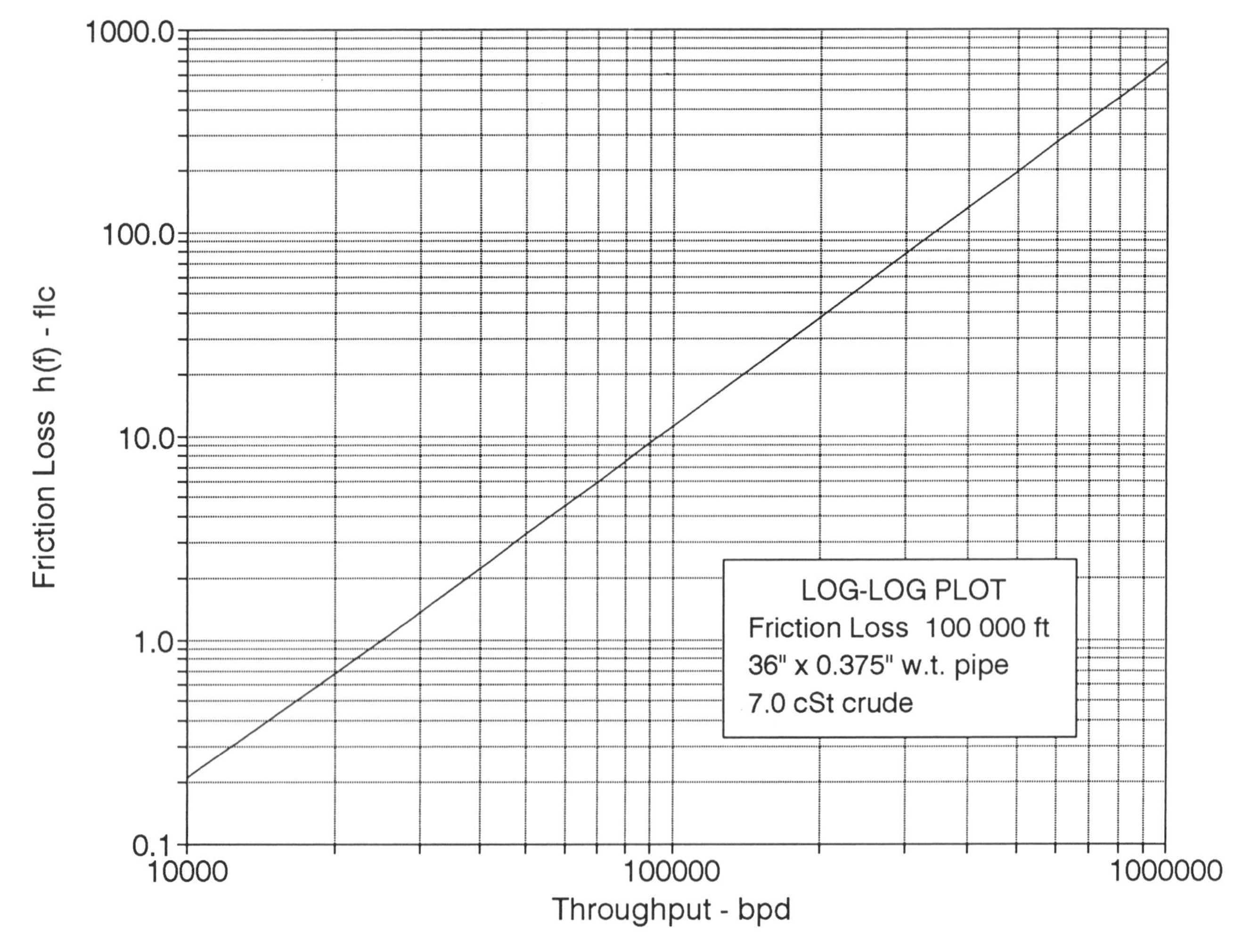

Figure 16-4.

These calculations yield h_f in terms of the pipe and fluid parameters and the flow rate. The inverse, finding the flow rate given h_f and the fluid and pipe parameters requires going to $\mathbf{Q_f}$ (or $\mathbf{Q_p}$ if p_f is given instead of h_f).

$$\mathbf{S_f} = \left(\frac{Ha - Hb}{L/1000}\right) \tag{16-3}$$

and

$$\mathbf{Q_f} = \sqrt{\frac{\mathbf{S_f} d^5}{C_2}} \times \left(-2\log\left(2.8250\frac{\nu}{C_1}\sqrt{\frac{C_2}{\mathbf{S_f} d^3}}\right)\right) \tag{16-4}$$

Thus

$$\mathbf{Q_f} = \sqrt{\frac{\left(\frac{Ha - Hb}{L/1000}\right)d^5}{C_2}} \times \left(-2\log\left(2.8250\frac{\nu}{C_1}\sqrt{\frac{C_2}{\left(\frac{Ha - Hb}{L/1000}\right)d^3}}\right)\right) \tag{16-5}$$

This is the simplest application of $\mathbf{Q_f}$, the flow rate equation. There are many problems that cannot be solved without iterative methods *without* using $\mathbf{Q_f}$. It is a complex equation, but once programmed, it is always there.

$\mathbf{Q_f}$ in this form is needlessly cumbersome and will not be used further in the text herein. $\mathbf{Q_f}$ should be found by first computing $\mathbf{S_f}$ from *Ha*, *Hb*, and *L*, and then substituting in the simpler $\mathbf{S_f}$ form of $\mathbf{Q_f}$. When $\mathbf{Q_f}$ is programmed into a computer, $\mathbf{S_f}$ can be computed first and then carried to the $\mathbf{Q_f}$ calculation as a parameter in the calling sequence for $\mathbf{Q_f}$. If symbolic manipulation is required, $\mathbf{Q_f}$ can be written in terms of natural logarithms as in Table 16-3.

Figure 16-5 shows a plot of $\mathbf{Q_f}$ vs $\mathbf{S_f}$ for a 24–inch pipe flowing 7.0 cSt crude. Note the axes for a plot of $\mathbf{Q_f}$ vs $\mathbf{S_f}$ are rotated 90° from that of a plot of $\mathbf{S_f}$ vs Q or of $h_f - Q$; in the first case the abscissas are flow rate and the ordinates are head loss, whereas in the second case the abscissas are unit head loss and the ordinates are flow rate.

CCT No. 16-1 gives the layout for computer solutions for both the head–unknown and flowrate–unknown problems. Neither problem requires an iterative solution.

Example 2: Pipelines in Series

If *n* pipelines are placed in series, by which is meant the output of the first line is connected directly into the input of the second, etc., the result is

$$h_{f1} + h_{f2} \ldots + h_{fn} = L_1 \mathbf{S_{f1}} + L_2 \mathbf{S_{f2}} \ldots + L_n \mathbf{S_{fn}} \qquad (16\text{-}6)$$

This is the basic equation for solving Example 2 on Figure 16-1, which is a pipeline composed of three pipes in series.

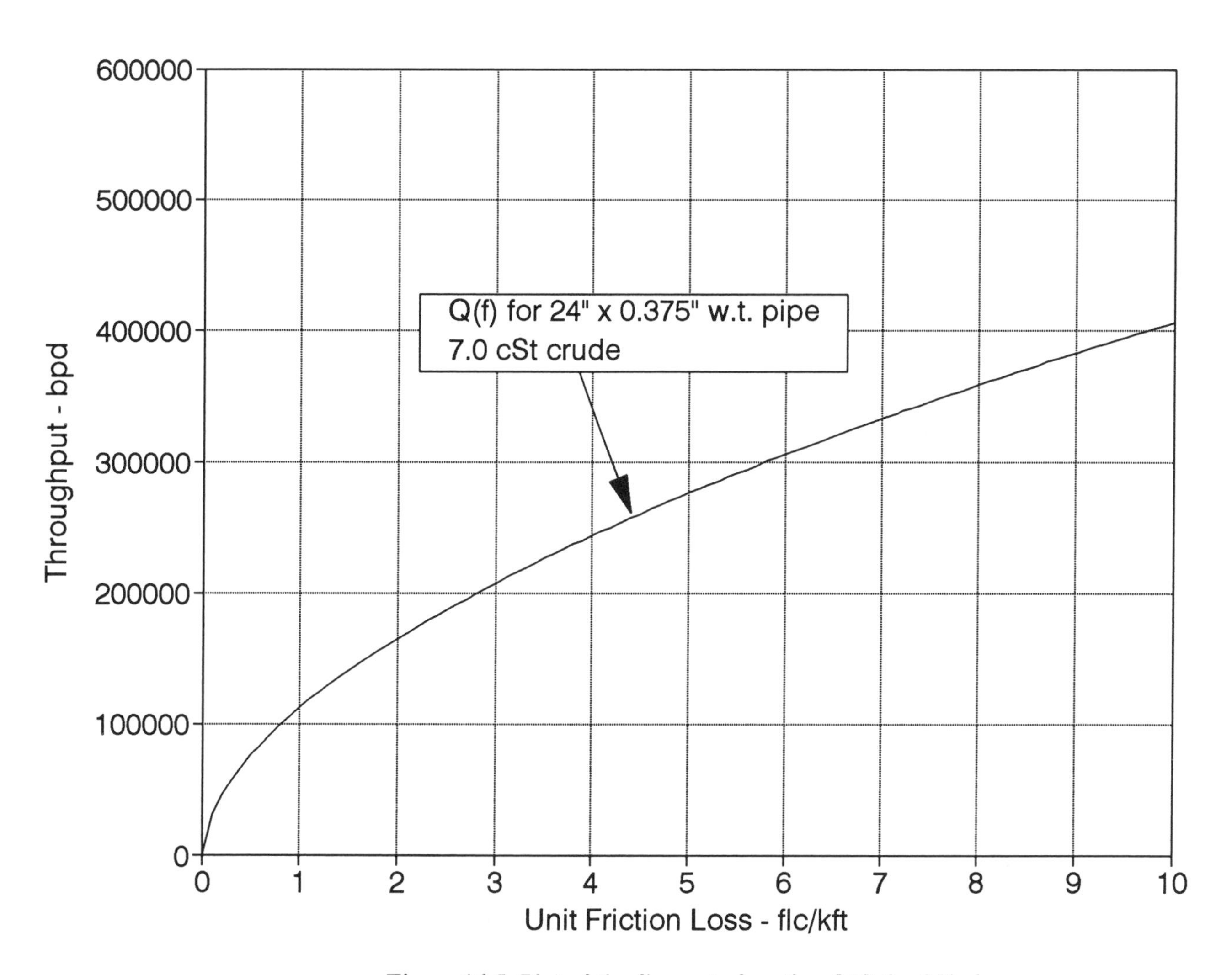

Figure 16-5. Plot of the flow rate function Q(f) for 24" pipe.

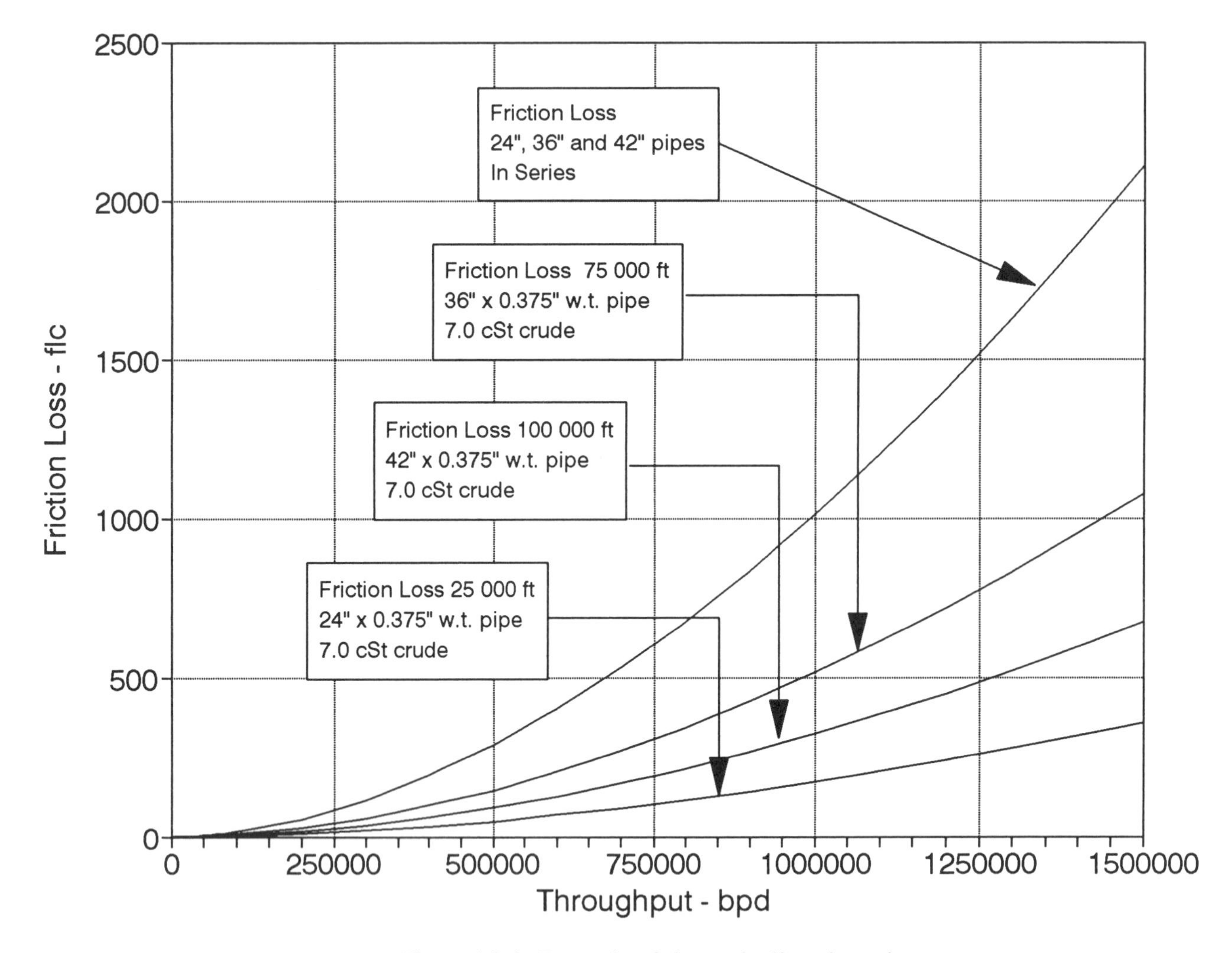

Figure 16-6. Example of three pipelines in series.

There are several parameters available aside from the pipe and fluid characteristics:

$$h_{f1} + h_{f2} + h_{f3} = L_1\mathbf{S_{f1}} + L_2\mathbf{S_{f2}} + L_3\mathbf{S_{f3}} = Ha - Hb \quad (16\text{-}7)$$

Figure 16-6 illustrates graphically the $h_f - Q$ relations of 25 kft of 24–inch in series with 75 kft of 36–inch and 100 kft of 42–inch and line. The top curve is exactly that of Equation 16-7; it is the sum of $h_{f24} + h_{f36} + h_{f42}$. Thus, the problem of finding the unknown total head given the flow rate is simple; add the individual head losses.

Finding the flow rate, given the total head, isn't as easy. Figure 16-7 shows the graphical solution; draw a vertical line along the line of constant Q and draw lines of constant h_f through the intersection of this line with the $h_f - Q$ curves of the individual lines. The extensions of these lines to the scale of ordinates yields the flow rates.

The analytical solution is a little more difficult, but given the *solve* functions available in both calculators and computers, it isn't too difficult. **CCT** No. 16-2 (lower solution) is the key; sum the values of $\mathbf{Q_f}$ against the value of $(Ha - Hb)$ and solve iteratively for $\mathbf{Q_f}$.

The top of **CCT** No. 16-2 is an illustration of the first described class of problem where Q and one value of head are known.

Example 3: Parallel Pipelines

The parallel pipelines of Figure 16-1, Example 3, either do or do not present a problem.

If *Ha* and *Hb* are known, a graphical solution along the lines of Figure 16-8 can be made; compute and plot $\mathbf{Q_f}$ against the unit friction loss *(Ha – Hb)/L* and sum the individual curves along the vertical line of constant h_f.

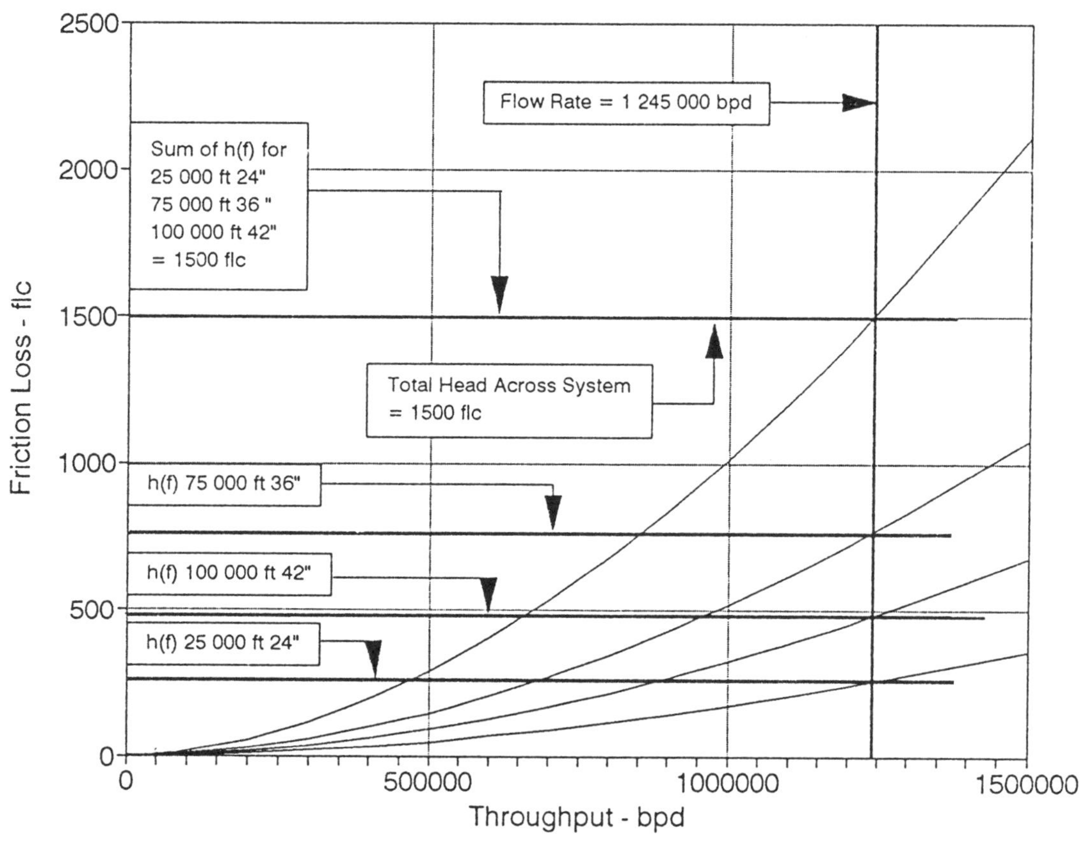

Figure 16-7. Solving for flow rate in three pipelines in eries. Total head = 1500 flc.

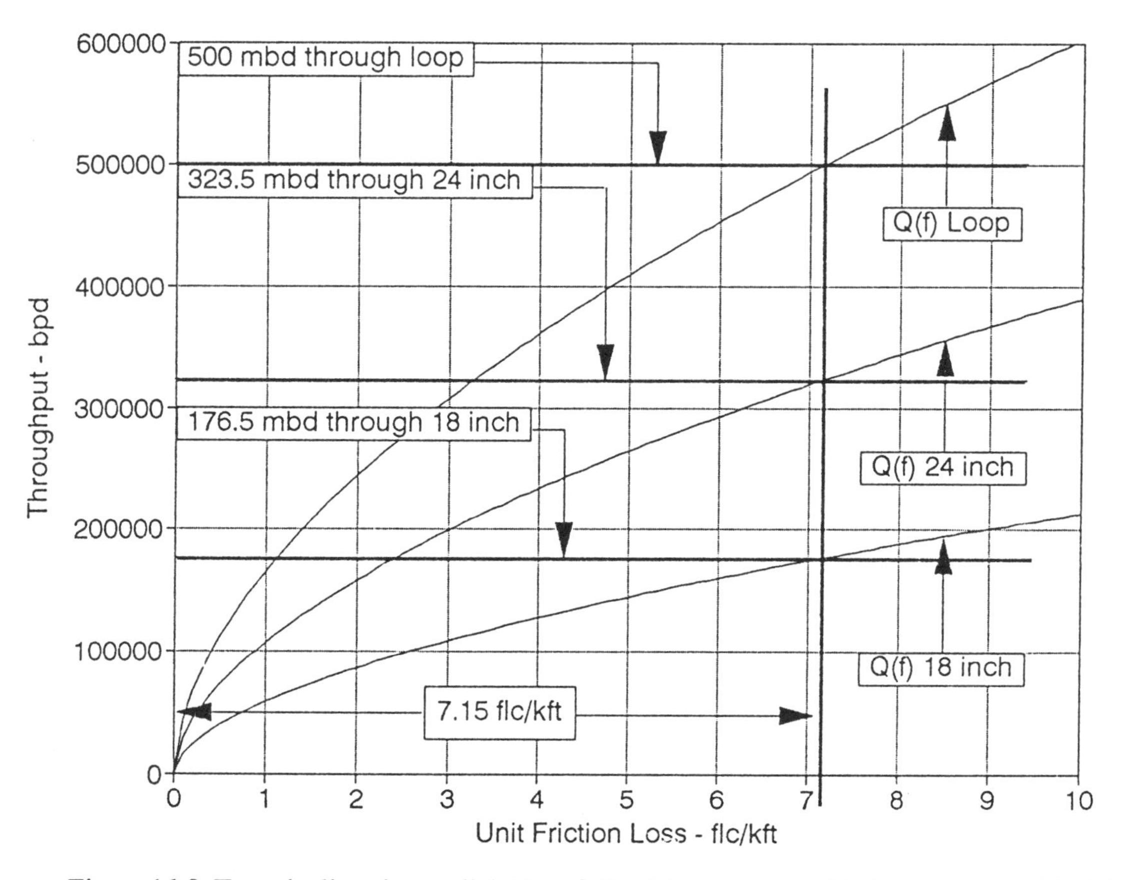

Figure 16-8. Two pipelines in parallel. Note Q(f) plots are rotated 90 degrees from S(f) plots.

The analytical method for solving this problem is on Page 1 of **CCT** No. 16-3. It is exactly as described in words in the above.

The other problem, where *Q* and only one head (either *Ha* or *Hb*) is known, is one of the piping configurations that requires a solution similar to that of pipe networks, i.e., by having recourse to both of Kirchoff's laws.

The solution on Page 2 of **CCT** No. 16-3 solves the problem on Page 1 backwards; given *Qa* and *Hb*, find *Ha* and the individual values of *Q1*, *Q2*, and *Q3*. In this solution the individual values of $\mathbf{Q_f}$ are summed to *Qa*; *Q1*, *Q2*, and *Q3* are defined in terms of *Qa* and each other; and the individual values of $\mathbf{S_f}$ are equated. This yields four equations in four unknowns, and the values of *Q1*, *Q2*, *Q3*, and *Ha* are the result. This problem can be solved by using any of the canned multi–equation solvers, or it can be programmed in any algorithmic computer language by setting up a solution using a version of the multi–variable Newton–Raphson method described in Chapter 13. As presented, however, **CCT** No. 16-3 is a *MathCad* template that will solve the problem directly.

Example 4: Partially Looped Pipelines

A common situation in pipelining is that shown in Figure 16-1, Example 4. This is a pipeline that did not have enough capacity so a parallel line—called a *loop*—was constructed along a part of its length.

The usual problem—how much loop to construct to achieve the desired increase in throughput—is simple if the loop line is the same diameter as the original, but is somewhat more complicated if the diameters differ.

If the diameters are the same, then $Q2 = Q3 = Qa/2$, and two equations in two unknowns can be set up and solved by substitution:

$$\begin{aligned} &L1 \cdot \mathbf{S_f}(Qa, D1) + L2 \cdot \mathbf{S_f}(Qa/2, D2) = Ha - Hb \\ &L1 + L2 = LT \end{aligned} \tag{16-8}$$

where $LT =$ total length of the pipeline

If the line diameters are *not* equal the solution begins to become cumbersome. **CCT** No. 16-4, Page 1, solves the problem where *Ha* and *Qa* are known, and *Hb*, *Hc*, *Q2*, and *Q3* are needed. Here there are four unknowns so four equations are required:

$$\begin{aligned} &Q2 + Q3 = Qa \\ &\frac{Ha - Hc}{L1} = \mathbf{S_f}(Qa, D1) \\ &\frac{Hc - Hb}{L2} = \mathbf{S_f}(Q2, D2) \\ &\frac{Hc - Hb}{L3} = \mathbf{S_f}(Q3, L3) \end{aligned} \tag{16-9}$$

This set of simultaneous non–linear equations can be solved by methods mentioned above.

The solution to the loop–length problem is shown on **CCT** No. 16-4, Page 2. Here the givens are *LT* and *Qa*, the total line length and the required flow rate, and *Ha* and *Hb*, the inlet and outlet heads. Here *L2* (and, of course, *L1* follows), *Q2* and *Q3* are wanted. *Hc* is wanted as well so that the solution may be checked: (*Hc* – *Hb*) is the total loss over the loop. This is a problem with five unknowns requiring five equations and three of them are not linear so one of the *solver* methods must be used or a solving algorithm written. The solution on Page 2 is the inverse of that on Page 1. Note that in the iterative solutions here only some 3+ significant figures are available. For pipeline flow problems, that is enough.

Example 5: Branching Pipelines

A *branching* pipeline is not a *looped* pipeline; a looped pipeline has only one source and one sink, whereas a branching pipeline has at least two of one or the other. In Figure 16-1, Example 5, there are two sinks.

[As an aside, this is the problem that appears in nearly every fluid mechanics text: a Y–shaped pipeline system connects three reservoirs, each of a different elevation (tanks, lakes, whatever), and the problem is to determine whether water flows to or from the intermediate reservoir. It is not a trivial problem.]

On **CCT** No. 16-5 the same equations as those on **CCT** No. 16-4 are used to solve the problem.

The solution on Page 1 is more of a pipeline problem; given *Ha*, *Qa*, *Q2*, and *Q3*, find the other three heads *Hb*, *Hc*, and *Hd*. This is the kind of problem involved in setting up a delivery into two tanks different distances from and different elevations in relation to the inlet point.

The solution on Page 2 is the three–reservoir problem; given *Ha*, *Hb*, and *Hc*, find *Qa*, *Q2*, and *Q3*. The equations would also yield *Hd* if required.

It should be mentioned that these equations will also solve the reverse kind of problem: two source pipelines feeding a single pipeline delivering to a single sink. None of the equations given have signs to indicate the direction of flow, and common sense will indicate where a flow is, say, *plus* or *minus*. The equations can be made to adjust for direction of flow automatically—most of the large, network-solving, computer programs do this—but for most purposes less complicated equations and more common sense will find the correct answer.

Example 6: Pipeline Networks

Example 6 on Figure 16-1 is a kind of generic pipeline network. It has a source, two sinks, four junctions, and five legs. While networks are treated quite thoroughly in some texts, among which Giles,[1] Daugherty, Franzini, and Finnemore,[2] Streeter and Wylie,[3] and Vennard and Street,[4] can be recommended, a small book—*Analysis of Flow in Pipe Networks* by Jeppson[5]—is distinguished in that the author not only covers the three modern methods of solving pipe networks with procedural computer languages (Hardy–Cross, Newton–Raphson, and Linear Theory methods) but also includes dozens of detailed numerical solutions as examples. I highly recommended this reference.

Oil pipeliners should realize that, without exception, all of the solutions for networks published in fluid mechanics texts are for water distribution networks in which the roughness (and, in most cases, even the equivalent inside diameter) of the pipes; the flow rates, and in fact, much of the time, even the directions of flows; and the internal junction pressures, are not known within engineering accuracy. Also, flow in these kinds of networks is never constant or steady–state; these networks are never, ever, quiescent.

Under such conditions it is not unusual for writers to recommend approximate methods of solution: if the inputs are known to no better than 20 or 30 percent, is it important to try to get an answer within a percent or two? The answer, of course, is that it isn't.

Oil pipeliners, on the other hand, usually don't have to deal with complicated pipe networks; very rarely will an oil pipeline network have more than three or four legs, and then only in exceptional cases. But when they need an answer they just can't get along with an accuracy of 20 percent. This section, therefore, is dedicated to solving *small* piping networks with *good* accuracy.

The sample network in Figure 16-9 shows Vennard and Street's[14] recommended nomenclature for designating the pipe and flow parameters of pipeline networks. *Junctions* are identified by capital letters (A, B, C...etc.), *pipes* (or legs) with Arabic numbers (1, 2, 3...etc.), and *loops* with Roman numerals (I, II...etc.). While this kind of setting out the parts of a network is not necessary, the use of this or some other consistent kind of identification is really important for large networks and is of considerable assistance in small ones. Figure 16-10 shows this kind of nomenclature applied to a simple two–loop network.

There are essentially an infinite number of network problems; if more complexity is desired, add more loops or legs. If more difficulty is wanted, leave out some of the important variables.

Figure 16-11, a small network of small diameter pipes, illustrates the kind of complexity that can be found in simple networks. Here none of the heads (or pressures) are known, only the flow into the network from the source and from the network at the two sinks. Yet all of the internal flows—the flows in the individual legs—can be found fairly simply.

CCT No. 16-6 shows the solution. Five flows are unknown, so five equations are required. Three of these can be written as continuity equations around three of the junctions (there are always (J − 1) continuity equations available in a network), and the other two are energy or head–loss equations written around the two loops. Using a canned solver the solution appears quite simply. If the algorithm has to be programmed, note that the loop equations are non–linear.

A much more complex, though simpler looking, problem is shown in Figure 16-12. Here there is one source, two sinks, three legs, three junctions, and one loop, with *Ha*, *Qb* and *Qc* known. The solution is tied to solving two equations, one describing the flow in *L1* in terms of the flow in *L2* and *Qb*, and the second describing the flow in *L2* in terms of *Q3* and *Qc*. In each case where flow in a leg is required it is described in terms of $\mathbf{Q_f}$ for that leg expanded so as to be expressed in terms of the junction heads *Ha*, *Hb*, and *Hc*. Solution of these two highly non–linear equations yields *Hb* and *Hc* directly. **CCT** No. 16-7 shows how the problem can be set up and solved with a solver; it can be programmed in any of the algorithmic languages using a Newton–Raphson kind of solution.

More complex networks can be solved in similar ways. The *MathCad* canned solver *minerr* used in most of the **CCT** solutions for network problems can actually handle

(text continued on page 318)

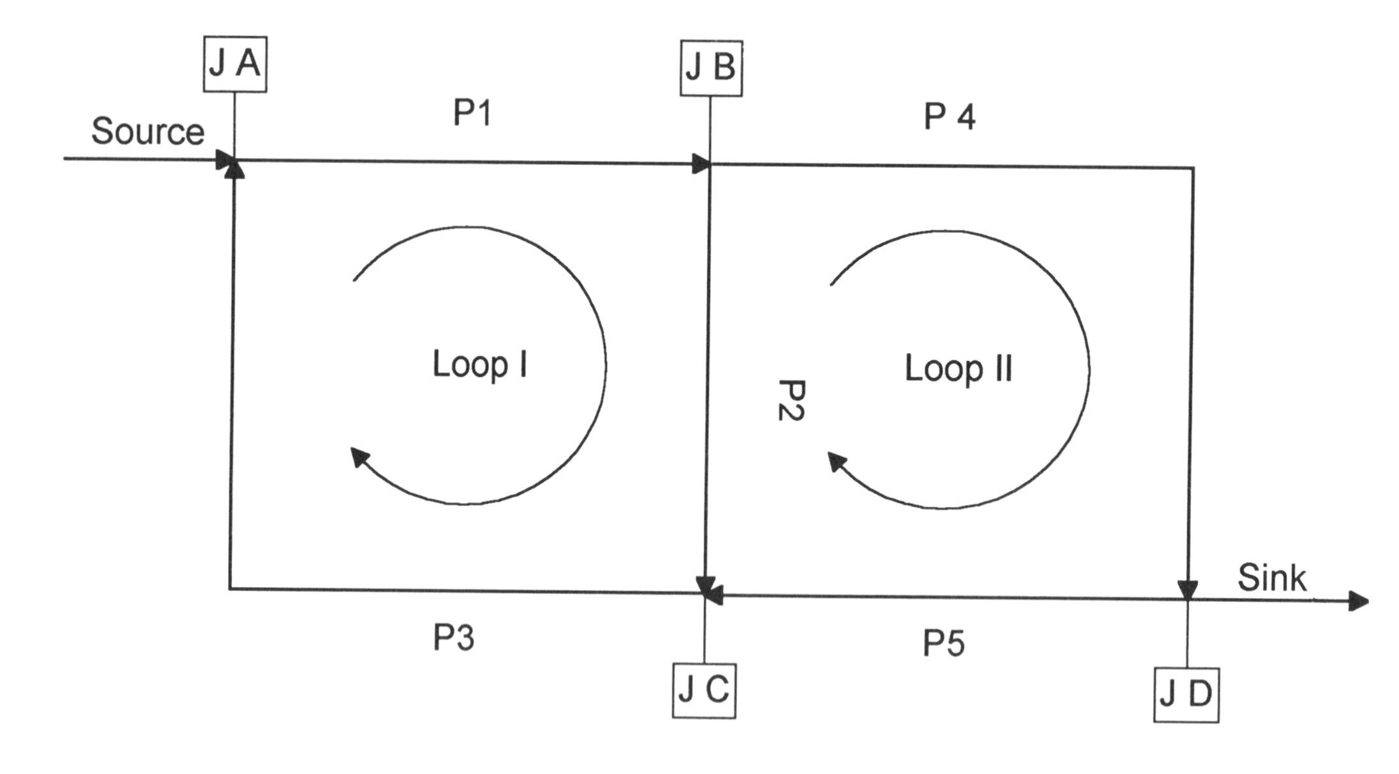

Figure 16-9. Recommended nomenclature for piping networks. **[J]**unctions are keyed to capital letters (A, B, etc.), [P]ipes to Arabic numbers (1, 2, etc.), and **[L]**oops to Roman numbers (I, II, etc.).

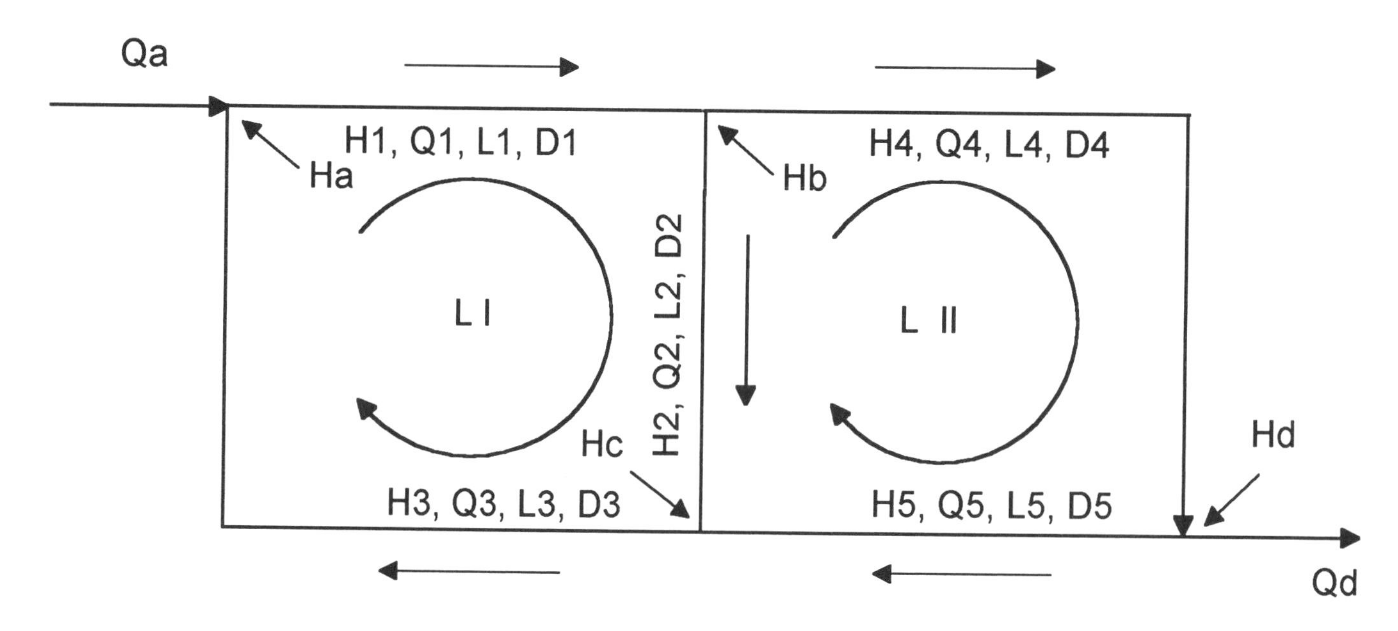

Figure 16-10. Recommended nomenclature applied to two–loop circuit with four nodes, five pipes, one source, and one sink.

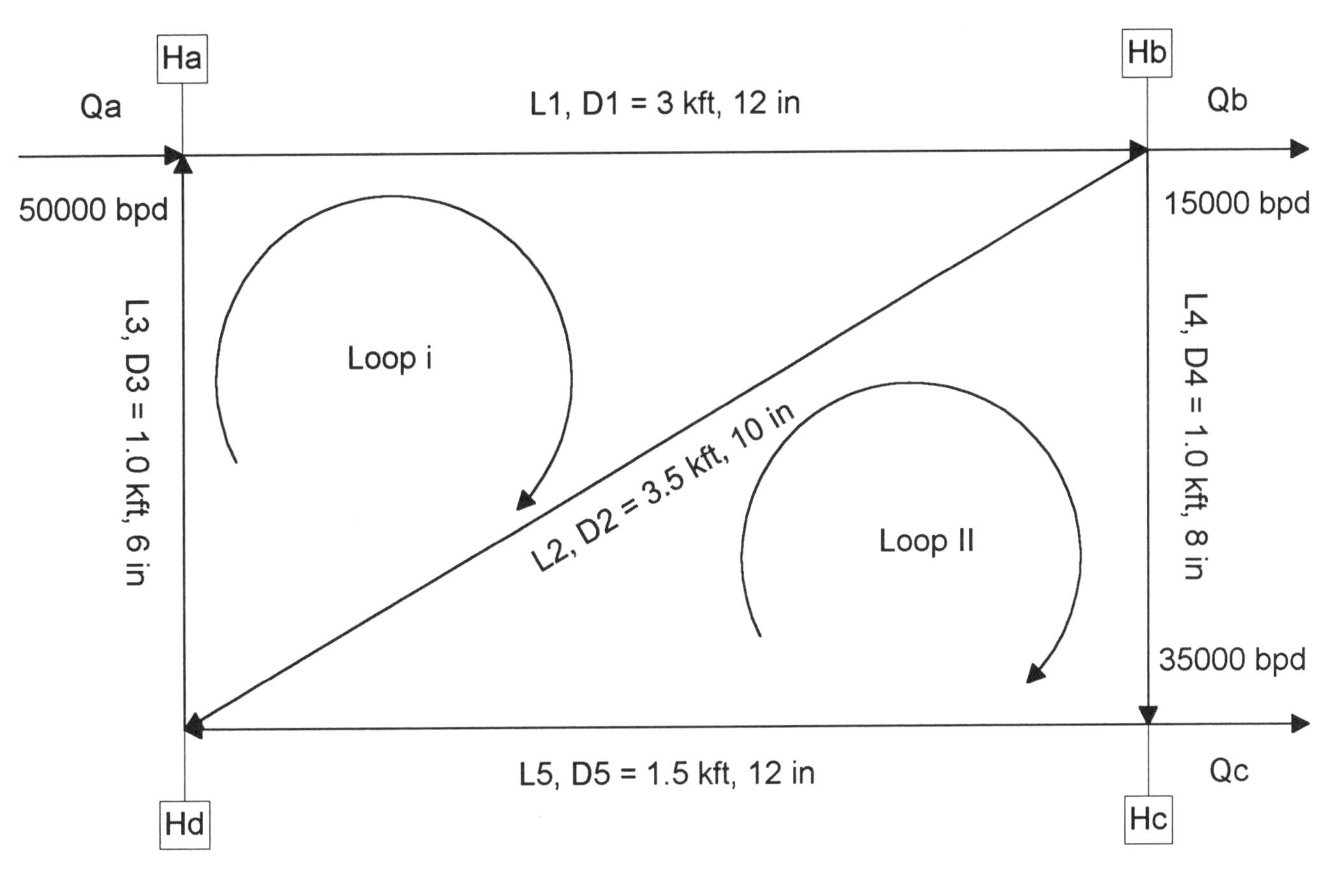

Figure 16-11. Network with one source, two sinks, two loops, four nodes, and five lines.

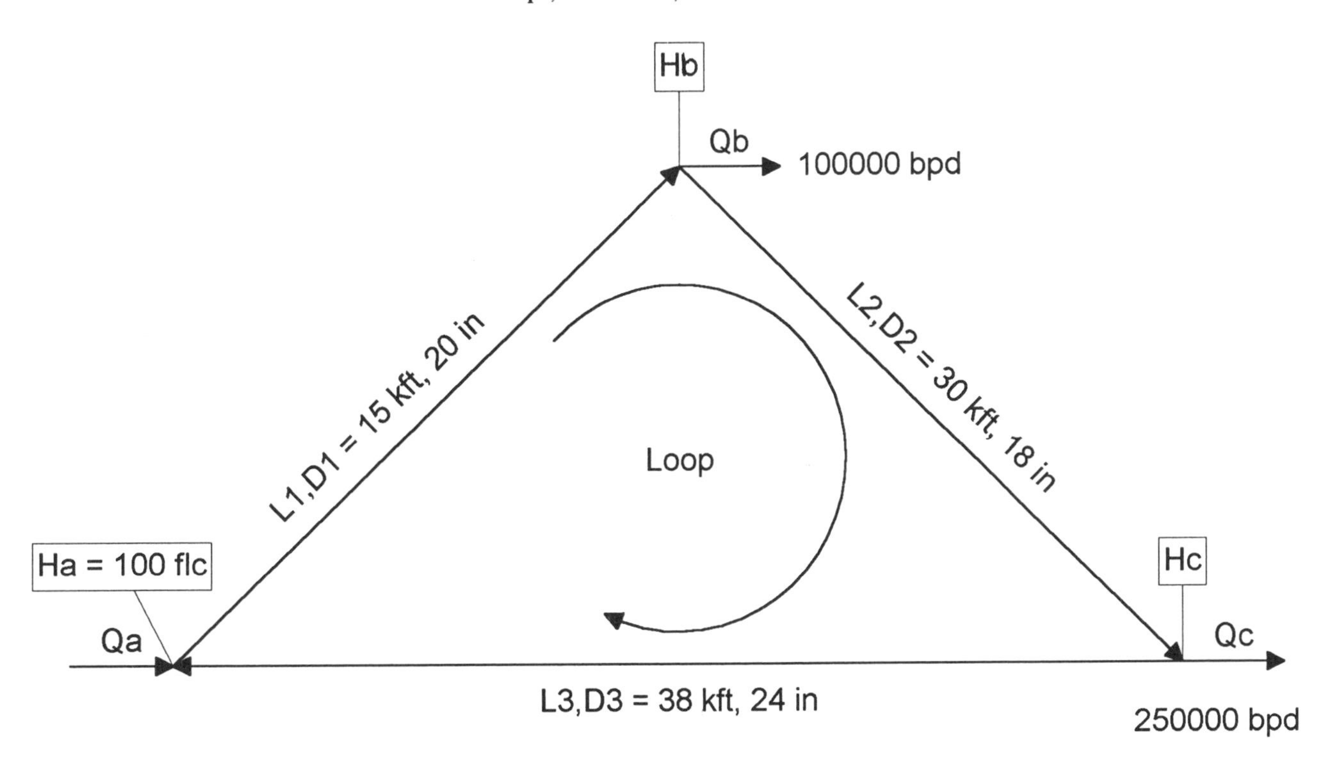

Figure 16-12. Three leg, three node, one loop network with one source and two sinks.

(text continued from page 315)

up to 50 equations, and other even more powerful solvers are commercially available. However, if a pipeliner ever has to solve a network that has to be described in 40 or 50 equations, then he isn't working on *oil* pipelines.

Conclusion

The six pipeline configurations of Figure 16-1 are typical of the kind of problems that need to be solved in routine pipeline work. Only in recent years, with the coming of the programmable calculator and the desktop computer, have seriously accurate solutions to some of them been possible. Many of them are not amenable to symbolic solution, and as Professor Fazarinc said (see Reference 2, Introduction):

> "We cannot solve algebraic equations of order higher than four. Consequently we cannot solve differential equations of order higher than four, either. Furthermore, we can solve only a handful of non–linear equations.... This leaves out the vast majority of interesting problems."

The numerical solution of engineering problems has, in the period commencing with the appearance of the Hewlett–Packard Model 65 programmable calculator in the mid–1970's, on up to today's powerful desktop machines with their built–in or *canned* mathematical solution applications, become so casual that Professor Fazarinc's recommendation to teach students to use the finite difference, or similar, forms of equations rather than those derived from the infinitesimal calculus has, in real fact, essentially been put in place.

The engineering world of *Hydraulics for Pipeliners*, Second Edition, is far different from that of the original *Hydraulics for Pipeliners*, with its graphical and 10–inch slide rule solutions (approximations?).

Whether it is a *better* engineering world is conjectural.

References

1 Giles, Ranald V., *Fluid Mechanics and Hydraulics,* Schaum's Outline Series. New York: McGraw–Hill Book Company, 1962.

2 Daugherty, R. L., Franzini, J. B., and Finnemore, E. J., *Fluid Mechanics With Engineering Applications*, 8th ed. New York: McGraw–Hill, Inc., 1985.

3 Streeter, V. L., and Wylie, B. J., *Fluid Mechanics,* 8th ed. New York: McGraw–Hill, Inc., 1985.

4 Vennard, J. K., and Street, R. L., *Elementary Fluid Mechanics*, 6th ed. New York: John Wiley & Sons, 1982.

5 Jeppson, R. W., *Analysis of Flow in Pipe Networks.* Ann Arbor, Michigan: Ann Arbor Science Publishers, Inc., fifth printing, 1982.

Computer Computation Template

CCT No. 16-1: Calculation of heads and flow rates for a single pipeline.

References: Chapter 16. Figure 16-1, Example 1. Equations 16-1 and 16-5.

Input

Factors $C_1 := 92.241232$ $C_2 := 0.02647271$ $C_3 := 0.01147629$ $K := \frac{3.6}{2.302585}$ $k := \frac{2.0}{2.302585}$

Input Data $Q := 1250000$ bpd $Ha := 3000$ flc $L := 100$ kft $d := 35.25$ in $\nu := 7.0$ cSt

Case I: Q and Ha known, find Hb

Equations

$$S_f := C_2 \cdot \left[\frac{4}{\left(K \cdot \ln\left(\frac{C_1 \cdot Q}{8 \cdot d \cdot \nu} \right) \right)^2} \right] \cdot \frac{Q^2}{d^5}$$

$h_f := L \cdot S_f$ $Hb := Ha - h_f$

Results

$S_f = 10.325$ flc / kft $h_f = 1032.5$ flc $Hb = 1967.5$ flc

Case II: Ha and Hb known, find Q

Equations

$$S_f := \frac{Ha - Hb}{L}$$

$$Q := \sqrt{\frac{S_f d^5}{C_2}} \cdot \left(-2 \cdot \log\left(2.8250 \cdot \frac{\nu}{C_1} \cdot \sqrt{\frac{C_2}{S_f d^3}} \right) \right)$$

Results

$Q = 1.2486 \cdot 10^6$ bpd

Computer Computation Template

CCT No. 16-2: Calculation of heads and flow rates for series pipelines.

References: Chapter 16. Figure 16-1, Example 2. Equations 16-6 and 16-7.

Input

Factors $C_1 := 92.241232$ $C_2 := 0.02647271$ $C_3 := 0.01147629$ $K := \frac{3.6}{2.302585}$ $k := \frac{2.0}{2.302585}$

Pipeline and Fluid Input Data $L_1 := 100$ kft $L_2 := 200$ kft $L_3 := 300$ kft

$d_1 := 23.25$ in. $d_2 := 29.25$ in. $d_3 := 35.25$ in. $\nu := 7.0$ cSt

Case I: Q and Hb known, find Ha, Hc and Hd.

Flow Input Data $Q := 250000$ bpd $Hb := 100$ flc

Equations

$$h_{f1} := C_2 \cdot L_1 \cdot \frac{4}{\left(K \cdot ln\left(\frac{C_1 \cdot Q}{8 \cdot d_1 \cdot \nu}\right)\right)^2} \cdot \frac{Q^2}{d_1^5} \qquad h_{f2} := C_2 \cdot L_2 \cdot \frac{4}{\left(K \cdot ln\left(\frac{C_1 \cdot Q}{8 \cdot d_2 \cdot \nu}\right)\right)^2} \cdot \frac{Q^2}{d_2^5} \qquad h_{f3} := C_2 \cdot L_3 \cdot \frac{4}{\left(K \cdot ln\left(\frac{C_1 \cdot Q}{8 \cdot d_3 \cdot \nu}\right)\right)^2} \cdot \frac{Q^2}{d_3^5}$$

$Hd := 100 + h_{f3}$ $Hc := Hd + h_{f2}$ $Ha := Hc + h_{f1}$

Results

$h_{f1} = 416.5$ flc $h_{f2} = 277.2$ flc $h_{f3} = 170.1$ flc $h_{f1} + h_{f2} + h_{f3} = 863.8$ flc

$Ha = 963.8$ flc $Hc = 547.3$ flc $Hd = 270.1$ flc

Case II: Ha and Hb known, find Q

Flow Input Data $Ha = 963.8$ flc $Hb = 100.0$ flc

Equations

Given

$$C_2 \cdot L_1 \cdot \frac{4}{\left(K \cdot ln\left(\frac{C_1 \cdot Q}{8 \cdot d_1 \cdot \nu}\right)\right)^2} \cdot \frac{Q^2}{d_1^5} + C_2 \cdot L_2 \cdot \frac{4}{\left(K \cdot ln\left(\frac{C_1 \cdot Q}{8 \cdot d_2 \cdot \nu}\right)\right)^2} \cdot \frac{Q^2}{d_2^5} + C_2 \cdot L_3 \cdot \frac{4}{\left(K \cdot ln\left(\frac{C_1 \cdot Q}{8 \cdot d_3 \cdot \nu}\right)\right)^2} \cdot \frac{Q^2}{d_3^5} = Ha - Hb$$

Results

$Minerr(Q) = 2.500 \cdot 10^5$ bpd

Computer Computation Template

CCT No. 16-3: Calculations of heads and flow rates for parallel pipelines

Page 1

References: Chapter 16. Figure 16-1, Example 3. Equation 16-5.

Input

Factors $C_1 := 92.241232$ $\quad C_2 := 0.02647271$ $\quad C_3 := 0.01147629$ $\quad K := \dfrac{3.6}{2.302585}$ $\quad k := \dfrac{2.0}{2.302585}$

Pipeline and Fluid Input Data $L_1 := 145.3$ kft $\quad L_2 := 145.3$kft $\quad L_3 := 145.3$ kft $\quad \nu := 7.0$ cst

$d_1 := 35.25$ in. $\quad d_2 := 29.25$in. $\quad d_3 := 23.25$ in

Flow Input Data $Ha := 1600$ flc $\quad Hb := 100$ flc

Variables $S_1 := \dfrac{Ha - Hb}{L_1}$ $\quad S_2 := \dfrac{Ha - Hb}{L_2}$ $\quad S_3 := \dfrac{Ha - Hb}{L_3}$

Case I: Ha and Hb known, Find Qa, Q1, Q2 and Q3 for three parallel pipelines.

Equations

$$Q1 := \sqrt{\frac{S_1 \cdot d_1^{\,5}}{C_2}} \cdot \left(-k \cdot ln \left(2.8250 \cdot \frac{\nu}{C_1} \cdot \sqrt{\frac{C_2}{S_1 \cdot d_1^{\,3}}} \right) \right)$$

$$Q2 := \sqrt{\frac{S_2 \cdot d_2^{\,5}}{C_2}} \cdot \left(-k \cdot ln \left(2.8250 \cdot \frac{\nu}{C_1} \cdot \sqrt{\frac{C_2}{S_2 \cdot d_2^{\,3}}} \right) \right)$$

$$Q3 := \sqrt{\frac{S_3 \cdot d_3^{\,5}}{C_2}} \cdot \left(-k \cdot ln \left(2.8250 \cdot \frac{\nu}{C_1} \cdot \sqrt{\frac{C_2}{S_3 \cdot d_3^{\,3}}} \right) \right)$$

Results

$Q1 = 1.249 \cdot 10^6$ bpd $\quad Q2 = 7.609 \cdot 10^5$ bpd $\quad Q3 = 4.132 \cdot 10^5$ bpd

$Qa := Q1 + Q2 + Q3$ bpd $\quad Qa = 2.423 \cdot 10^6$ bpd

Computer Computation Template

CCT No. 16-3: Calculations of heads and flow rates for parallel pipelines

Page 2

References: Chapter 16. Figure 16-1, Example 3. Equation 16-5.

Input

Flow Input Data $Qa := 2.423 \cdot 10^{6}$ bpd $Hb := 100$ flc $S_1 = 10.323$ $S_2 = 10.323$ $S_3 = 10.323$

Case II: Qa and Hb known, Find Ha, Q1, Q2 and Q3 for three parallel pipelines.

Equations

Given

$$\left[\sqrt{\frac{S_1 \cdot d_1^{5}}{C_2}} \cdot \left(-k \cdot \ln\left(2.8250 \cdot \frac{v}{C_1} \cdot \sqrt{\frac{C_2}{S_1 \cdot d_1^{3}}}\right)\right) + \sqrt{\frac{S_2 \cdot d_2^{5}}{C_2}} \cdot \left(-k \cdot \ln\left(2.8250 \cdot \frac{v}{C_1} \cdot \sqrt{\frac{C_2}{S_2 \cdot d_2^{3}}}\right)\right) \ldots + \sqrt{\frac{S_3 \cdot d_3^{5}}{C_2}} \cdot \left(-k \cdot \ln\left(2.8250 \cdot \frac{v}{C_1} \cdot \sqrt{\frac{C_2}{S_3 \cdot d_3^{3}}}\right)\right)\right] = Qa$$

$Qa = Q1 + Q2 + Q3$ $S_1 = S_2$ $S_2 = S_3$

Results

$$minerr(Q1, Q2, Q3, Ha) = \begin{bmatrix} 1.249 \cdot 10^{6} \\ 7.609 \cdot 10^{5} \\ 4.136 \cdot 10^{5} \\ 1600.000 \end{bmatrix}$$

$Q1 = 1.249 \cdot 10^{6}$ bpd

$Q2 = 7.609 \cdot 10^{5}$ bpd

$Q3 = 4.132 \cdot 10^{5}$ bpd

$Ha = 1.600 \cdot 10^{3}$ flc

Computer Computation Template

CCT No. 16-4: Calculations of heads and flow rates in pipelines with partial loops.

Page 1

References: Chapter 16. Figure 16-1, Example 4. Equations 16-8 and 16-9.

Input

Factors $C_1 := 92.241232$ $C_2 := 0.02647271$ $C_3 := 0.01147629$ $K := \frac{3.6}{2.302585}$ $k := \frac{2.0}{2.302585}$

Pipeline and Fluid Input Data $L_1 := 264$ kft $L_2 := 132$ kft $L_3 := 132$ kft $\nu := 7.0$ cSt

$d_1 := 19.25$ in $d_2 := 19.25$ in $d_3 := 29.25$ in

Flow Input Data $Qa := 250000$ bpd $Ha := 3000$ flc

Case I: Ha and Qa known, find Hb, Hc, Q2, and Q3

Seed Variables $Hc := 200$ flc $Hb := 100$ $Q2 := 100000$ bpd $Q3 := 150000$ bpd

Equations

Given

$$\frac{Ha - Hc}{L_1} = C_2 \cdot \left[\frac{4}{\left(K \cdot \ln\left(\frac{C_1 \cdot Qa}{8 \cdot d_1 \cdot \nu} \right) \right)^2} \right] \cdot \frac{Qa^2}{d_1^5}$$

$$\frac{Hc - Hb}{L_2} = C_2 \cdot \left[\frac{4}{\left(K \cdot \ln\left(\frac{C_1 \cdot Q2}{8 \cdot d_2 \cdot \nu} \right) \right)^2} \right] \cdot \frac{Q2^2}{d_2^5}$$

$$\frac{Hc - Hb}{L_3} = C_2 \cdot \left[\frac{4}{\left(K \cdot \ln\left(\frac{C_1 \cdot Q3}{8 \cdot d_3 \cdot \nu} \right) \right)^2} \right] \cdot \frac{Q3^2}{d_3^5}$$

$$Q2 + Q3 = Qa$$

Results

$$minerr(Q2, Q3, Hb, Hc) = \begin{bmatrix} 61261 \\ 188739 \\ 169 \\ 280 \end{bmatrix} \begin{matrix} \text{bpd} \\ \text{bpd} \\ \text{flc} \\ \text{flc} \end{matrix}$$

Computer Computation Template

CCT No. 16-4: Calculations of heads and flow rates in pipelines with partial loops.

Page 2

References: Chapter 16. Figure 16-1, Example 4. Equations 16-8 and 16-9.

Input

Flow Input Data $L_T := 396$ kft $Qa := 250000$ bpd $Ha := 3000$ flc $Hb := 169$ flc

Case II: L_T,Qa, Ha and Hb known, find L_1, L_2,Hc, Q2 and Q3

Seed Variables $L_2 := 100$ kft $L_1 := 296$ kft $Q2 := 50000$ bpd $Q3 := 200000$ bpd

Equations

Given

$$\frac{Ha - Hc}{L_1} = C_2 \cdot \left[\frac{4}{\left(K \cdot \ln\left(\frac{C_1 \cdot Qa}{8 \cdot d_1 \cdot v} \right) \right)^2} \right] \cdot \frac{Qa^2}{d_1^{\,5}}$$

$$\frac{Hc - Hb}{L_2} = C_2 \cdot \left[\frac{4}{\left(K \cdot \ln\left(\frac{C_1 \cdot Q2}{8 \cdot d_2 \cdot v} \right) \right)^2} \right] \cdot \frac{Q2^2}{d_2^{\,5}}$$

$$\frac{Hc - Hb}{L_3} = C_2 \cdot \left[\frac{4}{\left(K \cdot \ln\left(\frac{C_1 \cdot Q3}{8 \cdot d_3 \cdot v} \right) \right)^2} \right] \cdot \frac{Q3^2}{d_3^{\,5}}$$

$$L_1 + L_2 = 396 \qquad Q2 + Q3 = Qa$$

Results

$$minerr(L_1, L_2, Hc, Q2, Q3) = \begin{bmatrix} 264 \\ 132 \\ 280 \\ 61268 \\ 188732 \end{bmatrix} \begin{matrix} \text{kft} \\ \text{kft} \\ \text{flc} \\ \text{bpd} \\ \text{bpd} \end{matrix}$$

Computer Computation Template

CCT No. 16-5: Calculation of heads and flow rates for a branched pipeline.

Page 1

References: Chapter 16. Figure 16-1, Example 5. Equation 16-5.

Input

Factors $C_1 := 92.241232$ $C_2 := 0.02647271$ $C_3 := 0.01147629$ $K := \frac{3.6}{2.302585}$ $k := \frac{2.0}{2.302585}$

Pipeline and Fluid Input Data $L_1 := 264$ kft $L_2 := 132$ kft $L_3 := 66$ kft $\nu := 7.0$cSt

$d_1 := 19.25$ in $d_2 := 19.25$ in $d_3 := 29.25$ in

Flow Input Data $Ha := 3000$ flc $Qa := 250000$ bpd $Q2 := 100000$ bpd $Q3 := 150000$bpd

Case I: Ha, Qa, Q2 and Q3 known, find Hb, Hc, and Hd.

Seed Variables $Hb := 500$ flc $Hc := 500$ flc $Hd := 500$ flc

Equations

Given

$$\frac{Ha - Hc}{L_1} = C_2 \cdot \left[\frac{4}{\left(K \cdot ln\left(\frac{C_1 \cdot Qa}{8 \cdot d_1 \cdot \nu} \right) \right)^2} \right] \cdot \frac{Qa^2}{d_1^{\,5}}$$

$$\frac{Hc - Hd}{L_2} = C_2 \cdot \left[\frac{4}{\left(K \cdot ln\left(\frac{C_1 \cdot Q2}{8 \cdot d_2 \cdot \nu} \right) \right)^2} \right] \cdot \frac{Q2^2}{d_2^{\,5}}$$

$$\frac{Hc - Hb}{L_3} = C_2 \cdot \left[\frac{4}{\left(K \cdot ln\left(\frac{C_1 \cdot Q3}{8 \cdot d_3 \cdot \nu} \right) \right)^2} \right] \cdot \frac{Q3^2}{d_3^{\,5}}$$

$$Q2 + Q3 = Qa$$

Results

$$minerr(Hb, Hc, Hd) = \begin{pmatrix} 243 \\ 280 \\ 16 \end{pmatrix} \begin{matrix} \text{flc} \\ \text{flc} \\ \text{flc} \end{matrix}$$

Computer Computation Template

CCT No. 16-5: Calculation of heads and flows for a branched pipeline.

Page 2

References: Chapter 16. Figure 16-1, Example 5. Equation 16-5.

Input

Factors $C_1 := 92.241232$ $C_2 := 0.02647271$ $C_3 := 0.01147629$ $K := \frac{3.6}{2.302585}$ $k := \frac{2.0}{2.302585}$

Pipeline and Fluid Input Data $L_1 := 264$ kft $L_2 := 132$ kft $L_3 := 66$ kft $\nu := 7.0$ cSt

$d_1 := 19.25$ in $d_2 := 19.25$ in $d_3 := 29.25$ in

Flow Input Data $Ha := 3000$ flc $Hb := 243$ flc $Hd := 16$

Case II: Ha, Hb, and Hd known, find Hc, Qa, Q2, and Q3

Seed Variables $Qa := 250000$ bpd $Q2 := 100000$ bpd $Q3 := 150000$ bpd $Hc := 500$ flc

Equations

Given

$$\frac{Ha - Hc}{L_1} = C_2 \cdot \left[\frac{4}{\left(K \cdot ln\left(\frac{C_1 \cdot Qa}{8 \cdot d_1 \cdot \nu}\right)\right)^2}\right] \cdot \frac{Qa^2}{d_1^5}$$

$$\frac{Hc - Hd}{L_2} = C_2 \cdot \left[\frac{4}{\left(K \cdot ln\left(\frac{C_1 \cdot Q2}{8 \cdot d_2 \cdot \nu}\right)\right)^2}\right] \cdot \frac{Q2^2}{d_2^5}$$

$$\frac{Hc - Hb}{L_3} = C_2 \cdot \left[\frac{4}{\left(K \cdot ln\left(\frac{C_1 \cdot Q3}{8 \cdot d_3 \cdot \nu}\right)\right)^2}\right] \cdot \frac{Q3^2}{d_3^5}$$

$$Q2 + Q3 = Qa$$

Results

$$minerr(Hc, Qa, Q2, Q3) = \begin{bmatrix} 280 \\ 250015 \\ 99980 \\ 150035 \end{bmatrix} \begin{matrix} \text{flc} \\ \text{bpd} \\ \text{bpd} \\ \text{bpd} \end{matrix}$$

Computer Computation Template

CCT No. 16-6: Calculations for five leg, four node network with one source, two sinks, and five lines.

References: Chapter 16. Figure 16-1, Example 6 (general case). Figure 16-11 (case specific).

Input

Factors $C_1 := 92.241232$ $C_2 := 0.02647271$ $C_3 := 0.01147629$ $K := \frac{3.6}{2.302585}$ $k := \frac{2.0}{2.302585}$

Pipeline and Fluid Input Data $L_1 := 3.0$ kft $L_2 := 3.5$ kft $L_3 := 1.0$ kft $L_4 := 1.0$ kft $L_5 := 1.5$ kft

$d_1 := 12.0$ in. $d_2 := 10.0$ in. $d_3 := 6.0$ in. $d_4 := 8.0$ in $d_5 := 12.0$ in

$\nu := 7.0$ cSt

Flow Input Data $Qa := 50000$ bpd $Qb := 15000$ bpd $Qc := 35000$ bpd

Case I: Ha, Hb and Hc known, find Q_1,Q_2,Q_3,Q_4,Q_5.

Seed Variables $Q_1 := 40000$ bpd $Q_2 := 10000$ bpd $Q_3 := 20000$ bpd $Q_4 = 20000$ bpd $Q_5 := 15000$

Equations

Given

$$Q_1 + Q_3 = Qa \qquad Q_1 - Q_2 - Q_4 = Qb \qquad Q_4 + Q_5 = Qc$$

$$\left[C_2 \cdot L_1 \cdot \left[\frac{4}{\left(K \cdot \ln\left(\frac{C_1 \cdot Q_1}{8 \cdot d_1 \cdot \nu}\right)\right)^2}\right] \cdot \frac{Q_1^2}{d_1^5} + C_2 \cdot L_2 \cdot \left[\frac{4}{\left(K \cdot \ln\left(\frac{C_1 \cdot Q_2}{8 \cdot d_2 \cdot \nu}\right)\right)^2}\right] \cdot \frac{Q_2^2}{d_2^5}\right] - C_2 \cdot L_3 \cdot \left[\frac{4}{\left(K \cdot \ln\left(\frac{C_1 \cdot Q_3}{8 \cdot d_3 \cdot \nu}\right)\right)^2}\right] \cdot \frac{Q_3^2}{d_3^5} = 0$$

$$\left[C_2 \cdot L_4 \cdot \left[\frac{4}{\left(K \cdot \ln\left(\frac{C_1 \cdot Q_4}{8 \cdot d_4 \cdot \nu}\right)\right)^2}\right] \cdot \frac{Q_4^2}{d_4^5} - C_2 \cdot L_5 \cdot \left[\frac{4}{\left(K \cdot \ln\left(\frac{C_1 \cdot Q_5}{8 \cdot d_5 \cdot \nu}\right)\right)^2}\right] \cdot \frac{Q_5^2}{d_5^5}\right] - C_2 \cdot L_2 \cdot \left[\frac{4}{\left(K \cdot \ln\left(\frac{C_1 \cdot Q_2}{8 \cdot d_2 \cdot \nu}\right)\right)^2}\right] \cdot \frac{Q_2^2}{d_2^5} = 0$$

Results

$$minerr(Q_1, Q_2, Q_3, Q_4, Q_5) = \begin{bmatrix} 37860 \\ 8884 \\ 12140 \\ 13976 \\ 21024 \end{bmatrix} \begin{matrix} \text{bpd} \\ \text{bpd} \\ \text{bpd} \\ \text{bpd} \\ \text{bpd} \end{matrix}$$

Computer Computation Template

CCT No. 16-7: Calculations for three leg, three node network with one source, two sinks, and one loop.

References: Chapter 16. Figure 16-1, Example 6 (general case). Figure 16-12 (case specific).

Input

Factors $C_1 := 92.241232$ $C_2 := 0.02647271$ $C_3 := 0.01147629$ $K := \frac{3.6}{2.302585}$ $k := \frac{2.0}{2.302585}$

Pipeline and Fluid Input Data $L_1 := 15.0$ kft $L_2 := 30.0$ kft $L_3 := 38.0$ kft $\nu := 7.0$ cSt

$d_1 := 20.0$ in $d_2 := 18.0$ in $d_3 := 24.0$ in

Flow Input Data $Qb := 100000$ bpd $Qc := 250000$ bpd $Ha := 100$ flc

Case I: Ha, Qb and Qc known, find Hb and Hc

Seed Variables $Hb := 20$ flc $Hc := 80$ flc

Equations

Given

$$\left[\frac{-\sqrt{Ha-Hb}}{\sqrt{L_1}}\cdot\frac{d_1^{\left(\frac{5}{2}\right)}}{\sqrt{C_2}}\cdot k\cdot ln\left[2.825\cdot\frac{\nu}{C_1}\cdot\frac{\sqrt{C_2}}{\sqrt{Ha-Hb}}\cdot\frac{\sqrt{L_1}}{d_1^{\left(\frac{3}{2}\right)}}\right]\right]-\left[\frac{-\sqrt{Hb-Hc}}{\sqrt{L_2}}\cdot\frac{d_2^{\left(\frac{5}{2}\right)}}{\sqrt{C_2}}\cdot k\cdot ln\left[2.825\cdot\frac{\nu}{C_1}\cdot\frac{\sqrt{C_2}}{\sqrt{Hb-Hc}}\cdot\frac{\sqrt{L_2}}{d_2^{\left(\frac{3}{2}\right)}}\right]\right]-Qb=0$$

$$\left[\frac{-\sqrt{Hb-Hc}}{\sqrt{L_2}}\cdot\frac{d_2^{\left(\frac{5}{2}\right)}}{\sqrt{C_2}}\cdot k\cdot ln\left[2.825\cdot\frac{\nu}{C_1}\cdot\frac{\sqrt{C_2}}{\sqrt{Hb-Hc}}\cdot\frac{\sqrt{L_2}}{d_2^{\left(\frac{3}{2}\right)}}\right]\right]+\left[\frac{-\sqrt{Ha-Hc}}{\sqrt{L_3}}\cdot\frac{d_3^{\left(\frac{5}{2}\right)}}{\sqrt{C_2}}\cdot k\cdot ln\left[2.825\cdot\frac{\nu}{C_1}\cdot\frac{\sqrt{C_2}}{\sqrt{Ha-Hc}}\cdot\frac{\sqrt{L_3}}{d_3^{\left(\frac{3}{2}\right)}}\right]\right]-Qc=0$$

Results

$$minerr(Hb, Hc) = \begin{pmatrix} 44.2 \\ 14.0 \end{pmatrix} \begin{matrix} \text{flc} \\ \text{flc} \end{matrix}$$

Index